Guide to Wireless Mesh Networks

The **Computer Communications and Networks** series is a range of textbooks, monographs and handbooks. It sets out to provide students, researchers and non-specialists alike with a sure grounding in current knowledge, together with comprehensible access to the latest developments in computer communications and networking.

Emphasis is placed on clear and explanatory styles that support a tutorial approach, so that even the most complex of topics is presented in a lucid and intelligible manner.

For other titles published in this series, go to
http://www.springer.com/series/4198

Sudip Misra • Subhas Chandra Misra
Isaac Woungang
Editors

Guide to Wireless
Mesh Networks

Springer

Editors

Sudip Misra
School of Information Technology
Indian Institute of Technology
Kharagpur, India

Isaac Woungang
Department of Computer Science
Ryerson University
Toronto, Canada

Subhas Chandra Misra
Department of Industrial & Management
  Engineering
Indian Institute of Technology
Kanpur, India

Series Editor
Professor A.J. Sammes, BSc, MPhil, PhD, FBCS, CEng
Centre for Forensic Computing
Cranfield University
DCMT, Shrivenham
Swindon SN6 8LA
UK

CCN Series ISSN 1617-7975
ISBN: 978-1-84996-804-1   e-ISBN: 978-1-84800-909-7
DOI: 10.1007/978-1-84800-909-7

British Library Cataloguing in Publication Data
A catalogue record for this book is available from the British Library

Printed on acid-free paper

Springer Science+Business Media
springer.com

Dedicated to the newborns:
Tultuli (Subhas's daughter)
and Babai (Sudip's son)
and Isaac's grand ma:
Maria Happi

Preface

Overview and Goals

Wireless communication technologies are undergoing rapid advancements. The last few years have experienced a steep growth in research in the area of wireless mesh networks (WMNs). The attractiveness of WMNs, in general, is attributed to their characteristics such as the ability to dynamically self-organize and self-configure, coupled with the ability to maintain mesh connectivity leading, in effect, to low set-up/installation costs, simpler maintenance tasks, and service coverage with high reliability and fault-tolerance. WMNs also support their integration with existing wireless networks such as cellular networks, WLANs, wireless-fidelity (Wi-Fi), and worldwide interoperability of microwave access (WiMAX). WMNs have found useful applications in a broad range of domains such as broadband home networking, commercial/business networking, and community networking – particularly attractive in offering broadband wireless access with low initial installation and set-up costs.

Even though WMNs have emerged to be attractive and they hold great promises for our future, there are several challenges that need to be addressed. Some of the well known challenges are attributed to issues relating to scalability (significant drop in throughput with the increase in the number of nodes), multicasting, offering quality of service guarantees, energy efficiency, and security.

This handbook attempts to provide a comprehensive guide on fundamental key topics coupled with new ideas and results in the areas of WMNs. The book has been prepared keeping in mind that it needs to prove itself to be a valuable resource dealing with both the important core and the specialized issues in WMNs. We have attempted to offer a wide coverage of topics. We hope that it will be a valuable reference for students, instructors, researchers, and industry practitioners. We believe, this is particularly an attractive feature of this book, as the very limited selection of books available on WMNs we are aware of, are written primarily for academicians/researchers. We have attempted to make this book useful for both the academicians and the practitioners alike.

Organization and Features

The book is organized into 19 chapters, each chapter written by topical area experts. Chapter 1 is devoted to the basics of WMNs and their relationship between MANETs. Chapter 2 is devoted to issues concerning medium access. Chapters 3–8 discuss about various issues concerning routing and channel assignment in WMNs. Chapter 8 is worth mentioning, as it introduces a very interesting and important chapter on routing metrics. Chapters 9–11 center on congestion and other transport layer issues. The issues of multinetwork convergence and scalability are very key issues in WMNs and they are discussed in Chaps. 12 and 13. Chapter 14 is dedicated to the issues concerning mobility. The rest of the chapters, Chaps. 15–19, are devoted to some of the specialized topics relating to WMN such as the WiMAX metro area mesh networks, the symbiosis of cognitive radio with WMNs and the construction and evaluation of testbeds in WMNs.

We list below some of the important features of this book, which, we believe, would make this book a valuable resource for our readers:

- Most of the chapters of the book are authored by prominent academicians/ researchers/practitioners in WMNs that have been working with these topics for quite a few years now and have thorough understanding of the concepts.
- The authors of this book are distributed in a large number of countries and most of them are affiliated with institutions of worldwide repute. This gives this book an international flavor. The readers of this book can get absorbed by perspectives, suggestions, experiences and issues projected forward by authors from different countries.
- Almost all the chapters in this book have a distinct section providing *directions for future research*, which, particularly, targets researchers working in these areas. We believe, this section in each chapter should provide insight to the researchers about some of the current research issues.
- The authors of each chapter have also attempted to the extent possible to provide a comprehensive bibliography, which should greatly help the researchers and readers interested further to dig into the topic.
- Almost all chapters of this book have a separate section outlining *thoughts for practitioners*. We believe, this section in every chapter will be particularly useful for industry practitioners working directly with the practical aspects behind enabling these technologies in the field.
- Most of the chapters provide a list of important terminologies and their brief definitions.
- Most of the chapters also provide a set of questions at the end that can help in assessing the understanding of the readers.
- To make the book useful for pedagogical purposes, almost all chapters of the book also have a corresponding set of presentation slides. The slides can be obtained as a supplementary resource by contacting the publisher, Springer.

We have made attempts in all possible way we could to make the different chapters of the book look as much coherent and synchronized as possible. However, it cannot be denied that as the chapters were written by different authors, it was not fully possible to fully achieve this task. We believe that this is a limitation of most edited books of this sort.

Target Audience

The book is written by primarily targeting the student community. This includes the students of all levels – those getting introduced to these areas, those having an intermediate level of knowledge of the topics and those who are already knowledgeable about many of the topics. To keep up with this goal, we have attempted to design the overall structure and content of the book in such a manner that makes it useful at all learning levels. To aid in the learning process, almost all chapters have a set of questions at the end of the chapter. Also, in order that teachers can use this book for classroom teaching, the book also comes with presentation slides and sample solutions to exercise questions, which are available as supplementary resources.

The secondary audience for this book is the research community, whether they are working in the academia or in the industry. To meet the specific needs to this audience group, most chapters of the book also have a section in which attempts have been made to provide directions for future research.

Finally, we have also taken into consideration the needs to those readers, typically from the industries, who have quest for getting insight into the practical significance of the topics, i.e., how the spectrum of knowledge and the ideas are relevant for real-life working of WMNs.

Supplementary Resources

As mentioned earlier, the book comes with the following supplementary resources:

- Solution manual, having sample solutions to most questions provided at the end of the chapters.
- Presentation slides, which can be used for classroom instruction by teachers.

Teachers can contact the publisher, Springer, to get access to these resources.

Acknowledgments

We are extremely thankful to the 50 authors of the 19 chapters of this book, who have worked very hard to bring this unique resource forward for help of the student, researcher, and practitioner community. The authors were very much interactive at all stages of preparation of the book from initial development of concept to finalization. We feel it is contextual to mention that as the individual chapters of this book are written by different authors, the responsibility of the contents of each of the chapters lies with the concerned authors.

We are also very much thankful to our colleagues in the Springer publishing and marketing teams, in particular, Mr. Wayne Wheeler and Ms. Catherine Brett, who tirelessly worked with us and guided us in the publication process. Special thanks also go to them for taking special interest in publishing this book, considering the current worldwide market needs for such a book.

Finally, we would like to thank our parents, Prof. J.C. Misra, Mrs. Shorasi Misra, Mr. John Sime, Mrs. Christine Seupa, our wives Satamita, Sulagna and Clarisse, and our children, Babai, Tultuli, Clyde, Lenny, and Kylian, for the continuous support and encouragement they offered during this project.

Kharagpur, India   *Sudip Misra*
Boston, MA, USA   *Subhas Chandra Misra*
Toronto, ON, Canada   *Isaac Woungang*

Contents

Contributors

Fuad Abujarad Department of Computer Science and Engineering, 3115 Engineering Building, Michigan State University, East Lansing, MI 48824, USA, abujarad@cse.msu.edu

Dharma P. Agrawal Department of Computer Science, OBR Center of Distributed and Mobile Computing, University of Cincinnati, Cincinnati, OH 45221-0030, USA, dpa@cs.uc.edu

Mahesh Arumugam Cisco Systems, Inc., 170 W Tasman Drive, San Jose, CA 95134, USA, maarumug@cisco.com

Leonardo Badia IMT Lucca Institute for Advanced Studies, Piazza S. Ponziano 6, 55100 Lucca, Italy, l.badia@imtlucca.it

N. Balakrishnan Supercomputer Education and Research Center, Indian Institute of Science, Bangalore 560012, India, balki@serc.iisc.ernet.in

Rainer Baumann Computer Engineering and Networks Laboratory (TIK), ETH Zurich, Gloriastrasse 35, 8092 Zurich, Switzerland, baumann@tik.ee.ethz.ch

Raouf Boutaba David R. Cheriton School of Computer Science, University of Waterloo, 200 University Ave West, Waterloo, ON, Canada N2L 3G1, rboutaba@uwaterloo.ca

Paolo Bucciol DAUIN c/o Politecnico di Torino, C.so Duca degli Abruzzi 24, 10129 Torino, Italy, paolo.bucciol@polito.it

Daniel Câmara Institut Eurécom, 2229 Route des Crêtes, BP 193, 06560 Sophia-Antipolis, France, daniel.camara@eurecom.fr

Chunjie Cao Key Laboratory of Computer Networks and Information Security of the Ministry of Education, Xidian University, Xi'an 710071, People's Republic of China, chjcao@mail.xidian.edu.cn

Bin Chang Computer Science, Vanderbilt University, VU Station B 351824, Nashville, TN 37235, USA, bin.chang@vanderbilt.edu

Marco Conti Instituto di Informatica e Telematica (IIT), Italian National Research Council (CNR), Via G. Moruzzi, 1, 56124 Pisa, Italy, marco.conti@iit.cnr.it

Liang Dai Computer Science, Vanderbilt University, VU Station B 351824, Nashville, TN 37235, USA, liang.dai@vanderbilt.edu

Sajal K. Das Department of Computer Science and Engineering, Center for Research in Wireless Mobility and Networking (CReWMaN), The University of Texas at Arlington, Arlington, TX 76019-0015, USA, das@cse.uta.edu

Juan Carlos De Martin DAUIN c/o Politecnico di Torino, C.so Duca degli Abruzzi 24, 10129 Torino, Italy, demartin@polito.it

K.L. Eddie Law Department of Electrical and Computer Engineering, Ryerson University, 350 Victoria Street, Toronto, ON, Canada M5B 2K3, eddie@ee.ryerson.ca

Fethi Filali Institut Eurécom, 2229 Route des Crêtes, BP 193, 06560 Sophia-Antipolis, France, fethi.filali@eurecom.fr

Nikos Fragoulis Electronics Laboratory, Department of Physics, University of Patras, 26500 Patras, Greece, nfrag@upatras.gr

Dirk Grunwald Department of Computer Science, University of Colorado at Boulder, UCB 430, Boulder, CO 80309, USA, dirk.grunwald@cs.colorado.edu

Hossam Hassanein School of Computing, Queens University, Kingston, ON, Canada K7L 3N6, hossam@cs.queensu.ca

Simon Heimlicher Computer Engineering and Networks Laboratory (TIK), ETH Zurich, Gloriastrasse 35, 8092 Zurich, Switzerland, heimlicher@tik.ee.ethz.ch

Feiyi Huang Department of Electronic and Electrical Engineering, University College London, London WC1E 6BT, UK, f.huang@ee.ucl.ac.uk

Brent Ishibashi David R. Cheriton School of Computer Science, University of Waterloo, 200 University Ave West, Waterloo, ON, Canada N2L 3G1, bkishiba@uwaterloo.ca

S.S. Iyengar Louisiana State University, 298 Coates Hall, Tower Dr, Baton Rouge, LA 70803, USA, iyengar@csc.lsu.edu

Arshad Jhumka Department of Computer Science, University of Warwick, Coventry CV4 7AL, UK, arshad@dcs.warwick.ac.uk

Merkourios Karaliopoulos Computer Engineering and Networks Laboratory (TIK), ETH Zurich, Gloriastrasse 35, 8092 Zurich, Switzerland, karaliopoulos@tik.ee.ethz.ch

Sandeep Kulkarni Department of Computer Science and Engineering, 3115 Engineering Building, Michigan State University, East Lansing, MI 48824, USA, sandeep@cse.msu.edu

Luciano Lenzini Department of Information Engineering, University of Pisa, via Diotisalvi 2, 56122 Pisa, Italy, l.lenzini@iet.unipi.it

Frank Yong Li Department of Information and Communication Technology, University of Agder, 4898 Grimstad, Norway, frank.li@uia.no

Jianfeng Ma Key Laboratory of Computer Networks and Information Security of the Ministry of Education, Xidian University, Xi'an 710071, People's Republic of China, jfma@mail.xidian.edu.cn

Robert McTasney Department of Computer Science, University of Colorado at Boulder, UCB 430, Boulder, CO 80309, USA, robert.mctasney@colorado.edu

Ulrich Meis Department of Computer Science, RWTH Aachen University, Ahornstrasse 55, 52074 Aachen, Germany, meis@cs.rwth-aachen.de

Vinod Mirchandani Faculty of Information Technology, University of Technology, Sydney, P.O. Box 123, Broadway, NSW 2007, Australia, vinodm@it.uts.edu.au

Georgios Parissidis Computer Engineering and Networks Laboratory (TIK), ETH Zurich, Gloriastrasse 35, 8092 Zurich, Switzerland, parissid@tik.ee.ethz.ch

Bernhard Plattner Computer Engineering and Networks Laboratory (TIK), ETH Zurich, Gloriastrasse 35, 8092 Zurich, Switzerland, plattner@tik.ee.ethz.ch

Ante Prodan Faculty of Information Technology, University of Technology, Sydney, P.O. Box 123, Broadway, NSW 2007, Australia, aprodan@it.uts.edu.au

Bahareh Sadeghi Intel Corporation, 2111 NE 25th Ave, JF3-206, Hillsboro, OR 97124, USA, bahareh.sadeghi@intel.com

Douglas Sicker Department of Computer Science, University of Colorado at Boulder, UCB 430, Boulder, CO 80309, USA, douglas.sicker@colorado.edu

Habiba Skalli IMT Lucca Institute for Advanced Studies, Piazza S. Ponziano 6, 55100 Lucca, Italy, h.skalli@imtlucca.it

Thrasyvoulos Spyropoulos Computer Engineering and Networks Laboratory (TIK), ETH Zurich, Gloriastrasse 35, 8092 Zurich, Switzerland, spyropoulos@tik.ee.ethz.ch

S. Srivathsan Louisiana State University, 164 Coates Hall, Tower Dr, Baton Rouge, LA 70803, USA, ssrini1@lsu.edu

Ahmed Iyanda Sulyman Department of Electrical and Computer Engineering, Royal Military College of Canada, P.O. Box 1700, Station Forces, Kingston, ON, Canada K7K 7B4, ahmed.sulyman@rmc.ca

Junfang Wang Department of Computer Science, OBR Center of Distributed and Mobile Computing, University of Cincinnati, Cincinnati, OH 45221-0030, USA, wangjf@cs.uc.edu

Jonathan Wellons Computer Science, Vanderbilt University, VU Station B 351824, Nashville, TN 37235, USA, jonathan.wellons@vanderbilt.edu

Martin Wenig Department of Computer Science, RWTH Aachen University, Ahornstrasse 55, 52074 Aachen, Germany, wenig@cs.rwth-aachen.de

Bin Xie Department of Computer Science, OBR Center of Distributed and Mobile Computing, University of Cincinnati, Cincinnati, OH 45221-0030, USA, xieb@cs.uc.edu

Yuan Xue Computer Science, Vanderbilt University, VU Station B 351824, Nashville, TN 37235, USA, yuan.xue@vanderbilt.edu

Yang Yang Department of Electronic and Electrical Engineering, University College London, London WC1E 6BT, UK, y.yang@ee.ucl.ac.uk

Alexander Zimmermann Department of Computer Science, RWTH Aachen University, Ahornstrasse 55, 52074 Aachen, Germany, zimmermann@cs.rwth-aachen.de

Chapter 1
Journey from Mobile Ad Hoc Networks to Wireless Mesh Networks

Junfang Wang, Bin Xie, and Dharma P. Agrawal

Abstract A wireless mesh network (WMN) is a particular type of mobile ad hoc network (MANET), which aims to provide ubiquitous high bandwidth access for a large number of users. A pure MANET is dynamically formed by mobile devices without the requirement of any existing infrastructure or prior network configuration. Similar to MANETs, a WMN also has the ability of self-organization, self-discovering, self-healing, and self-configuration. However, a WMN is typically a collection of stationary mesh routers (MRs) with each employing multiple radios. Some MRs have wired connections and act as the Internet gateways (IGWs) to provide Internet connectivity for other MRs. These new features of WMNs over MANETs enable them to be a promising alternative for high broadband Internet access. In this chapter, we elaborate on the evolution from MANETs to WMNs and provide a comprehensive understanding of WMNs from theoretical aspects to practical protocols, while comparing it with MANETs. In particular, we focus on the following critical issues with respect to WMN deployment: *Network Capacity, Positioning Technique, Fairness Transmission* and *Multiradio Routing Protocols*. We end this chapter with some open problems and future directions in WMNs.

1.1 Introduction

A pure MANET is dynamically established by mobile devices grouped together as needed without any support from existing infrastructure as shown in Fig. 1.1. The mobile devices in the network communicate with each other through single or multi hop wireless links. The key benefit of MANET communication is that it enables

J. Wang (✉), B. Xie and D.P. Agrawal
Department of Computer Science, OBR Center of Distributed and Mobile Computing,
University of Cincinnati, Cincinnati, OH 45221-0030, USA
e-mail: {wangjf, xieb, dpa}@cs.uc.edu

S. Misra et al. (eds.), *Guide to Wireless Mesh Networks*, Computer Communications
and Networks, DOI 10.1007/978-1-84800-909-7_1,
© Springer-Verlag London Limited 2009

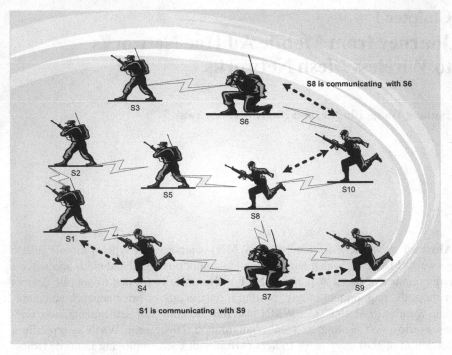

S8 is communicating with S6

S3
S6
S2
S5
S10
S8
S1
S4
S7
S9
S1 is communicating with S9

Fig. 1.1 An example MANET

us to form a network spontaneously without the need of having any infrastructure, which is both expensive and time-consuming.

The first MANET was initialized around 30 years ago by Defense Advanced Research Projects Agency (DARPA). Despite of peculiar advantages associated with MANETs, they have not been widely accepted for civilian applications. This could be primarily because of two reasons (1) some limitations of MANETs such as the security and limited throughput hinder MANETs from civilian applications and (2) the military and emergence applications dominate the research direction in MANETs so that most of the works target how to meet the unique requirements for these applications, such as the dynamic topology, which may not be solidly necessary for civilian applications.

In the recent years WMNs emerge as pragmatic multihop ad hoc networks to provide the high bandwidth Internet service to communities, enterprises, or entire cities. A WMN is a particular multihop ad hoc network, consisting of two parts: mesh backbone and mesh clients as shown in Fig. 1.2. The stationary wireless mesh routers (MRs) interconnecting through single/multi hop wireless links form the backbone. A few MRs with the wired connections act as the IGWs to exchange the traffic between the Internet and the WMN. The mesh clients can be the mobile wireless devices such as cell phones and laptops. The mobile clients connect to any MRs to access the Internet via the IGWs in a multihop fashion. Compared to pure MANETs, a WMN has a hierarchical structure and the topology of the wireless

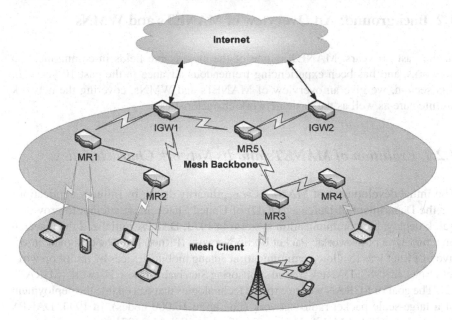

Fig. 1.2 A typical architecture of a WMN

backbone is relatively stable. These facilitate their deployment and application for Internet accessibility. Compared to the traditional wired and wireless networks (e.g., Wi-Fi), WMNs provide a cost-efficient way to support Internet services because of the reduced dependence on infrastructure by taking the advantage of multihop Internet access.

In this chapter, we first take an overview of the history of MANETs and WMNs. Then from theoretical aspects to practical protocols, we provide a comprehensive understanding of the evolution from MANETs to WMNs, concentrating specifically on the following issues:

- *Network capacity*. From the perspective of theoretical analysis, we present the capacity of MANETs (throughput per MR), and the capacity gain of WMNs in comparison with MANETs.
- *Positioning technique*. We detail the positioning technique of MRs and IGWs in WMNs that needs to satisfy various design constraints imposed by the system requirements.
- *Fairness transmission*. We explain the unfairness problem that exists in both MANETs and WMNs. We give special attention to end-to-end unfairness problem on WMNs compared to MANETs and provide an overview of the current approaches in WMNs.
- *Multiradio routing*. We study the routing issues in MANETs, including the design requirement and existing routing protocols. We also specify the requirements of routing protocols for WMNs in contrast to MANETs and then elaborate on some new routing metrics for WMNs.

1.2 Background: An Overview of MANETs and WMNs

In the past 30 years, MANET is one of the most active fields in communication networks, and has been experiencing tremendous advance in the past 10 years. In this section, we give an overview of MANETs and WMNs, covering the network architecture as well as the basic network characteristics.

1.2.1 Evolution of MANETs and Its Network Characteristics

The initial development of MANETs was primarily driven by military applications for the Department of Defense (DOD) of United States, in an attempt to provide a quick deployable communication system. In 1973, DARPA initialized the research on a new type of networks: Packet Radio Network (PRnet) [1]. The designed objective of PRnet was to allow communication among mobile devices by multihop wireless links. In 1983, DARPA started sponsoring Survivable Radio Network (SURAN) [2]. The goal of SURAN was to explore technologies that can enable the deployment of a large-scale packet radio network (i.e., up to 10,000 nodes). In 1994, DARPA launched the Global Mobile Information Systems (GloMo) [37] program, intending to develop new wireless ad hoc networking technology compatible with the Internet technology. At the same time, MANETs have received extensive attention from both the academia and industries because of explosive growth of personal wireless communication devices in the 1990s. In 1997, the Internet Engineering Task Force (IETF) established the Mobile Ad Hoc Network (MANET) Working Group to standardize MANET routing protocols such as Ad hoc On-Demand Distance Vector Routing (AODV) [3], Dynamic Source Routing (DSR) [4], etc.

The most prominent characteristic of a MANET is its infrastructureless. In other words, a MANET has no special centralized authority like cellular base station (BS) in support of wireless communication and is useful for a variety of communication applications where the infrastructure is unavailable for users. In brief, the main characteristics of MANETs can be summarized as follows:

- *Self-forming, self-configuring, and self-healing.* The multihop communication and its management in a MANET are automatic and spontaneous without any centralized network authority. In most scenarios, a MANET has no Internet accessibility for nodes.
- *Dynamic topology.* The locations of nodes in a MANET keep on changing because of node mobility and new node enrollment and leave without prior notification. Therefore, the network topology changes in an unpredictable manner.
- *Constrained resources.* MANETs suffer from limited energy and network bandwidth. The mobile nodes are usually powered by battery for transmitting or receiving packets to/from other nodes. Thus, the node can only work for a limited period because of limited power. In addition, all the nodes in a MANET usually operate in a shared wireless channel. Thus, the network bandwidth is limited and nodes compete with each other for accessing the medium.

The network performance (e.g., packet loss) of a MANET is significantly affected by path loss, interference, and other factors like other wireless networks. There are some new challenges in MANETs. For example, time-varying topology causes frequent link broken that needs path reconstruction to maintain the communication. The quality of service (QoS) of an application may be significantly impacted by link breakage and channel contention. Furthermore, the energy constraint requires the MANET design to be scalable for the purpose of energy-efficiency. Although MANETs offer the benefits such as self-forming, self-configuring, and self-healing so that each node is free to move while maintaining its communication, the challenging problems limit the application of MANETs. Currently, MANETs are still largely deployed for military networks rather than civilian applications.

1.2.2 Evolution from MANETs to WMNs

Since the late 1990s, the world has been experiencing both the prosperity of the Internet and the popularization of personal wireless devices. People always strive to design a low-cost and ubiquitous wireless environment that can allow their personal wireless devices to access Internet anywhere and anytime. WMNs are therefore proposed to satisfy the civilian application requirements while addressing the limitation of MANETs. In contrast to peer to peer structure of a MANET, a WMN consists of MRs forming a wireless backbone. As shown in Fig. 1.2, a number of MRs are interconnected by wireless links so that mobile devices are connected with the Internet via the wireless backbone. A MR not only handles the traffic from its associated mobile devices but also relays the traffic from other MRs. Similar to MANETs, a WMN still has the ability of self-configuring and self-healing, which keep some advantages of MANETs in a WMN. For example, the Internet access of a mobile device can be implemented in a multihop manner, which reduces installation cost of the overall system. The main improvements of WMNs over MANETs are as follows:

- *Infrastructure support of MANET*. In a MANET, the mobile device is usually not able to access the Internet. On the contrary, the IGWs of a WMN integrate with the Internet infrastructure, which provides the Internet accessibility for mobile clients that may not directly connected to the Internet. Internet accessibility is critical for civilian applications so that the Internet services such as email, Web browser, etc., can be enabled in WMNs. When the mobile device is moving in a WMN, it can handover its services from one MR to another. If it moves out the range of the WMN, the service continues if any other network is available. This may require an internetworking scheme for such mobility support.
- *Minimal mobility of MR*. In a WMN, each MR is situated in a fixed position, such as the roof of a building, and has minimal mobility. Therefore, the wireless backbone has a fixed topology unless some MRs or interfaces are added into or removed from the network. This is important to reduce the probability of link breakage and improve the network throughput.

- *Rich energy of MR.* In contrast to a MANET where the nodes are subject to energy constraint, every MR is usually connected to a rich power supply.
- *Multiradio and multichannel.* The network bandwidth is improved in WMNs because each MR can use multiple orthogonal channels. To manipulate multiple channels, a MR is configured with multiple radios (e.g., interfaces), which can simultaneously transmit/receive packets in multiple orthogonal channels without interference with each other. For example, the network capacity can be almost doubled if each MR has two radios to operate in two separate channels. As we discussed in the previous section, all nodes in a MANET only work on a shared channel with a single radio at each node. Table 1.1 lists the number of channels in IEEE 802.11s, the operating frequency, and their available data rate per channel.

These improvements with respect to the network architecture resolve some of the limitations and also improve the network performance as compared to the traditional wireless network such as MANET, cellular network, and WLAN. In terms of commercial implementation and application requirements, WMN offers many benefits and some of them are:

- *High-speed wireless system.* The wireless backbone of a WMN provides wireless accessibility for mobile users to communication with the Internet. IEEE 802.11g works in the 2.4-GHz band (like 802.11b) with a maximum raw data rate of 54 Mbps. With IEEE 802.11n products, the data rate between 100 and 200 Mbps becomes feasible for the mass market, which is not achievable in earlier wireless technology. The high data rate is enabled by high-speed Physical Layer (PHY) technologies such as sensitive Modulation Coding Schemes (MCS) and Multiple Input/Output (MIMO). Therefore, wireless backbone can be used to provide a platform for supporting various real-time commercial applications and bandwidth-consuming communication needs. The network capacity (throughput per MR) can be further improved by deploying multiple orthogonal channels over multiple interfaces. Meanwhile, the infrastructure nodes (i.e., IGWs) serve as the traffic sinks for the wireless backbone, and improve the network capacity when the number of IGWs asymptotically grows up with the number of MRs.
- *Promising coverage and connectivity.* The Internet accessibility is not available in a pure MANET because of its infrastructureless nature. In a WLAN, each access point is connected to the Internet and mobile devices are connected to the access point directly by using a centralized medium access. Therefore, the

Table 1.1 Multiple channels in IEEE 802.11

Protocol	Number of channels	Frequency (GHz)	Data rate (Mbps)	Transmission range (m)
IEEE 802.11a	12/13 indoor 4/5 outdoor	5	23–54 (2 stream)	30–100
IEEE 802.11b	14	2.4–2.5	5–11 (2 stream)	35–110
IEEE 802.11g	14	2.4–2.5	19–54 (2 stream)	35–110
IEEE 802.11h	12/13	5	≤100	30–100
IEEE 802.11n	14 in 2.4 GHz	2.4, 5	74–248 (2 stream)	70–160

coverage is largely limited and the connectivity is only available in a single hop range. A wireless personal area network (WPAN) provides high-speed data transmission, but it is only available for a short range because of limited transmission power and its operating frequency band. In a WMN, each MR not only operates as a host to aggregate data from its associated mobile clients, but also as a router to forward packets on behalf of other MRs. In this manner, the coverage and connectivity are significantly extended in a multihop fashion. Furthermore, the mobile device can fast and seamlessly handoff from an accessing MR to a neighboring MR at a faster rate by employing an appropriate migrating scheme.

• *Flexible and cost-efficient deployment.* WMN can be commercially operated in many ways. At first, it can be constructed and managed by a single Internet Service Provider (ISP). In such a case, each MR is under the control of an ISP. Secondly, a WMN can be semi-managed, meaning that the core MRs in a network are controlled by a single or several ISPs whereas the other MRs can be added by any users, subjecting to some operation and payment model. Moreover, a commercial WMN can be operated in unmanaged way that every MR is managed by an independent entity. For example, a user in a community can independently install a MR on the roof of its house and share the connection with a near-by ISP. Whichever commercial operating models, a MR is able to flexibly add or uninstall from a WMN because of self-configuration and self-forming. By taking advantage of IGWs, the Internet connectivity of a MR can be achieved by a multihop wireless connection. The loose integration of WMNs with the Internet significantly reduces the required wired links, which are expensive in link connection and hardware. Therefore, deployment of a WMN becomes a cost-efficient solution and is beneficial for mobile users.

Because of promising popularity in the market, several IEEE standard groups have been established to define specifications for WMN techniques in terms of different network types. In 2003, IEEE 802.15 and IEEE 802.11 each established a new subgroup, i.e., 802.15.5 and 802.11s, to standardize mesh network with their respective devices. IEEE 802.16a group and IEEE 802.20, the working group established in 2002 for mobile broadband wireless access networks, both include mesh into their standards as well. In the following subsection, we illustrate some example WMNs.

1.2.3 Free-For-Use and Commercial WMN Examples

Unlike MANETs, early stage of research works of WMNs have been performed on actual test beds or free-to-use networks. The first free-to-use WMN is Roofnet [5], which is an 802.11b/g community-oriented WMN developed by the Massachusetts Institute of Technology (MIT) to provide broadband Internet access to the users in Cambridge, MA. Some other free-to-use WMNs [6] have Champaign-Urbana Community Wireless Network (CUWiN), Seattle Wireless, Broadband and Wireless Network (BWN), and Southampton Open Wireless Network (SOWN), Technology For

All (TFA) [7]. These WMNs are typically implemented with open source software and free of addition of new nodes.

These test beds fostered tremendous advances and consequently built up the confidence for commercial applications in the design of architectures, protocols, algorithms, services, and applications of WMNs. Some of the commercial WMN solutions [8] include Tropos, BelAir, Cisco, Nortel, Microsoft, Firetide, Sensoria Corporation, PacketHop, MeshDynamics, and Radiant Networks. We illustrate them by using the example of Cisco.

Cisco [9] has commercial WMN solution by using its Aironet products to allow government, public safety, and transportation organization to build cost-effective outdoor WMNs for private or public use. The Aironet MR products are designed to provide secure, high-bandwidth, and scalable networks to enable access to fixed and mobile applications across metropolitan areas. The products use an 802.11 radio to provide network connectivity to the end users and a separate radio that is used for communication with the other MRs on the backbone. For instance, Aironet 1500 works as a MR that uses 802.11g/b for connecting the end users while using 802.11a radio to connect with neighboring MRs. The Aironet 1500 series support 16 broadcast identifiers to create multiple Wireless LANs so that the accessing network can be segmented to provide services to multiple user types. The IGW node is able to connect 32 other Cisco MRs (i.e., Aironet 1500).

A routing algorithm based on Adaptive Wireless Path Protocol (AWPP) allows remote MRs to dynamically select the multihop path toward the destination or the IGW. If new MRs are added to the network, each MR self-adjusts to ensure networking capability. The Cisco Aironet 1500 Series interacts with Cisco wireless LAN controllers and Cisco Wireless Control System (WCS) Software, having centralized key functions, scalable management, security, and mobility support. The security solution is compliant with 802.11i, Wi-Fi Protected Access (WPA2), and Wired Equivalent Privacy (WEP), which provide authentication for various WAP types and ensure data privacy with necessary encryption. The MR joins the network using X.509 digital certification and the wireless backbone uses the hardware-based Advanced Encryption Standard (AES) encryption. The Cisco solution based on the dual-radio approach raises the question of scalability and capacity in the infrastructure mesh, where all the MRs use 802.11a and the clients use 802.11b/g.

1.3 MANET and WMN Network Capacity

1.3.1 Capacity of Pure MANETs

Gupta and Kumar [10] derived the network capacity of a pure MANET. Intuitively, the network capacity is expected to increase with the network size because of spatial reuse of wireless channel. In fact, the theoretical analysis [10] has proved that the capacity of MANETs decreases with the growth of the network size. We consider a MANET with n MR nodes arbitrarily located in a disk R. Each node chooses an arbitrarily destination to which it wishes to send traffic at an arbitrary rate by single hop or multihops. If the channel has the capacity of w bits/s, the throughput per node in the MANET is $\Theta(w/\sqrt{n})$ bits/s [10]. In other words, the throughput per node decreases with the increase of the number of nodes. The main reasons of limited node capacity are:

- *Multihop communication*. The traffic of a node has to be forwarded in a multihop fashion that repeatedly consumes radio resource.
- *Co-channel interference*. Concurrent transmission is not allowed by two nodes that are in the interference range. The wireless medium can only be spatially reused when the co-channel interference can be avoided.
- *Asynchronous sending and receiving*. The nodes usually cannot receive and send simultaneously at the same channel.

1.3.2 Capacity Gain of WMNs

In this part, we study the capacity gain of the WMN. Compared to MANET, WMN increases its network capacity in two ways: the addition of infrastructure nodes [11] and the usage of multiradio for operating multichannel [12].

1.3.2.1 Capacity Gain of Infrastructure

Liu et al. [11] studied the capacity gain by adding the infrastructure to a pure MANET, where access points like IGWs are placed either in a random or arbitrary manner. For instance, as shown in Fig. 1.3a, two IGWs are added in the MANET. It assumes that these IGWs are interconnected to the Internet with wired network of infinite bandwidth. Similar to an ad hoc node, each IGW has a data rate of w_1 bp over the common channel. The bandwidth w_1 is ivided into three parts: intracell w_1^1, uplink subchannel w_1^2, downlink subchannel w_1^3, where $w_1^2 = w_1^3$. Note that a

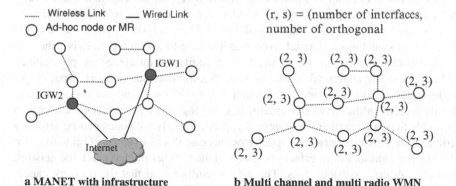

a MANET with infrastructure   b Multi channel and multi radio WMN

Fig. 1.3 (a) MANET with infrastructure. (b) Multichannel and multiradio WMN

deterministic routing scheme is used for a node to communicate. If the destination is outside the IGW in which the source node is located, the traffic is sent through the IGWs. Otherwise, the traffic is forwarded to the destination node hop by hop. Liu et al. [11] showed that the number of IGWs involved in multihop forwarding impacts the network capacity. If the number of IGWs (i.e., m) grows slower than $\sqrt{n}$, the network capacity per node scales as $\theta(w_1^1/\sqrt{n^* \log(n/m^2)})$. In this case, as a pure MANET, the network capacity decreases as the network size (i.e., n) increases. The improvement of network capacity because of IGWs is thus insignificant. On the contrary, if the IGWs are added at the rate faster than $\sqrt{n}$, the network capacity per node scales as $\theta(w_1^2 {}^* m/n)$, meaning that the network capacity per node increases linearly as more IGWs are added. The IGWs reduce the number of wireless transmissions between nodes. Hence, IGWs can accommodate more traffic and effectively improve the network capacity. In order to achieve a nonnegligible capacity gain in an infrastructure-based MANET, the investment of IGWs should be high enough, i.e., the number of IGWs should grow at a rate faster than $\sqrt{n}$, as the number of ad hoc node increases. Thus, n node network with growing traffic requires a minimal increase in the number of IGWs such that the network capacity does not drop.

1.3.2.2 Capacity Gain of Multichannel and Multiradio

Kyasanur and Vaidya [12] investigated the gain of the network capacity because of multiply channels. They theoretically showed how multiple channels and interfaces can improve the network capacity. In a multichannel network, all nodes are equipped with multiple interfaces (r) to operate on s orthogonal channels, $1 \leq r \leq s$. In order to show the impact of multichannel, they consider a network without IGW nodes. As shown in Fig. 1.3b, each node has at most three wireless interfaces ($r = 3$) to operate on three orthogonal channels ($s = 3$). A MR is capable of concurrently transmitting or receiving traffic on multiple available interfaces at a given time. It assumes that the total data rate possible by simultaneously using all orthogonal channels is w_2. It is noted that $w_2 = s^* w$, if every channel has the data rate of w. If the number of channels is equal to the number of interfaces, all the channels can be fully utilized. Otherwise, depending on s/r, the network capacity may suffer capacity loss because of inefficient channel utilization. The analysis assumes that there is no delay in switching a channel from one interface to another. Otherwise, the network capacity will be further reduced, if no additional interfaces are provisioned at the nodes. The analytical results indicate that the maximal network capacity per node scales as $\theta(w_2/\sqrt{n^* \log n})$ bp when $1 \leq s/r \leq \log n$ and the nodes are randomly located in the network domain. If $s/r \geq \log n$ (i.e., interfaces are less than channels than a rate), there is always a capacity loss. If the nodes can be arbitrary situated, the maximal network capacity per node scales as $\theta(w_2/\sqrt{n})$ bp with $r = s$. If $s > r$ (i.e., number of interfaces is less than the number of channels), the network capacity suffers a capacity loss. The above results reveal that the network capacity is improved in comparison with pure MANETs because of the use of multiple interfaces and multiple channels (i.e., $w_2 > w$). The actual improvement depends on the number of channels and interfaces, and the routing and transmission scheduling protocols used [12]. In order to achieve the maximal capacity improvement, all channels have to be effectively utilized, possibly by dynamically switching channels at the interface.

1.4 WMN Positioning Technique

As given in Sect. 1.2, the node in a MANET is randomly located and movable depending on the mobility pattern. On the contrary, the location of the MR in a WMN is stationary with minimal mobility. Again, the placement of the MRs and the IGWs is one of critical factors that determine the WMN performance [10]. If the MRs and the IGWs are randomly situated as in the MANET, it encounters the following problems:

- *Unbalanced load distribution.* The low traffic areas in a WMN domain may have many MRs whereas the heavy traffic areas could be covered with a few MRs. Thus, congestion may occur in the hotspots, resulting poor network performance in these areas. On the contrary, the MRs in the free traffic remain underutilized. Moreover, in a WMN, the traffic is primarily oriented towards the IGWs and the areas close to the IGWs have the high traffic. Hence, it is necessary to place more MRs or interfaces near the IGWs so as to alleviate congestion and achieve better network performance. The random deployment policy cannot take this factor into account and MRs around the IGWs may be the bottleneck of the traffic flow.
- *Uncontrolled interference.* Independent of its neighboring and global network information a random placement may create a network topology with high degree of interference, which could significantly deteriorate the network performance. For example, in an IEEE 802.11 wireless network, because of hidden terminal problem, the behavior of a node is decided not only on by its own capability, but also by the behavior of neighboring nodes and its hidden nodes. The achievable throughput for a MR decreases when the transmission rate of its neighboring nodes and hidden nodes increases. It is because of the fact that this MR may have a reduced chance to use channel/subchannel because of the presence of contention and interference. In contrast, a well positioned network can help in mitigating the interference by selecting the optimal position for each MR.
- *Unreliable architecture.* A random placement fails to consider the connectivity degree, which determines the fault-tolerance of the network. Therefore, the random placement approach results in an unreliable WMN architecture without considering fault-tolerance in the presence of link failures. It is possible that the failure of a link can disconnect the network so that all MRs, which connects to the IGWs by multi hop wireless links, may lose the Internet connectivity. For example, in a random constructed WMN, if an IGW fails, all MR connected to it may lose the Internet connectivity if there is no other neighboring IGWs that can provide an alternate connection.

Different locations of MRs and IGWs lead to different network topologies and architectures with distinct performance. Positioning technique in WMNs is required for appropriately placing the IGWs and MRs to achieve desired network performance. The WMN positioning technique can be defined as physical configuration of the IGWs and MRs including their locations and the number of interfaces on them. The configuration of IGWs and MRs is subjected to some constraints like geographical constraint, maximum number of channels, and the traffic demand. The positioning technique in WMNs can be further divided into two issues namely: IGW placement and MR placement as we study in detail.

1.4.1 Positioning Technique for IGWs

Positioning technique for IGWs targets to minimize the number of IGWs while meeting the bandwidth requirement. Because IGWs serve as the gateway to provide the Internet connectivity to MRs, IGWs usually are equipped with wired connections to the Internet, which increases the installation cost of a WMN. Therefore, to reduce the cost of the network, we need to keep the number of IGWs installed as small as possible. On the other hand, WMNs require enough IGWs to satisfy the network capacity needed by MRs. Therefore, the positioning technique for IGWs concerns where the IGWs should be located and how to minimize the number of IGWs while satisfying the MR Internet throughput demand. The critical questions should be answered while deploying IGWs:

- How many IGWs are needed in WMN?
- How many interfaces should be configured in the IGWs?
- Where the IGWs should be placed?
- How many and which MRs should be served by which IGW?

In the following section, we investigate Existing IGW selection algorithms that address these questions.

Cluster-based IGW selection. This deployment approach is proposed in [13], which is based on TDMA (Time division multiple access) MAC technology. In this scheme, the network nodes are divided into several disjoint clusters and in each cluster one IGW is deployed to serve the MRs in the cluster. To satisfy the minimal bandwidth requirement imposed by the QoS constraint, they assume each MR has a weight, which represents the bandwidth requirement. On the other hand, the total weight of all cluster nodes is also bounded because of limited capacity of the IGW. Then, the IGW deployment problem is abstracted as a clustering problem of minimizing the number of clusters that could satisfy the QoS constraints. Two polynomial time approximation clustering algorithms are proposed. The first algorithm, called shifting strategy, is a divide-and-conquer method and can have different computational complexity depending on different qualities of solutions. The shifting strategy is able to find a near optimal solution at the expense of high computational complexity. The second approach is a greedy domination-independent-set approach. This greedy algorithm selects a dominant-independent-set of the power graph as the set of cluster-headers, and guarantees an approximation ratio that is linear with the maximal radius R (i.e., the maximum number of hops from a MR to the IGW), independent of the network size. To satisfy the weight, depth and replay-load requirements, clusters created by either method is further refined based on the weight partition algorithm. To meet the assumption of the TDMA scheme, a special adaptive delivery mechanism is designed. Because of TDMA, this approach is not applicable for a generic WMN that is based on IEEE 802.11s.

Multihop WLAN-oriented IGW selection. The IGW positioning algorithm proposed in [14] is on the basis of IEEE 802.11 multihop WLAN network architecture. In their scheme, the constraints of wireless channel capacity, wireless interference, fault-tolerance, and variable traffic demands are all considered. With respect to co-channel interference, two coarse-grained interference models that capture the trend of throughput degradation because of wireless interference are proposed: the bounded hop-count model and smooth throughput degradation model. The IGW positioning problem is modeled as a capacitated facility location problem. There are two steps in their IGW positioning scheme: Given a set of potential IGW locations, they first prune the search space by grouping points into equivalent classes. Second, they use a greedy placement approach by which they iteratively pick an IGW that maximizes the total demands. In this step, the search on the IGWs is on the equivalence classes created in step one. To decrease the computing complexity, some approximations such as traffic demand are also used in the greedy IGW selection phase. In a WLAN, each node only has one interface and the drawback of this approach is that it fails to consider the issues of multichannel and multiradio.

OPEN/CLOSE Heuristic IGW selection. The problem of optimum IGW selection specifically for a mesh network has been analyzed in [15] recently. They have developed two approaches to select the optimal number of IGWs and determine the placement of IGWs. Given a network with MRs, the IGW selection approach determines the appropriate location for an IGW. The first approach is based on an integer linear programming model to greedily select IGWs from a set of potential alternatives and calculates all possible solutions of IGW placement in term of establishment cost and communication capacity between IGWs and MRs. But, the computational complexity of this approach increases with the number of potential IGWs, which limits its effectiveness for a large mesh network. The second approach is an OPEN/CLOSE heuristic to find a sub-optimal solution. The heuristic scheme starts from any feasible solution and repeatedly decreases the investment cost by a certain percentage. If no more solution can be found, the current solution is claimed to be the best approximation.

QoS-based IGW selection. A QoS-based IGW selection approach for WMNs in [16] developed a recursive algorithm with the purpose of minimizing the number of IGWs and satisfying the QoS requirement. In this approach, an one-hop dominating set of original graph will be greedily found first and this result will be used as the input of next recursion. The one-hop dominating set means that every node is connected to a clusterhead directly (i.e., single hop). The greedy dominating-set searching operation continue until the cluster radius reaches R, which is the predefined upper bound of cluster radius (i.e., maximum number of hop from a MR to the IGW). Finally, the clusterhead is the selected IGW. Because the cluster formulation fails to consider the hop length from each individual MR to the IGW, many faraway MRs may be attached to the clusterhead (i.e., IGW), and thus could not minimize the number of hop from the MR to the IGW.

Tree-based IGW selection. In [17], A WMN is modeled as a tree-based network architecture. In this architecture, the IGW is the root of the tree and all MRs are attached to the tree. The author first formulates the IGW selection as the problem of selecting multiple trees from an initial network graph by a linear program (LP). Then, they developed two heuristic algorithms: Degree-based Greedy Dominating Tree Set Partitioning (Degree-based GDTSP) and Weight-based Greedy Dominating Tree Set Partitioning (Weight-based GDTSP), for the purpose of cost-effective IGW selection. Both algorithms target to partition an initial network graph into multiple trees while considering the MR and IGW capacity. However, the tree-based WMN network architecture fails to consider the multipaths that allows each MR to connect the IGW using multiple paths.

1.4.2 Positioning Technique for MRs

The MR positioning problem can be described as a way to cover a given region with a minimum number of MRs and interfaces while meeting the traffic demand. According to this definition, the goal for MR positioning includes:

- Provide enough network capacity for satisfying the traffic demand
- Minimize the number of MRs and their configured interfaces, thereby minimizing the cost of deploying the network
- Avoid congestion incurred by balancing the traffic
- Maximize the network capacity with a certain number of MRs and their configured interfaces
- Provide fault-tolerance

The efficient placement of MRs is a challenging issue because of many practical constraints and contradictory requirements such as cost, link capacity, wireless interference, and varying traffic demands. For example, sparse MR placement is favored in terms of cost whereas dense placement provides better fault-tolerance. The main factors that should be considered in the MR placement are as follows:

- *Geographical restriction.* The MR placement should satisfy the coverage requirement expected by users and should also meet the location limitations imposed by the geographical terrain.
- *Connectivity degree.* The WMN consisting of MRs should be fully connected, meaning every MR is at least connected to an IGW by a path. On the other hand, a

higher connectivity degree is beneficial to satisfy the variable traffic demand and is helpful in maintaining fault-tolerant links to provide resilient to MR failures.

- *Traffic information and link capacity.* The traffic information should be considered carefully for placing MR. In a heavy traffic area, more MRs or interfaces should be placed to meet the high traffic demand. The potential relaying path should be explored to avoid the congestion problem.
- *Co-channel interference.* The physical parameters of wireless link, such as fading and Signal-to-Noise Ratio (SNR), vary considerably in different geographical environments. The MR placement has to minimize the co-channel interference.

Compared to the research work on IGW placement, the study pertain to MR placement is still in the infancy time. The two network architectures commonly discussed in the most previous approaches are grid-based or tree-based [18]. In both these approaches, each MR has been assumed to be equipped with the same number of interfaces. In a grid-based mesh network, each node interface (e.g., 2, 3, or 4) connects to four neighboring nodes; whereas in a tree-based mesh network, each node connects to three neighboring nodes in which two nodes are descendent nodes in the tree. Both deployments are too simple for the real world. In addition, uniform interface configuration may again lead to poor performance and low utilization efficiency for the network architecture. For instance, in the tree-based architecture, a MR at a higher level needs to carry more traffic than its descendent nodes. They are supposed to require more radios. Otherwise, they might suffer from short of capacity. In contrast, those nodes far away from the IGWs may need fewer interfaces because they have less relaying traffic.

One of the elementary explorations about MR placement is presented in [19]. The authors discussed the MR placement on the condition that MRs can only be placed in the predecide candidate positions while considering the coverage, connectivity, and traffic demand constraints. They proposed a two-phase heuristic algorithm to find the optimal MRs and their positions. In phase I, the algorithm greedily excludes the candidate nodes that do not cause the uncovered hole by testing all the candidate nodes. The remaining node set, called coverage set, can satisfy the network coverage requirement but not the connectivity. For example, some nodes still have no route to the IGW. The nodes in the coverage set that can connect to each other form a cluster. In the other words, a cluster in the coverage set is the nodes they are connected at least by a path. In phase II, the algorithm adopted an add-and-merge procedure to select minimal number additional candidate nodes and add these nodes into clusters so that clusters can merge together. The mergence happened only if the aggregated traffic of the resulting cluster will not violate the traffic demand constraints.

Figure 1.4 illustrates an example how the algorithm proceeds. Initially, there are 43 candidate nodes as Fig. 1.4a shows. After removal of unnecessary nodes in phase I, it generates 11 nodes marked as black points and 10 corresponding clusters (i.e., Δ_i denotes cluster i as shown in Fig. 1.4a). It is because among 11 nodes, only nodes v_5 and v_6 are within the transmission range of each other and can be connected wirelessly. It can be seen that the nodes provide full coverage, but they still are not a connected graph. Figure 1.4b illustrates one step of add-and-merge procedure. Each of the dash lines represents that a new node has been added to a cluster. For example

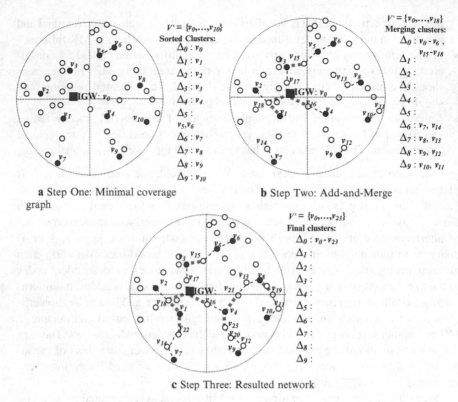

a Step One: Minimal coverage graph

b Step Two: Add-and-Merge

c Step Three: Resulted network

Fig. 1.4 Heuristic algorithm for MR placement

v_{11} is introduced into Δ_9 and v_{15} into Δ_5. When a node is added, two clusters may be able to merge as one cluster. In Fig. 1.4, the square dot dash line represents the mergence of two connected-clusters. For instance, when v_{15} is added into Δ_5, Δ_5 is then able to merge with Δ_3 because v_{15} and v_3 are neighboring nodes. The final deployment showed in Fig. 1.4c includes 24 nodes after more nodes are added with a similar procedure.

1.5 Fairness Transmission in 802.11-Based MANETs and WMNs

IEEE 802.11 Distributed Coordination Function (DCF) is the most used MAC protocol in both MANETs and WMNs. DCF implements wireless medium sharing among a number of devices (i.e., nodes) through the use of carrier sense multiple access/collision avoidance (CSMA/CA) technology with a random back-off policy. A node first senses status of the channel for ongoing transmissions before sending a packet over a channel. If the channel is already in use, the node defers its transmission. It waits for a random time and re-attempts to sense the channel. On the other hand, if the channel is currently free, the node starts transmission. Such a mechanism is very effective when the channel is not overloaded, because it allows a node to transmit the packet immediately with a minimum delay. On the other hand, it always has a chance of collision if multiple nodes sense the channel free and begin transmission simultaneously. Although CSMA/CA solves the problem of channel contention, this MAC protocol, together with traditional transport protocols result in severe unfairness problems, which degrade the network performance:

- *Local unfairness* among the nodes that are within the interference range
- *End-to-end unfairness* between end-to-end multiple-hop flows

1.5.1 Local Unfairness

The phenomenon of local unfairness is that some flows dominate the transmission for a long period time while the other flows have no chance to seize the channel for their transmission. The main reasons that cause the problem are hidden and exposed terminal conditions and 802.11 MAC (i.e., CSMA/CA) backup policy.

If some devices within the interference range compete the medium for packet transmission, only one node can use the channel for transmission at a given time, while other node have to defer their transmission and enter into backup status as MAC protocol requires. Because the Transmission Control Protocol (TCP) also has the backup policy, the suspending transmission in MAC layer causes TCP to further back off. Thus, unfairness occurs and the losing flow suffers from a low transmission rate.

Figure 1.5 shows the two unfair node topologies: the hidden terminals and the exposed terminals. In Fig. 1.5a, S0 and S1 are hidden nodes and packets sent by the sender S0's can be corrupted by S1's signals. In other words, the reception at the receiver R0 fails because of S1's simultaneous transmission. On the contrary, packets sent from S1 to R1 are immune to S0's interference. Such a scenario causes unfairness because S1 has more chance for packet transmission than S0. In Fig. 1.5b, sender S2 can sense the other two senders but unfairness exists as following. In this

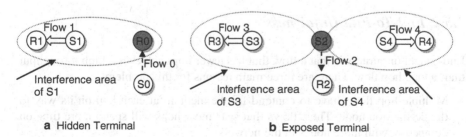

a Hidden Terminal   **b** Exposed Terminal

Fig. 1.5 Local unfairness

scenario, S2 is exposed to two senders (i.e., S3 and S4) whereas nodes S3 and S4 are only exposed to one sender (i.e., S2). Therefore, S2 is forced to back off more often, and thus flow 2 (S2—>R2) has less chance to transmit its packets than the other two flows. In order to evaluate the unfairness in the above scenarios, simulations are carried out for considering a case that the total bandwidth of a channel is 20 Mbps. The simulation results [20] have shown when the total bandwidth of a channel is 20 Mbps, Flow 0 and Flow 1 achieve 1 Mbps to 20 Mbps bandwidth allocation, respectively, in the scenario of Fig. 1.5a and the three flows in the scenario of Fig. 1.5b have the bandwidth of 18 Mbps, 3 Mbps, and 18 Mbps, respectively.

Local fairness has a similar impact on either 802.11-based MANETs or 802.11-based WMNs. For the purpose of providing fairness medium access in the 802.11-based network, some approaches have been proposed, and they can be classified into two categories:

General fair queuing. Some fair queuing algorithms are evaluated in [21]. The key point of wireless fair queuing algorithms is to monitor and predict the flow status and channel condition. The network then assigns more network resource to the node that experiences bursting and location-dependent errors in wireless channels, which compensates the losing flows. This approach requires a centralized coordinator such as access point that is able to schedule the wireless traffic in the network.

Self-adoption protocol. Self-adaptation in 802.11 networks has been recently addressed to provide a way for fair network bandwidth allocation. This approach includes altering the MAC back-off durations [22], switching from sender-initiate mode to receiver-initiate mode [23], and adapting the transmission rate and time scheduling [24]. For example, in the above hidden-terminal scenario in Fig. 1.5a, if the receiver R0 in Flow 0, rather than the sender S0, initiates the transmission by sending a request-to-receive (RTR) packet, then the sender S1 would back off while S0 is transmitting. Thus, Flow 0 no longer experiences packet losses [23]. On the other hand, in the exposed-terminal scenario, if the sender S2 shortens its random back-off window to a smaller value than the other two senders, then S2 will have a higher chance to use the shared medium than the other two senders [24]. These adaptation approaches, however, require modifications of the existing 802.11 MAC protocol.

1.5.2 End-To-End Unfairness

End-to-end unfairness is that a flow that has fewer hops achieves a high throughput than a long hop flow. There are three main reasons for this problem:

- Multiple-hop flows have to contend for the medium at each hop on its way to the destination node. Thus, flows that span more hops will spend more time on competing with more nodes in the network.

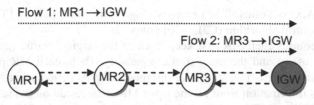

Fig. 1.6 End to end unfairness

- The packets forwarded on a longer path are more likely to be dropped than that of a shorter path.
- The end-to-end unfairness is getting even worse when the transport protocol, i.e., TCP, cannot distinguish the reasons of packet drop such as network congestion, packet collision, or stale route. Instead, it equally treats all packet losses as an indicator of network congestion and the sender will consequently reduce its congestion window size and lower the sending rate. These increases in round trip time (RTT) and packet loss lead to a lower throughput for the longer flows.

Figure 1.6 shows such an unfair example consisting of a three-hop TCP flow (i.e., Flow 1) and a one-hop TCP flow (i.e., Flow 2). Flow 1 and Flow 2 intend to transmit the same amount of packets. The experiments in [25,26] show Flow 1 always get far less share of the bandwidth than Flow 2 and starve most of time. The reason is that Flow 2 has an increased number of channel contention, an increased probability of packet loss, and more TCP backup times.

End-to-end unfairness problem in the WMN exhibits some unique characteristics compared to the MANET. The premise of MANET is that every node is interested in communicating with any other node in the network, and thus the flows could be well distributed in the network following different multihop paths. However, in a WMN, most of the packets are IGW-oriented, meaning that the packets are predominately transmitted between the MRs and the IGWs. In other words, the packets aggregated by a MR mostly lead to an IGW for the Internet, and the IGW sends the packets to the MR, upon receiving these packets from the Internet. Because of the fact that the flows originated by a distant node away now are more likely to share the same MRs with those flows closer to the IGWs, the queue delay accumulation will further slow the long-distance flows down. In other words, the default first in first out (FIFO) queue in each MR will make the packets of long-distance flows wait for a longer time while they are traveling a long distance toward the IGW. On the contrary, the traffic from the MRs closer to the IGWs can be forwarded to the IGW fast with a low packet loss. The following part is some approaches regarding to the end-to-end unfairness problem.

1.5.2.1 Global Fair Bandwidth Allocation

This approach requires a global view of the network topology and traffic load to make the decision on global fair network bandwidth allocation, exemplified by the

Inter-Transit Access Points (TAP) Fairness Algorithm (IFA) [25] and Co-ordinate Congestion Control algorithm (C3L) [26] protocols.

The IFA requires each MR first keep track of the original traffic (generated by its associated users) and the relay traffic separately. Then, each MR periodically broadcasts and forwards this statistical information to its neighboring MRs. Upon collecting this information from all the other MRs that reside on the same routing branch to an IGW, a MR is able to locally compute its own fair network share and correspondingly limit its sending rate according to the calculated result.

The C3L protocol is designed to achieve max–min per-flow fairness in the network. It first divides the whole network into several collision domains based on node topology; the network capacity of each domain is estimated by measuring the senders' queue length change while varying the traffic load pushed into that domain. It then allocates the network capacity to each node/link in that domain, proportionally to the number of flows going through it. The throughput of a multihop flow that crosses multiple collision domains will be determined by the smallest bandwidth granted by these domains.

1.5.2.2 Local Unfairness Measurement and Adjustment

Protocols such as ad hoc transport protocol (ATP) [27] and an end to end rate-based flow control scheme (EXACT) [28] only need local adjustments at each MR to help multihop flows to achieve the max-min fairness. These protocols first estimate the available bandwidth of each mesh link by measuring the queue, transmission, and contention delay at each MR. Then, each MR divides its bandwidth (or equivalently, the total service time) equally among all flows going through it. Thus, a multihop flow's throughput will be determined by the smallest bandwidth granted by the MRs along its routing path to the IGW. The difference between ATP and EXACT is that ATP calculates the "average packet service time" at each node, and the sender of each flow will adjust its transmission according to the maximum of these estimated service time. On the contrary, in EXACT each node divides the measured bandwidth among all the flows going through it.

1.6 Multihop and Multiradio Routing

1.6.1 Multihop Routing in MANETs

In a MANET, each node with a limited transmission range acts as traffic source, destination, or a router to collaboratively forward data packets for other nodes without a priori knowledge of the network topology. Therefore, a MANET routing protocol is required for a node to find the communication path in a dynamic network environment. In general, a routing protocol includes two parts: routing discovery and routing maintenance. In the routing discovery stage, the node discovers the path in a self-configurable manner before sending traffic. The basic idea of routing discovery is that the node initially advertises its existence and listens to the advertisement from its neighboring nodes. The node then knows the presence of its neighboring nodes as well as the way to reach them. The node can further find out the nodes that are outside its transmission range after its neighboring nodes have been discovered. In the stage of routing maintenance, a route can be reconstructed once the using path is broken. The routing reconstruction can be performed by a global or local routing discovery procedure to find the new path to the destination.

The design of a MANET routing protocol is a complex problem that has to consider many performance requirements (1) fast routing discovery, (2) fast routing recovery, (3) small communication overhead, (4) low computational overhead, and (5) efficiency and scalability for a large-scale network. These requirements are mostly imposed by the application requirements such as low traffic delay, node constraints such as limited power and memory, and node mobility. In the past few years, a variety of MANET routing protocols have been developed, and they can be classified into proactive, reactive, or hybrid, depending on the reaction of node in the routing determination process. On the other hand, they can be also categorized into flat or hierarchical protocols according to the logical network structure.

1.6.1.1 Proactive vs. Reactive

The proactive protocol maintains the routes for all destinations at every node up to date. Because every node always keeps the routes to all destinations, the communication connection can be immediately established any time by a node for reaching a destination, resulting in a minimal routing delay. In response to topology change because of node mobility, nodes have to periodically exchange the routing and link information between them even if the communication is not required. The proactive routing maintenance results in a heavy communication overhead in maintaining all routes up to date. Consequently, the network size and the node mobility are two hindrances in designing a scalable proactive routing protocol. The typical proactive protocols include Destination Sequenced Distance-Vector (DSDV) [29], Wireless Routing Protocol (WRP) [30], etc.

In contrast to proactive routing protocols, a reactive (or on-demand) routing protocol discovers or updates a multihop path from a source to the destination only when the communication is required. The multihop routing process is initiated only when a node has a packet to be delivered to a destination, and the path is maintained until the session is finished. Consequently, a reactive protocol avoids the prohibitive cost of routing maintenance as required in a proactive protocol, and thus achieves less communication overhead and scalability. On the other hand, it always suffers a delay on the path establishment before packet delivery if there is no fresh route to the destination. The typical reactive routing protocols include DSR [4], AODV [29], etc. The detailed information can again be found in [31].

1.6.1.2 Flat vs. Hierarchical

In a flat routing protocol, the MANET does not need to maintain any specific structure such as hierarchy, and every node performs the same functionalities. Upon receiving a routing request from a node, all neighboring nodes response by forwarding it to its neighboring nodes if itself is not the destination node. The flat protocol saves the communication overhead caused by maintaining a hierarchical network structure. However, the scalability and complexity may be problems in a large-scale network. For example, AODV is a flat routing protocol.

Rather than a flat network structure, a hierarchical routing protocol maintains a hierarchical network structure, which offers scalability and reduces the complexity in routing computation as well. In general, a hierarchical routing protocol organizes the network as a hierarchy, consisting of certain number of clusters, from the lowest level 0 to the top level $L - 1$. In each level, it may have multiple clusters and each cluster elects certain nodes as the leaders. The cluster leaders maintain network state information at multiple levels of granularity. A level i $(0 \leq i < L)$ cluster consists of level $(i + 1)$ clusters. By taking advantage of multilevel network structure, the hierarchical routing protocol implements a hierarchical addressing. The main idea of a routing discovery is that a routing request, which is initiated by a node at level i, is successively forwarded to its lower level $(i - 1)$, until the request reaches the level that contains the destination. The hierarchical routing protocols differ in the approaches that are used for clusters to collect network state information, and the particular path used in the routing process. The hierarchical network structure reduces the number of participating nodes and the communication overhead for routing discovery and maintenance. However, these protocols suffer the communication overhead on maintaining the network hierarchy. Hierarchical state routing (HSR) [32] is a typical hierarchical routing protocol that maintains a hierarchical topology. A cluster elects a clusterhead at the lowest level and again the clusterhead is a member of the next higher level. On the higher level, superclusters are formed, and so on. A multihop route can be directly established within the cluster. On the other hand, if the destination is outside the same cluster, the node requests the clusterhead, which is able to forward the packet to the next level until to the clusterhead that the destination belongs to. The packet from this clusterhead then travels down to the destination node.

1.6.2 Multihop and Multiradio WMN Routing

The stationary MRs renders the node topology of a WMN to be fixed. On the other hand, each MR is connected to an external power supply so that it has no energy constraint. Thus, quite different from the routing design objectives in MANETs, the routing protocol in WMNs more focuses on how to determine the routing paths that can maximize the network throughput, taking advantage of multiratio. The routing protocols for MANETs have the following problems when they are directly applied in WMNs:

- *Minimum hop routing metric.* The minimum hop routing metric results in the poor throughput in WMNs. The minimum hop routing metric has been extensively used in MANETs because a route with smaller number of hops involves fewer forwarding nodes, and consequently fewer transmissions and lesser energy expense. Furthermore, a high packet delivery ratio can be guaranteed with fewer hop routes in a highly dynamical wireless environment. However, the minimum hop routing metric does not perform well in WMNs. The primary reason is that a higher packet loss or lower throughput link may be selected because of less transmission hops. The link quality is affected by two factors: the distance between the transmitter and the receiver, and the interference at the receiver. The minimum hop routing always choose the node that is furthest away as the next hop node. However, as the distance between the two nodes increases, the SNR decreases, which results in that the receiver may not be able to correctly decode the data. In addition, minimum hop metric may also select the link having the heavy interference, which also results in a low SNR in the receiver. As a result, the packet error rate and loss rate increase, which reduces the network throughput.
- *Multiradio and multichannel.* The routing protocols in MANETs are not able to take advantage of the channel diversity in WMNs. As the discussion in Sect. 1.3, multichannel and multiradio techniques improve the network capacity. However, the existing protocols in MANETs have no functionality that evaluates the impact of multiple channels or radios. In a WMN, the routing protocol has to choose the best channel or radio in terms of link quality.
- *Load-balancing route.* There is no load-balancing strategy in the MANET routing protocols. A MANET typically is dominated by peer-to-peer traffic so that the source and the destination are randomly distributed in the network. However, WMNs has to support rich Internet services and so the traffic volume may be very high and is oriented toward the IGWs. Therefore, the traffic is concentrated on the paths between the MRs and the IGWs. Furthermore, the traffic burst may occur unpredictably, depending on the applications. The MRs close the IGWs and the paths directed to the IGWs are easily turned into the hotspots because they have to deliver more Internet traffic. The minimum hop routing protocol routes the data by using the shortest path without considering load-balancing on the nodes and the paths. In this manner, when more and more packets continue arriving on a node or a path, it is overburdened, which increases the number of dropped packets and transmission delay. Moreover, additional retransmitted packets generate more traffic. As a consequence, the network performance further deteriorates.

Therefore, new routing metrics have to be designed in the WMN routing protocol that considers multiple radios and the traffic property in WMNs. In the following part, we discuss three new developed routing metrics for WMN routing.

Expected transmission count (ETX) [33] is one of the earliest new metric for WMNs. ETX takes into account asymmetry link loss in the two direction of each link and uses the fixed-size probing packets to evaluate d_f, the forward delivery

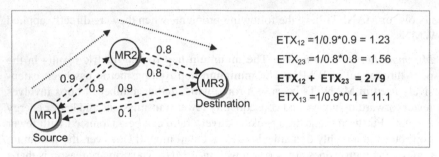

Fig. 1.7 An example of ETX

ratio, i.e., the probability that a data packet successfully arrives at the receiver, and d_r, the reverse delivery ratio, i.e., the probability that the ACK packet is successfully received. Then, the ETX value of each link is equal to $1/(d_f \times d_r)$. The EXT value of a path is the sum of the link ETX. Instead of choosing the path with minimum hops, the routing algorithm selects the path with the minimum ETX. Let us consider the source and the destination: MR1, MR3, respectively, as shown in Fig. 1.7. If a routing protocol uses the minimum hop metric as in the MANET, it chooses the direct link between MR1 and MR3 as the communication path. However, as shown in Fig. 1.7, the reverse delivery ratio from MR3 to MR1 is only 0.1, which is indicated by the successful reception probability of ACK packets. Compared to other links (i.e., 0.8), the link bandwidth in this direction is extremely low. In contrast, if the path (1–2–3) is selected using ETX as routing metric, the path can achieve a high throughput in both directions.

The drawback of EXT is that it does not identify the different throughput of the links and thus their metric may still suffer from a low throughput. On the other hand, ETX fails to consider the multichannel and traffic load balancing. These limitations motivate the development of new protocols as illustrated below.

Weighted cumulative expected transmission time (WCETT) [34] is a routing metric that takes into account both link throughput and channel diversity. It first extends the link ETX to the expected transmission time (ETT) with which to address the link bandwidth. The ETT is defined as: ETT = ETX × size of probe packet/current bandwidth. Given a path with n hops, k involved channels, it defines W_i as the sum of ETT of hops working on channel i in the path:

$$W_i = \sum \text{ETT}_j \quad \text{Hop } j \text{ on channel } i, \quad 1 \le i \le k. \tag{1.1}$$

Then, it defines α as the maximal value W_i of k channels

$$\alpha = \text{Max } W_i, \quad 1 \le i \le k. \tag{1.2}$$

Furthermore, let β be the sum of ETT of all the hops of the path

$$\beta = \sum_{j=1}^{n} \text{ETT}_j. \tag{1.3}$$

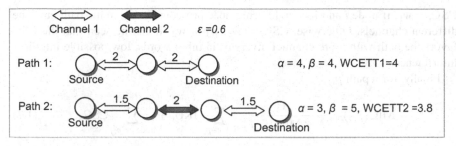

Fig. 1.8 An example of WCETT

Consequently, WCETT of a path can be formulated as

$$\text{WCETT} = (1 - \varepsilon) \times \beta + \varepsilon \times \alpha, \tag{1.4}$$

where ε is a coefficient between 0 and 1.

The path with the least value of WCETT will be selected by the source node for packet transmission. The path having more channel diversity tends to have less intra interference (packets of the same flow contend with each other at different hops). Therefore, WCETT use α to capture the channel diversity. At the same time, the path with a high link quality has the lower value of β. In (1.4), ε specified the trade-off between α and β. If $\varepsilon = 0$, WCETT is similar with ETX. Figure 1.8 is an example that explains WCETT. In the example network, it has two channels, i.e., channel 1 and channel 2. Considering $\varepsilon = 0.4$, Path1 uses only one channel and has $\text{WCETT}_1 = 4$. Path 2 uses two channels. $W_{\text{channel 1}} = 1.5 + 1.5 = 3$, $W_{\text{channel 2}} = 2$. Thus $\alpha = 3$, $\beta = 5$ and so $\text{WCETT}_2 = 5 \times 0.4 + 3 \times 0.6 = 3.8$. Thus, path 2 will be selected for the source although it is a longer than path than path 1 and has the similar link quality with path 1.

However, WCETT ignores the interflow interference and the routing still cannot bypass the nodes with heavy load intentionally. How to calculate of WCETT with low expense is still an open problem.

Interference and channels switching (MIC). MIC [35] is proposed to capture interflow and intraflow interference in WMNs. The MIC of a path includes two submetrics: Interference aware resource usage (IRU) and channel switching cost (CSC). IRU evaluates the interflow interference, link quality of different bandwidth. The CSC evaluates the intraflow interference. Given a link e: it has

$$\text{IRU}_e = \text{ETT}_e \times N_e, \tag{1.5}$$

where N_e is the number of the links that can interfere e. IRU gives a larger weight to the link with more interflow interference source.

For any node i: it has

$$\text{CSC}_e = \begin{cases} w_1 & \text{No common channels} \\ w_2 & \text{Having common channels} \end{cases} \quad 0 \leq w_1 < w_1. \tag{1.6}$$

$CSC_i = w_1$, if node i and its neighboring node previous node i in the path have the different channels. Otherwise, $CSC_i = w_2$, $w_1 < w_2$. It can be seen that the CSC favors the path with more channel diversity, in other words, low possible intraflow interference.

Finally, for a path p_{ij},

$$\text{MIC}(p_{ij}) = \frac{1}{N \times \text{Min(ETT)}} \sum_{\text{link } e \in p_{ij}} \text{IRU}_e + \sum_{\text{node } x \in p_{ij}} \text{CSC}_x, \qquad (1.7)$$

where N is the total number of nodes in a WMN and Min(ETT) means the smallest ETT in the network.

1.7 Thoughts of Practitioners

WMNs demonstrate a strong potential of supporting high bandwidth Internet access. On the other hand, WMNs reduce the construction cost because of less dependence on the infrastructure. These critical features foster the reality civilian applications in developing cost-efficient WMNs.

The key application of WMNs is to offer Internet access at the low income community areas in a city or the rural areas to promote a variety of Internet services with a low price. An example for such kind of WMNs is TFA [7]. TFA started in 2004 to provide Internet service for the low income area of the southeast Houston. TFA covers $3\,\text{km}^2$ with one IGW and 18 MRs to provide 1 Mbps minimum access rate. The monthly payment for a TFA user is approximately half of the cable or DSL connections. In the early 2007, it serves 2,000 users.

Municipal high-speed Internet service is allowable by designing a WMN in the hotspot area such as mall. Users are able to access the Internet by using their laptops or PDAs. Desirable mobility can be supported by WMNs anywhere in the municipal area whether in the car or on the street. One of leading projects for such a WMN design is Chaska Net [36] being deployed in Chaska, Minnesota, USA. A number of MRs are densely mounted on the street lamps. Chaska Net plans to have 200 MRs to cover 15 square miles in the city area and maintains a low monthly charge.

Some other promising WMN applications include enterprise wireless network, metropolitan area networks, transport information system, etc.

1.8 Direction for Future Research

WMNs provide a new paradigm for high bandwidth wireless network that tightly integrates multiradio and multichannel MANET with the Internet. On the contrary to the limited civilian application of MANETs, in the past few years not only many nonprofit WMNs have been deployed, but also many industrial giants have released their commercial WMN solutions. However, there are still a number of open issues before the advantages of WMNs can fully take effect. These challenging problems involve all seven Open Systems Interconnection (ISO) protocol layers. Specifically, the critical problems related to above discussion are summarized as below:

- *Capacity improvement.* The current implemented WMNs are still far from the theoretical capacity because these implementations could not effectively combat the interference problem, channel assignment problem, etc. In order to improve the network capacity, the following issues are critically expected:
  - In the stage of network setup, the IGW and MR configuration strategy is necessary to provide the sufficient radio and channel resource for the MR, which has a high traffic demand.
  - In the stage of network deployment, the approach of distributed channel assignment is needed to dynamically assign the orthogonal channel to interfaces of MR in such a way to minimize the interference and maximize the network performance.

- *Efficient routing protocols.* Multihop, Multiradio, and Multipath routing protocols are required for effectively deploying a WMN. The key issues are:
  - WMN routing algorithm should take into account, the availability and the diversity of multiple radios and multiple channels.
  - Reliable routing metrics should be developed for efficiently identifying the link quality with the consideration of load-balancing.

- *Fairness.* Most of works pertain to fairness are still in the experimental phase and require further evaluation:
  - Enhanced or new MAC protocols should be proposed to increase the local fairness.
  - Enhanced or new transport protocol should be designed with which to effectively distinguish the packet loss because of congestion from other possible reasons in wireless networks.

1.9 Conclusions

A MANET is usually spontaneously deployed in an area for supporting peer-to-peer communication. A WMN inherits some characteristics of MANETs, such as self-origination, self-healing, and multihop communication. On the other hand, a WMN has a hierarchical architecture, i.e., static backbone and mobile clients, which facilitates the civilian application of the WMN to provide ubiquitous high-bandwidth Internet accessibility. In this chapter, we present the fundamental concepts with respect to the MANET and the WMN. In particular, we study the network design issues of a WMN from the following aspects: network capacity, IGW and MR positioning technique, unfairness transmission problem, and routing issues, by comparing them with MANETs. From these aspects we investigate the current approaches in the literature.

1.10 Terminologies

1. *Ad hoc routing*. In a MANET, each node with a limited transmission range acts as traffic source, destination, or a router to collaboratively forward data packets for other nodes without a priori knowledge of the network topology. A MANET ad hoc routing protocol is required for a node to find the communication path in a dynamic network environment. It includes two parts: routing discovery and routing maintenance.
2. *End to end unfairness*. End-to-end unfairness is that a flow that has fewer hops achieves a high throughput than a long hop flow.
3. *Internet gateway*. A few MRs with the wired connections act as the Internet gateway (IGW) to exchange the traffic between the Internet and the WMN.
4. *Local unfairness*. The phenomenon that some flows dominate the transmission for a long period time while the other flows have no chance to seize the channel for their transmission is called local unfairness.
5. *MANET*. MANET is the abbreviation of mobile ad hoc network. A pure MANET is dynamically established by mobile devices grouped together as needed without any support from the existing infrastructure. The mobile devices in the network communicate with each other through single or multihop wireless links.
6. *Mesh backbone*. The stationary wireless mesh routers (MRs) interconnecting through single/multi hop wireless links form the backbone.
7. *Network capacity*. The maximal amount of traffic load that a network can support at a time.
8. *Positioning technique*. The WMN positioning technique can be defined as physical configuration of the IGW and MRs including their locations and the number of interfaces on them.
9. *Routing metric*. Routing metric is used to differentiate the quality of different routing paths.
10. *WMN: Wireless mesh network*. A WMN is a particular multihop ad hoc network, consisting of two parts: mesh backbone, and mesh clients. The stationary wireless mesh routers (MRs) interconnecting through single/multi hop wireless links form the backbone. The MR with the wired connections acts as the IGW to exchange the traffic between the Internet and the WMN. The mesh clients can be the mobile wireless device such as cell phones and laptops. The mobile clients connect to any MRs to access the Internet via the IGW in a multihop fashion.

1.11 Questions

1. What is the main difference between MANETs and WMNs?
2. What is the main characteristic of traffic load distribution in MANETs and WMNs?
3. What is the theoretical throughput per node in a MANET of a random network model? How is the scalability of MANETs?

4. How does the network throughput improve if some infrastructure nodes are added into a MANET?
5. What are the benefits for adopting positioning technique in WMNs?
6. What is Max-flow Min-cut theorem?
7. What is the exposed terminal problem and how does the local unfairness happen in such circumstance?
8. What are the main reasons that lead to end-to-end unfairness problem?
9. What are advantages and disadvantages of reactive routing protocols in MANETs?
10. Please calculate the WCETT of the following paths. Which one will be chosen in terms of WCETT?

References

1. J. Jubin and J. D. Turnow, The DARPA packet radio network protocols, Proceedings of IEEE, 75(1), 21–32, (1987).
2. D. A. Beyer, Accomplishments of the DARPA SURAN program, Proceedings of the Military Communications Conference (MILCOM), Sep (1990).
3. C. E. Perkins and E. M. Royer, Ad-hoc on-demand distance vector routing, Proceeding of the Second IEEE workshop Mobile Computing System and Applications, 90–100 Feb (1999).
4. D. B. Johnson and D. A. Maltz, Dynamic Source Routing in Ad-Hoc Wireless Networks, Mobile Computing, T. Imielinski and H. Korth, Eds., Kluwer, Dordrecht, 153–181 (1996).
5. J. Bicket, D. Aguayo, S. Biswas, and R. Morris, Architecture and evaluation of an unplanned 802.11b mesh network, Proceedings of the Eleventh Annual International Conference on Mobile Computing and Networking (Mobicom), Aug (2005).
6. S. A. Mahmud, S. Khan, S. Khan, and H. Al Raweshidy, A comparison of MANETs and WMNs: Commercial feasibility of community wireless networks and MANETs, Proceeding of the First International Conference on Access Networks (AccessNet), Athens, Greece, Sep (2006).
7. J. Camp, E. Knightly, and W. Reed, Developing and deploying multihop wireless networks for low-income communities, Proceedings of Digital Communities, Jun (2005).
8. http://www.dailywireless.org/2007/04/23/belair-live-in-london/
9. http://www.cisco.com/en/US/products/ps6548/prod_brochure0900aecd8036884a.html
10. P. Gupta and P. R. Kumar, The capacity of wireless networks, IEEE Transactions on Information Theory, 46(2), 388–404, (2000).
11. B. Liu, Z. Liu, and D. Towsley, Capacity of a wireless ad hoc network with infrastructure, Computer Science Dept. University of Massachusetts Amherst, Technical Report, (2004).
12. P. Kyasanur and N. H. Vaidya, Capacity of multi-channel wireless networks: Impact of number of channels and interfaces, Proceedings of the Eleventh Annual International Conference on Mobile Computing and Networking (Mobicom), Aug (2005).
13. Y. Bejerano, Efficient integration of multihop wireless and wired networks with QoS constraints, IEEE/ACM Transaction on Networking, 12(6), 1064–1078, (2004).
14. L. Qiu, R. Chandra, K. Jain, and M. Mahdian, Optimizing the placement of integration points in multi-hop wireless networks, Proceeding of the Twelfth IEEE International Conference on Network Protocols (ICNP), Oct (2004).
15. R. Prasad and H. Wu, Minimum-cost gateway deployment in cellular WiFi networks, Proceeding of Consumer Communications and Networking Conference(CCNC), Las Vegas, Jan (2006).
16. B. Aoun, R. Boutaba, Y. Iraqi, and G. Kenward, Gateway placement optimization in wireless mesh networks with QoS constraints, IEEE Journal on Selected Areas in Communications, 24(11), 2127–2136, (2006).

17. B. He, B. Xie, and D. P. Agrawal, Optimizing the internet gateway deployment in a wireless mesh network, Proceeding of the Fourth IEEE International Conference on Mobile Ad-Hoc and Sensor Systems(MASS), Pisa, Italy, Oct (2007).
18. A. Raniwala and T-c. Chiueh, Architecture and algorithms for an IEEE 802.11-based multi-channel wireless WMN, IEEE INFOCOM, Mar (2005).
19. J. Wang, B. Xie, K. Cai, and D. P. Agrawal, Efficient mesh router placement in wireless mesh networks, Proceeding of the Fourth IEEE International Conference on Mobile Ad-Hoc and Sensor Systems(MASS), Pisa, Italy, Oct (2007).
20. K. Cai, M. Blackstock, R. Lotun, M. Feeley, C. Krasic, and J. Wang, Wireless unfairness: Alleviate MAC congestion first!, Proceeding of the Second ACM International Workshop on Wireless Network Testbeds, Experimental Evaluation and Characterization (WiNTECH) in Conjunction with MobiCom, Montreal, Canada, Sep (2007).
21. T. Nandagopal, S. Lu, and V. Bharghavan, A unified architecture for the design and evaluation of wireless fair queueing algorithms, Proceedings of the Fourth Annual International Conference on Mobile Computing and Networking (Mobicom), Oct (1998).
22. N. H. Vaidya, P. Bahl, and S. Gupta, Distributed fair scheduling in a wireless LAN, Proceedings of the Sixth Annual International Conference on Mobile Computing and Networking (Mobicom), Aug (2000).
23. F. Talucci, M. Gerla, and L. Fratta, Macabi (maca by invitation): A receiver oriented access protocol for wireless multiple networks, in PIMRC 1997, 1994, pp. 1–4.
24. B. Sadeghi, V. Kanodia, A. Sabharwal, and E. Knightly, Opportunistic media success for multirate ad hoc networks, Proceedings of the Eighth Annual International Conference on Mobile Computing and Networking (Mobicom), Sep (2002).
25. V. Gambiroza, B. Sadeghi, and E. Knightly; End-to-end performance and fairness in multihop wireless backhaul networks, Proceedings of the Tenth Annual International Conference on Mobile Computing and Networking (Mobicom), Sep (2004).
26. A. Raniwala, P. De, S. Sharma, R. Krishnan, and T. Chiueh, End-to-End flow fairness over IEEE 802.11-based wireless mesh networks, INFOCOM, May (2007).
27. K. Sundaresan, V. Anantharaman, H. Y. Hsieh, and R. Sivakumar; ATP: A reliable transport protocol for ad hoc networks, Proceeding of the Fourth ACM Interational Symposium on Mobile Ad Hoc Networking and Computing(MobiHoc), Jun (2003).
28. K. Chen, K. Nahrstedt, and N. Vaidya; The utility of explicit rate-based Flow control in mobile ad hoc networks, Proceeding of IEEE Wireless Communications and Networking Conference (WCNC), Mar (2004).
29. C. E. Perkins and P. Bhagwat, Highly dynamic destination-sequenced distance-vector routing (DSDV) for mobile computers, Computer Communication Review, 24(4), 234–244, (1994).
30. S. Murthy and J. J. Garcia-Luna-Aceves, An efficient routing protocol for wireless networks, ACM Mobile Networks and Appilications Journal, Special Issue on Routing in Mobile Communication Networks, 1, 183–197, (1996).
31. D. P. Agrawal and Qing-An Zeng, Introduction to Wireless and Mobile Systems, Chapter 13, Brooks/Cole (Thomson Learning), Pacific Grove, CA, ISBN No. 0534-40851-6
32. A. Iwata, C.-C. Chiang, G. Pei, M. Gerla, and T.-W. Chen, Scalable routing strategies for ad hoc wireless networks, IEEE Journal on Selected Areas in Communications, Special Issue on Ad-Hoc Networks, 17(8), 1369–1379, (1999).
33. D. De. Couto, D. Aguayo, J. Bicket, and R. Morris, High-throughput path metric for multihop wireless routing, Proceedings of the Ninth Annual International Conference on Mobile Computing and Networking (Mobicom), Sep (2003).
34. R. Draves, J. Padhye, and B. Zill, Routing in multi-radio, multi-hop. wireless mesh networks, Proceedings of the Tenth Annual International Conference on Mobile Computing and Networking (Mobicom), Sep (2004).
35. Y. Yang, J. Wang, and R. Kravets, Designing routing metrics for mesh networks, Proceeding of IEEE Workshop on Wireless Mesh Networks (WiMesh), Sep (2005).
36. http://www.chaska.net/
37. B. M. Leiner, R. J. Ruth, and A. R. Sastry, Goals and challenges of the DARPA GloMo program [global mobile information systems], IEEE Personal Communications, 3(6), 34–43, (1996).

Chapter 2
Medium Access Control in Wireless Mesh Networks

Feiyi Huang and Yang Yang

Abstract In wireless mesh networks, ad hoc and infrastructure modes are both used to support multihop data transmission from mesh clients. Because of the hybrid network architecture, existing problems and challenges in centralized or distributed networks become even worse and severely damage the network performance. In this chapter, we first analyze the network architecture and identify some technical challenges on the design of medium access control protocols in wireless mesh networks. The corresponding solutions are reviewed, followed by a brief discussion of some open issues.

2.1 Introduction

In recent years, wireless mesh networks (WMN) [1,2], together with related applications and services, have been actively researched. New applications include digital home, broadband, and wireless home Internet access, community and neighborhood networking, enterprise networking, metropolitan area networks, building automation, health and medical systems, public safety and security surveillance systems, intelligent transportation systems, emergency and disaster networking, etc. Generally speaking, a WMN is a group of self-organized and self-configured mesh clients and mesh routers interconnected via wireless links (Fig. 2.1). Mesh clients can be different kinds of user devices with wireless network interface cards (NIC), such as PCs, laptops, PDAs, and mobile phones. They have limited resources and capabilities in terms of energy supply, processing ability, radio coverage range, etc. Wireless mesh routers can be access points (AP) of wireless local area network (WLAN), sink nodes of wireless sensor network, base stations (BS) of cellular network, or

Y. Yang (✉)
Department of Electronic and Electrical Engineering, University College London,
London WC1E 6BT, UK
e-mail: y.yang@ee.ucl.ac.uk

S. Misra et al. (eds.), *Guide to Wireless Mesh Networks*, Computer Communications and Networks, DOI 10.1007/978-1-84800-909-7_2,

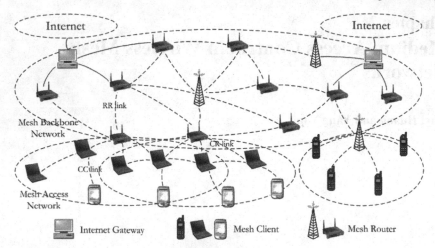

Fig. 2.1 Infrastructure of wireless mesh networks [1]

furthermore, a special kind of hardware device that has multiple types of radio tech-nologies and able to work properly in each of these networks. Mesh routers are usually much more powerful than clients in terms of computation and communi-cation capabilities, and have continuous power supply. They usually stay static and supply connections and services for mesh clients.

Ad hoc mode interconnections via wireless meshing among mesh routers (router-to-router RR links) construct the wireless mesh backbone network. There are several Internet gateways located at the edge of the backbone network so as to provide Inter-net access for the mesh network. For efficiency, wired line connections are usually used for these gateways between a gateway and the Internet, as well as between a gateway and mesh routers. A universal radio technology is usually used for the entire backbone network although there might be a number of heterogeneous networks, e.g., WLAN, cellular, WPAN. To join in an existing mesh network that is using a different radio technology, one additional wireless interface has to be equipped on the mesh router that does not use the common radio. On the other hand, multiple radio technologies coexisting in one wireless mesh backbone network are also pos-sible. However, difficulty in implementation and costly requirements of hardware make it less attractive in a real network. When a new/existing router joins/leaves the backbone, the network will be self-organized and self-configured accordingly. As the mesh routers usually stay static, and the backbone topology changes only when new routers join in and existing routers fail or leave.

2.2 Background

The structure of a wireless mesh access network is very different from the back-bone network. Figure 2.2 illustrates the conventional centralized network structure wherein clients access to WMN through a client-to-router (CR) wireless link. The

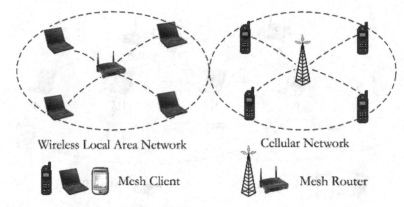

Fig. 2.2 Infrastructure of centralized networks

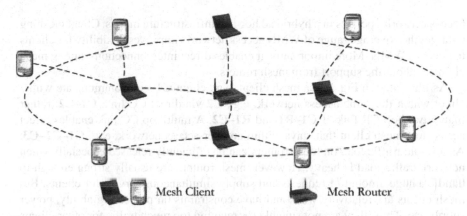

Fig. 2.3 Client meshing in wireless mesh access networks

mesh router manages all access requests from clients within that access network, and supplies an Internet connection for them. The access procedure from clients to the router follows the corresponding access mechanisms of that access network, e.g., CSMA/CA.

On the other hand, a wireless mesh access network enables ad hoc mode peer-to-peer interconnections among mesh clients, namely "client meshing." It can be achieved among any type of client that shares the same radio technology. As shown in Fig. 2.3, with "client meshing" mesh clients that stay outside of the radio coverage range of a mesh router can rely on other intermediate clients to relay packets for them to get WMN access network connections. Thus, packets from a mesh client that stay far away from the mesh router have to travel through a multihop hybrid client-to-client (CC) and client-to-router (CR) wireless link before reaching its destination. The number of hops is determined by the geographic position of the mesh client and organization structure of the access network. In this case, a wireless mesh

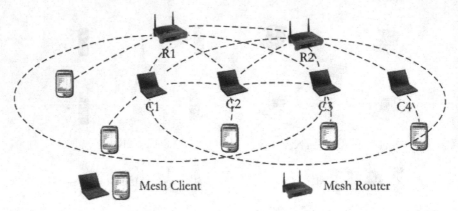

Fig. 2.4 Wireless client meshing

access network operates in a hybrid ad hoc and infrastructure modes. Client meshing enlarges the coverage range of WMN access network, improves flexibility for clients to access a WMN. More importantly, it enables direct interconnections among mesh clients without the support from mesh routers.

As illustrated in Fig. 2.4, a mesh client, e.g., client **C1**, can communicate with a client within the same access network, e.g., **C2** via direct CC link, **C1–C2**, rather than a two hop CR link of **C1–R1** and **R1–C2**. A multihop CC link enables direct access to a mesh client that stays within another access network, e.g., **C1–C2–C3**. As a result, traffic load on mesh routers can be efficiently released especially when network traffic load is heavy. However, mesh routers are usually strong enough to handle a huge amount of requests and supply simultaneous service for clients. But mesh clients are relatively weak and have constraints on processing ability, power supply, etc. Thus, they are not suitable for relaying too much traffic for other clients unless client meshing connection has fewer hops, e.g., traffic from client **C1** to **C2**. According to the research in [3], transport capability, especially the TCP throughput, drops dramatically when the number of CC hop increases. Therefore, too long a multihop CC connection in WMN is not attractive at all. In other words, within a wireless mesh access network, the mesh router still takes the major role to manage the access and packet transmission procedures for mesh clients. Client meshing is an efficient supplementary access method for neighboring clients, e.g., **C1** and **C2** or **C2** and **C3** in Fig. 2.4. Pure client meshing constructs a complete distributed network that has the identical characteristic with the conventional ad hoc network and is less attractive in WMN.

The WMN network structure and connection characteristics determine a number of problems and challenges in medium access control (MAC) process. In WMN backbone network, mesh routers are usually powerful enough and have no constraints on computation, communication, and energy supply. Thus, multiradio, cognitive radio, and multichannel MAC methodologies [4–6] are possible to be implemented to supply flexible and robust WMN backbone interconnection with good fault tolerant and QoS capabilities. In WMN access network, MAC protocols

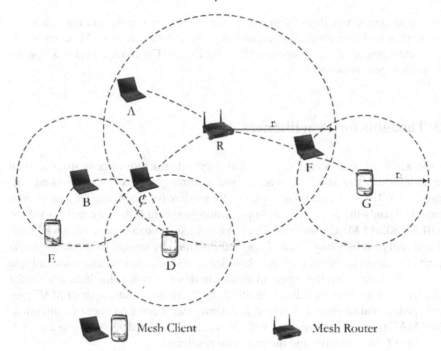

Fig. 2.5 Infrastructure of wireless mesh access networks

have to manage both CC and CR access procedures with as less cost as possible. Generally speaking, mesh clients are relatively weaker than mesh routers, so that simple but efficient medium access solutions with collision avoidance and energy conservation capabilities are more attractive in WMN access networks. Figure 2.5 illustrates a typical wireless mesh access network with seven mesh clients (**A–G**) and one mesh router (**R**). The radio coverage area of a mesh router is larger than that of a mesh client, which are bounded by circles with radius r_1 and r_2 ($r_1 \geq r_2$), respectively. The two-hop transmission path from client **D** to router **R** contains one CC and one CR connection. Therefore, the intermediate mesh client, i.e., client **C**, on the path has to relay traffic for its neighboring clients, as well as transmitting its own packets. During the CC communication procedure, the "hidden terminal" problem still has not been well addressed and severely degrades the access successful probability and system throughput. Terminals that stand within the radio coverage range of the receiver (client **C**) but out of the coverage of the sender (client **D**), are hidden terminals of the sender (client **B**). Transmission from client **D** to **C** is interrupted if client **B** transmits packets to client **C** during the same period. On the other hand, traffic will be accumulated from multiple directions when it traverses through the network until it reaches the mesh router. This phenomenon makes the clients closer to the router become traffic hot-spots and thus induces some critical problems such as unbalanced resource use and unsustainable system connectivity. Together with the hidden terminal problem, constraints of energy supply and limitations on computation and communication capabilities on mesh clients make MAC protocols quite a challenge to address these problems for CC and CR communications in wireless mesh access networks.

2.3 Thoughts for Practitioners

As the wireless mesh access network has a hybrid structure of centralized and ad hoc architecture, the MAC layer access mechanisms proposed for wireless ad hoc, sensor and WLAN are potentially suitable for mesh networks. There are a number of papers that study the possibility of implementing existing MAC protocols to WMN, e.g. IEEE 802.11 MAC protocols. The existing MAC protocols have been well studied and analyzed by many researchers, and classified by several methods. From the aspect of channel division, they are classified into single channel and dual/multiple channels, whereas from the aspect of session initiator, they are classified into sender initialized and receiver initialized. In this chapter, we study the targets of MAC protocols and provide a general solution to achieve that. Then a number of contention-based MAC protocols are reviewed, discussed, and classified according to the functionalities of the protocols and the problems resolved.

2.3.1 Collision Avoidance

On MAC layer, packet level collisions usually happen when more than one packet arrives on the same channel, at the same receiver and at the same time (or within a certain period of time). The basic functionality of a MAC protocol is to avoid and resolve packet collisions as much as possible. As the "hidden terminal problem" is the major collision causer, most of existing protocols attempt to resolve this problem, e.g., RTS/CTS handshake-based and busy-tone-based mechanisms.

To illustrate the access mechanism clearly, we take a hypothetic conversation scenario as an example and describe the general access procedure according to this example. In Fig. 2.6, when more than one speaker (**A, B**, and **C**), who speak the same language (packet transmitted on the same channel) try to speak to the same listener (**R**), MAC protocols coordinate their transmissions and try to avoid collisions among them.

When **A, B**, and **C** intend to speak to **R**, they listen to the environment for any ongoing conversations (carrier sense). If any ongoing conversation is heard they know some other person is speaking, and they will keep silent until the environment become silent again (the CSMA/CA [7] mechanism). When **A** is talking to **R, B** can rely on this mechanism to prevent from interrupting the conversation from **A** to **R**. Because of limited coverage range of voice, **C** cannot hear both **A** and **B** as it

Fig. 2.6 Contending environment

Speaker A

Listener R

Speaker C

Speaker B

stay's quite far away from them, and vice versa (the "hidden terminal problem"). To resolve the "hidden terminal problem," packets are separated into control message (RTS and CTS) and data payload (DATA). To reserve the channel, RTS is usually transmitted without enough protection and easy to fail. In this example, the control message refers to the "hello" word, and the data payload refers to the content of conversations. The "hello" word contains the brief information of the outgoing conversation content, such as location of the speaker and listener (source and destination address), conversation duration (packet length), etc. There are a number of protocols that operate, based on the RTS/CTS handshake, to reserve the channel for safe DATA transmission.

2.3.1.1 RTS/CTS Handshake-Based MAC

The Multiple Access Collision Avoidance (MACA) [8] MAC protocol was proposed in the early 1990s. The RTS/CTS handshake process was proposed to improve the payload transmission successful probability in packet radio networks. The general access procedure of this series of protocols is illustrated in Fig. 2.7. When a speaker talks, he transmits a "hello" message (RTS) first of all instead of the conversation content directly. This "hello" message is used to contend for the conversation opportunity. In Fig. 2.7, speaker **C** succeeds in the contending phase and gets the right to talk to the listener **R**. He then receives a short "yes" message (CTS) from **R**. The CTS contains similar information with RTS such as source and destination address and duration of the transmission. When the CTS packet arrives at other contending senders (speaker **A** and **B**), they are aware of the busy period length of that listener and keep silence for corresponding duration. When the CTS arrives at the successful sender (speaker **C**), he regards the "yes" message as the permission of talking and start transmitting his DATA. A confirmation message (ACK) is sent back from the listener when the transmission is finished.

The existing MAC protocols with sender initialized RTS/CTS-based handshake process follow the general procedure described above. The multiple access collision avoidance wireless (MACAW) [9] protocol tries to adapt the MACA in the

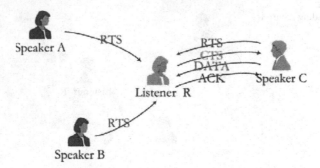

Fig. 2.7 RTS/CTS handshake

unreliable wireless network by introducing an acknowledgement control message, ACK. The ACK is transmitted back from the DATA receiver to the sender to confirm a successful payload transmission. The access procedure of MACAW is improved to a RTS–CTS–DATA–ACK four-way handshake. Then, the floor acquisition multiple access (FAMA) [10] further improves this four way handshake by perform the CSMA before transmitting the RTS packet.

The separation of control message and data payload greatly improves the successful probability of data payload. The system performance is then improved as the payload is usually longer in transmission duration and more vulnerable in collisions. Vital drawbacks still exist in the RTS/CTS-based MAC protocols as RTS packets are usually transmitted without efficient protections. Collision among RTS packets is very difficult to avoid. According to the analysis in [11], frequent RTS collisions can also severely degrade the system performance. How to avoid or alleviate RTS–RTS collision as much as possible becomes an open issue of MAC protocol design and optimization.

2.3.1.2 Receiver Initialized MAC

There are a number of protocols that let the receiver initialize the communication process rather than the sender. When the receiver becomes free, a control message named RTR (ready to receive) is transmitted to encourage potential senders to access the receiver one by one. These polling-based protocols follow a general access procedure as illustrated in Fig. 2.8 as follows.

The invitation message is dedicated to one of the neighboring speakers of the listener, say RTR-C in Fig. 2.8. Non-intended speakers will hear the invitation message and then keep silence according to the information contained, e.g., speaker A. After exchanging DATA between the speaker and the listener, an acknowledgement message is sent to the speaker and the listener will continue to invite other speakers to talk.

The multiple access collision avoidance by invitation (MACA-BI) [12] proposes the receiver initialized mechanism based on the MACA protocol. This polling-based

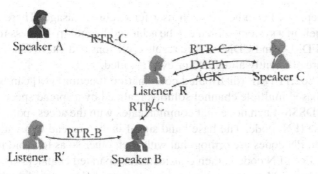

Fig. 2.8 Receiver initialized protocols

access mechanism has excellent collision avoidance capability. Data payload collision is completely avoided and control message collision is also efficiently alleviated. Unfortunately, the collision among control messages (RTR) still exists. In Fig. 2.8, if listener **R** and **R'** transmit invitation packet at the same time, the RTR packets (RTR-B and RTR-C) will collide with each other at speaker **B** and failed. The RIMA-SP/DP/BP (receiver initialized multiple access – simple polling/dual polling/broadcast polling) [13] follow the same line of receiver invitation and improve this kind of mechanism one step further. The RIMA-SP mechanism employs a RTR–DATA–ACK handshake. And the RIMA-DP mechanism further improves the protocol by change handshake process to RTR–DATA–DATA–ACK or RTS–CTS–DATA–ACK. The first DATA transmission is sent from the RTR receiver while the second DATA is sent from the RTR sender. More payload transmission is enabled by allowing a reverse payload transmission from the RTR sender to the RTR receiver. If the invited RTR receiver does not have packet to transmit, it replies with a CTS packet to wait for the upcoming DATA. RIMA-BP utilizes a broadcast polling mode and the RTR receivers have to transmit RTS to further reserve the channel. When RTS collision happens, a No-Transmission-Request (NTR) is broadcasted to forbid the subsequent DATA transmission.

2.3.1.3 Dual/Multiple-Channel-Based MAC

Compared with single-channel-based MAC protocols discussed above, there are a number of protocols that use more than one wire less channel for each wireless user to exclusively transmit their own packets. Two-channel separation is another important implementation issue wherein both channels are used to transmit control messages and data packets separately and are shared by each terminal. This type of protocol has inherent collision avoidance ability between control and data packets. By implementing random access MAC protocols on the control channel only, the contending process happens on control channel only, so as to achieve collision free data payload transmission. One typical multichannel implementation issue is assigning a separated channel to each user for exclusive usage. There are usually several channels in a system, which can be achieved by multiple access mechanisms like TDMA, FDMA, and CDMA. As a result, collisions and interference among different users are efficiently alleviated or even avoided.

The IEEE 802.11 DCF (distributed coordination function) [14] can be classified to a CDMA-based multiple channel solution realized by a spread spectrum mechanism namely DSSS. Each node that communicates with the access point is assigned a pseudo noise (PN) code. The base band signal is then spread to a wide transmission band. The PN codes are orthogonal with each other so as to avoid interference among users. Each PN code is then considered a separated channel.

Busy-tone-based MAC protocols are usually regarded as a special type of multiple channel issue. Busy-tone signals occupy a separated and small (compared to the data transmission channel bandwidth) range of total available frequency bandwidth. As a result, busy-tone detection time is on the microsecond (ms) level. Busy-tone signals can be transmitted with very simple methods, for example a power impulse to indicate a busy tone is on without the conventional modulation and coding process. Thus, busy tone setting up and detecting time is considered as the major cost of busy-tone-based MAC protocols. Busy-tone signals have only two statuses "on" and "off" so as to indicate the status of the channel(s) or indicate the state of individual terminals, "busy" and "idle." In Fig. 2.9, a general access procedure of busy-tone base random access MAC mechanism is provided. Speakers still use "hello" messages (RTS packets) to contend the listener. When one of them succeeds, instead of CTS, the listener broadcasts a busy-tone signal, which can be considered as "wave hand" by the listener. This action indicates the RTS transmission is successful and the listener is waiting for the conversation. Note that the "wave hand" action can be performed along with the packet transmission. Thus, busy tone is able to be continuously broadcasted when the sender (speaker C) transmitting DATA until finished. Any other potential speakers (A or B) will then recognize the "wave hand" action and gets aware of the busy status of the listener. They will keep silence and try to send RTS again until the busy tone is set off.

One of the most successful busy-tone-based MAC protocol is known as the dual busy tone multiple access (DBTMA) protocol [15]. In DBTMA, the control and data packets are transmitted on a single shared wireless channel. Two out-of-band

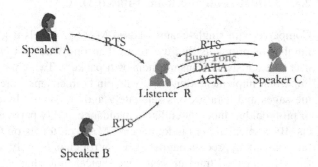

Fig. 2.9 Busy-tone-based protocols

busy-tone signals are employed to indicate the status of terminal. $BT_t \psi$ indicates a terminal is transmitting RTS message and is broadcasted along with RTS transmission; whereas the $BT_r \psi$ indicates the terminal is receiving the data payload and is broadcasted when receiving DATA. According to the analysis, it is shown that the performance of DBTMA is much better than the RTS/CTS-based protocols.

Unfortunately, busy-tone-based MAC protocols have not been widely implemented so far. This is because how to define the out-of-band channel for busy tones is not well addressed. Furthermore, busy-tone-based MAC protocols usually require full-duplex communication. But this functionally is not enabled on every NIC card. However, good performance of the busy-tone-based MAC protocols makes the implementation issue become an attractive topic.

2.3.2 Energy Conservation

In the wireless communication environment, mobile terminals always have limited energy supply, which makes the energy conservation become a continuous requirement. Generally speaking, the energy conservation capability is able to be optimized on each layer of protocol stack. Here we consider the energy conservation only from the MAC/physical layer's perspective.

On mesh clients, energy is mainly spent on computation and communication, wherein the communication energy consumption takes the major part. The medium access procedure should be carefully designed to reduce the energy cost during the communication process: first of all, packet collision is the major energy waster, especially when traffic load is heavy. If collision happens, all involved packets are failed when capture effect [16] is not taken into account (from the pure MAC layer's perspective). They are scheduled for retransmission after a random back-off delay and contend to access the channel again. This involves a complicated process of sending, receiving, rescheduling, and resending, which wastes a large amount of energy at both sides of senders and receivers. Thus, the packet collision is expected to be avoided as far as possible.

Second, mesh clients have to keep active and continuously sense carrier at all times to avoid missing any possible incoming packet (control or data). Thus, the idle listening procedure consumes a large amount of communication energy, e.g., the Idle:Receive:Send ratio is measured by 1:1.05:1.4 in [17]. The idle listening energy waste can be tackled by making mobile terminals wake up and sleep alternatively [18]. Thus, energy conservation can be alleviated with the cost of increasing access delay. Time slot scheme [19] is also able to reduce the energy cost on idle listening by dividing the time into equal sized slots and let the transmission be conducted at the beginning of each time slot. Then, packet receivers do not have to keep active all the time, but only at the beginning of each time slot. Time slot scheme requires the system to be perfectly synchronized. Thus, it is more suitable for the centralized structure networks rather the distributed ones.

Fig. 2.10 Packet overhearing problem

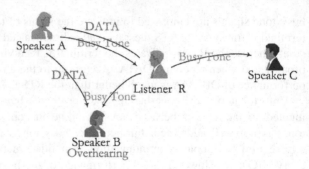

Third, the packet overhearing problem is another energy waster illustrated in Fig. 2.10. When speaker **A** is talking to the listener **R**, speaker **B** stays within the radio coverage range of both the sender and the receiver. Therefore, it has to keep silence during the conversation period of **A** and **R**. However, it is quite unnecessary for **B** to keep active during this period, which would make it overhear the packet sent by **A** and waste energy on packet receiving. The power aware medium access with signaling (PAMAS) [20] protocol is designed to avoid overhearing unnecessary packet transmission by letting mobile terminals switch off according to their own judgment: if a terminal can hear both the busy tone from the listener and the DATA from the transmitter, he can then confirm he is not the intended receiver of the DATA packet. That terminal will then power itself off until the DATA transmission is finished so as to solve the overhearing problem.

From the physical layer's perspective, if the transmission power level is appropriately set, the packet is then able to arrive at the intended terminal with minimum power level required, then the energy spent on packet transmission is reduced and at the same time the interference to other terminals, which might induce collisions is reduced as well. With this methodology, MAC protocols are usually designed along with physical layer parameters, such as capture effect, signal to interference plus noise ratio (SINR), taken into account. This type of MAC protocol is summarized and discussed in the following section.

2.3.3 Interference Resistance

From the pure MAC layer's perspective, any simultaneous transmission in which more than one packet arrives at the same receiver at the same time on the same channel will induce a packet collision. The unwanted packet is considered as the interference of the intended packet as illustrated in Fig. 2.11. This phenomenon will make the transmission from speaker **A** to listener **R1** fail. However, while considering the radio propagation characteristics, the power level of packets and signals will

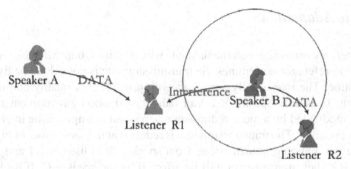

Fig. 2.11 Transmission interference

progressively fade when transmitted further. Thus, if the speaker **B** that transmits interfering packets stays far away from the listener **R1**, the interference is somehow tolerable. The interference tolerance capability highly depends on the SINR. Interference is regarded as tolerable if the ratio between the intended signal and the interference plus noise has a larger value than the required threshold.

According to this characteristic, carefully designing the transmission power level can reduce the interference among mobile terminals. One typical implementation is use of the power adaptive/interference aware mechanism over the pure MAC protocol. In Fig. 2.11, when the listener **R1** is receiving DATA from speaker **A**, he is able to hear another DATA transmission from speaker **B**. If speaker **B** carefully adjusts his volume to make sure that only the listener **R2** can hear his speech clearly rather than being too loud so as to interfere with listener **R1**. This type of design methodology is based on the assumption that $R_{rec}\psi$ (the transmission range of a packet to be correctly received), $R_{intf}\psi$ (the range to cause interference), and $R_{sen}\psi$ (the range to be sensed) follow a general equation $R_{rec} \leq R_{intf} \leq R_{sen}$. The power control medium access control (PCM) [21] protocol is designed based on the MACAW RTS–CTS–DATA–ACK four-way handshake mechanism. At the sender side, the RTS is transmitted at the maximum power level. This power level is reduced to a necessary level when transmitting DATA according to information included in the CTS packet sent back. The receiver transmits the CTS with the maximum power level as well and transmits ACK with reduced level according to the information included in DATA packet. This mechanism let the packet listener feedback to the speaker to inform minimum power volume required, and vice versa. As a result, radio transmission energy consumption is efficiently controlled with interference reduced at the same time. In [22], the DBTMA protocol is improved by selecting the appropriate transmission power level for busy-tone signals and data packets. Busy-tone signals broadcasted by the receiver are transmitted on the maximum power level to provide good protection of DATA protection during the packet receiving phase. The RTS and DATA packets are transmitted on the minimum power level to avoid interrupting other transmissions.

2.3.4 Rate Adaptation

When packets are transmitted on the unstable wireless links, unpredictable link fluctuation and interference determines the transmission rate is not able to be fixed to a
constant value. The rate usually varies according to the link quality and network
environment. Generally speaking, a bad radio propagation environment requires
packets be modulated by a more redundant mechanism to improve the interference
resistance capability. The transmission rate is reduced at the same time. In Fig. 2.12,
if there are two ongoing transmissions from speaker **A** to listener **R1** and from **B**
to listener **R2**, their transmission will be affected by the speaker **C**. If packets are
transmitted on the same power level, the listener **R1** suffers more serious interference as it stays nearer to the interferer (speaker **C**) and farther to the intended sender
(speaker **A**). As a result, speaker **A** has to transmit packet with much lower rate than
speaker **B** to reduce error rate and retransmission times.

To deal with this problem, a number of solutions try to adapt the transmission rate
according to the real-time SINR ratio at the receiver side. By monitoring the SINR,
the optimal transmission rate can be selected to maintain the system throughput and
minimize the transmission error rate. In [23], the authors propose a power control
mechanism (distributed power control with active link protection) combined with
the rate control (adaptive probing) and flow control (pipelining) mechanism. The
first control frame is transmitted with an initial power level, which is progressively
upgraded until reach the required SINR ratio. Then the transmission data rate is
selected according to the SINR determined.

Rate adaptive MAC protocols try to reflect the physical layer radio channel quality to the MAC layer and adjust the modulation method to satisfy physical layer
requirements. Using a more redundant modulation mechanism will increase the
packet transmission delay at the same time. As a result, the vulnerable period of

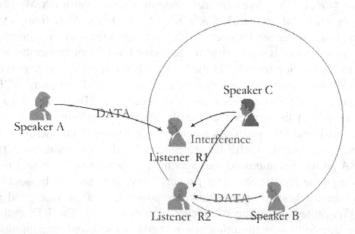

Fig. 2.12 Rate adaptive protocols

packet transmission increases when transmission error rate decreases. Therefore, to balance transmission error rate and access delay, an appropriate modulation and coding mechanism should be carefully selected.

2.4 Directions of Future Research

In wireless mesh access networks, ad hoc and infrastructure modes are usually both used to support multihop data transmission from mesh clients to a mesh router. However, in case that all mesh clients stay within the radio coverage of the mesh router, they can communicate with the mesh router directly. The network topology is then a pure centralized architecture, as illustrated in Fig. 2.13, rather than the hybrid structure. If the transmission power of some clients, say client **C3**, is lowered down, the number of reachable neighboring terminals of **C3** is decreasing. If the communication range of **C3** is not able to cover the mesh router, some intermediate clients have to relay a packet for it. Then, the network topology becomes a hybrid structure. By carefully adjust the power level of several clients within a wireless mesh access network, the network topology and packet transmission route are indirectly affected and controlled.

A number of benefits can be provided by the topology and routing control. As an example in Fig. 2.13, client **C3** and **C4** stay quite far away from the mesh router. Thus, their packet transmissions are quite sensitive to interference and noise at the receiver side and they usually have lower transmission data rate. If they transmit their packet through a multihop path of **C3–C1–R** and **C4–C2–R** respectively with radio coverage reduced, several benefits can be achieved: the transmission rate is improved with the interference reduced, the transmission successful rate is enhanced, and the traffic load on the mesh router is released.

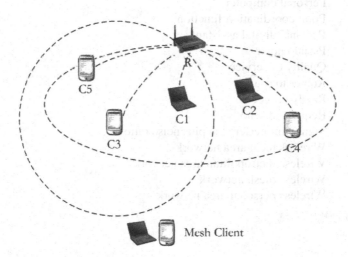

Fig. 2.13 Topology control

2.5 Conclusions

In WMN, because of the special network architecture, it is quite a challenge to manage MAC process for mesh clients and mesh routers. To avoid packet collisions among contending terminals, not only data payload, but also control messages should be protected against packet collisions as far as possible. Energy conservation capability is preferred by any mesh client related communications. Power aware, interference aware, and rate adaptive capabilities are usually considered at the same time by combining the MAC layer and physical layer characteristics together. To achieve high rate, energy saving and collision free transmissions, a number of open issues described in this chapter should be addressed.

2.6 Terminologies

ACK	Acknowledgment
AP	Access points
BS	Base stations
CC	Client to client
CDMA	Code division multiple access
CR	Client to router
CSMA	Carrier sense multiple access
CSMA/CA	Carrier sense multiple access/collision avoidance
CTS	Clear to send
DCF	Distributed coordination function
MAC	Medium access control
NIC	Network interface card
PC	Personal computer
PCF	Point coordination function
PDA	Personal digital assistant
PN	Pseudo noise
QoS	Quality of service
RR	Router to router
RTR	Ready to receive
RTS	Request to send
SINR	Signal to interference plus noise ratio
WLAN	Wireless local area network
WMAN	Wireless metropolitan area network
WMN	Wireless mesh network
WPAN	Wireless personal area network

2.7 Questions

1. What is the infrastructure of wireless mesh backbone network?
2. What is the infrastructure of wireless mesh access network?
3. What is the major difference between wireless mesh network and wireless ad hoc/sensor network?
4. What is the hidden terminal problem?
5. How does the RTS–CTS handshake-based MAC solve/alleviate the hidden terminal problem?
6. Please specify the major characteristics of busy-tone signals.
7. Please specify the major energy consumer and waster during the MAC protocol operation process.
8. Please specify the difference between packet collision and interference.
9. Please specify the relationship between SINR and transmission rate.
10. Please provide one promising topology control based MAC solution.
11. Please specify the major difference of MAC protocols that used in wireless mesh access network and in wireless mesh backbone network.

References

1. I. F. Akyildiz, X. Wang, and W. Wang, Wireless mesh networks: A survey, Computer Networks Journal, 47(4), 445 487, (2005).
2. A. Raniwala and T.-C. Chiueh, Architecture and algorithms for an IEEE 802.11-based multi-channel wireless mesh network, Proceedings of IEEE Infocom, 3, 2223–2234, (2005).
3. A. Jain, A. Pruthi, R. C. Thakur, and M. P. S. Bhatia, TCP analysis over wireless mobile ad hoc networks, Proceedings of IEEE International Conference on Personal Wireless Communications, 95–99, Dec. (2002).
4. N. Choi, Y. Seok, and Y. Choi, Multi-channel MAC protocol for mobile ad hoc networks, Proceedings of IEEE Vehicular Technology Conference, 2, 1379–1382, (2003).
5. A. Adya, P. Bahl, J. Padhye, A. Wolman, and L. Zhou, A multi-radio unification protocol for IEEE 802.11wireless networks, Proceedings of IEEE First International Conference on Broadband Networks, 344–354, Oct. (2004).
6. N. Choi, M. Patel, and S. Venkatesan, A full duplex multi-channel MAC protocol for multi-hop cognitive radio networks, Proceedings of First International Conference on Cognitive Radio Oriented Wireless Networks and Communications, 1–5, June (2006).
7. A. Colvin, CSMA with collision avoidance, Computer Communication, 6(5), 227–235, (1983).
8. P. Karn, MACA – a new channel access method for packet radio, Proceedings of ARRL/CRRL Amateur Radio Ninth Computer Networking Conference, 134–140, (1990).
9. V. Bharghavan, A. Demers, S. Shenker, and L. Zhang, MACAW: A medium access protocol for wireless LANs, Proceedings of ACM SIGCOMM 94 Conference, 24(4), 212–225, (1994).
10. C. L. Fullmer and J. J. Garcia-Luna-Aceves, Floor acquisition multiple access (FAMA) for packet radio networks, Proceedings of ACM SIGCOMM 95 Conference, 25(4), 262–273, (1995).
11. Y. Yang, F. Huang, X. Gu, M. Guizani, and H.-H. Chen, Double sense multiple access for wireless ad hoc networks, Computer Networks, 51(14), 3978–3988, (2007).

12. F. Talucci, M. Gerla, and L. Fratta, MACA-BI (MACA By Invitation)-a receiver oriented access protocol for wireless multihop networks, Proceedings of the Eighth IEEE International Symposium on Personal, Indoor and Mobile Radio Communications, 2, 435–439, (1997).
13. J. J. Garcia-Luna-Aceves, and A. Tzamaloukas, Receiver-initiated collision avoidance in wireless networks, Wireless Networks, 8(2/3), 249–263, (2002).
14. ANSI/IEEE Std 802.11, Edition: Wireless LAN Medium Access Control (MAC) and Physical Layer (PHY) Specifications, (1999).
15. Z. J. Haas and J. Deng, Dual busy tone multiple access (DBTMA) – a multiple access control scheme for ad hoc networks, IEEE Transactions on Communication, 50(6), 975–985, (2002).
16. R. L. Borchardt and T. T. Ha, Power capture ALOHA, Proceedings of IEEE Military Communications Conference, 2, 703–707, (1988).
17. M. Stemm and R. H. Katz, Measuring and reducing energy consumption of network interfaces in hand-held devices, IEICE Transactions on Communications, E80-B(8), 1125–1131, (1997).
18. W. Ye, J. Heidemann, and D. Estrin, An energy-efficient MAC protocol for wireless sensor networks, Proceedings of IEEE Infocom, 1567–1576, June (2002).
19. L. Roberts, Aloha packet system with and without slots and capture, ACM SIGCOMM Computer Communication Review, 5(2), (1975).
20. S. Singh and C. S. Raghavendra, PAMAS – power aware multi-access protocol with signaling for ad hoc networks, ACM SIGCOMM Computer Communication Review, 28(3), 5–26, (1998).
21. E.-S. Jung and N. H. Vaidya, A power control MAC protocol for ad hoc networks, Wireless Networks, 11(1–2), 55–66, (2005).
22. S.-L. Wu, Y.-C. Tseng, and J.-P. Sheu, Intelligent medium access for mobile ad hoc networks with busytones and power control, IEEE Journal on Selected Areas in Communications, 18(9), 1647–1657, (2000).
23. J.-W. Kim, and N. Bambos, Power-efficient MAC scheme using channel probing in multirate wireless ad hoc networks, Proceedings of IEEE Vehicular Technology Conference, 4, 2380–2384, (2002).

Chapter 3
Hierarchical and QoS-Aware Routing in Multihop Wireless Mesh Networks

Paolo Bucciol, Frank Y. Li, Nikos Fragoulis, and Juan Carlos De Martin

Abstract This chapter presents a novel approach to provide scalable and service-oriented communications over multihop mesh networks. In such a scenario, the performance of existing routing algorithms can be impaired by problems related to scalability, stability, and service-awareness. We start with a survey on existing solutions including commercial proposals. Then a novel framework, targeted at providing a reliable set of services in multihop mesh networks, is presented. Within this framework, several extensions to the OLSR routing protocol are proposed, such as hierarchical routing, power aware routing, multihoming, load balancing and multiple interface support. A cross-layer design technique is described to take the advantage of additional information, such as link-layer notification. A heuristic, service-oriented Quality of Service (QoS) approach is also proposed, aiming at providing QoS among both different nodes (via IEEE 802.11e standard) and different flows within a single node (via the Hierarchical Token Bucket mechanism).

3.1 Introduction: Motivation and Challenges

Today's broadband communication systems are more robust and far more ubiquitous than they used to be 5 or 10 years ago. The increment in bandwidth allows spread delivery of high bit-rate multimedia content and high Quality of Service (QoS) demanding services, such as Internet Protocol Television (IPTV) and Voice over IP (VoIP). However, remote areas of the planet – such as rural and mountainous areas – still cannot benefit from the advantages offered by broadband access. This usually happens in areas where broadband connections (or even low-speed connections) would not be deployed because of technical and/or economical reasons.

P. Bucciol (✉)
Department of Control and Computer Engineering, Politecnico di Torino, Corso Duca degli Abruzzi 24, 10129 Torino, Italy
e-mail: paolo.bucciol@polito.it

S. Misra et al. (eds.), *Guide to Wireless Mesh Networks*, Computer Communications and Networks, DOI 10.1007/978-1-84800-909-7_3,
© Springer-Verlag London Limited 2009

.The resulting so-called "digital divide" prevents the inhabitants of such areas from accessing many services, also impairing the types of services that they can receive and how fast these services can be accessed.

To provide broadband access to residential customers, various technologies, such as optical fiber, twisted pair cables, cable TV, digital subscriber line (DSL), satellite communications, and various kinds of wireless networks can be used, depending on a specific service provider and on the location of the end users. Although the DSL technology appears as probably the most popular technology for broadband access in urban areas, its deployment in rural and mountainous areas appears problematic because of its very limited coverage in terms of maximum distance between the Internet gateway and the user premises. Wireless networks, for instance Wireless Local Area Networks (WLANs) based on the IEEE 802.11 standard protocol [1], are promising to provide access in rural areas. At the same time, WLANs can also be attractive in urban scenarios because of lower infrastructure costs and greater ubiquity [2, 3].

However, one-hop wireless networks are either costly and usually require channel licenses (e.g., 3G networks and licensed WiMAX) or have limited coverage (e.g., 802.11 WLANs). The challenge of designing wireless networks aimed at providing broadband access and QoS has then to be revisited by resorting to multihop broadband wireless networks. Such networks, when connected to the Internet, and designed in a mesh network form, are called multihop Wireless Mesh Networks (WMNs). Multihop WMNs can be considered as a promising solution to bridge the digital divide [4], and to provide ubiquitous access in a rural scenario in a cost-effective way.

Current deployments of mesh networks are mainly targeting at urban areas and/or university campuses (please refer to Sect. 3.3 for more details). Compared to rural regions, these environments are more friendly to network deployments, maintenance and operations, compared to their rural counterparts. They are for instance characterized by spatial node proximity, easier node accessibility, better weather conditions, shorter links, different electromagnetic scenarios, smaller network size, and typically higher cost. Furthermore, the hardware, the routing protocols, the software and the security strategies employed in these networks are not suitable for the deployment of WMNs in remote areas – where the diversity in terms of environmental conditions is much greater than the urban scenario. Last – but not least – particular care shall be taken in the choice of the routing protocol when the number of end-users is high. Routing protocols, which are currently being deployed, are in fact not scalable and not reliable when running over a huge WMN with thousands of nodes.

Many of the technologies described in this chapter has been investigated, deployed, and tested in the context of the ADHOCSYS research project. ADHOC-SYS [4] is an Information Society Technologies (IST) Specific Targeted Research Project (STREP), supported by the European Commission in the context of Sixth Framework Programme (FP6), under the strategic objective "Broadband for All." This 2-year long project started in November 2005, and aimed at providing reliable broadband access in rural and mountain regions via multihop WMNs.

The rest of this chapter is organized as follows. In Sect. 3.2, background information on multihop mesh networks and the OLSR routing protocol are presented. In Sect. 3.3, a survey of the already existing mesh networks – both academic and commercial – is presented. Section 3.4 illustrates a proposed framework for WMN deployment in rural areas, which has been implemented in the context of the ADHOCSYS project. In Sect. 3.5, a few enhancements to the standard OLSR protocol, aimed at improving network scalability and reliability, are proposed. Section 3.6 describes how to guarantee different levels of QoS to the main services of the network. Section 3.7 proposes some directions for future research, whereas Sect. 3.8 presents some thoughts to practitioners of WMNs. Finally, conclusions are drawn in Sect. 3.9.

3.2 Background Information

3.2.1 Multihop Mesh Networks

WMNs are an emerging architecture based on multihop wireless transmission. It is a key technology for next generation wireless networks, showing rapid progress and many new inspiring applications. WMNs seem to be significantly attractive to network operators for providing new applications that cannot be easily supported by other wireless technologies. The persistent driving force in the development of WMNs comes from their envisioned advantages, including extended coverage, robustness, self-configuration, easy maintenance, and lower costs.

WMNs are expected to solve the current wireless network limitations and improve their performance. However, despite the wide scale research on this topic and a large number of experimental results, a few critical aspects of this architecture still remain open. These aspects include scalability, mesh connectivity, QoS, ease of use, security, compatibility, and interoperability. Researchers and industries are proposing modifications to existing protocols or designing completely new protocols. Several working groups of standardization bodies, such as IEEE 802.11, IEEE 802.15, IEEE 802.16, and IEEE 802.20 are working actively toward this direction.

3.2.2 The OLSR Routing Protocol

Optimized link state routing (OLSR) [5] is the most representative proactive routing protocol for ad hoc networking, and it has been extensively studied both in theory and in practice. Inherited from open shortest path first (OSPF), it is a link state routing protocol, where each router keeps the topology information of the entire network, and where the routes to all other nodes are always available, no matter whether there is any ongoing traffic or not.

To minimize protocol overhead introduced by link state protocols, OLSR is built based on the concept of Multipoint Relay (MPR). The MPR of a node is one of its one-hop neighbors that has been selected as the next hop to reach a maximum number of its two-hop neighbors. The node that has selected its MPRs is referred to as an MPR selector. The set of MPRs, which is a subset of all one-hop neighbors of the MPR selector, can therefore reach all two-hop neighbors of the MPR selectors.

The optimization of protocol overhead is achieved through the use of MPRs in a distributed network. Firstly, only MPRs with nonempty MPR selectors can generate Topology Control (TC) messages. Secondly, when TC messages are received by a node, only those nodes that are MPRs will further forward TC messages to other nodes inside the same network. Thirdly, an MPR node may choose to report partial link state.

OLSR operates in three main steps:

- *Neighbor sensing*. This is achieved by exchanging HELLO messages between all one-hop neighbors in a network. Through periodic HELLO messages received from its one-hop neighbors, a node is able to select its MPRs. Correspondingly, a link state database and a neighborhood database are established by each node based on neighbor sensing.
- *Topology control messages dissemination*. Each node, through its MPRs, periodically advertises its link information to all other nodes inside the network. As a consequence, all nodes inside a network have necessary topology information for all links between any two nodes inside the same network.
- *Routing table calculation*. Based on TC messages received from other nodes, a node is able to compute its shortest-path routes to all reachable nodes in the network, by using an algorithm similar to the Dijkstra's algorithm. According to RFC 3626 [5], the shortest-path in terms of the number of hops is used for route calculation in OLSR.

3.3 Survey of Existing Multihop Mesh Networks

To date, a number of WMN systems have been deployed enabling a variety of applications. These applications range from broadband home networking to provisioning of services for metropolitan areas to solve public safety problems *et similia* [6].

3.3.1 Academic Multihop Mesh Networks

Many academic research testbeds have been established and actively tested to further the development of WMNs. Some of the more relevant projects include:

- *MIT*: *Roofnet*. Roofnet [7] is an experimental multihop IEEE 802.11b/g mesh testbed providing broadband Internet access to Cambridge, MA. The major feature of Roofnet is that it is does not rely on a priori configuration.

- *Georgia Institute of Technology*: *BWN-Mesh*. BWN-Mesh [8] is a mesh test-bed at Broadband and Wireless Network (BWN) Lab at Georgia Institute of Technology. BWN-Mesh consists of 15 IEEE 802.11b/g mesh routers, some of which are connected via gateway/bridge to other future generation test-beds, e.g., wireless sensor networks (WSNs). Other nodes residing in BWN-Mesh are laptops and desktops.
- *State University of New York*: *Hyacinth*. Hyacinth [9] is a test-bed at Experimental Computer Systems Lab (ECSL) at State University of New York. In Hyacinth, each node uses multiple IEEE 802.11 radios. So Hyacinth is a multi-channel WMN. Hyacinth is intended to be readily built using IEEE 802.11a/b/g, and 802.16a technology.
- *University of Illinois*: *Net-X*. University of Illinois at Urbana-Champaign has presented a 4-node multichannel 802.11b testbed called Net-X [10]. Each node is equipped with two cards whose channels were determined based on the load-aware channel assignment algorithm. The multichannel network achieves 2.63 times higher throughput as compared to that of the single channel network.
- *Carnegie-Mellon University*: One of the earliest mesh network testbeds is Carnegie-Mellon University's mobile ad hoc network (MANET) test-bed [11]. It consists of seven nodes: two stationary nodes, five car mounted nodes that drive around the testbed site, and one car mounted node that enters and leaves the site.

3.3.2 Commercial Multihop Mesh Networks

Many companies throughout the world have explored WMNs and have put products into real-world deployments, based on their proprietary solutions.

3.3.2.1 US Companies

- *MeshNetworks*, developed by Motorola, is devoted on mobile broadband Internet access, which provides high-speed access to mobile users [12]. MeshNetworks provides innovative solutions, such as Quad-Division Multiple Access (QDMA) radio technology and adaptive transmission protocol. Installations include the city of Buffalo, Minnesota, the city of Ripon, California, and Las Vegas.
- *SkyPilot Networks* provides broadband Internet access using WMNs [13]. Features include eight directional antennas, high power radios and dynamic bandwidth scheduling. SkyPilot's products are equipping installations in California and elsewhere.
- *Tropos Networks* provides metro-scale Wi-Fi mesh networking services [14]. Their MetroMesh architecture is aimed at providing high speed and affordable broadband data communication, with focus on VoIP services. Cities such as San Francisco, Chaska in Minnesota, have adopted Tropos solutions.

- *Firetide* provides applications on indoors and outdoors Layer 2 connectivity [15]. It offers products with 2.4 GHz and 5 GHz radio technologies, with Advanced Encryption Standard (AES), Wired Equivalent Privacy (WEP) security measures, and network management software. Firetide has deployed many installations for municipalities, warehouses, hospitals, and educational institutes.
- *Intel* has been conducting research on WMNs since 2002 [16]. Their Berkeley Research Lab has been focused on issues such low power and traffic balancing.
- *Microsoft* has been focusing on community WMNs [17]. Their software, called mesh connectivity layer (MCL) aims at routing and link quality. Modifications are transparent to other layers.
- *Meraki* provides low-cost solutions for indoor and outdoor networks, and covers applications scaling from a small home and hotel networks, to large city networks [18]. Meraki provides *Meraki Mini* and *Meraki Outdoor* products, which facilitate easy and low cost large scale WMN deployment. Meraki has also deployed a WMN for the city of Prestonsburg KY.
- Other companies, like *Nortel, Packet Hop, Ricochet Networks, Strix Systems* [19], and *SkyPilot Networks* also offer exciting products with cutting-edge technologies and have deployed many WMN systems.

3.3.2.2 European Companies

- *LamTech*, formally known as *Radiant Networks*, focuses on broadband Internet access [20]. Their main product, MESHWORK, uses asynchronous transfer mode (ATM) switching in wireless routing, four directional mobile antennas therefore making links directional.
- *NOW.co.uk*, a division of the *NOW Wireless* group, offers Mesh4G, and has deployed networks in many cities in Great Britain (Portsmouth, Hampshire, Glascow, and others) [21].
- *Locust World* focuses on community networking, featuring MeshAP hardware [22].
- *Wilibox*, based in Lithuania, provides a Willi platform running a Linux based OS [23]. Wilibox provides WILI MESH product, which is a secure, QoS capable, portable Linux based OSI layer 2 wireless mesh networking software platform targeting at enterprise, campus, WISP networks covering significant areas with 802.11 wireless access.
- *Essentia Wifless*, based in Italy, provides a complete solution for wireless connectivity through the Wifless^{TM} ESS family [24]. The main product is ESS 2456x, which can operate in every possible configuration, acting as a simple stand-alone Access Point, or as a node in a BGP-4 routed PTP link, or as a base station in a PMP coverage, as a simple node in a WDS Mesh Network as well as in more complex OLSR Mesh Networks.
- *Wi-Next*, based in Italy, provides the N.A.A.W. (acronym for autoreconfigurable network device), a software technology able to transform a wide range of devices into networks nodes that can collaborate for the creation and optimization of Wireless organic networks, both outdoor and indoor [25].

3.4 The Proposed Framework

A single WMN may cover a large area, which in addition might also be densely populated. Simulations performed to study such challenging case scenario show that the standard routing protocols (such as OLSR and Ad Hoc On-Demand Distance Vector (AODV)) fail to guarantee scalability and reliability to the network [26]. To allow better scalability, the whole mesh network can then be split into a hierarchical structure, composing several "clouds" interconnected by a wireless backbone. The resulting *two-tier* hierarchy represents a good tradeoff between complexity, performance, and scalability. An example of such hierarchical network is depicted in Fig. 3.1.

The *first tier* network (backbone) is represented in *deep dark*, and can be a standard multihop wireless network consisting of several long wireless links. When designing such long distance links, special care should be taken in adding the appropriate redundancy to the network, to avoid single points of failure (SPOF). Long distances and short delays between transmitters and receivers can be achieved by means of fine tuning of IEEE 802.11 protocol parameters (such as those related to the exponential backoff mechanism) and by employing directional antennas.

The strength of the signal that will be received can be calculated according to (3.1), which represents the *link budget*:

$$P_r = P_t - L_t + G_t - \text{PL} - \text{FL} + G_r - L_r, \tag{3.1}$$

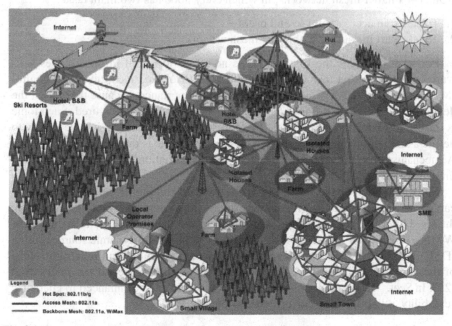

Fig. 3.1 An example of hierarchical (two-tier) multihop wireless mesh network

where the parameters are expressed in dB and have the following meaning:

- P_r is the estimate of the received power level at the receiver
- P_t is the power generated by the transmitter
- L_t, L_r represent the attenuation of the transmitter cable and receiver cable, respectively, to which the antennas are connected (~ 0 dB for integrated antennas)
- G_t, G_r represent the transmitter antenna gain and the receiver antenna gain, respectively
- FL is the *fade loss* takes into account the effects of multipath, fading, and other propagation issues. Many models exist in literature, which estimate the fade loss depending on the actual physical environment. As a rule-of-thumb, 10-dB fade loss is an estimate for short IEEE 802.11a/b/g communication whose line-of-sight has at least 60% Fresnel Zone clearance
- PL is the *path loss*, which represents the attenuation of the signal strength caused by the communication medium. *Path loss* can be calculated from (3.2), which corresponds to a simplified model based on the ideal propagation condition with only one clear line-of-sight link between the transmitter and the receiver. In (3.2), f indicates the operating frequency expressed in Hertz and d is the distance in meters between the sender and the receiver antennas:

$$PL = 32.4 + 20\text{Log}(f) + 20\text{Log}(d). \tag{3.2}$$

The *second tier* network is represented in *light dark* in Fig. 3.1. It is composed of clouds of smaller mesh networks, in which every node has two main tasks:

1. To act as an access point to end users
2. To take part in routing within the mesh network

For what concerns the first task, it is necessary to install in those nodes software components that take care of the physical (PHY) and medium access control (MAC) layer issues. MAC issues are of extreme importance, because the QoS of the designed network will strongly depend on how the MAC is configured. Such topic is discussed in details in Sect. 6.2. To allow service-oriented QoS it is necessary to define a taxonomy among the service classes. This topic is discussed in Sect. 3.6.1. A mechanism for allowing effective service differentiation within a single node is then analyzed in Sect. 3.6.3.

For what concerns the second task, the standard routing protocols designed for cabled networks, such as OSPF v2 (the most widely deployed routing protocol in the fixed Internet), experience problems that prevent their usability in the context of WMNs. Protocols for MANETs such as OLSR, AODV, and dynamic source routing (DSR), on the other side, still exhibit problems in scalability and reliability. None of the aforementioned protocols can be selected to run on second tier nodes *as is* [27]. In Sect. 3.5, we explain how the OLSR protocol can be enhanced to guarantee scalability, reliability, and performance within the WMN.

3.5 Extensions to the OLSR Routing Protocol

The legacy OLSR protocol is designed for mobile ad hoc networks, not for WMNs. Consequently, much effort has been devoted to reduce protocol overhead. In large-size WMNs, however, the router nodes are static. Therefore, other aspects shall be considered instead, for instance reliability, scalability, throughput maximization, load balancing among routers, and cross-layer link notification.

In this section, six enhancements of OLSR are presented for achieving higher network reliability through multipath and multihoming, better scalability with hierarchical topology, higher multihop throughput through multiple interfaces, cross-layer optimization with link notification and load balancing, metric-based route table calculation, and power-aware routing. These enhancements have been implemented and tested in a real-life mesh network (in the context of the ADHOCSYS project) with considerable performance improvement over the legacy OLSR protocol.

3.5.1 Hierarchical Routing

The use of the concept of "cluster," is introduced in this section. A network is divided into several clusters and different clusters are connected by border gateways. More specifically, a two-level hierarchy is preferable in a medium- or large-sized mesh network, where the access network nodes (interconnected mesh routers in a cluster) form the level-2 tier, whereas the backbone nodes that connect several access networks form the level-1 network. Additionally, gateways to the Internet can be connected directly either to the level-1 tier or to the level-2 tier, depending on the locations of the gateways.

The level-2 tier is composed by one or more access subnetworks (the light dark clouds in Fig. 3.1). An access subnetwork, which is connected to other access subnetworks, is referred to as a cluster. A backbone node serves as the cluster-head and advertises its reachability to other clusters periodically. The cluster-heads are predefined, and in this way there is no need to develop an algorithm for their selection. The cluster-heads are aware of the existence of each other through periodic handshake messages, and are connected to each other, either directly or via multihop relays.

The main idea behind our hierarchical OLSR solution is to achieve the goal through enhanced Host and Network Association (HNA) messages and address aggregation. More specifically, each cluster head uses HNA to advertise its reachability for both sides:

- *Intercluster HNA message* advertises a cluster head's connectivity of all nodes, including both level-2 nodes and Internet gateway nodes inside the same cluster, to other clusters. This message is sent to all other connected cluster heads using *unicast* packets (note this is different from the standard version of OLSR), or

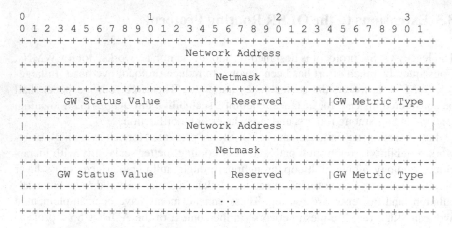

Fig. 3.2 Illustration of the extended HNA message format

subnet-directed-broadcast packets. Both the level-2 nodes and the gateways are advertised as connected subnets, specified by the netmask field in HNA. If a level-2 node or a gateway is advertised as an individual node, then the netmask is set as /32 because the length of the IPv4 address is 32 bits.

– *Intracluster HNA message* advertises a cluster-head's connectivity to other clusters, including also Internet gateways from another cluster. This message is sent to all nodes inside the same cluster. Both level-2 nodes and gateways from another cluster are advertised as connected subnets, and are specified by the netmask field in HNA.

– For both intercluster and intracluster HNA messages, an *extended HNA message format* has been defined, so that either hop-count or other metrics can be used for gateway selection of level-2 nodes. Figure 3.2 illustrates the extended format for HNA messages.

3.5.2 Multihoming with Load Balancing

OLSR HNA messages allow gateway nodes to announce their connection (network address and netmask) with the Internet to other OLSR nodes. In case of multihoming, the gateway, which is closest to the end-user by means of the number of hops, is always chosen as the default gateway by the legacy OLSR. The other gateway will be used only if the default gateway is down, and the process of finding another gateway may take up to a few seconds.

With the proposed multihoming enhancement, a node uses a metric-based policy to select the best gateway. These metrics include for example link and path capacity, traffic load and other QoS parameters, in addition to the number of hops. Three

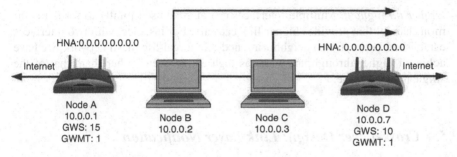

Fig. 3.3 A simple example of multihoming with load balancing

types of load balancing are considered here, namely load balancing among channels, paths and gateway nodes. Given that two or more channels coexist between a pair of nodes, if one channel is close to congestion, another channel should be used. Similarly, if one path is over-loaded, the routing table calculation process will recalculate a new path. This task is triggered by including the traffic load information in a newly defined LINKINFO message, which has been implemented as a plug-in to OLSR. For multihomed networks, the traffic load status is monitored at each gateway and is disseminated to other nodes inside the network, using a modified HNA message. Once this information is available at each router, the router is able to reroute its traffic toward a lighter-loaded gateway. This process needs to be carried out periodically so that the traffic load through the whole network is balanced among available gateways. In Fig. 3.3 an example of the implemented multihoming mechanism with load balancing is presented. In the example, B has chosen gateway D as its gateway based on the received load status, even through B is closer to gateway A in number of hops.

3.5.3 Multiple Interface Support

In the legacy OLSR, the multiple interface declaration (MID) message is used when a node has several interfaces, but only a single interface is selected as the main (working) interface. That is, only one interface will be used for path establishment.

With the implemented multiple interface enhancement, each individual interface is treated independently and multiple interfaces can work at the same time. That is, more than one link can be established between two neighbour nodes. As a consequence, the following benefits are achieved:

– *Higher reliability.* If one link is down, a node with multiple interfaces could still provide routing path for the end users. For instance, with two interfaces between a pair of nodes, the link between these two nodes is still available even if one of the two channels is broken.

– *Higher throughput.* Multiple interfaces can also be used jointly to form a common channel that provides higher link capacity. For instance, with two interfaces established between two neighboring nodes in a real-life mesh network, we have achieved higher throughput, twice as high as, or even higher than, that of the single interface case.

3.5.4 Cross-Layer Design: Link Layer Notification

When a link breaks, the legacy OLSR protocol reacts to this change by exchanging HELLO and TC messages, which may take up to a few seconds. With the link layer notification enhancement, a new path, if it exists, will be available immediately after a link break. With this enhancement, noninterrupted access services can be provided to the end-users.

The basis for the link layer notification enhancement is to utilize link break information gathered at the MAC layer to impose OLSR routing table recalculation. More specifically, the MAC layer detects the link break and sends a notification to the protocol layer, and upon receiving such a notification, which is treated as a topology or neighbor change, OLSR shall immediately conduct routing table recalculation.

3.5.5 Metric-Based Routing Table Calculation

With this enhancement, a routing algorithm similar to the Dijkstra's algorithm has been implemented. As the input of this algorithm, the "cost" of each link within the network will be advertised throughout the whole network so that each router has the topology information needed for its routing calculation. The link cost could be data rate, delay, load status, or any other metrics of interest. Based upon this link cost information, a router is able to build its routing table according to the minimal path cost criterion.

To implement the metric-based OLSR routing table calculation, a new metric, referred to as "link cost" has been introduced for each link state. Depending on the scenario, the network administrator can choose which metric(s) to use for a specific network, by specifying corresponding parameters in the modified OLSR configuration file.

3.5.6 Power Aware Routing

The power aware extension to the OLSR routing protocol aims at avoiding connectivity problems in case of node failures because of low battery power. Problems related to power consumption include, for instance, the possibility that node batteries get drained; the choice of the best transmitting power to minimize power consumption without affecting network connectivity; the choice of the best interval to send special advertisement packets to refresh network topology information, by finding a fair tradeoff between power consumption and network topology awareness.

Within this section, a solution to avoid the aforementioned problems by including power awareness in the OLSR routing protocol is proposed. This solution has also been implemented as an OLSR enhancement in the context of the ADHOCSYS project.

3.5.6.1 Obtaining Node and Link Conditions

The first step toward power awareness is to obtain information about the battery status of the nodes participating in the network. This information will be then used in the proposed power aware enhancements to the OLSR routing algorithm, which are described in the following subsections.

The battery level on each node can be obtained via system calls based, for instance, on the advanced power management (APM) or advanced configuration power interface (ACPI) software interfaces, when the nodes run an operating system supporting them. Otherwise, custom hardware and software must be used.

The obtained battery state information needs then to be disseminated and stored. The OLSR routing algorithm [5] already provides mechanisms for disseminating and storing information about individual nodes and links, which are based, respectively, on HELLO messages, topology sets, and link sets.

The HELLO messages, in fact, contain information about neighbor interface addresses, and are used, upon reception, to fill the link set and the topology set, in each node. The link set maintains the set of all local links, i.e., pairs of interfaces, where the first is a local interface and the second a remote interface. The topology set maintains information of all possible destinations. Therefore a possible implementation of a mechanism to diffuse and store additional information about nodes and links could be based on an extension of the HELLO message and a corresponding extension of topology and link sets.

Another possible implementation could be the definition of one or more custom messages containing all the additional information needed by the network. This method brings some advantages, including the fact that any standard OLSR node could still work in this network, by simply ignoring custom messages. Such type of extensions is compatible with other extensions to OLSR and can interoperate with them. However, one disadvantage is that new message types are introduced, leading to more protocol overhead, and higher power consumption.

This technique is already implemented in the *powerinfo* plug-in [5], an example of OLSR plug-in, in which a new type message type called POWERINFO has been defined to diffuse information about the power source and battery state of each node. The proposed implementation of the enhancement is based upon this example, which however needs improvement because it only provides power information for each node, but does not specify how a node, which has received such information, would react.

In the *powerinfo* plug-in example, the POWERINFO message is sent as a regular message, at an interval of every 2.5 s. Nodes that do not support this message simply use the default flooding algorithm to forward this message, whereas nodes equipped with this plug-in can then take other necessary actions.

Each node, in OLSR, maintains a set of link tuples and topology tuples, which store information about all local links, i.e., pairs of interfaces, and on all possible destinations. These tuples are updated upon reception of HELLO messages, or after expiration of the validity period of the maintained information.

In our enhancement, we define two additional information bases to store information about nodes and links. The first step toward this implementation is the definition of two new information bases, defined within the modified version of core OLSRD. One information base is used to store the "cost" of each node, which can be used within a route. This means, for example, that nodes with nearly drained battery should get a very high cost. Another information base is used to store the "cost" of each link. These new information bases are then used in the routing table computation algorithm, which is described in Sect. 3.5.6.2.

3.5.6.2 Routing Table Computation

The computation of the routing table is very important from the viewpoint of power awareness. Intuitively, the idea is to choose routes that allow for less power consumption, and which do not involve nodes that have power supply problems.

Several algorithms can be applied to achieve these goals, and they could lead to the choice of different routes. The best route will be computed using all of these algorithms at the same time. In this way, each algorithm is assigned a weight that reflects the importance of the considered metric and each algorithm assigns a score to each route. The final score of each route is given by the weighted sum of the scores assigned by all algorithms and the route with the highest score is selected. This technique for the selection of the optimal route allows adopting several algorithms to assign scores to available routes. The algorithms chosen in the proposed implementation are the following two:

1. *Battery level at each node.* The basic idea behind this algorithm is that nodes with nearly drained batteries should not be involved in routing, to prolong their lifetime. The algorithm should choose a route for each flow that maximizes the battery lifetime of involved nodes, thus discarding routes that involve nodes with nearly drained batteries. This algorithm can be easily implemented considering the average battery level of each node in the involved route. For example, in Fig. 3.4, the route via Nodes 1 and 2 has been selected as it has an average battery level of $(50 + 70)/2 = 60\%$, whereas the route via Node 3 is not selected because it has an average battery level of 10% even though this path is shorter in

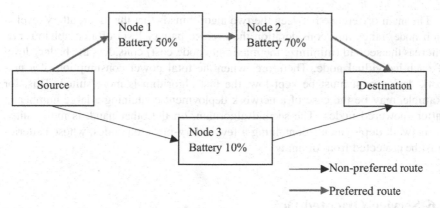

Fig. 3.4 Power aware routing – battery level at each node

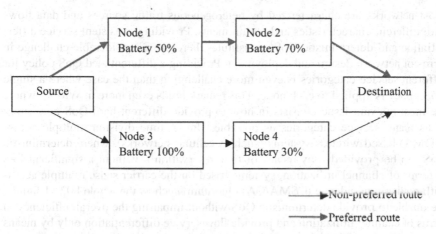

Fig. 3.5 Power aware routing – minimize the maximum node cost

number of hops. In this algorithm, nodes, which are not battery powered, should always be considered with a battery level of 100%.

An alerting mechanism in case of drained batteries can also be implemented by means of this algorithm. Nodes with nearly drained batteries send a message to their neighbors, so that they are advised to adopt alternative routes. This alert mechanism can be implemented exploiting the WILLINGNESS parameter of OLSR, which describes how much a node is "willing" to contribute to routing of messages. The WILLINGNESS value will then be set to WILL_NEVER in case the battery level of a given node is below a given threshold.

2. *Minimize the maximum node cost.* This algorithm is similar to the previous one, with the difference that only the node that has the minimum battery level is considered, instead of the sum of the battery level of all nodes involved in the route. This means, routes that involve the node that has the lowest battery levels will be discarded, as shown in Fig. 3.5.

The main difference between the two algorithms is that the first (battery level at each node) takes into consideration the average battery level of a complete route, whereas the second (minimize the maximum node cost) considers the battery level of each individual node. Therefore, when the total power consumption of a network deployment must be kept low, the first algorithm is more suited. This, for example, may be the case of a network deployment containing a large number of battery-powered nodes. The second algorithm, on the other hand, is more suited for network deployments containing a few battery-powered nodes, whose batteries must be protected from draining.

3.6 Service-Oriented QoS

Most networks are characterized by heterogeneous traffic sources and data flows with different characteristics and requirements. Providing consistent service differentiation and deterministic QoS[1] constitutes therefore a considerable challenge in terms of network design and deployment. Providing a differentiated QoS policy for different service categories is even more challenging than the case where a single QoS policy is applied to each node. This remark holds even more in WMNs, where the most relevant issue consists in how to provide different hard QoS guarantees to as many service categories as possible. Unlike time division multiple access (TDMA)-based wireless systems (such as cellular networks), where deterministic QoS can be provided – by means of time reservation – without a significant loss in terms of channel utilization, systems based on the carrier sense multiple access with collision avoidance (CSMA/CA) algorithm, such as the whole 802.11 family, are unable to provide deterministic QoS without impairing the overall efficiency in terms of channel utilization and provide flow/service differentiation only by means of probabilistic QoS.

The 802.11e standard [28], ratified in 2005, provides service differentiation for up to four different "Access Categories" (AC). In the same year the Wi-Fi Media Alliance published a white paper based on a prior draft version of the 802.11e standard, the so-called Wi-Fi for MultiMedia (WMM) [29], with the goal to promote interoperability among different hardware producers and allow hardware QoS-enabled 802.11e MACs to enter the market without having to wait the time required to finalize 802.11e. WMM basically provides a subset of the 802.11e functionalities. However, both 802.11e and WMM have been designed to provide probabilistic QoS. Deterministic QoS shall then be achieved by implementing different or complementary solutions within the mesh network, such as the Hierarchical Token Bucket (HTB) mechanism [30].

---

[1] Deterministic QoS, also referred to as hard QoS, means that QoS requirements for each flow/service are *always* satisfied ("I need 100 kbit/s for flow A, I will get 100 kbit/s for flow A"). Probabilistic QoS (soft QoS), on the other side, means that QoS requirements for each flow/service will be satisfied on average over a relatively wide time interval ("I need 100 kbit/s for flow A, I will get an average of 100 kbit/s for flow A over a 10-s time window").

In this section, we will briefly classify several application classes – or *services* – and analyze their QoS requirements (Sect. 3.6.1). We will then proceed to examine how different levels of QoS can be provided to those services in the mesh network (probabilistic QoS, via the 802.11e/WMM protocol[2]) (Sect. 3.6.2) and within a single mesh node (deterministic QoS, via the HTB mechanism) (Sect. 3.6.3).

The proposed QoS mechanism is presented in Fig. 3.6. As illustrated, when a packet P is received by a generic node N, belonging to the two-tier WMN, its source is checked. This process can be done, for instance, by examining from which interface packet P has been received. If P comes from an external source (Internet gateway or end user), or it has not been classified yet, it is analyzed and classified by means of the flow/application to which it belongs to.

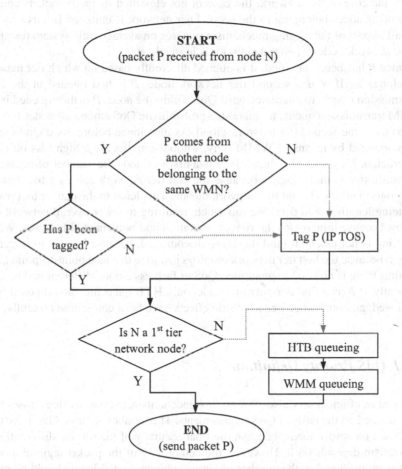

Fig. 3.6 Example of packet tagging within a two-tier hierarchical wireless mesh network

---

[2] It is worth noting that full interoperability is guaranteed between WMM- and 802.11e-enabled devices only when the set of common features is employed [27].

The classification process tags all packets belonging to the same application class with the same "label," stored in the Type Of Service (TOS) field of the IP header (for instance, by employing the "layer-7 filter" application [31]). The main goal of this process is to allow interworking between the various QoS mechanisms, which will be implemented via a unique IP TOS value for each information element (data packet). Similar mechanisms (using 802.1Q VLAN tags) have been considered in [32].

The discussion on the techniques used to determine the application class is outside the scope of the chapter. Refer to [33] and [34] for more details. The tagging mechanism can be computationally intensive if it is running on every network node. However, in a stable WMN, because the classification is made when the packet enters the core of the network, the case of not classified (tagged) packets coming from other nodes belonging to the second tier network is unlikely. In this way, the overall impact of the tagging mechanism requires on average little system resources in the network nodes when the network is in steady state.

Once P has been classified, it is queued differently based on which tier network N belongs to. If N is a second tier network node, P is first queued in the HTB transmission queue, to guarantee hard QoS within the node. P is then queued in the WMM transmission queue, to guarantee probabilistic QoS among all nodes belonging to the same second tier network cloud. As mentioned before, hard QoS cannot be guaranteed by means of WMM only, without introducing a high loss on channel efficiency. On the other hand, if N is a first tier node the number of packets to be transmitted is much higher because the first tier network acts as a backbone to interconnect all the second tier network clouds. This leads to the danger that packet queueing length would decrease too much, resulting in lower overall network efficiency, because it increases the risk of saturation and bottleneck. Moreover, WMM queueing is not essential (and therefore discouraged, because it leads to increased delays) because the first tier network employs just long distance point-to-point links, and thus there is no need to guarantee QoS or fairness among different nodes. Consequently, if N is a first tier network node, only HTB queueing should possibly be employed, and only when negative side effects have been considered carefully.

3.6.1 QoS Priority Definition

To adopt an efficient service differentiation mechanism, various service classes have to be defined on the basis of QoS requirements. The number of classes has to be chosen based on actual needs, because the final accuracy of the service differentiation mechanism depends on it. However, the complexity of the packet tagging mechanism is proportional to the number of service classes, so the latter should be maintained as low as possible. Standard application classification techniques for fixed networks are not suitable for a large multihop hierarchical WMN, because the latter case requires, for example, specific treatment for services essential to the network, such as routing information.

Table 3.1 Definition of service classes based on application requirements

Service class	Applications
I	Strong latency constraint, small bandwidth (VoIP, chat)
II	High throughput (transaction processing, file transfer)
III	Interactive, best-effort (Web browsing, e-mail)
	Essential set of services for the users
IV	Routing, battery information
	Essential set of services for the network
V	Emergency calls
VI	High throughput and latency constraint (streaming video)
VII	P2P applications
VIII	Unclassified traffic

In Table 3.1, an example of application classification is presented. The QoS definition for application classes I, II, III is based on the conventional QoS classification, which relies mainly on the delay and bit error rate requirements of different service classes. Classes from IV to VII have also been defined to allow finer service differentiation policies. Because the proposed QoS definition is not node-based, but flow-based, various traffic flows generated or received by a given node may belong to different classes, at different times. In this perspective, the service differentiation mechanisms can adapt to the actual network conditions.

3.6.2 Service Differentiation Among the Mesh Network: The 802.11e/WMM Medium Access Control

The new Enhanced DCF Channel Access (EDCA) mechanism defined by 802.11e implements up to eight different queues at MAC layer, corresponding to a different Traffic Category. The WMM MAC, as shown in Fig. 3.7, is limited to four different MAC queues, each corresponding to a different access category. Service differentiation is achieved by means of different values of the MAC layer parameters, which regulate the contention mechanism of the wireless channel (CWmin, CWmax, AIFSn) [35].

The standard classification of traffic typologies into WMM ACs and 802.11e TCs is presented in Table 3.2. In WMM, allocation of a given packet into a determined AC depends on the actual value of the IP TOS field. Particular care should then be made when setting those values, to maintain the coherency between them and the chosen service differentiation strategy (defined previously in Sect. 3.6.1).

Under ordinary conditions, the WMM MAC can be used to provide unequal error protection to different traffic flows, thus guaranteeing soft QoS among different nodes [36]. However, it is not generally true that higher AC priority means better performance. For instance, if one of the four ACs transports a much higher bitrate than the other ACs, the channel becomes saturated. Under saturation conditions, the

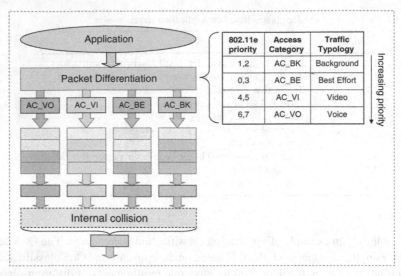

Fig. 3.7 Schematic representation of the WMM MAC layer

Table 3.2 Matching between traffic typologies, WMM ACs and 802.11e TCs

Traffic typology	WMM access category	802.11e Traffic category
Background	AC_BK (Lowest Priority)	0,1
Best effort	AC_BE	2,3
Video	AC_VI	4,5
Voice	AC_VO (Highest Priority)	6,7

packet loss rate and delay of such AC increase, and network efficiency and reliability may drop to unacceptable levels. To deterministically avoid this unwanted behaviour, other mechanisms have to be implemented to perform smart service differentiation and to feed the four AC queues properly. One of those mechanisms is the HTB mechanism, described in detail in Sect. 3.6.3.

3.6.3 Service Differentiation Within a Single Node: The Hierarchical Token Bucket Mechanism

The WMM-based prioritization mechanism, as pointed out in Sect. 3.6.2, is implemented to guarantee soft QoS among different nodes of the same two-tier mesh network, and allows high efficiency levels in wireless channel utilization. Additional mechanisms have however to be employed to guarantee hard QoS and avoid saturation. Taking into consideration nodes based on the Linux operating system, for instance, deterministic QoS can be provided by using the QoS features of the Linux kernel.[3]

By using software level (Linux) queueing mechanisms, queue parameters can be set to provide deterministic QoS. The HTB mechanism [30], part of the Linux kernel, can be installed on every mesh node and is able to effectively manage the node outbound policy [37]. It is currently used with success in many open source and commercial implementations [38], and it has been proved to successfully work in conjunction with WMM in various projects [39]. Thus, HTB can be used to guarantee hard service differentiation within a single mesh node, whereas WMM will perform traffic prioritization among different nodes.

To exploit the functionalities of the HTB mechanism, we need to define properly its tree structure. For the sake of comprehension, we suppose that the same structure will be implemented in each two-tier wireless node. Some *application categories*, which include one or more of the *application classes* defined in Sect. 3.6.1, have to be defined. This categorization strongly depends on the user and network needs. For instance, if the final goal is to reduce the digital divide of a given area, we prefer to privilege a small set of *essential services*, such as Web browsing and e-mail reading, over more complex services such as video streaming and peer-to-peer downloading. It is worth noting that this case study is different from the common QoS approach, which prioritizes applications with stringent delay/jitter/bandwidth requirements.

Let us now proceed with the chosen case study. The steps to be completed to have a fully working, HTB and WMM enabled two-tier network, are the following: definition of appropriate application categories, creation of the HTB tree structure, dimensioning of the HTB parameters and allocation of the application categories to the WMM access categories.

3.6.3.1 Definition of the Application Categories

If we target our QoS approach to a small set of *essential services*, three application categories can be defined:

- *Category A* (priority) groups essential services for both users and the network;
- *Category B* (delay) includes application with strict delay constraints;
- *Category C* (throughput) includes both high throughput (but not essential) services and uncategorized flows.

The three aforementioned application categories are listed in decreasing order of priority. If we had selected a traditional QoS approach, category A would have included low delay and low bandwidth services, category B would have included low delay, bandwidth-demanding services, whereas category C would have included other services. In this vision, *essential services* would have fallen into category C.

---

[3] A commonly used distribution for wireless mesh nodes is the OpenWRT distribution (URL: http://openwrt.org/).

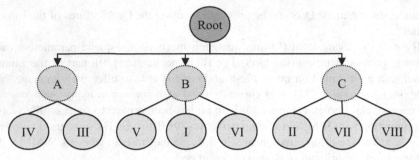

Fig. 3.8 Proposed structure for the Hierarchical Token Bucket tree

Table 3.3 HTB parameters for the various application categories

Category	HTB parameters		Priority
	Rate	Ceiling	
A	0.7 BW	BW	2
B	0.2 BW	0.9 BW	1
C	0.1 BW	0.9 BW	3

3.6.3.2 Creation of the HTB Tree Structure

The HTB mechanism is based on a *tree structure*. HTB can guarantee a fixed bandwidth for each application category (hard QoS), also defining a priority order based on which the several application categories are served. Figure 3.8 illustrates the resulting structure of the HTB tree, where the application categories have been defined as in Sect. 3.6.3.1. The application classes, defined in Sect. 3.6.1, which are represented at the bottom of the figure, are connected to the three application categories represented in the second level in the figure. The root of the HTB tree is represented at the top of the figure.

The main parameters to be defined for each application category are: *rate R* (kb s^{-1}), which defines the allocated bandwidth; *ceiling C* (kb s^{-1}), which is the maximum bandwidth that can be used when the other categories are not using their full bandwidth; and *priority*, which determines the priority order on which the categories will be served. It is strongly suggested to determine the *rate* and *ceiling* values based on the bandwidth actually available (*BW* parameter[4]). Example values are shown in Table 3.3.

In the example, Category A has been given the maximum steady-state bandwidth (70%, against 20% of B and 30% of C), whereas Category B has been given the

---

[4] BW states which fraction of the nominal bandwidth can be used to successfully transmit data ("net payload"). It can be approximated, in first instance, as a fraction of the physical connection speed of the specific wireless medium (e.g., 50% for 802.11b, 60% for 802.11e).

Table 3.4 Relationship between application classes, application categories, and WMM ACs

HTB application category	Application class	WMM access category
C	II, VII, VIII	0 (Best effort)
B	I, VI	1
A	III	2
A, B	IV, V	3 (Highest priority)

maximum priority (because of delay constraints). It is worth noting that, in this case, ceiling of B shall be maintained lower than BW, to avoid the situation where A gets no transmission opportunities, even over a short period of time.

3.6.3.3 Allocation of the Application Categories to the WMM Access Categories

The third step to be performed is to determine the relationship between the application categories defined in the HTB structure, the application classes and the WMM Access Categories. For what concerns our specific case study, a possible classification is shown in Table 3.4. Because WMM provides probabilistic QoS, it is necessary to put the privileged services into the two highest WMM ACs, to avoid unwanted side effects caused by the soft QoS mechanism.

3.6.3.4 Dimensioning of the HTB Parameters for the Application Classes

The last step of the QoS configuration concerns the dimensioning of the most significant HTB parameters for various application classes. They have to be calculated for each application class and depend strongly on the actual configuration of the wireless network and on the expected usage of each class. In our case, these values are shown in Table 3.3 (*rate, ceiling, priority*) with the addition of the *child queueing discipline* (qdisc) parameter. The latter determines how the flows belonging to the same application class have to be scheduled.

The two most common values for the *child qdisc* parameter are SFQ and PFIFO. SFQ is the acronym for statistically fair queueing. This queueing discipline is self-explaining and schedules the flows to guarantee fairness over time. We suggest use SFQ when dealing with high bandwidth services. PFIFO, on the other side, is a more efficient implementation of the first-in-first-out queueing discipline. It should be used for low delay services, such as emergency calls (Application Class V).

3.7 Directions for Future Research

WMNs have been a hot research topic in very recent years, and numerous test-beds and commercial networks have been deployed and tested. However, many aspects relating to WMNs still deserve for further investigation, such as:

- How to achieve higher throughput, e.g., up to 100 Mbp, over multihops with vast geographic coverage?
- How to achieve balanced traffic load in the presence of multiple gateways?
- How to achieve optimized tradeoff between reliability and redundancy?
- How to ensure QoS for a set of end-users when the channel is close to congestion, given a shared medium?
- How to achieve better performance for multimedia services over multi-hop WMNs?
- How to improve the scalability of the routing protocol when the number of routers (not the number of end-users) exceeds 1,000?
- How to support node mobility in WMNs and how to achieve seamless handoff between WMNs and cellular networks?
- How to handle and implement metric-based routing with multiple metrics for different traffic classes?

3.8 Thoughts for Practitioners

To deploy successfully a real-life WMN, a number of questions have to be taken into account carefully, from design, implementation, to test and installation. In the following, we give a nonexhaustive list of these questions that may be helpful for readers who are interested in developing WMN technologies:

- What is the main purpose of the WMN to be deployed? In which region will the WMN be deployed?
- How many end-users are expected for the to-be-deployed WMN? How many mesh routers do we need? In hierarchical or flat topology? At which locations will the mesh routers be placed? Do they provide full coverage for the end-users?
- Can all mesh routers be installed with AC-power supply? If not, how to select a power supply panel that provides noninterrupted service?
- Where is/are the nearest Internet gateway(s) and which types of the gateways are they?
- Which routing protocol to use in the to-be-deployed WMN? Does it provide all expected features? If not, how to we extend the protocol to support other designed features?
- Does the selected routing protocol provide open-source codes?
- Which hardware and Linux platform will be used for our mesh routers? Which operation system will be used in the selected hardware?

- Which tool will be used for remote monitoring and autoconfiguration of the mesh routers?
- Do we need to provide QoS in the to-be-deployed WMN? If yes, which QoS mechanisms will be used? How to classify traffic classes and how to prioritize high-priority traffic class(es)?
- Which type of security and authentication mechanism will be used in the to-be-deployed WMN?
- Is the to-be-deployed WMN extendable, in terms of hardware updating, software upgrading, new services provisioning and network capacity expansion?
- How about the cost of the to-be-deployed WMN? Is there any trade-off between equipment redundancy and reliability of the network?
- Which business models will be used by the WISP of the to-be-deployed WMN?

3.9 Conclusions

In this chapter, we have given a brief introduction to WMNs and a short survey of existing academic and commercial mesh networks. With focus on probably the most promising application scenario of WMNs, where a multihop WMN is used for providing broadband Internet access, we have proposed a framework with scalable, reliable, metric-based routing and QoS-aware features. Two important aspects within this framework, multihomed routing with load balancing and QoS provisioning, have been described in more details. Finally, based on our expertise and experience in this area, a few potential research topics and thoughts for practitioners within WMNs have been outlined, which may help interested readers identify their areas of interest for further research.

3.10 Questions

1. What are the main differences between cabled networks and WMNs?
2. What is a hierarchical WMN?
3. What are the main application scenarios for WMNs?
4. What is *link budget*? Calculate the link budget for a 2.4-GHz wireless link where: the two communicating devices are distant 150 m and are in Line of Sight, the transmitted power is equal to 18 dBm, the power loss of the coaxial cable at both sender's and receiver's end is equal to $0.2 \, \text{dB m}^{-1}$, the length of the sender's coaxial cable is 2 m, the length of the receiver's coaxial cable is 1 m and the antenna gain at both sender's and receiver's length is 8 dB.
5. What are link state routing protocols?
6. What is OLSR? How does OLSR work? How is optimization achieved in OLSR?

7. What is metric-based routing? How to implement metric-based routing calculation in OLSR?
8. Explain the need of introducing a power aware enhancement in OLSR.
9. In the power aware routing OLSR enhancement, which are pro's and con's of selecting the "Battery level at each node" or the "Minimize the maximum node cost" algorithm?
10. What do "Probabilistic QoS" and "Deterministic QoS" mean? What is the drawback of providing hard QoS with the WMM mechanism?
11. Explain how the WMM and HTB mechanisms work.

References

1. ISO/IEEC 8802-11. Wireless LAN Medium Access Control (MAC) and Physical Layer (PHY) Specifications. ANSI/IEEE Std 802.11, 1999
2. I. F. Akyildiz and X. Wang, A survey on wireless mesh networks, IEEE Communications Magazine, 43(9), 23–30, 2005
3. H. Moustafa, U. Javaid, T. M. Rasheed, S. M. Senouci, and D. Meddour, A Panorama on Wireless Mesh Networks: Architectures, Applications and Technical Challenges, First International Workshop on "Wireless Mesh: Moving Towards Applications," (WIMESHNETs) 06, Waterloo, Canada, August 2006
4. The ADHOCSYS Project. URL: http://www.adhocsys.org
5. T. Clausen and P. Jacquet, Optimized Link State Routing Protocol (OLSR), IETF RFC 3626, Oct. 2003. URL: http://www.ietf.org/rfc/rfc3626.txt
6. L. Chen, Wireless Mesh Networks. URL: http://www.cs.wustl.edu/~jain/cse574-06/ftp/wireless_mesh.pdf
7. MIT Roofnet. URL: http://pdos.csail.mit.edu/roofnet/doku.php
8. Georgia Institute of Technology BWN-Mesh. URL: http://www.ece.gatech.edu/research/labs/bwn/mesh/
9. State University of New York Hyacinth. URL http://www.ecsl.cs.sunysb.edu/multichannel/
10. University of Illinois Net-X. URL: http://www.crhc.uiuc.edu/wireless/netx.html
11. D. A. Maltz, J. Broch, and D. B. Johnson, Experiences Designing and Building a Multi-Hop Wireless Ad Hoc Network Testbed, URL: http://reports-archive.adm.cs.cmu.edu/anon/1999/CMU-CS-99-116.pdf
12. MeshNetworks. URL: http://www.meshnetworks.com/
13. SkyPilot Networks. URL: http://www.skypilot.com/
14. Tropos Networks. URL: http://www.tropos.com/
15. Firetide. URL: http://www.firetide.com/
16. Intel Research. URL: http://www.intel.com/research/exploratory/heterogeneous.htm
17. Microsoft Mesh Networks Research. URL: http://research.microsoft.com/mesh/
18. Meraki networks. URL: http://www.meraki.com
19. Strix Systems. URL: http://www.strixsystems.com/
20. Radiant Networks. URL: http://www.radiant-networks.com/
21. NowWireless. URL: http://www.now.co.uk/
22. Locust world. URL: http://www.locustworld.com/
23. WiliBox. URL: http://www.wilibox.com/
24. Essentia Wifless. URL: http://www.wifless.com/pagina.asp?id=39
25. Wi-Next. URL: http://www.winext.eu/
26. L. Leschiutta et al. Specification of Algorithms for Static Reconfiguration of Ad-Hoc Network, and Multimedia Services, final version. ADHOCSYS project, Deliverable D14, IST-2004-026548, project restricted (available on request, please contact authors), delivered to the EC, May 2007

27. F. Y. Li et al. Implications on the Use of MANETs Protocols. ADHOCSYS project, Deliverable D3, IST-2004-026548, project restricted (available on request, please contact authors), delivered to the EC, May 2006
28. IEEE 802 Committee. Wireless LAN Medium Access Control (MAC) and Physical Layer (PHY) Specifications – Amendment 8: Medium Access Control (MAC) Quality of Service Enhancements. IEEE Std 802.11e, September 2005
29. Wi-Fi alliance. Support for Multimedia Application with Quality of Service in Wi-Fi Networks. URL: http://www.wi-fi.org, Sep. 2004
30. M. Devera. Hierarchical Token Bucket Theory, Technical Report. URL: http://luxik.cdi.cz/~devik/qos/htb/manual/theory.htm
31. Application Layer Packet Classifier for Linux (l7-filter). URL: http://l7-filter.sourceforge.net
32. H. Zhu et al. A survey of quality of service in IEEE 802.11 networks, IEEE Wireless Communications, 11(4), 6–14, 2004
33. M. Crotti, M. Dusi, F. Gringoli, and L. Saltarelli, Traffic classification through simple statistical fingerprinting, ACM SIGCOMM Computer Communication Review, 37(1), 7–16, 2007
34. M. Crotti, F. Gringoli, P. Pelosato, and L. Saltarelli, A Statistical Approach to IP-level Classification of Network Traffic, The 2006 IEEE International Conference on Communications (ICC), Vol. 1, pp. 170–176, Istanbul, Turkey, 11–15 June 2006
35. G. Bianchi, I. Tinniriello, and L. Scalia, Understanding 802.11e contention-based prioritization mechanisms and their coexistence with legacy 802.11 stations, IEEE Network, 19(4), 28–34, 2005
36. G. Bianchi and I. Tinnirello, Remarks on IEEE 802.11 DCF performance analysis, IEEE Communications Letters, 9(8), 765–767, 2005
37. D. Ivancic, N. Hadjina, and D. Basch, Analysis of Precision of the HTB Packet Scheduler, 18th International Conference on Applied Electromagnetics and Communications (ICECom), 2005
38. Arcturus Networks, SIPjack[TM] series, URL: http://www.arcturusnetworks.com
39. T. Sprull and J. Lockwood, Extensible Network Configuration and Communication Framework, URL: http://www.arl.wustl.edu/~todd/sproull_iwan_05.pdf

Chapter 4
Stabilizing Interference-Free Slot Assignment for Wireless Mesh Networks

Mahesh Arumugam, Arshad Jhumka, Fuad Abujarad, and Sandeep S. Kulkarni

Abstract In this chapter, we focus on stabilizing interference-free slot assignment to WMN nodes. These slot assignments allow each node to transmit its data while ensuring that it does not interfere with other nodes. We proceed as follows: First, we focus on infrastructure-only part where we only consider static infrastructure nodes. We present three algorithms in this category. The first two are based on communication topology and address centralized or distributed slot assignment. The third focuses on slot assignment where infrastructure nodes are deployed with some geometric distribution to cover the desired area. Subsequently, we extend this protocol for the case where there are mobile client nodes that are in the vicinity of the infrastructure nodes. And, finally, we present algorithm for the case where a client node is only in the vicinity of other client nodes.

4.1 Introduction

Wireless mesh networks (WMNs) are one type of wireless networks constructed with mesh routers and mesh clients. Mesh routers are the backbone of the WMNs and form a mesh network within themselves. Mesh clients can serve as hosts to their application(s) and at the same time serve as routers to other clients. Mesh clients can also form WMNs within themselves to provide connections between nodes that are not within the transmission range of each other.

Mesh routers are typically stationary. They form the infrastructure of the WMNs. They are different from traditional wireless routers in that they are often equipped with multiple wireless interfaces. This increases their transmission compatibilities and capabilities. The mesh routers are connected with each other in a way to form an

S.S. Kulkarni (✉)
Department of Computer Science and Engineering, 3115 Engineering Building,
Michigan State University, East Lansing, MI 48824, USA
e-mail: sandeep@cse.msu.edu

S. Misra et al. (eds.), *Guide to Wireless Mesh Networks*, Computer Communications
and Networks, DOI 10.1007/978-1-84800-909-7_4,
© Springer-Verlag London Limited 2009

infrastructure through which their clients can connect to other larger networks such as the Internet. One of the advantages the mesh routers have is that they require less transmission power, that they can use the multihop connections.

Mesh clients can be of several types such as desktops, laptops, phones, sensors, etc. They are mostly mobile and have wireless interfaces to connect them to mesh routers and other clients. Although mesh clients without wireless interfaces can connect to wireless mesh routers via Ethernet connections, we do not consider them in this chapter that our focus is on MAC layer for wireless channel. Clients in WMNs can form a network within themselves without mesh routers. The clients, while serving their users application, also serve as routers and forward other clients messages to the requested destination. The difference between the mesh routers and the mesh clients is that mesh clients do not have the gateway or bridge functions.

There are a wide variety of applications that can benefit from WMNs. These applications include: broadband home networking, community and neighborhood networking, enterprise networking, metropolitan area networking, transportation systems, building automation, health and medical systems, and security surveillance systems. To illustrate the use of WMNs consider for example the community and neighborhood networking: The typical way of setting up community and neighborhood networks is by connecting a wireless router to the Internet though cable or DSL modem. These types of networks include many points of access to connect their clients to the Internet. This architecture suffers from many limitations: communication between clients in the same network have to go through the Internet, expensive and high bandwidth routers are required to cover the neighborhood, and many dead zones may exist in between homes. WMNs can overcome these problems. Another example of the applications of WMNs is the enterprise networking. Although there was a significant increase in the use of wireless networks in enterprise networking, wireless units are being used as isolated groups with no link between them except Ethernet connections, which are expensive to setup. WMNs will eliminate the need for any Ethernet connections between wireless units by using wireless mesh routers.

WMNs are different from traditional wireless networks. In WMNs, clients are connected to more than one point of access. However, in traditional wireless networks, nodes are connected to single point of access. Providing direct connections to every node can be expensive and impractical, because it requires setting up many points of access. WMNs use their routers and clients transmission power to overcome such problems. The advantage of the WMNs over the conventional wireless networks is that, it provides reliable, cost effective, robust, self-configuring, and self-organizing networks.

Organization of the chapter. The rest of the chapter is organized as follows: In Sect. 4.2, we identify different classifications of WMNs. In Sect. 4.3, we present the model and assumptions made in this chapter. We also define the problem of slot assignment and what it means to be stabilizing. In Sect. 4.4, we present our algorithm for the case where only infrastructure nodes are considered. In Sect. 4.5, we extend it to deal with the case where client nodes are present but only in the vicinity of infrastructure nodes, i.e., they do not form a client WMN by themselves. In Sect. 4.6, we extend it to deal with the case where all client WMNs are permitted.

In Sect. 4.7, we identify thoughts for practitioners and in Sect. 4.8, we identify future directions for research. Finally, we summarize in Sect. 4.9.

4.2 Background

WMNs are broadly classified in terms of Infrastructure/backbone WMN, Client WMNs, and Hybrid WMNs depending upon the types of nodes participating in them [1].

- *Infrastructure/backbone WMNs*. In infrastructure/backbone WMN, networks are built with routers only. Routers are connected to each other with many links to form backbone network that can be utilized by clients. Routers in this group use two types of communications: infrastructure communication, to connect within themselves and other networks, and user communication, to connect with their clients. Also, this group uses the gateway and bridge functionality of the routers to connect clients to each other and to existing networks such as the Internet.
- *Client WMNs*. This group consists of peer-to-peer networks among clients. Nodes can serve as hosts to the user application(s), and can provide routing functionality to connect with other nodes. Therefore, it is possible for the client WMNs to be built without routers. In this group, packets will be transmitted from the source to the destination though multiple nodes. Clients will forward packets from one node to another until the packet reaches its destination. Nodes in this group are equipped with routing and transmission capabilities to help them perform their tasks.
- *Hybrid WMNs*. This type of network combines both the infrastructure and the client WMNs together in one network. Both, routers and clients will provide point of access to the network. In the hybrid WMNs, clients will provide more capabilities to the network. They will be used to connect other clients, who are not in any transmission range of any router, to the network.

Our focus is on MAC layer for WMNs. The computation of the MAC needs to be distributed using a collaborative protocol among the WMN nodes. Furthermore, the MAC layer needs to adapt to the mobility of the nodes. However, whenever possible, such mobility should be assisted with static infrastructure nodes that are typically present in WMNs. Given these requirements, there is a need to design a new MAC layer for WMNs that meets them. In this chapter, we propose a TDMA-based MAC protocol that meets such requirements.

In particular, in this chapter, we propose interference free slot assignment for three types of WMNs. Our first algorithm focuses on the infrastructure nodes. This algorithm focuses on the case where nodes are static and (relatively) stable. Being relatively stable and static allows one to implement efficient slot assignment algorithms from traditional networking. In the second algorithm, we extend it to deal with the client nodes that are mobile but are in the vicinity of infrastructure nodes. In the third algorithm, we present an algorithm for the case where client nodes are not in the vicinity of infrastructure nodes and, hence form client WMN.

4.3 Model and Problem Statement

In this section, we describe the WMN model and the assumptions in our algorithms. We also identify why these assumptions are reasonable or how they can be met using existing techniques. In other words, many of the assumptions are made for brevity of presentation and can be removed by incorporating existing techniques.

Existence of a leader. We assume that there exists a unique mesh router, denoted as the leader that initiates the TDMA slot assignment. There are several ways in which this leader could be chosen. In particular, one may use any of the existing leader election algorithms [2–6]. The use of these algorithms also ensures that if the leader fails then another unique leader would be elected. Moreover, the leader election algorithm can be tuned to identify the leader most appropriate for a given WMN. For example, one may prefer the leader to be a node with high degree or a node that is centrally located or a node that is expected to have high availability (e.g., bridge node provided by a service provider). Furthermore, leader election could be done independent of the normal node operation that infrastructure nodes are typically equipped with multiple wireless interfaces. Also, if the network provides it, the leader could also be chosen from available centralized servers (e.g., for authentication). Finally, the exact leader and its location are not crucial in that the amount of work that the leader is performing is limited to the initial slots assignment and to the reorganization cycles.

Faults/recovery of infrastructure. The leader makes slots assignment based on its knowledge about working nodes in the network. After slot assignment, if a mesh router fails, then it will be eliminated from the list of infrastructure nodes in the next reorganization cycle. All the slots that were assigned to faulty nodes will be reclaimed and reassigned to the other active routers. If a new mesh router joins the network it will be treated as mesh client until the next organization cycle is started. Hence, it will run the same algorithm that the client node would use until it receives its new slots in the next reorganization cycle.

Neighborhood discovery. There are no requirements on the way in which the mesh routers will be arranged in their network space. For simplicity, we assume that each node knows its neighbors. This could be implemented in several ways. For example, each node could maintain a list, say listen-from, of nodes it can hear from. It can communicate this list to the leader during the leader election process and before the initial slots are assigned. As a result, the leader can compute the network topology. All the nodes in the network will send a message to the leader containing the list of the nodes they can listen to. The leader then will be able to compute two important sets: The first set is the listen-from set, which contains the list of all nodes that a specific node can listen to. The second set is the talk-to set, which contains the set of nodes that are in the transmission range of a specific node. After computing these two sets the leader will share this information with all other nodes. Therefore, each mesh router will be able to compute its communication range and with which group of mesh routers its messages may collide. Also the nodes will know the network they are in, and the routing layer can use this information to compute the best route.

Note that in the first two cases (infrastructure-only and clients in the vicinity of infrastructure nodes), only infrastructure nodes need to know other infrastructure nodes that it can communicate with. Clients (in the case where they are in the vicinity of infrastructure nodes) only need to know the infrastructure nodes that they are close to. For the third case (clients forming WMN by themselves), clients need to know their neighbors but this discovery can be easily achieved using standard techniques such as those in [7–9].

4.3.1 Problem Statement

In this section, we precisely define the problem of slot assignment. For sake of simplicity, first, we consider the case where only one frequency is available and, hence, all nodes are transmitting on the same frequency. With a single frequency, the problem of slot assignment is the problem of time division multiple access (TDMA) where each node is assigned a set of slots in which it can transmit.

Now, consider the case where two nodes, say a and b are transmitting and both messages could be received by a common neighbor c. Now, if both a and b transmit simultaneously then there will be a collision at c thereby preventing c from receiving either message. The goal of the slot assignment algorithm is to assign slots to each node so that no two nodes transmit simultaneously if their messages will collide at some node. To define the problem statement, we view the WMN as a graph $G - (V, E)$ where V consists of all nodes (infrastructure nodes, clients, etc.) and E denotes the links between them. The pair (v_1, v_2) is in E if v_1 can communicate with v_2. Note that the relation E is reflexive, i.e., for any node v_1, $(v_1, v_1) \in E$. It may not be symmetric, i.e., it is possible that $(v_1, v_2) \in E$ and $(v_2, v_1) \notin E$.

First, we define the notion of collision group in WMN; the collision group of a node, say j, includes those nodes that should not transmit when j is transmitting. Based on the above discussion, we define collision group as follows:

Definition 4.1 (Collision group). The collision group of j is $CG(j)$, where

$$CG(j) = \{k \mid \ni l : (j, l) \in E \land (k, l) \in E\} - \{j\}.$$

Now, using the collision group, we can define the problem of slot assignment for a single frequency (i.e., TDMA). The slot assignment problem is to assign each node a set of slots such that two nodes transmitting simultaneously are not in the collision group of each other. Thus, the problem of slot assignment is as follows:

Using this definition, the problem statement for slot assignment is as shown in Fig. 4.1.

Note that this definition of collision group and TDMA takes into account the unidirectional nature of the links. In other words, if there are two nodes j and k such that l can communicate with j and k although neither j nor k can communicate with l then the above problem statement allows j and k to transmit simultaneously. Ideally, one should solve the problem of slot assignment by considering the existence

> **Problem statement for slot assignment.** *Assign a set of slots,*
> *say* slot$_j$ *to each node* j *such that*
> $$\forall j,k : j \in CG(k) \Rightarrow slot_j \cap slot_k = \phi$$

Fig. 4.1 Problem statement for slot assignment

> **Problem statement for slot assignment with symmetric**
> **links.** *The problem of slot assignment with bi-directional links*
> *is to assign a set of slots, say slot$_j$ to each node j such that*
> $$\forall j,k : j \in SCG(k) \Rightarrow slot_j \cap slot_k = \phi$$

Fig. 4.2 Problem statement for symmetric slot assignment

of unidirectional links. However, in certain cases, for sake of simplicity, an algorithm may treat all links as bidirectional and assign slots accordingly. For such an algorithm, we define the notion of symmetric collision group,

Definition 4.2 (Symmetric collision group). The symmetric collision group of node j is SCG(j) where

$$\text{SCG}(j) = \left\{ k \,\middle|\, \exists l : (j,l) \in E' \wedge (k,l) \in E' \right\} - \{j\},$$

where

$$E' = \{(j,k),(k,j) \,|\, (j,k) \in E\}.$$

Using this definition, the problem statement for slot assignment is as shown in Fig. 4.2

In case of WMNs, a node may be able transmit on multiple frequencies. Hence, the slot assignment not only has to deal with assignment of timeslots but assignment of frequencies as well. In such a model, we view each slot assigned to a node to be of the form (f, t), where (f, t) denotes that the node is allowed to transmit at time t on frequency f. With such a definition of slots, the problem of slot assignment remains the same as above with the exception that the slot identifies both frequency and time slot.

4.3.1.1 Additional Restrictions on Slot Assignments in WMNs

Note that the above problem statement imposes no restrictions on which frequencies should be assigned to which nodes. In practice, however, such restrictions could be in place because of hardware limitations. To illustrate this, consider the case where there are two infrastructure nodes, say x_1 and x_2 that communicate with their respective clients y_1 and y_2. Furthermore, in such a network, assume that the infrastructure nodes use one frequency to communicate with other infrastructure node and another frequency to communicate with clients. Then, such a network could use one frequency, say f_1, for communication between x_1 and x_2, another, say f_2, for communication between x_1 and y_1 and a third frequency, say, f_3, for communication between x_2 and y_2. Observe that in this scenario, x_1 does not (respectively, cannot) communicate with frequency f_3. Hence, it should not be assigned slots of the form $(f_3, -)$. Likewise, all slots assigned to y_1 would be of the form $(f_2, -)$.

Furthermore, in our solution for the case where clients are always in the vicinity of an infrastructure node, the infrastructure nodes are assigned their slots and the clients *borrow* these slots from them. Given such a model and the scenario in the previous paragraph, x_1 would be assigned all slots in frequency f_2. In this case, we say that frequency f_2 is assigned to node x_1.

4.3.1.2 Unidirectional Antennas

The above problem statement assumes that communication is omnidirectional, i.e., when a node sends a message, it is broadcast to all neighbors that can listen to it. The problem (and the solutions in this chapter) could be easily modified to deal with the case where communication uses unidirectional antennas. However, this issue is outside the scope of this chapter.

4.3.2 Defining Self-stabilization of Slot Assignment

A solution to the slot assignment problem from the previous Sect. 4.3.1 would ensure that when a node transmits, its message would not collide with other messages and would be received by intended receiver(s). However, if certain faults occur then the interference freedom property may be violated. Examples of such faults include clock drift, variable communication characteristics, etc. Particularly, if clocks drift then two slots of neighboring nodes may correspond to the same time. Or, if communication characteristics change, e.g., if the slots are assigned under the assumption that j is not in the collision group of k. Now, if communication range of j increases (because of hardware changes or variable nature of communication characteristic) or a new node is present in the area covered by communication ranges of j and k then the time slots must be reevaluated so that the collision freedom is guaranteed.

Although certain changes in topologies would be handled explicitly, e.g., when nodes move their slots would be recomputed to ensure collision freedom, unexpected faults could cause the system to reach states where collision freedom is violated. In these situations, it is necessary to restore the system to a legitimate state. In particular, we argue that the slot assignment algorithms should be self-stabilizing [10, 11].

Definition 4.3 (Self-stabilization). A system is self-stabilizing if starting from an arbitrary state, it (1) recovers to legitimate state, and (2) upon recovery continues to be in legitimate states forever.

Thus, a self-stabilizing slot assignment algorithm would ensure that even if faults cause corruption of slots assigned to nodes, the network would eventually recover to states where correct slot assignment is reestablished.

Our approach for providing stabilization is based on periodic update. In particular, in periodic update, each node would update its slots so that they are conflict free with other nodes in its collision group. The update would also ensure that the number of unused slots is reduced/eliminated. Thus, the maximum time for restoring time slots is directly proportional to the time between these updates. And, the overhead of stabilization is inversely proportional to the time between updates. Hence, the value of the period should be chosen based on system needs, level of acceptable overhead, probability of faults, etc. We note, however, that while worst case depends upon the period, many faults would be handled locally whenever feasible. For example, in most situations, node failure, repair, or movement would be handled locally and immediately.

4.4 Self-stabilizing Frame Assignment for Infrastructure Network

In this section, we present algorithms for assigning frames to infrastructure nodes of a WMN. To achieve collision free communication, we need to ensure that the frames assigned to a node are unique within its distance-2 neighborhood. This can be achieved using distance-2 coloring. As an illustration, consider the communication topology shown in Fig. 4.3. Figure 4.3a considers bidirectional links among the nodes. In this example, if nodes a and b transmit simultaneously then it causes a collision at node c. Hence, a and b should transmit in different frames. In other words, a and b should get different colors. Similarly, node c cannot transmit simultaneously with neighbors of a or b as it leads to collision at a or b. Figure 4.3b considers some unidirectional links among the nodes. In this example, if two nodes can send messages to a common neighbor then they should get different colors to ensure collision free communication. On the other hand, node c may transmit simultaneously with nodes that can send messages to a or b directly. This does not cause a collision at a or b as c cannot talk to a or b directly.

Distance-2 coloring. Given a communication graph $G = (V, E)$ for the infrastructure network, compute E' such that two distinct nodes a and b are connected in E' if they are connected in E or if they share a common neighbor c that can hear from both a and b. To obtain distance-2 coloring, we require that $\forall (i, j) \in E' ::$ color.$i \neq$ color.j. Formally, the problem statement is defined in Fig. 4.4.

Fig. 4.3 Distance-2 coloring

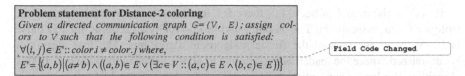

Problem statement for Distance-2 coloring
Given a directed communication graph $G= (V, E)$; *assign colors to* V *such that the following condition is satisfied:*
$\forall (i, j) \in E' :: color.i \neq color.j$ *where,*
$E' = \{(a,b) | (a \neq b) \wedge ((a,b) \in E \vee (\exists c \in V :: (a,c) \in E \wedge (b,c) \in E))\}$

> Field Code Changed

Fig. 4.4 Problem statement for distance-2 coloring

Frame assignment. The algorithms presented in this section assign complete frames to each node. Each frame consists of x (≥ 1) slots. The infrastructure node will choose the first slot within its assigned frames for sending messages. The node will assign a subset of other slots to the clients attached to it. (We refer the reader to Sect. 4.5 for a discussion on the algorithm that assigns slots to the clients.)

We present two methods for frame assignment in infrastructure networks. The first method considers a network with arbitrary topology where each node is aware of only its local neighborhood (i.e., the nodes that it can directly communicate with). This method is suitable for the case where the application can tolerate network initialization time (to setup the interference-free frames). Moreover, this method does not require any global knowledge and location information. By contrast, the second method considers a network where the locations of the nodes are known up front and the network is deployed in some geometric topology to cover a given region. In such networks, the frames can be assigned offline to a *location* and each node can statically determine its frames by virtue of where it is located. This method is suitable for the case where each node is equipped with GPS for determining its location (i.e., its global coordinates). Such networks allow the nodes to start functioning immediately (without significant network initialization overhead) that the frames are computed statically and time synchronization is achieved using GPS. Moreover, addition of new nodes to the network is much faster.

With this introduction, we present our algorithms in more detail, next.

4.4.1 Frame Assignment Algorithms Using Graph Coloring

In this section, we present two frame assignment algorithms in infrastructure network with arbitrary topologies. We use graph coloring to assign frames to the nodes. Specifically, we obtain distance-2 coloring of nodes that identifies the initial frame assignments. Distance-2 coloring ensures that the colors assigned to nodes i and j are different if i is in the collision group of j (or vice versa). This will ensure that the frame assignment (and, therefore, the slot assignment) would meet the problem statement in Fig. 4.1. Once a node determines its color (i.e., the initial frame), it can compute subsequent frames assigned to it by using the number of colors required to obtain distance-2 coloring. Suppose $color_i$ is the color assigned to node i. Node i gets $\forall c : c \geq 0 : color_i + c \times K$ frames, where K is the number of colors required to obtain distance-2 coloring. In this chapter, K is also referred as the frame period.

Based on the model in Sect. 4.3, there is a leader in the network that is responsible for frame assignment. The leader could be chosen by algorithms such as those in [2–6]. Now, we present the frame assignment algorithms. The first algorithm is centralized where the leader assigns colors to all the nodes in the network. In this algorithm, the leader can optimize the number of colors required. However, in this algorithm, the leader has to learn the network topology before it can assign colors. The second algorithm is distributed in the sense that each node chooses its color depending on the colors chosen by its neighborhood. The main advantage of this algorithm is that, unlike the centralized algorithm, addition of new nodes does not involve the leader. Additionally, the distributed algorithm does not require a node to learn the entire network topology.

Next, we discuss these two algorithms in detail.

4.4.1.1 Algorithm 1: Centralized Coloring

In this algorithm, colors are assigned to the nodes in a centralized fashion. This is achieved using the following three step process (1) computing the global network topology, (2) coloring the nodes such that two nodes that are within distance-2 of each other have unique colors, and (3) distributing the colors and the frame period to all the nodes in the network. Next, we discuss these steps in detail.

Step 1: Computing the Network Topology

As mentioned earlier, each node is aware only of its local neighborhood. As discussed in Sect. 4.3, all nodes communicate their local neighborhood to the leader. That many leader election algorithms actually construct a spanning tree that is rooted at the leader, these messages could be sent along this tree. Alternatively, similar to algorithms in [7,8], these messages could be sent using broadcast primitives. Furthermore, by allowing nodes to combine messages of different nodes, number of messages could be reduced. Thus, once the leader election is complete and a leader is decided, the leader will be aware of the entire network topology.

Step 2: Distance-2 Coloring of the Nodes

The leader can then apply [12] to obtain distance-2 coloring of the network. Specifically, in [12], Lloyd and Ramanathan present *minimum degree last* algorithm for distance-2 coloring. First, the algorithm assigns a unique label to each node in a *progressive* fashion. Suppose node i is labeled p. The next node the algorithm chooses to label will be the one with the least number of neighbors in the subgraph formed by all unlabeled nodes. The label of that node is $p + 1$. Once all the nodes are labeled, the algorithm then colors the nodes starting with the highest labeled node. When a node is selected for coloring, the algorithm assigns the lowest numbered color that does not conflict with previously colored nodes. Lloyd and Ramanathan show that ordering obtained using the labeling of nodes is crucial in bounding the worst-case performance of the algorithm. Also, they prove that obtaining an optimal distance-2 coloring of planar graphs is NP-complete (even in an offline setup).

Note that since WMN nodes could transmit on multiple frequencies, the algorithm in [12] would be repeated for each frequency. Moreover, the graph being considered for each frequency would be different based on the nodes that can actually transmit on that frequency. Thus, the number of colors required for each frequency may be different. It follows that the frame period used for different frequencies may be different.

Step 3: Distributing Colors and Frame Period to the Nodes

Once the leader computes the colors of the nodes, it distributes them to the nodes. Towards this end, the leader communicates the color assignments and the frame period (which is equal to the number of colors required to obtain distance-2 coloring) in the slots allocated to it. Whenever a node receives color assignments, it does the following (1) determines its initial frame assignment (from the color assigned to it), (2) computes its subsequent frame assignments using the frame period, and (3) communicates the color assignments it received to its neighbors in slots assigned to it. Continuing in this fashion, the color assignments and frame period are distributed to the nodes and each node determines its frames.

Self stabilization

We sketch the outline of how self-stabilization is achieved. As mentioned in Sect. 4.3.2, the stabilization is provided by periodic revalidation of frames. This revalidation ensures that frames remain collision free and in case of (controlled) topology change such as addition or removal, frames are recomputed. In case of arbitrary failures, the validation messages may collide preventing a node from receiving its revised frame assignment. In this case, after a timeout, the node reverts to using CSMA and restricts application traffic. This minimizes network traffic and helps the revalidation messages to succeed. Once the revalidation of frames is complete, the node subsequently resumes application traffic. The value of the timeout depends on the frequency of update messages and number of nodes. The details of computing this timeout are available in [13].

4.4.1.2 Algorithm 2: Distributed Coloring

In this section, we present our distributed coloring algorithm for frame assignment in infrastructure network. We propose a layered architecture that includes (1) distance-2 neighborhood layer, (2) token circulation layer, (3) distance-2 coloring layer or the TDMA layer, and (4) application layer. The distance-2 neighborhood layer is responsible for maintaining distance-2 neighbor path-list at each node. Path-list of a node identifies the path to all the distance-2 neighbors of the node. The token circulation layer is responsible for circulating a token in such a way that every mesh router is visited at least once in each circulation. The token-circulation layer assumes that the subgraph obtained using only the bidirectional links of the network is connected. The token circulation algorithm uses only the bidirectional links to circulate the token. The distance-2 coloring layer is responsible for determining the initial frame of the node. Whenever a node receives the token, it can choose or validate its color. As before, the color of the node identifies the initial frame of the node. Finally, the application layer is where the actual application resides. All application message communication goes through the TDMA layer. Next, we discuss the first three layers in more detail.

Distance-2 Neighborhood Layer

As mentioned before, this layer is responsible for maintaining distance-2 neighbors of a node. Towards this end, each node sends distance-2 neighborhood discovery messages. More specifically, each node communicates the information about its immediate neighbors (i.e., nodes that can send messages to this node and nodes that can hear from this node) up to certain distance in the network. The distance up to which a node forwards the information is a tunable parameter. Before the frames are assigned to each node, nodes communicate using CSMA mechanism and rely on back-off schemes for reliability. Once the frames are assigned, this layer sends distance-2 neighborhood discovery messages in the slots assigned to a node.

Token Circulation Layer

The token circulation later is responsible for maintaining a spanning tree rooted at the leader and traversing the graph infinitely often. The leader initiates the token circulation in the network. As mentioned earlier, we assume that the subgraph formed with bidirectional links in the network is connected. The token traverses the network using bidirectional links (as it provides acknowledgment to the node that forwards the token). In this section, we do not present a new algorithm for token circulation. Rather, we only identify the constraints that this layer needs to satisfy. This layer should recover from token losses and presence of multiple tokens in the networks. Existing graph traversal algorithms [14–17] satisfy these constraints and, hence, any of these can be used.

Distance-2 Coloring/TDMA Layer

We use the token circulation protocol in designing a distance-2 coloring algorithm. As before, each node is aware of its local neighborhood. Whenever a node receives the token (from the token circulation layer), it chooses its color. Towards this end, node j first computes the set $used_j$, which contains the colors used in its distance-2 neighborhood. Once it determines this set, it chooses its color such that $color_j \notin used_j$. Subsequently, it reports its color to its distance-2 neighbors (using the slots assigned to it). This action is important that it lets the nodes in the distance-2 neighborhood of j that are not yet colored to compute their *used* sets appropriately. Finally, j forwards the token to one of its distance 1 neighbors (using the token circulation layer).

As an example, consider Fig. 4.5. Each node first computes the path-list to their distance-2 neighbors. Table 4.1 identifies the distance-2 neighbors of the nodes for the topology shown in Fig. 4.5a. The token circulation layer maintains a depth first tree rooted at the leader, i.e., node r. An example depth-first tree is shown in Fig. 4.5b. Whenever a node receives the token, the distance-2 coloring/TDMA layer computes the colors used in its distance-2 neighborhood.

Suppose in Fig. 4.5b colors assigned to nodes r, a, and c be 0, 1, and 2, respectively. The colors of other nodes are not yet assigned (i.e., undefined) that the token has not reached them yet). When b receives the token, it knows the colors that have been taken by nodes in its distance-2 neighborhood. The colors assigned to nodes in the distance-2 neighborhood of b are $\{1, 2\}$. Now, b chooses a color that does not conflict with this set. In the example shown in Fig. 4.5, b sets its color to 0, the

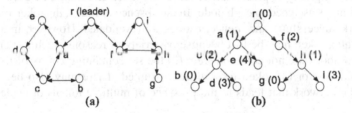

(a)   (b)

Fig. 4.5 Color assignments using distributed coloring. **a** Topology of the network and **b** traversal and color assignments

Table 4.1 Distance-2 neighborhood for the topology shown in Fig. 4.5

Node	Distance-2 neighborhood
r	$\{a, c, e, f, i, h\}$
a	$\{r, b, c, e, d, f\}$
b	$\{c, a, d\}$
c	$\{r, a, b, d, e\}$
d	$\{a, b, c, e\}$
e	$\{r, a, c, d\}$
f	$\{r, a, i, g, h\}$
g	$\{f, h, i\}$
h	$\{r, f, g, i\}$
i	$\{r, f, h\}$

minimum color that does not conflict with the colors used in its distance-2 neighborhood. Similarly, other nodes choose their colors.

Once a node determines its color, it can compute frames assigned to it. As discussed in Sect. 4.4.1.1, color of a node identifies the initial frame. A node can compute the subsequent frames assigned to it using the frame period (which is equal
to the number of colors required to obtain distance-2 coloring). Once the token
circulation is complete, the leader knows the number of colors required to obtain
distance-2 coloring in the network. In the subsequent token circulation, it forwards
this information to the nodes and each node computes its frames accordingly. More
precisely, the color of the node identifies the initial frame assigned to it and subsequent frames are computed using the frame period (as discussed earlier). In the
example shown in Fig. 4.5, the frame period is 5. Hence, frames assigned to node b
are: $\forall c : c \geq 0 : 0 + c \times 5$. Similarly, other nodes determine their frames.

Self-stabilization

We sketch the outline of how self-stabilization is achieved in this algorithm. In the
above algorithm, if the nodes are assigned correct frames then validating them is
straight-forward. For example, we can use a simple diffusing computation to report
the colors to distance-2 neighborhood and ensure the frames are consistent. In this
chapter, for simplicity of presentation, we let the token circulation be used for validation of frames assigned to each node. In the absence of faults, the token circulates
the network successfully and, hence, frames are revalidated. However, in the presence of faults, token may be lost because of variety of reasons, such as (1) frames
assigned to nodes are not collision-free, (2) the set containing colors of neighbors
is corrupted, and/or (3) token message is corrupted. There may also be transient
faults in the network that leads to the presence of multiple tokens or cycles in the
network.

Dealing with cycles. To deal with the issue of cycles, we add a time-to-live (TTL)
field to the token. Whenever the leader initiates token circulation, it sets TTL to the
number of hops the token traverses during one circulation. That the token traverses
an edge twice (once during visiting a node and once during backtracking), the leader
sets TTL to $2 \times |E_t|$, where $|E_t|$ is the number of edges traversed in one circulation.
Remember that token uses only the bidirectional links and the network formed by
the bidirectional links is connected. At each hop, the token decrements its TTL
value. When it reaches zero, the token circulation is terminated. Thus, this ensures
that when the token is caught in a cycle, token circulation terminates and the token
is lost.

Dealing with multiple tokens/lost tokens. This is achieved by keeping a timeout at
the leader. The value of the timeout is chosen in such a way that any token sent by the
leader would return back before the expiry of the timeout. The value of this timeout
depends upon the number of nodes in the network. For detailed analysis of the timeout computation, we refer the reader to [13]. Thus, if a token is lost then the leader

can generate it by sending another token. If there are multiple tokens then either they will get lost (because of expiry of TTL) or they will return to the leader before the expiry of the timeout. If the leader receives multiple tokens before the expiry of timeout then it implies that there were several tokens in the network. The leader can destroy them. Finally, each node also keeps a timeout to deal with the possible loss of token. Upon expiry of this timeout, similar Sect. 4.4.1.1, it reverts to using CSMA and blocking the application traffic so that the new token circulation would succeed. It follows that upon expiry of the timeout when the leader sends a token, there is only one token in the system and this token would circulate to reestablish the frames assigned to each node.

4.4.1.3 Addition/Removal of Nodes in the Network

In this section, we address the issue of addition/removal of mesh routers to/from the infrastructure network. Dealing with removal of nodes is straight-forward. Whenever a node is removed or fails, the frames assigned to other nodes still remain collision-free and, hence, normal operation of the network is not interrupted. However, frames assigned to the removed/failed node are wasted.

New mesh routers are typically added to the network to improve the footprint of the network and to reduce the load on mesh routers. To address the issue of frame assignment to the newly added nodes, we discuss two approaches. The first approach requires the new nodes to behave like client nodes. The second approach requires the nodes to choose conflict-free frames by listening to token circulation and distance-2 neighborhood discovery messages.

Approach 1: Adding new mesh routers as clients. In this approach, whenever a new node is added to the network, it becomes a client of one of its neighbors. Frames of the added node are assigned by its parent infrastructure node using the approach presented in Sect. 4.5.

Approach 2: Passive addition of new mesh routers. This approach requires that whenever a node forwards the token (as part of the revalidation process to verify the colors assigned to the nodes), it includes its color and the colors assigned to its distance-1 neighbors. Suppose a new mesh router, say, q is added to the network. Before q joins the network and starts communicating application messages, this approach requires q to learn the colors assigned to its distance-2 neighborhood. One way to achieve this is by listening to token circulation of its distance-1 neighbors. Once q learns the colors assigned to the nodes within distance 2, it selects its color. Thus, q can subsequently determine frames assigned to it. Now, when q sends a message, it announces its presence to its neighbors.

With this approach, if two or more nodes are added simultaneously in the same neighborhood then these new routers may choose conflicting colors and, hence, collisions may occur. However, that the distributed coloring algorithm is self-stabilizing, the network self-stabilizes to states where the color assignments are collision-free. Thus, controlled addition of new routers can be achieved.

4.4.1.4 Claiming Unused Frames

The algorithms discussed in this section assign uniform bandwidth to all nodes. In this section, we discuss an extension where nodes can claim unused frames/slots in the network, if available. This approach embeds information about the frames/slots that a node requests in the token and relies on the token circulation layer.

Each node is aware of the frames used by the nodes in its distance-2 neighborhood. Hence, a node can determine the unused slots and if necessary request for the same. A node (say, j) that requires additional bandwidth does the following. Suppose j requires unused slots identified by the set request$_j$. Upon receiving the token, j embeds request$_j$ to the token along with a timestamp that indicates when j made the request. Towards this end, the token contains three tuple information in the set token.requestSet (1) ID of the node, (2) unused slots requested by it, and (3) the timestamp. To request for unused slots, j sets token.requestSet = token.requestSet $\cup$ (j, request$_j$, timestamp$_j$).

Now, when a node receives the token, it checks token.requestSet to determine if there are any requests for unused slots by nodes in its distance-2 neighborhood. Suppose node k receives the token and finds request from a neighbor j that is within distance 2 of it. Before k decides about the fate of this request, it checks token.requestStatus to determine if other neighbors within distance 2 of j have accepted or denied this request. Towards this end, the token contains three tuple information about the status of each request in token.requestStatus (1) ID of the node, (2) timestamp when the request was made, and (3) status (accept or deny). If (j, timestamp$_j$, deny) is already present in token.requestStatus then k simply ignores j's request as the request cannot be satisfied by some neighbor within distance 2 of j. Otherwise, k proceeds as follows. If there are no other conflicting requests or j's timestamp is earlier than other requests then k lets j claim the slots. To accept j's request, k sets token.requestStatus = token.requestStatus $\cup$ (j, timestamp$_j$, accept) if no other neighbor within distance 2 of j already added this information. If k finds j's request conflicting then it updates token.requestStatus with (j, timestamp$_j$, deny).

When the token reaches j in the next token circulation round, it contains the status of its request. It checks token.requestStatus to determine the status of its request. It maps the timestamp$_j$ to request$_j$ to identify the slots it has been allowed to use. Once j identifies the additional slots, it removes its request and status information from the token. Specifically, if j is allowed to use slots it requested in request$_j$ then it sets:

1. token.requestStatus = token.requestStatus − (j, timestamp$_j$)
2. token.requestSet = token.requestSet − (j, request$_j$, timestamp$_j$).

Now, j can start using the slots in request$_j$. Thus, infrastructure nodes can request for unused slots when necessary using the token circulation layer. Furthermore, when a node requests unused slots, it learns the status of its request within one token circulation round. Additionally, to deal with starvation, we can use lease mechanisms (e.g., [18]) where a node is required to renew the additional slots within a certain period of time.

4.4.2 Frame Assignment in Infrastructure Network with Known Locations

In this section, we consider infrastructure networks where the locations of the mesh routers are known up front. In this algorithm, frames are assigned offline and each node statically identifies the frames assigned to it by determining its physical location. To obtain a TDMA schedule, we proceed as follows (1) impose a (virtual) grid on top of the deployed region, (2) compute the interference range of the network (in terms of grid distances), (3) determine the initial frame assignment, and (4) compute the frame period.

4.4.2.1 Step 1: Impose a Virtual Grid

As mentioned earlier, in this algorithm, we assume that the infrastructure network is deployed in some geometric topology. Each node determines its physical location using some mechanism (e.g., GPS). This assumption is reasonable that the infrastructure nodes are powerful with sophisticated hardware. The node can then map this physical location in to virtual grid coordinates. That the node knows the physical location where the virtual grid origin is located and the grid dimensions, it is straight-forward to calculate the virtual grid coordinates. Figure 4.6 shows an example of imposing a virtual grid on top of the deployed network. Observe that more than one node may be present in a given grid location.

4.4.2.2 Step 2: Compute the Interference Range

In this algorithm, communication ranges are defined in terms of grid distances. For example, in Fig. 4.6, distance between neighbors *a* and *b* is 2 grid hops. Furthermore, in this algorithm, we restrict nodes to communicate only with its grid neighbors, i.e., nodes that are 1 grid hop away. Now, we define the notion of interference range in the context of this algorithm. The maximum grid distance up to which a

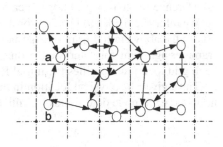

Fig. 4.6 Imposing a virtual grid over the deployed network

node (say, j) can successfully communicate is called the *interference range* of j. The interference range of the entire network is the maximum of interference ranges of all nodes in the network. This value can be computed either statically (before deployment) or dynamically (after deployment).

An infrastructure network with known locations will be typically used to cover a geometric area, it is expected that the deployment is performed systematically, i.e., it is known up front. As a result, the interference ranges of the nodes can be computed easily. Also, ideally, the interference range should be as close to 1 as possible. This could be achieved by using appropriate signal strength at each node. Using the interference range of each node, the interference range of the network is determined statically. Alternatively, similar to previous subsection, the interference range of the network could be computed using a leader. In particular, in this case, each node is aware of its local topology and can compute its own interference range during network initialization. This value would then be communicated with the leader who would determine the overall network interference.

4.4.2.3 Step 3: Assign Initial Frame to Grid Locations

Once a node determines its interference range it knows the frames it is allowed to use. The frames are assigned to grid locations rather than nodes. To ensure connectivity with this algorithm, we need to ensure that the network formed by links that are 1 grid hop distance is connected. This can be achieved by fine-tuning the virtual grid distances used in imposing a grid on top of the network.

Interference range of 1 grid hop. Let us know consider frame assignment to the grid locations for interference range of 1 grid hop. Suppose grid location $\langle 0,0 \rangle$ is assigned frame 0. Locations $\langle 1,0 \rangle$ and $\langle 0,1 \rangle$ will hear messages from a node in $\langle 0,0 \rangle$. Without loss of generality, $\langle 1,0 \rangle$ is assigned frame 1. Location $\langle 0,1 \rangle$ cannot be assigned frame 1 as it will cause a collision at location $\langle 1,1 \rangle$. Therefore, $\langle 0,1 \rangle$ is assigned frame 2. In general, for interference range of 1 grid hop, location $\langle i,j \rangle$ is assigned frame $i+2j$.

Interference range of y grid hops. Now, we consider the general case where the interference range is y grid hops. Suppose grid location $\langle 0,0 \rangle$ is assigned frame 0. Locations $\langle 1,0 \rangle$ and $\langle 0,1 \rangle$ will hear messages from a node in $\langle 0,0 \rangle$. Additionally, nodes in locations that are within y grid hops may also receive the message. Suppose $\langle 1,0 \rangle$ is assigned frame 1. Location $\langle 0,1 \rangle$ is assigned frame $y+1$. In general, for interference range of y grid hops, location $\langle i,j \rangle$ is assigned frame $i+(y+1)j$.

We refer the reader to [19, 20] for detailed discussion on the collision-free property of the above initial frame assignment algorithm. Figure 4.7 illustrates the initial frame assignment for interference range of 2 grid hops. The numbers marked in each grid location identifies the initial frame assigned to that location.

Thus, given its physical location in the network and the interference range of the network, each node determines its initial frames statically.

Fig. 4.7 Initial frames of some grid locations for the network with interference range of 2 grid hops

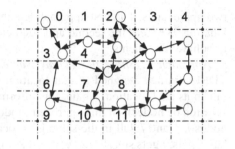

4.4.2.4 Step 4: Compute Frame Period

To compute TDMA frames, we need to determine the period between successive frames assigned to a location. As mentioned before, two locations $\langle i_1, j_1 \rangle$ and $\langle i_2, j_2 \rangle$ are assigned same frame if nodes located at these locations do not interfere the communication of each other. In this context, we use the notion of *collision-group* (cf. Definition 4.1). As defined in Sect. 4.3, collision-group of a location (say, $\langle i, j \rangle$) identifies the locations that could potentially affect the communication of $\langle i, j \rangle$.

Consider the example shown in Fig. 4.7. The collision-group of a node at $\langle 0, 0 \rangle$ includes nodes located at

$$\{\langle 0,0 \rangle, \langle 1,0 \rangle, \langle 2,0 \rangle, \langle 3,0 \rangle, \langle 0,1 \rangle, \langle 1,1 \rangle, \langle 2,1 \rangle, \langle 0,2 \rangle, \langle 1,2 \rangle, \langle 0,3 \rangle\}.$$

As a result, $\langle 0, 0 \rangle$ gets another slot only after the locations identified in its collision-group have been assigned a slot. In this example, maximum frame assigned to locations in the collision-group of $\langle 0, 0 \rangle$ is 9. Hence, $\langle 0, 0 \rangle$ can choose to transmit in the next frame, i.e., frame 10. In other words, the TDMA frame period for the network shown in Fig. 4.7 is 10.

In general, if the interference range of the network is y then the TDMA frame period is $(y + 1)^2 + 1$. Therefore, location $\langle i, j \rangle$ is assigned the following frames: $\forall c : c \geq 0 : t_{\langle i,j \rangle} + c \times P$, where $t_{\langle i,j \rangle}$ is the initial frame assigned to that location and $P = (y + 1)^2 + 1$ is the TDMA frame period. (We refer the reader to [19, 20] for a formal proof of correctness of this algorithm.)

Thus, each node statically determines its initial TDMA frame and the TDMA frame period.

4.4.2.5 Dealing with Multiple Nodes in One Grid Location

When a virtual grid is imposed over the deployed network it is possible that multiple nodes fall into a single grid location. As discussed before, the algorithm presented in this section assigns TDMA frames to a grid location rather than the nodes. To deal with the case where multiple nodes are present in a single grid location, we discuss two approaches. The first approach lets the nodes share the frames assigned to that location. Whereas the second approach elects a leader and requires the other nodes to behave like clients.

Approach 1: Share the TDMA frames. In this approach, the nodes share the TDMA frames assigned to that location. Each node is aware of other nodes that fall in its grid location. Again, this information is statically available (based on the location of nodes). In this approach, nodes share the frames uniformly. Suppose 3 nodes i, j, and k fall in the same grid location and $i < j < k$. Assuming the frame has x slots, i gets slots $0 \ldots ((x/3) - 1)$, j gets slots $(x/3) \ldots ((2x/3) - 1)$, and k gets slots $(2x/3) \ldots (x - 1)$.

Approach 2: Elect a leader. In this approach, each node is aware of other nodes that fall in its grid location. The nodes, however, do not share the frames assigned to that location. Instead, they elect a leader for the location. Existing leader election algorithms [2–6] can be used to elect a leader. The leader becomes the mesh router of that location. All the other nodes in that location become clients of that leader.

4.5 Slot Assignment for Infrastructure Nodes and Clients in their Vicinity

As discussed in Sect. 4.2, WMNs are classified in terms of Infrastructure WMN, Client WMN, and Hybrid WMN. In Sect. 4.4, we presented a slot assignment scheme for infrastructure network. In this section, we extend it to deal with hybrid WMNs where clients are in the vicinity of infrastructure nodes. We visualize such a network as a set of clusters. Each cluster consists of one infrastructure node that is the cluster head for that network and a collection of clients that communicates with it. Of these, infrastructure nodes are static and reliable, i.e., they rarely crash. Client nodes, however, are mobile and are subject to crash failures.

That a client node may be close to multiple infrastructure nodes, it can be part of multiple clusters. To deal with this situation, we let the client consist of several *virtual* clients, each of which is in one cluster. In particular, each virtual client would receive slots based on the cluster it is in. The client can use any of the slots assigned to its virtual client(s). As far as the application is concerned, it can send/receive its messages through any virtual client. However, certain control messages (such as for allocating slots, requesting slots etc) would be forwarded to the appropriate virtual client based on the infrastructure node involved. It follows that each virtual client will only get control messages from the cluster it is in. For brevity of presentation, whenever it is clear from the context, we drop the word *virtual* in the presentation of the algorithm and use the *client* to mean *virtual client*. The algorithm presented in this section would in fact be executed by each virtual client.

We consider two types of mobility for a client node (1) intracluster mobility and (2) intercluster mobility. To accurately capture the notion of cluster mobility, we, first, define the notion of *cluster state*.

Definition 4.4 (Cluster state). The state of a cluster $C_i = (V_i, E_i)$ at instant $t \geq 0$, denoted by $S(C_i, t)$, is a tuple consisting of the set of nodes and links in C_i at t, i.e., $S(C_i, t) = (V_i^t, E_i^t)$, where V_i^t (respectively. E_i^t) is the set of nodes (respectively. links) in C_i at t.

Intracluster mobility captures the fact that, even though nodes are mobile, the mobility pattern is not totally random. The nodes, rather, move in a somewhat coordinated way. Note that when nodes are mobile, if they remain within their assigned cluster, the set of nodes within that cluster remains unchanged, though the links between nodes can change. We now define intracluster mobility.

Definition 4.5 (intracluster mobility). Given two consecutive states of a cluster C_i, i.e., given $S(C_i, t)$ and $S(C_i, t+1)$, we say that there is intracluster mobility in cluster C_i at instant $(t+1)$ if $V_i^t = V_i^{t+1}$ and $E_i^t \neq E_i^{t+1}$.

Intercluster mobility can occur if the mobility pattern of the nodes is random. When a node leaves its cluster, it is because it has either crashed or has joined a new one because of mobility, causing the set of nodes in its original cluster to change (contrast with intracluster mobility). Further, when a cluster head crashes or joins a new cluster, its original cluster ceases to exist, and a new cluster needs to be generated. Using our notation for cluster states, we denote the remaining set of nodes by $(\phi_i^{t+1}, \phi_i^{t+1})$. Note that this set of nodes is not a cluster that it does not have a head.

Definition 4.6 (intercluster mobility). Given cluster C_i with two consecutive states $S(C_i, t)$ and $S(C_i, t+1)$, we say that a node p has left cluster C_i at instant $(t+1)$ if $\exists p \in V_i^t$ such that $p \in V_i^t \wedge p \notin V_i^{t+1}$. We also say that node p has joined cluster C_i at instant $(t+1)$ if $\exists p \in V_i^{t+1}$ such that $p \notin V_i^t \wedge p \in V_i^{t+1}$. We say that there is there is intercluster mobility from cluster C_i to cluster C_j if there exists instants t_1 and t_2, and a node p such that p has left C_i at t_1, and joined C_j at t_2.

Note that there need not be any relationship between t_1 and t_2 that a cluster head may detect the presence of a new node n before the previous head detects n's absence.

Recall that a client consists of several virtual clients. Consider the case where a client is moving from an infrastructure node A to another infrastructure node B. In this case, initially, it would have only one virtual client that receives slots from A. When it is in the overlapping region between A and B, it has two virtual clients and will receive slots from both A and B. Once it is out of contact with A, the virtual client corresponding to A would be terminated and it would receive slots only from B. Thus, the *hand-off* between clusters can be handled smoothly.

Our approach is as follows: First, we observe that the infrastructure nodes are more powerful than client nodes. Hence, given two clients say c_1 and c_2, if c_1 is in the collision group of c_2 then the infrastructure node(s) associated with c_1 are also in the collision group of the infrastructure node(s) associated with c_2. Hence, to provide collision-free slot assignment, we can rely on the slot assignment to infrastructure nodes.

Hence, for the scenario where clients are in the vicinity of infrastructure nodes, we first run the frame assignment algorithm from Sect. 4.4 to assign individual frames to each infrastructure node. Now, client nodes can borrow slots from these frames. To enable such *borrowing* of slots by clients, we adopt a service-oriented perspective for the slot assignment problem: Client nodes can either request slots assigned to their respective cluster head and they can return them. In turn, the cluster head (an infrastructure node) allocates slots to requesting nodes, or returns an updated schedule after nodes have relinquished some slots. To achieve this, the cluster head offers four methods, namely:

1. Request_slot(id).
2. return_slots().
3. Allocate_slot(id).
4. send_schedule().

First, notice that if every node in the cluster is assigned a different set of slots, then collision-freedom is ensured within the cluster. The algorithm in Sect. 4.4 assigns each cluster head a set of frames. The head then allocates slots from this frame in a nonoverlapping manner. That the original frames are interference-free, it follows that assignment of slots to clients also preserves this interference-freedom.

Program for cluster head. We present the program for the cluster head in Fig. 4.8. The cluster head is responsible for scheduling access to the medium. It services requests for slots, and also de-allocates slots when nodes do not use them. Also, this code is executed only in the frames assigned to it.

The first event is a timing event, whereby after each clock tick, the head updates the current slot. After a given number of slots, i.e., max_slots, which is decided by the size of the frame, the value of the current_slot is reset to 1. In the first slot of every frame, the head sends the schedule for the current frame, which every node in its cluster follows. Then, from slot 2 onwards, the head listens for messages from clients. If it does not hear from an expected client for a threshold number of frames, it decides that the client is no longer in its cluster, and reclaims its slot. In the last slot of the frame, the head listens to requests for slots from new nodes and allocates a slot to the node, by updating the schedule for the next frame.

Program for clients. We now develop the program for the client nodes. Similar to the cluster head, the client nodes need to keep track of slot count to determine when they can transmit (Figs. 4.9 and 4.10).

Since the cluster head sends schedule information in the first slot of the frame, client nodes wait to hear the message to determine their slot assignment. Because of the interference freedom property of the solution in Sect. 4.4, the message of the cluster head does not collide with other infrastructure nodes. Hence, every correct node will hear it. However, that the head is also susceptible to transient failures, the nodes may end with corrupted information.

Furthermore, because nodes are mobile, they may end up hearing from a different head, and they keep track of the assigned slots with each head. Note that $\oplus$ denotes overwrite (Figs. 4.11).

```
variables:
% Timing information
current_slot int init 0;

%State information
node_id: int; % Personal head information
max_slots: int;
TYPE node_access: int×int×int×int; % (id, slot, start frame, rate)
schedule[1...max_slots]: array of node_access sequences
frame_schedule: array of int % array indexed by slot number returning id.

algorithm M_TDMA:
    do forever{
    case ⟨event⟩ of

% updating timing information.
    1. ⟨tick()⟩
        current_slot:=current_slot + 1;
        if (current_slot mod max_slots = 1) then
            current_slot := 1;fi

% first slot of frame reached; head transmits cluster information.
    2. ⟨current_slot = 1⟩
        send_schedule();fi;

% head does not hear an expected message from slot 2 onwards.
    3. ⟨(3 ≤current_slot ≤ max_slots)⟩
        if (frame_schedule[current_slot - 1] ≠ ⊥) ∧
        not rcv(⟨frame_schedule[current_slot - 1], payload⟩)) then
            frame_schedule[current_slot - 1] := return_slots();fi;

% head receives a slot request.
    4. rcv⟨request_slot(id)⟩) then
        frame_schedule:=allocate_slot(id);fi
```

Fig. 4.8 Program for the cluster head (infrastructure node)

```
variables:
% Node information
node_id, head_id: int, node_slot: int ∪{⊥};
node_slots: head_id → int

% Timing information
current_slot:int init 0;


algorithm M_TDMA:
    do forever{
    case ⟨event⟩ of
```

Fig. 4.9 Program for client

% timing information being updated at client nodes.
1. ⟨tick()⟩
        current_slot:=current_slot + 1;

Fig. 4.10 Program for client (continued)

% a non-head receive cluster information from head
2. ⟨rcv(⟨hd_id, rnd_sched, curr_slot⟩)⟩
        current_slot := curr_slot;% slot synchronization with head.
        head_id := hd_id;
        node_slot := lookup(rnd_sched, node_id);
        if (hd_id ∈ dom(node_slots) ∧ (node_slot ≠ ⊥)) then
            node_slots := node_slots ⊕ {hd_id ↦ node_slot};
        else
            node_slots := node_slots ⊕ {hd_id ↦ ⊥};fi;

Fig. 4.11 Program for client (continued)

% a non-head transmits in its slot.
3. ⟨current_slot = node_slot⟩
        bcast(⟨node_id, payload⟩);

Fig. 4.12 Program for client (continued)

% A client does not hear its own message.
4. ⟨not rcv(node_id, payload)⟩
        node_slots := node_slots \ {head_id ↦ node_slot};

Fig. 4.13 Program for client (continued)

When a node reaches its transmission slot, it sends it payload with its id tagged to it (Fig. 4.12).

In the case the message has collided then the node does not hear its own message. It then knows that its head will remove it from its cluster by removing its slot entry (Fig. 4.13).

In slot (max_slots), i.e., the one before last, a new node to a cluster requests slots in the next frame. It checks whether it already has a slot assigned with the current head, and if not, makes the request (Fig. 4.14).

After the end of the final slot, i.e., the start of the first slot of the next frame, if a requesting node does not hear its own request, then it knows it has collided, and execute exponential back off when making another request (Fig. 4.15).

```
% head does not hear an expected message
    5. ⟨(current_slot = max_slots)⟩
        if (node_slots(head_id) = ⊥) then
            request_slot(node_id);fi;
```

Fig. 4.14 Program for client (continued)

```
% after nonhead transmits a request
    6. ⟨(current_slot > max_slots) or (current_slot = 1)⟩
        if not (rcv⟨request_slot(id)⟩) then
            execute exponential backoff();fi;
```

Fig. 4.15 Program for client (continued)

Self-stabilization. Next, we sketch the proof of self-stabilization. Observe that if the state of a client node is corrupted, then a collision may occur when the client transmits its payload message. The failure to detect the message will cause the cluster head to remove the node's entry for the following frame. The client will thus have a assigned slot ⊥. The client will thus need to request for new slots again. Hence, the fault can be corrected. In case the state corruption does not lead to collision, the head will still detect an expected message, and will remove the node's entry. Ultimately, the node will need to request slots again, leading to correction.

Likewise, if the state of the head is corrupted such that nodes have conflicting information, collisions will occur. The head will reassign new slots to the requesting nodes, thus correcting earlier faults.

4.5.1 Example

In Fig. 4.16, there are two infrastructure nodes, namely head i, and head j, whose region coverage overlap. Each client node belonging to a unique head has a unique slot number. For example, nodes belonging to node i's cluster only will have the following slot information: $(i, 2)$, meaning the node belongs to cluster i, and can transmit in slot 2 in the cluster.

Now, consider a node A belonging to cluster j and having slot 5. In Fig. 4.17, node A has moved, and is now both within the communication range of heads i and j. When it reaches the cluster j, it requests a new slot (event 5), which is allocated by head j. Thus, after its request for slots have been serviced, the slots for node A can be $\{(j,5),(i,6)\}$, where slot 6 has been allocated to it in cluster i. Thus, node A now has two virtual clients, one is cluster i, and another in cluster j. Each virtual client can transmit within its cluster's timeframe, in its assigned timeslot.

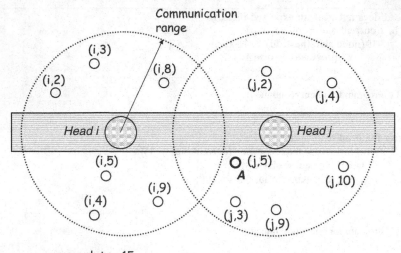

max-slots = 15

Fig. 4.16 State of hybrid WMN at time t_0

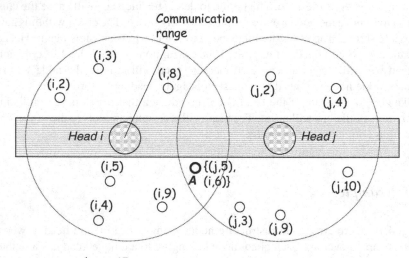

max-slots = 15

Fig. 4.17 State of a hybrid WMN at time $t_1 > t_0$

4.6 Slot Assignment Without Infrastructure Node Support

In this section, we extend the previous algorithms in Sects. 4.4 and 4.5 to deal with the case where client WMNs are not close to infrastructure nodes and, hence, form an ad hoc network. In particular, in Sect. 4.4, we had focused on the infrastructure WMN where only infrastructure nodes are considered. In Sect. 4.5, we considered the case where client nodes are introduced but only in the vicinity of the infrastructure nodes. Then, the network was partitioned into clusters where each cluster consisted of one infrastructure node and a collection of client nodes. That the cluster head was an infrastructure node, we could assume that the possibility of their failure/movement is low and, hence, such crash/movement was ignored. In this section, we further extend the setup to permit client nodes that are not in the vicinity of infrastructure nodes. Hence, a cluster may no longer contain an infrastructure node, i.e., the cluster head could be a client node. With this change, all nodes are subject to crash or move and the mobility can be intracluster or intercluster.

Based on the above discussion, in this section, we present a MAC protocol that guarantees collision freedom in spite of node mobility. Now, consider the case where nodes are GPS-enabled. In this situation, we could apply the algorithm in Sect. 4.4.2; clustering would be done in such a way that cluster heads are equipped with GPS devices and determine the frames assigned to them using their location. For networks where clients do not have GPS devices, we can perform hierarchical clustering using algorithms in [21–24]. For example, the algorithm in [23] establishes hierarchical clusters and identifies gateway nodes between cluster heads. Such hierarchical clustering could be used along with algorithm in Sect. 4.4.1 and allow the cluster head at one level in the hierarchy to assign frames to cluster heads at lower level. Finally, the cluster heads would assign the corresponding slots to clients within their cluster in a manner similar to the algorithm in Sect. 4.5. Another scenario in this category is where the clusters are far apart from each other and, hence, can have overlapping slots. Hence, when nodes move across cluster, collisions could occur. That the scenario where nodes are equipped with GPS devices is straightforward, in this section, we consider the other two scenarios.

We describe the setup first: A clustering algorithm is executed to create clusters. (Note that a hierarchical clustering algorithm may be executed, if needed.) Then, each node within a cluster is assigned a unique slot. Further, we split the time frame into two parts, namely (1) data part and (2) a control part, see Fig. 4.18. In the control part, the head and nodes execute only control events, whereas in the data part, the nodes send their payload. The MAC protocol we develop guarantees collision-freedom in the data part of the cluster schedule. Although collisions are

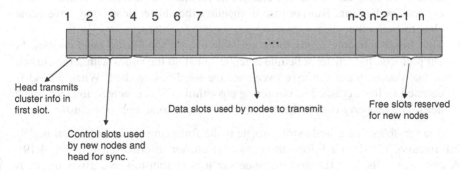

Fig. 4.18 A frame consisting of control slots and data slots

possible in the control part, such collisions are unlikely except in case of high number of nodes moving into the given cluster. Specifically, if more than one node moves into a cluster within any time frame, then collision will occur. And, in these situations, the existing cluster nodes will continue to function even though new nodes may not be able to participate in the cluster immediately.

Before presenting the algorithm, and the assumptions, we present two definitions that we will use in the algorithm:

Definition 4.7 (i-band). A node i is said to be in the i-band of node j if i is in the communication range of j.

Definition 4.8 (o-band). A node i is said to be in the o-band of node j if i can communicate with low probability with j.

Before presenting the MAC protocol, we make the following assumptions:

1. Nodes within a cluster have nonoverlapping slots.
2. All nodes including the cluster head(s) are (potentially) mobile. They can exhibit both intra and intercluster mobility.
3. Every node, including the head, can crash.
4. Any node remains within a cluster for a least one round, unless it crashes. This is to capture the fact that nodes cannot move beyond a certain speed. What this also implies is that nodes will get the chance to transmit. If a node does not satisfy this mobility constraint, then the algorithm we propose will not guarantee that the node can transmit its messages.
5. If in a given round r, a node is in the i-band of its cluster head, then in round $r+1$, it will at most be in the o-band of the head, unless the head crashes. Again, this is used to capture the fact that nodes cannot move beyond a certain speed. What it also allows is to distinguish between message collision and a node being far away.
6. No more than 1 round elapses for a node to not hear from a head i.e., a node may not go for more than 1 round without hearing from a head. (Thus could be trivially extended to k rounds where k is an arbitrary number.)
7. Messages sent by cluster heads do not collide on $k \geq 2$ consecutive rounds. This is reasonable because either the clusters are far apart or the slots assigned to them are not overlapping. But because of mobility the cluster heads may have come close to each other.
8. The cluster schedule needs to be available even if the cluster head crashes. In our protocol, the cluster schedule is replicated at all the nodes within the cluster, so that one of them can take over once the head has crashed. When a head is detected to have crashed, a clustering algorithm [25] is executed to elect a new head. The property then is that client nodes belong to at most one cluster.

The variables that a node stores relate to the following (1) an individual node's information, (2) timing information, and (3) cluster information (cf. Fig. 4.19). A node stores its own ID, and the node slot it is to transmit in a given frame. It

```
on every client node
variables:
% Node information
node_id, node_slot: int; head:{0,1, ⊥};
cluster_id: int; band:{i,o}; status:{⊥, waiting, ⊤, ?} init ⊤;
node_access: int × int × int × int; %(node_id,node_slot, start,rate)
% timing information
current_slot, next_slot init 0;  round init 1;
%cluster information
max_slots: int; new_slot: int init ⊥;
schedule[1...max_slots]: array of node_access sequences
algorithm M_TDMA:
    do forever{
    case ⟨event⟩ of
```

Fig. 4.19 Algorithm at each client node: state information

```
1. ⟨tick()⟩
      current_slot:=current_slot + 1;
      if (current_slot mod max_slots = 1) then
          round := round + 1;fi
```

Fig. 4.20 Algorithm at each client node: event 1

```
2. ⟨current_slot = (max_slots * (round - 1) + 1)⟩
      if (head) then
          bcast(⟨node_id, head, schedule[],round⟩); fi
```

Fig. 4.21 Algorithm at each client node: event 2

also keeps track of whether it is a cluster head or not. It also keeps track of its transmission parameters, i.e., its node access rights. A node keeps track of timing information by determining the current slot and the current round. Furthermore, every node keeps track of cluster information such as cluster schedule and the maximum number of slots in a frame.

The first event (cf. Fig. 4.20) that a node listens to is a timing event. It increments the current slot number with every slot interval. Then, once the current slot number reaches a predefined value, the slot number wraps around, and the node increments the round value, i.e., once the current slot reaches a predefined value, the node knows that it has reached the next round of transmission, and increments the "round" variable accordingly. When the first slot of a round is reached (second event), a cluster

```
   3. ⟨rcv(⟨hd_id, hd, sched[], rnd⟩)⟩
        if (head ∧ status = ?) then
            status := ⊤;
        if (head != 1 ∧ cluster_id = hd_id) then
            head, schedule,cluster_id, band, round, status := 0, sched,
hd_id, i|o, rnd, ⊤; fi
        if (head != 1 ∧ cluster_id ≠ hd_id) then
            head, schedule,cluster_id, band, round, status := 0, sched,
hd_id, i|o, rnd, ⊥; fi
        if (status = ⊤) then
            node_access := look_up(schedule,node_id); fi
```

Fig. 4.22 Algorithm at each client node: event 3

```
   4. ⟨current_slot = (max_slots * (round - 1) + 2)⟩
        if (head ∧ not rcv(⟨cluster-id,1,sched[], current_round⟩)) then
            status := ?; fi
        if (status = ⊤∧ not rcv(⟨cluster_id,1,sched'], current_round)) ∧ band=o)
            head,cluster_id,node_access, band, status:= ⊥, _, _, ⊥, ⊥; fi;
        if (status = ⊤∧ not rcv(⟨cluster id,1,sched'], current round)) ∧ band=i)
            status:= ?; fi;
        if (status =? ∧ not rcv(⟨cluster id,1,sched'], current round)) ∧ band=i)
            status:= _; cluster id, head := run cluster [9];fi;
        if (status = ⊥) then
            bcast(⟨node_id⟩);
            status := waiting; fi
```

Fig. 4.23 Algorithm at each client node: event 4

head will broadcast the current schedule for the cluster, together with the value of the current round (Fig. 4.21). This is the second event a node waits for.

When a head node hears its own message (Fig. 4.22), it knows there has been no collision at the sender, and it sets its status as T. When a nonhead node receives the message, its sets itself as a nonhead, and updates its version of the cluster schedule. It also keeps track of whether it is in the head's i-band or o-band, depending on the strength of the signal. It also sets its status as T.

If the second slot is reached (Fig. 4.23), and a cluster head has not yet heard its own message, it knows that there is something wrong, such as a collision, and it sets is status to undecided ("?"). On the other hand, if a nonhead node does not hear the message, and it was initially in the o-band of the head, then it assumes that it is now too far from the cluster head, and it resets all of its cluster information, i.e., it assumes it is no longer part of the cluster. If the nonhead node was in the i-band, then it assumes that some problem, such as collision, occurred, and sets its status to "?." On the other hand, if a node had its status as "?," and did not receive the message from the head, it concludes that the head as crashed. So, it resets its status to ⊥, and the node will take part in re-electing a new cluster head. Furthermore, if a node has status $\perp$, then it broadcasts a message, informing the head that it is a new node, and sets its status to "waiting" for transmission rights.

When the head receives a request for slots from a new node (Fig. 4.24), it does a noncolliding allocation. If there are empty slots, then the node is assigned one of these slots in every round. However, if there is only one empty slot, then, to be able to tolerate future mobility, the remaining bandwidth is halved by allowing the new node to transmit at half the rate of the empty slot.

When the third slot in a given round is reached (Fig. 4.25), the head then broadcasts the new transmission information, which is then picked up by the requesting node. Note that, because of assumption 4 above, the requesting node will still be in the cluster at when the head broadcasts its new transmission information.

---

5. $\langle \text{rcv}(\langle \text{new\_id} \rangle) \rangle$
   if (head $\wedge | \{k:4..\text{max\_slots}| \text{ schedule}[k] = \langle \rangle\}| \geq 2$) then
       new_slot:=choose$\{k:4..\text{max\_slots}| \text{ schedule}[k] = \langle \rangle\}$
       rate:=1;% rate=1 denotes every round
       schedule[new_slot]:= $\langle(\text{new\_id}, \text{new\_slot}, \text{round}, \text{rate})\rangle$; fi

   if (head $\wedge | \{k:4..\text{max\_slots}| \text{ schedule}[k] = \langle \rangle\}| = 1$) then
       new_slot:=choose$\{k:4..\text{max\_slots}| \text{ schedule}[k] = \langle \rangle\}$
       rate:=2;% rate=1 denotes every second round
       schedule[new_slot]:= $\langle(\text{new\_id}, \text{new\_slot}, \text{round}, \text{rate}), \perp\rangle$; fi

   if (head $\wedge | \{k:4..\text{max\_slots}| \text{ schedule}[k] = \langle \rangle\}| = 0$) then
       new_slot:=choose$\{k:4..\text{max\_slots}| |\text{schedule}[k]| > 1\}$;
       schedule:=update_schedule(schedule,new_id,new_slot);
       % rate increases exponentially

Fig. 4.24 Algorithm at each client node: event 5

---

6. $\langle \text{current\_slot} = (\text{max\_slots} * (\text{round} - 1) + 3) \rangle$
   if (head $\wedge$ new_slot $\neq \perp$) then
       bcast$(\langle \text{new\_id,new\_slot,round,rate} \rangle)$;
       new_slot:= $\perp$; fi

Fig. 4.25 Algorithm at each client node: event 6

---

7. $\langle \text{rcv}(\langle \text{id,slot,start,rate} \rangle) \rangle$
   if (node_id $= \text{id}$) then
       node_access:= (id,slot,start,rate);
       next_slot:= (start - 1)*max_slots + slot;
       status:= $\top$; fi

Fig. 4.26 Algorithm at each client node: event 7

8. ⟨current_slot = (max_slots * (round - 1) + 4)⟩
   if (not rcv(⟨(id,slot,start,rate)⟩) ∧ status = waiting) then
       exponential_backoff();

Fig. 4.27 Algorithm at each client node: event 8

9. ⟨current_slot = next_slot⟩
   if (status ≠ ?) then
       bcast(⟨node_id, cluster_id, payload⟩); fi
   next_slot:= current_slot + (max_slots * rate);

Fig. 4.28 Algorithm at each client node: event 9

10. ⟨(max_slots * (round - 1) + 4) ≤ current_slot ≤ (max_slots * round)⟩
    if (head) ∧ not (rcv(⟨(id, cluster, payload)⟩)) ∧ status ≠ ? then
        schedule:= reclaim_slots(schedule); fi

Fig. 4.29 Algorithm at each client node: event 10

When the requesting node receives its transmission information (Fig. 4.26), it calculates its set of assigned slots, and sets its status as T.

If the fourth slot is reached (Fig. 4.27), and a new node does not receive its new transmission information, then it concludes that either (1) the head has crashed or (2) its request has collided with another potential new node's request. Hence, the new nodes will execute exponential backoff, within the current cluster, before requesting transmission slots again.

After the fourth slot (Fig. 4.28), any node will transmit its payload in an allowed slot. Note that a node is only allowed to transmit when its status is T. A node will still retain its transmission information even in the presence of head crashes.

If the head does not hear from a node (Fig. 4.29), it assumes the node to have crashed, and it reclaims the slots allocated to the node.

4.7 Informal Proof of Correctness

Under a fault free scenario, the correctness proof can be reasoned as follows: A head *knows* that node m has joined its cluster at instant $n + 1$ when it assigns a slot to m, i.e., m has an entry in schedule. This occurs in the third slot of a round.

Under M_TDMA, when a node hears from a new head, it knows that it is in a cluster (without joining). This occurs in the first slot of a round, i.e., instant $n - 1$, and it updates its state. In the next slot, i.e., instant n, it broadcasts its ID, and sets it status to "waiting." When the head receives a message about a new node (hence there is no collision), event 5 is triggered. It assigns a slot to the new node according to whether there is a free slot or whether the last slot is to be shared. Because the remaining bandwidth is always halved, it means that, theoretically, there is always available bandwidth, half of which is then allocated.

In the third slot, i.e., instant $n+1$, it broadcasts the new node access information to the new node, indicating which slot is to be accessed when and what rate. At this point, the node joins the cluster. The cluster remains collision-free that, in any given round, no slot is shared by two nodes.

When a nonhead node crashes, the head detects it when no message is obtained in a given allocated slot. In this case, only the allocated slots are reclaimed. That the slot assignment was noncolliding, reclaiming unused slots maintain the noninterference property of the assignment. If a head is absent (crashes or move beyond its cluster), a clustering algorithm is executed, and a new head is elected. That the new head has the cluster schedule, it can maintain it. Later, once it detects that the previous head is absent, it will reclaim the unused slots. Again, noninterference is preserved in the data part of the schedule.

4.7.1 Example

In Fig. 4.30, two clusters are shown at some point during the execution of the protocol. Within each cluster, no pair of nodes has overlapping slots. However, nodes across clusters may have overlapping slots. Node H_1 (respectively, H_2) is the head of cluster C_1 (resp. C_2). Nodes shown in green are in the i-band of the respective head, whereas other nodes are in the o-band of the head. The numbers by the side of the nodes represent the node's assigned slot number. Also, we assume that a frame is 15 slots long (3 control slots and 12 data slots). That there are 11 nodes in the cluster, it means that H_1 cannot assign the remaining bandwidth (1 slot) completely to a new node.

If nodes display intracluster mobility only, then no collision will occur that nodes do not have overlapping slots. If a node moves from the i-band to the o-band of the head, then the probability of correct message transmission to the head is low, and the node may be incorrectly interpreted as crashed or not present. In either case, this will cause the node to lose its slots (event 10), which will preserve the collision-freedom property of the protocol. On the other hand, if a node moves from the o-band to the i-band of its cluster head, then two situations arise (1) if the node still has its slot, no new slots are assigned to it, thus collision-freedom is ensured, or (2) if the node did not have a slot, it will eventually be assigned a slot that do not overlap within the cluster, thus collision-freedom is ensured.

So, we will focus on intercluster mobility. In Fig. 4.31, a node moves from the o-band of the head of cluster C_2 to the i-band of the head of cluster C_1.

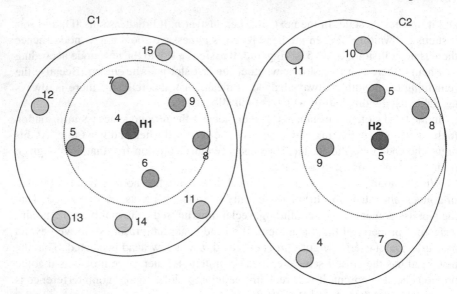

Fig. 4.30 Setup: Two clusters, with each node in each cluster having nonoverlapping slots. Nodes across clusters can have overlapping slots

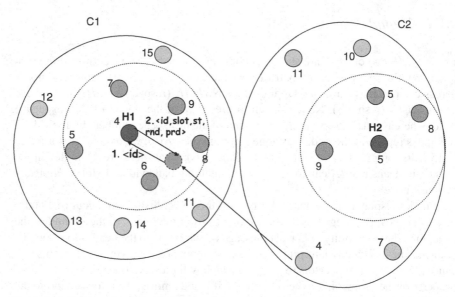

Fig. 4.31 A node moving from the *o*-band of the head of cluster C_2 to the *i*-band of the head of cluster C_1

When the node reaches its new cluster C_1, it broadcasts its ID (arrow 1). When the head receives this message in slot 2, it sends to the node its transmission information, viz its slot number, rate, and the starting round (arrow 2). Here, the head will detect that only slot 10 is free. Upon detecting that there is only one slot available, the head halves the bandwidth by allowing the requesting node to access the medium every second round, according to event 5 (second if statement). That the head calculates the transmission information in such a way that there is no slot overlap, collision-freedom is ensured. Thus, assume the execution is in round 2. The slot number in the current round will range from 16 to 30. Assume that node sid (read some id) makes a request in slot 17 (second slot in round 2), and is the only requesting node, then sid receives the following information in the third slot of the round: $\langle sid, 10, 2, 2 \rangle$. Thus, node sid will start transmitting when the slot reaches $(2 - 1) \times 15 + 10 = 25$, and it will also calculate its next transmitting slot by $((2 - 1) \times 15 + 10) + (2 \times 15) = 55$ (i.e., in slot 10 of round 4). On the other hand, when the head of C_2 does not hear sid transmit in the fourth slot of round 2, according to event 10, it reclaims the assigned slot.

In general, it can be shown that the algorithm tolerates node crashes, unrestricted node mobility, as well as new external nodes joining clusters.

4.8 Thoughts for Practitioners

In this section, we discuss how practitioners could use the algorithms presented in this chapter. First, we discuss the issue of computing interference range of the network. One of the important concerns to the practitioners is computing the interference range of the network for the algorithm discussed in Sect. 4.4.2. As mentioned in Sect. 4.4.2, the interference range can be computed either statically or dynamically. One approach to dynamically determine the ideal interference range of the network is as follows. Initially, all nodes assume some value as the interference range of the network. And, they compute their frames based on this value. A node, say, j, is aware of the mesh routers that it can communicate with and their locations, j can compute its interference range. Once all nodes compute their interference range, they broadcast this value to all the nodes in the network (in a collision-free manner using the frames assigned to them initially). Then, each node computes the maximum interference range of the network. Once the maximum interference range is determined, all nodes switch to the new schedule (that is computed using the interference range just computed) at a predetermined frame. Thus, interference range can be computed dynamically.

Second, an important concern in the distributed coloring algorithm is the token circulation reliability. The token may be lost because of message corruption, whenever a node forwards it. To ensure that it is not lost, the node expects its successor to forward the token within a certain interval. If it fails to receive such an *implicit acknowledgment* from its successor, it retransmits the token for a threshold number of times before entering recovery mode. With this mechanism, a node may receive duplicate tokens. Such duplicate messages are ignored.

Finally, an important concern about the algorithms presented in this chapter is scalability. Based on the discussion in [24], time required for stabilization of frame assignments to mesh routers is within application-tolerable limits. Additionally, we note that each mesh router form a cluster with their respective clients. Hence, the algorithms presented in Sect. 4.5 independently and concurrently for each cluster. Based on this discussion, the algorithms discussed in this chapter are highly scalable to large number of mesh routers and clients.

4.9 Directions for Future Research

There are several possible future directions for this work. First, the distributed coloring algorithm shows the feasibility of a stabilizing interference-free communication in a WMN where the mesh routers are aware of only their local neighborhood. However, the coloring algorithm is sequential in nature. Although the recovery time is acceptable for typical deployments, one interesting future research direction is to investigate deterministic concurrent coloring. We expect that such algorithms allow the routers to recover from arbitrary state corruption quickly.

Second, in the slot assignment algorithm where the routers are aware of their locations, an important concern is how to deal with the errors in the location. This may result in discrepancies in calculating in the interference range of the network and, hence, affect the interference-free nature of the TDMA schedule obtained. Therefore, another future direction is to investigate the feasibility of error-tolerant location-specific slot assignment for mesh routers.

Third, in this chapter, for simplicity of presentation, we considered the case where the number of slots assigned to all nodes is the same (or close). An interesting future work is to extend the algorithms to the case where some nodes are given larger bandwidth than others. For example, in case of solutions with graph coloring, such nodes could be assigned multiple colors thereby providing them more bandwidth. Letting such nodes have higher priority in claiming unused slots could also result in preferential treatment for some nodes.

4.10 Conclusions

In this chapter, we presented stabilizing algorithms for interference-free slot assignment in WMNs. First, we considered WMNs that only consist of infrastructure nodes. We presented three algorithms in this category. The first two algorithms relied on knowledge of local communication topology to determine how slots should be assigned to each node. Of these, one relied on centralized calculation of slots and the other relied on distributed calculation. These approaches provide a tradeoff between the time to add a new node to the network and bandwidth utilization. The third algorithm focused on situations where nodes are deployed to cover a geometric region. In this case, the knowledge about the node location was utilized to assign bandwidth efficiently as well as permitting quick addition of nodes in the network.

Subsequently, we extended this algorithm to deal with the case where mobile client nodes are added although they are close to infrastructure nodes. In this case, we first assigned slots to infrastructure nodes and provided an approach for clients to *borrow* these slots as needed.

We also presented an algorithm for slot assignment for clients that are not close to infrastructure nodes and, hence, form a client WMN. This algorithm allows client nodes to form clusters and permits cluster heads to assign slots to different nodes within the cluster. That an arbitrary hybrid WMN can be viewed as a union of client WMNs and a WMN where some clients are in the vicinity of the infrastructure nodes, the last two solutions can be applied for such WMNs.

The use of such slot assignment algorithms would be especially valuable when some quality of service needs to be provided to applications and where the data rate is moderate. By ensuring that communication of one node does not collide with that of other nodes allows one to provide guarantees on communication delay and guarantees on successful delivery. By contrast, if the load is low, CSMA-based approach may work better that low load makes it less likely that collision would occur.

4.11 Terminologies

1. *Infrastructure/backbone WMN.* In infrastructure/backbone WMN, networks are built with routers only. Routers are connected to each other with many links to form backbone network that can be utilized by clients. Routers in this group use two types of communications: infrastructure communication, to connect within themselves and other networks, and user communication, to connect with their clients. Also, this group uses the gateway and bridge functionality of the routers to connect clients to each other and to existing networks such as the Internet.
2. *Client WMNs.* This group consists of peer-to-peer networks among clients. Nodes can serve as hosts to the user application(s), and can provide routing functionality to connect other nodes to each other. Therefore, it is possible for the client WMNs to be built without routers. In this group, packets will be transmitted from the source to the destination though multiple nodes. Clients will forward packets from one node to another until the packet reaches its destination. Nodes in this group are equipped with routing and transmission capabilities to help them perform their tasks.
3. *Hybrid WMNs.* This type combines both the infrastructure and the client WMNs together in one network. Both, routers and clients will provide point of access to the network. In the hybrid WMNs, clients will provide more capabilities to the network. They will be used to connect other clients, who are not in any transmission range of any router, to the network.
4. *Time-division multiple access (TDMA).* Time-division multiple access is a medium access control layer that divides the time spectrum among the nodes in the network. Specifically, it assigns communication time slots to the nodes

such that messages transmitted from two nodes assigned the same slot do not interfere with each other.

5. *Frequency-division multiple access (FDMA)*. Frequency-division multiple access is a medium access layer that divides the frequency spectrum among the nodes in the network. In particular, FDMA assigns a frequency to each node such that messages transmitted by two nodes that are assigned the same frequency do not interfere with each other.

6. *Distance-2 coloring*. Distance-2 coloring is the problem of assigning colors to the nodes in the network such that two nodes within communication distance of 2 are colored differently. A solution to distance-2 coloring of a network topology provides a TDMA or a FDMA schedule for the nodes.

7. *Self-stabilization*. A system is self-stabilizing if starting from an arbitrary state, it (1) recovers to legitimate state and (2) upon recovery continues to be in legitimate states forever.

8. *Collision group*. Collision group of a node, say, j, is the set of nodes that share one or more nodes in their talk-to sets with that of talk-to set of j. In other words, collision-group of j identifies the nodes that could affect its communication.

9. *Symmetric collision group*. Symmetric collision group of a node, say, j, is the set of nodes that have one or more common neighbors with that of j.

10. *Interference range*. Interference range of a node is defined as the maximum (grid) distance up to which the node can successfully communicate.

11. *Intracluster mobility*. A node, say, j, displays intra cluster mobility if, within a given time frame, it remains within the same cluster but its neighborhood has changed.

12. *Intercluster mobility*: A node, say, j, displays inter cluster mobility if, at the start of a given time frame, it is in a cluster C_s and it is in another cluster C_e at the end of the time frame.

4.12 Questions

1. State the problem of distance-2 coloring in terms of collision group (as defined in Sect. 4.3.1).
2. Consider the following topology for infrastructure network.

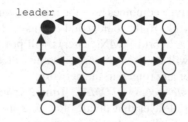

(a) Compute a frame assignment for the nodes using centralized coloring algorithm discussed in Sect. 4.4.1.1. Determine the time bound for distributing the frame assignments to all the nodes.
(b) Now, consider the distributed coloring algorithm discussed in Sect. 4.4.1.2. Compute a frame assignment for the nodes using this algorithm. Show the token traversal (assuming a depth first traversal).
(c) Assume that frames are assigned using the distributed coloring algorithm and nodes start transmitting application messages. Suppose node $\langle 1,1 \rangle$ fails while holding the token. Explain how the lost token is recovered.

3. State the advantages and disadvantages of using centralized coloring algorithm with respect to the distributed coloring algorithm.
4. Consider the following topology for infrastructure network.

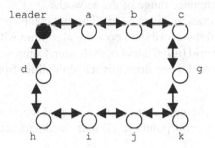

Let the frame assignments for the nodes are as follows.

Node	Initial frame
Leader	0
A	1
B	2
C	3
D	2
G	0
H	1
I	0
J	2
K	1

(a) What is the TDMA frame period for the above schedule?
(b) List down the nodes that can potentially get additional frames. Explain.
(c) Suppose nodes d and j request for frame #51 at frame #6. Determine whether both their requests be satisfied using the approach proposed in Sect. 4.4.1.4. If not, determine the node whose request will be satisfied. Is it the ideal solution? Explain.

5. Consider the following topology where each node knows its location (and the coordinates in a virtual grid as drawn in dotted lines).

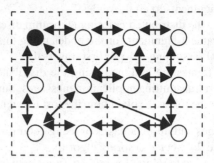

(a) What is the interference range of the network?
(b) Assign initial frames to the grid locations and compute the frame period.
(c) Suppose we used the distributed coloring algorithm with the node at the top-left corner of the grid (filled dark) as the leader. Compute the frame assignment and the frame period. How does this schedule differ from the one obtained in problem 4(b)?

6. Suppose the initial frame assigned to virtual grid location $\langle 0,0 \rangle$ is 0. Compute the initial frames of grid points in all four quadrants such that the frames are collision-free.

```
        North-west quadrant  │  North-east quadrant


  ──────────────────────────┼──────────────────────────
                             │<0,0>

        South-west quadrant  │  South-east quadrant
                             │
```

7. In the algorithm discussed in Sect. 4.4.2, the TDMA frame period is $(y+1)^2+1$, where y is the interference range of the network. Explain how the TDMA frame period is related to collision-group (defined in Sect. 4.3.1).
8. In a client mesh network, there can be several clusters. Using the MAC protocol for client mesh networks, explain what problem can happen if two clusters are close to each other. How can this problem be handled by the algorithm?
9. It has been shown that the MAC algorithm for a client mesh network tolerates inter cluster and intra cluster mobility. Prove that the MAC algorithm also tolerates the random waypoint mobility model (Note: the random waypoint mobility model is similar to the random motion, with the addition of pauses being introduced).

10. One assumption behind the MAC algorithm for client mesh network is that every node in a cluster has a distinct slot at system start up. Explain what can happen if such a requirement is violated.

References

1. I.F. Akyildiz, X. Wang, and W. Wang, Wireless mesh networks: A survey, Computer Networks, 47, 445–487, 2005.
2. A. Arora, Efficient reconfiguration of trees: A case study in the methodical design of nonmasking fault-tolerance, Proceedings of the Third International Symposium on Formal Techniques in Real Time and Fault-Tolerance, 110–127, 1994.
3. S. Bagchi and P. Das, A round-2 randomized leader election algorithm and latency for MDVM system, NWESP '05: Proceedings of the International Conference on Next Generation Web Services Practices, p. 103, 2005.
4. S. Dulman, P. Havinga, and J. Hurink, Wave leader election protocol for wireless sensor networks, International Symposium on Mobile Multimedia Systems and Applications (2003).
5. F. Gartner, A survey of self-stabilizing spanning tree construction algorithms, Swiss Federal Institute of Technology (EPFL) (2003).
6. N. Malpani, J. Welch, and N. Vaidya, Leader Election Algorithms for Mobile Ad Hoc Networks, Workshop on Discrete Algorithms and Methods for Mobile Computing and Communications (2000).
7. G. Chakrabarti and S.S. Kulkarni, Load balancing and resource reservation in mobile ad-hoc networks, Ad Hoc Networks, 4(2), 186–203, 2006.
8. D.B. Johnson and D.A. Maltz, Dynamic source routing in ad hoc wireless networks, Mobile Computing, vol. 353, 153–181, Kluwer, Dordrecht, (1996).
9. M.R. Pearlman, Z.J. Haas, P. Sholander, and S.S. Tabrizi, On the impact of alternate path routing for load balancing in mobile ad hoc networks, Proceeding of 2000 First Annual Workshop on Mobile and Ad Hoc Networking and Computing, Mobihoc 2000, Boston, MA, USA, 3–10 August (2000).
10. E.W. Dijkstra, Self-stabilizing systems in spite of distributed control, Communications of the ACM, 17(11), 643–644, (1974).
11. S. Dolev, Self-Stabilization, MIT, Cambridge, MA, (2000).
12. E.L. Lloyd and S. Ramanathan, On the complexity of distance-2 coloring, Proceedings of the International Conference on Computing and Information, (1992).
13. M. Arumugam and S.S. Kulkarni, Self-stabilizing deterministic time division multiple access for sensor networks, AIAA Journal of Aerospace Computing, Information, and Communication (JACIC), 3, 403–419, 2006.
14. A.K. Datta, C. Johnen, F. Petit, and V. Villain, Self-stabilizing depth-first token circulation in arbitrary rooted networks, Distributed Computing, 13, 207–218, (2000).
15. C. Johnen, G. Alari, J. Beauquier, and A.K. Datta, Self-stabilizing depth-first token passing on rooted networks, Proceedings of the Workshop on Distributed Algorithms, (1997).
16. F. Petit, Fast self-stabilizing depth-first token circulation, Proceedings of the Workshop on Self-Stabilizing Systems, Springer, vol. LNCS:2194, 200–215, (2001).
17. F. Petit and V. Villain, Color optimal self-stabilizing depth-first token circulation, Proceedings of the Symposium on Parallel Architectures, Algorithms, and Networks, (1997).
18. A.S. Tanenbaum, Computer Networks Prentice Hall, Upper Saddle River, NJ, (2003).
19. S.S. Kulkarni and M. Arumugam, SS-TDMA: A self-stabilizing MAC for sensor networks, Sensor Network Operations, Wiley-IEEE, New York, NY, (2006).
20. S.S. Kulkarni and U. Arumugam, Collision-free communication in sensor networks, Proceedings of the Sixth Symposium on Self-Stabilizing Systems (SSS), June (2003).

21. A.D. Amis and R. Prakash, Load-balancing clusters in wireless ad hoc networks, ASSET, 2000.
22. S. Basagni, Distributed and mobility-adaptive clustering for multimedia support in multi-hop wireless networks, Proceedings of Vehicular Technology Conference, 1999.
23. J. Gao, L.J. Guibas, J. Hershberger, L. Zhang, and A. Zhu, Discrete mobile centers, Proceedings of 17th ACM Symposium on Computational Geometry, (2001).
24. A.B. McDonald and T. Znati, A mobility based framework for adaptive clustering in wireless ad-hoc networks, IEEE Journal on Selected Areas in Communications, 17(8), 1466–1487, (1999).
25. M. Demirbas, A. Arora, V. Mittal, and V. Kulathumani, A fault-local self -stabilizing clustering service for wireless ad hoc networks, IEEE Transactions on Parallel and Distributed Systems, 17(9), 912–922, (2006).

Chapter 5
Channel Assignment Techniques for 802.11-Based Multiradio Wireless Mesh Networks

Ante Prodan and Vinod Mirchandani

Abstract This chapter gives an in-depth coverage of the area of channel assignment in 802.11-based multiradio wireless mesh networks (MR-WMN). Multiple channels in a MR-WMN can substantially increase the aggregate capacity of the Wireless Mesh Networks (WMN) if the channels are assigned to the nodes such that the overall interference is limited. To this end, use of graph theory to understand and address this problem facilitated by different representations of MR-WMNs is explained. Further, the inherent properties of the 802.11 radio's physical layer are identified to explain their influence on channel interference. We also examine the ways by which the emerging 802.11k standard will help to carry out an effective channel assignment. The usefulness of this chapter is made complete by giving a taxonomy of existing solutions, which is used as a preview to provide an extensive survey of the key research approaches proposed in the literature for channel assignment. We have contributed by way of putting together a comprehensive overview of the work in this area, which we believe does not exist with a similar scope. This chapter has been written such that it can be enjoyed and grasped by students as well as professionals.

5.1 Overview

This chapter is organized as follows: Sect. 5.2 provides an introduction to the chapter based on the terminology that is pertinent to the concepts discussed. In Sect. 5.3, we provide background information on the core problem in channel assignment for multiradio wireless mesh network (MR-WMN), and then explain the use of graph theory for channel assignment along with its aspects for MR-WMN model representation. Section 5.4 describes the key issues that are associated with

A. Prodan (✉)
Faculty of Information Technology, University of Technology, Sydney, P.O. Box 123, Broadway, NSW 2007, Australia
e-mail: aprodan@it.uts.edu.au

S. Misra et al. (eds.), *Guide to Wireless Mesh Networks*, Computer Communications and Networks, DOI 10.1007/978-1-84800-909-7_5,
© Springer-Verlag London Limited 2009

the physical layers of 802.11a/b/g radios. These are relevant for investigation because the channel assignment problem stems from the limiting attributes of the radio's physical layer. Further in Sect. 5.4, we describe the spectrum management capabilities of 802.11a/b/g radios and review the radio resource management mechanisms, which are proposed in 802.11k (draft 7). The taxonomy of the approaches to address the channel assignment problem in MR-WMN along with some of the key algorithms proposed in this regard are *detailed* out in Sect. 5.5. The thoughts for practitioners and directions for future research are provided in Sects. 5.6 and 5.7, respectively. Conclusions that can be drawn from this chapter are provided in Sect. 5.8.

5.2 Introduction

The throughput obtained at the physical layer of a wireless network is largely dependant on following functional procedures that are associated with the wireless medium:

- Channel selection
- Link adoption
- Transmit power control (TPC)
- Error correction schemes

In this chapter, we study the problem of channel assignment for 802.11-based multiradio wireless mesh networks (MR-WMN). MR-WMNs discussed are multihop, multiradio wireless networks based on IEEE[1] 802.11 suite of standards. The "multiradio" feature means that each wireless router has two or more radio interfaces that operate independently on different channels. Therefore, each router is capable of transmitting and receiving data simultaneously, albeit on different channels, as well as it can communicate with one or more neighbours. Further, the "multihop" feature signifies that each router, which is also called a node, can relay traffic from other nodes towards its final destination.

In this introductory section, we are going to familiarize the reader first with the basic terms and concepts that are necessary to understand the material provided in following sections. We will initiate with the definitions of terms such as *radio spectrum, channel and band*, which will be followed by an explanation of *radio interference*. We then provide a brief introduction to spectrum regulation and graph theory and end this section with an overview of IEEE 802.11 suite of standards.

5.2.1 What is a Radio Spectrum, Radio Channel, and Radio Band?

The term *radio spectrum* is broadly used to describe the collection of electromagnetic wave frequencies within the range of approximately 3 Hz–300 GHz. A *radio*

---

[1] Institute of Electrical and Electronics Engineers.

channel represents the radio spectrum within a limited range (e.g., 2.47–2.55 GHz), which is used to create a communications link between a transmitter and receiver. In this text, we also use the term *band* to describe spectrum range that includes more than one channel (e.g., 2–2.5 GHz; 802.11b/g band). Further on in this chapter, we will omit the term radio and use only terms such as *spectrum, channel, and band*.

5.2.2 Spectrum Regulation

Governmental regulatory bodies create policies that control as to who can do what, when, and how with a certain band of the radio spectrum. In United States this responsibility is carried out by an independent government agency called Federal Communication Commission (FCC).

There are many different bands, and within bands different classes of applications (e.g., analogue and digital TV and radio). Communication transmissions from services such as TV, radio broadcasting, and cellular phone get an exclusive access to their specified bands, whereas others such as those used for 802.11 are shared with other transmission systems. The regulators determine factors such as: transmit power levels, exact frequency range occupied by a particular band, and several other technical parameters. The general trend in recent years has been to use an auction process to lease the available spectrum to the highest bidder for example, the spectrum for new cellular services. This process is accompanied with less regulation of specific technologies and other technical details. However, some portions of spectrum have been made available to the public free of charge for limited range applications, like cordless phones and wireless LANs. These, so called, Industrial, Scientific and Medical (ISM) radio bands were originally reserved in most countries for industrial, scientific, and medical purposes rather than for general public communications. Further information about ISM bands in relation to IEEE 802.11 standards will be provided in the following sections.

Spectrum regulation is a very complex topic and for those who are interested in enhancing their knowledge in this area, we recommend a book authored by Nuechterlein and Weiser [1].

5.2.3 Channel Interference

In this chapter, we will examine *interference* as an important factor that influences the ability of two linked nodes to reliably communicate with each other at the desired communication rate. Within the scope of 802.11-based wireless networks, we distinguish between two types of interference: *radio interference* and *channel contention interference*.

Radio interference represents a physical interference that influences the entire spectrum of electromagnetic waves. Physical interference has two distinctive subtypes: *destructive interference and constructive interference. Destructive interference* is depicted in Fig. 5.1.[2] where three waveforms can be seen: the first waveform on the top represents message signal, the second waveform in the middle represents an unwanted interference signal and the bottom waveform represents the resulting signal waveform. As it can be observed in Fig. 5.1, the destructive interference has a detrimental effect on the transmitted radio signal, which often results in a loss of transmitted data. On the other hand, in the second type of interference i.e., the constructive interference depicted in Fig. 5.2 (waveforms are in the same order as in Fig. 5.1) has the converse effect on the signal as it results in an increase of its amplitude. However, the waveforms used in these two examples are synchronized

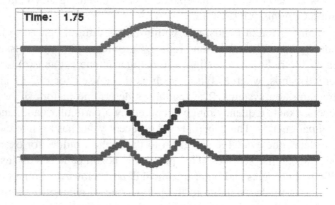

Fig. 5.1 Destructive interference

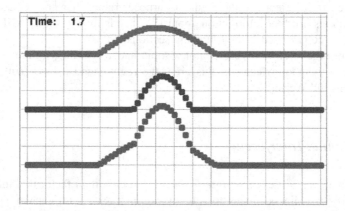

Fig. 5.2 Constructive interference

---

[2] The figures used to illustrate destructive and constructive interference are frames from the animation that is available on the website of Physics Department of Boston University: http://physics.bu.edu/~duffy/semester1/c21_interference.html.

in time, which is very rarely the case in reality. This means that in most of the cases both forms of interference have a detrimental effect on the signal.

The second type of interference mentioned at the beginning of this section, *contention interference*, is not a physical interference as in the previous cases although it also produces a negative effect. *Contention interference* stems from the Carrier sense multiple access with collision avoidance (CSMA/CA)-based MAC layer of 802.11 protocol that defines the behaviour of the 802.11 station, which has to wait until a channel is free to commence its transmission. Therefore, the channel may be occupied by transmission from any node that uses the same channel within the communication range. Because of the license free availability of 802.11 band this issue is even more pronounced in urban areas that contain a large number of 802.11-based networks. Thus, in urban areas all the co-located networks compete for the use of limited number of channels. In Sects. 5.3 and 5.4, we will examine ways of modelling and measuring interference that can provide the information necessary for an effective channel assignment process.

In addition to the term *interference* it is important to clearly define two related terms: *communication range* and *interference range*. *Communication range* is the range in which a reliable communication between two nodes is possible and *interference range* is the range in which transmissions from one node can detrimentally affect the transmissions from other nodes on a same or partially overlapping channel. It is important to note that the interference range is always larger than the communication range as shown in Fig. 5.3.

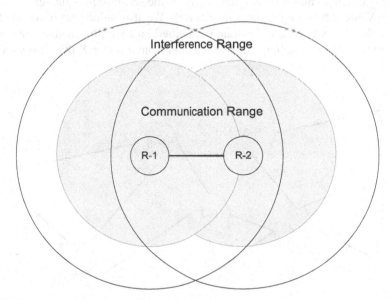

Fig. 5.3 Communication range and interference range

5.2.4 Graph Theory: Fundamentals

To model the behavior of interconnected structures mathematicians have developed the realm of *graph theory*. The fundamentals of this theory have their roots in the paper about the bridges of Königsberg by Leonhard Euler presented to the St. Petersburg Academy on August 26, 1735 [2]. In this subsection, we describe the basic elements that are necessary to understand the graph-based models, which are discussed further on in this chapter.

At any point in time all important elements of a WMN can be represented as a graph. A *graph, G*, is defined as a set of vertices V and a set of edges E and can be denoted as $G = (V,E)$. The sets V and E have to be nonempty and finite. An edge is a link between two vertices, which joins the vertices i and j, and is denoted by (i, j). The vertices i and j are the end-vertices of this edge. If an edge exists between two vertices, then these two vertices are called adjacent or neighbouring vertices of G. Two edges are called adjacent if they have one common end-vertex. An example of a graph is depicted in Fig. 5.4. To the edges of a graph, specific values or weights may be assigned for example to represent an interference level, in which case the graph is called a weighted graph [3].

For a graphical representation of MR-WMN, we can use the following simple mapping: Each vertex in a graph represents a router in a MR-WMN and each edge between two vertices represents a radio link between a pair of peer interfaces of two routers. See Fig. 5.4; routers are labeled from R-1 to R-10.

In addition to the mapping, we can also define sets of important elements. For example, we define the set of neighbours N_X as the set of routers that can be connected to X and that belong to the same network. We also define a set of routers IR_X that have the ability to receive a signal transmitted on a specific channel but which can be influenced by simultaneous transmission from a router X that also uses the

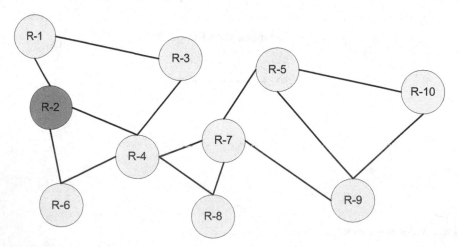

Fig. 5.4 The graph – representation of a mesh network

same or a partially overlapping channel. It is important to note that router X does not have to belong to the same network as the affected routers, but it can belong to any of different co-located networks.

In this model, we assume that the link between a pair of nodes is bidirectional (undirected). A link is said to be operational between two nodes if the signals transmitted from one node can be received above a minimum required power threshold by the peer node and vice versa. We define that two nodes are connected if there is a link between them.

This basic graph model enables us to explore in details the fundamental properties of mesh networks, including the connectivity, mutual interference, degree distribution and hop count.

For a further insight into graph theory and its applications in computer science refer to [4–6].

5.2.5 Introduction to IEEE 802.11 Set of Standards

The IEEE 802.11 family of standards was conceived in 1997, and since then it has been gradually evolving. For example, the 802.11b supplement standard defines the high rate direct sequence spread spectrum (HR/DSSS) transmission mode with a chip rate of 11Mchip/s, which provides the same occupied channel bandwidth and channelization scheme as Direct-Sequence Spread Spectrum (DSSS) in legacy standard. A higher data rate is achieved through a transmission mode, which is based on eight chip complementary code keying (CCK) modulation. The code set of complementary codes is richer than the set of Walsh codes that are used in DSSS [7].

In the period from 2000 to 2003, the 802.11b extension was the first technology that captured a significant market share. When it was initially released in 1999 the wireless networks based on physical layer augmentations of 802.11b and 802.11a were meant to be deployed as an extension to the existing wired LANs. However, as new wireless networks rapidly gained popularity the need for an amendment and further development of standards has arisen. Initially, networks based on 802.11a extension did not achieve the transmission rates of 802.11b networks. This was mainly because of higher prices and somewhat reduced range because of the use of 5 GHz spectrum. However, vendors continued with the refinement of the chipset technology, which has resulted in an improvement of 802.11a radio features. Today, this standard has captured a significant share of the corporate market. Another spinoff of the 802.11 standard is the 802.11g standard, which is fully backward compatible with 802.11b but offers higher data rates to be achieved through the use of orthogonal frequency-division multiplexing (OFDM). OFDM is the scheme that is also used in the physical layer of the 802.11a. In 2003 a third amendment for 802.11g was ratified.

Further increase in the number of deployed wireless LAN networks has lead to a higher level of interference in the congested 802.11b/g band that contains 11–14 channels (depending on regulatory domain). This is mainly because of the limited

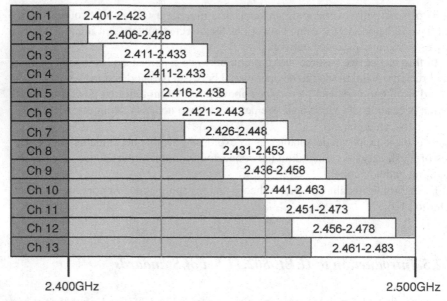

Ch 1	2.401-2.423
Ch 2	2.406-2.428
Ch 3	2.411-2.433
Ch 4	2.411-2.433
Ch 5	2.416-2.438
Ch 6	2.421-2.443
Ch 7	2.426-2.448
Ch 8	2.431-2.453
Ch 9	2.436-2.458
Ch 10	2.441-2.463
Ch 11	2.451-2.473
Ch 12	2.456-2.478
Ch 13	2.461-2.483

2.400GHz   2.500GHz

Fig. 5.5 Available channels in 2.4-GHz ISM band that are used for 802.11b/g

number of nonoverlapping[3] channels. It can be inferred from Fig. 5.5 that only channels 1, 6, and 11 are not overlapping. Further discussion on the limitations of 802.11 physical layers will be provided in Sect. 5.4.

Readers interested in a detailed analysis of IEEE 802.11 family of standards should refer to [7, 8].

5.3 Background

In this section, we will provide a detailed insight into the problem of channel assignment in 802.11-based mesh networks. We will also introduce few network models from three different perspectives of link, connectivity and interference.

5.3.1 Channel Assignment Problem Definition

For two 802.11-based interfaces to communicate with each other, they need to be assigned to a common channel. In a nutshell, a solution to a channel assignment problem determines which one of all available channels should be assigned to a

---

[3] The terms nonoverlapping and orthogonal have the same meaning and will be used interchangeably in this chapter.

given 802.11 interface. However, the number of available channels is limited and as more interfaces within the same interference range are assigned to the same radio channel or a partially overlapping channels, the effective bandwidth available to each interface decreases. Therefore, a good channel assignment algorithm needs to effectively balance between the goals of maintaining connectivity and increasing aggregate bandwidth. The problem definition will increase in complexity when we combine the constraints associated with routing and topology control along with the channel selection problem.

5.3.2 Different Models, Representations and Perspectives used for the Channel Assignment Problem

In this subsection, we will (1) Introduce channel selection problem models based on graph theory; (2) review few network representations from three different aspects: link, connectivity and interference and (3) we will examine in details IEEE 802.11 standard physical layers and their limitations that are pertinent to the channel assignment problem.

5.3.2.1 Models Based on Graph Theory

The graph coloring theory is used as a base for the theoretical modelling of channel assignment problem. In the early days of mobile telephony the channel assignment problem was modelled as an ordinary graph coloring problem, and graph coloring algorithms were used to solve it. Practical experience revealed that these solutions have a number of deficiencies. In the following few paragraphs, we provide an introduction to this type of modelling accompanied with an overview of its use in the literature.

In this type of model there is a vertex on a graph corresponding to each node (e.g., a mesh router) on a wireless network. An edge between two vertices on the graph represents the link between two nodes. This model does not contain an explicit representation for interfaces and assumes one interface per router; consequently it cannot be used to model MR-WMN. The color of each vertex represents a nonoverlapping channel and the goal of the channel assignment is to cover all vertices with the minimum number of colors such that no two adjacent vertices use the same channel. This model is illustrated in Fig. 5.6 – fill patterns are used instead of colors to represent different channels.

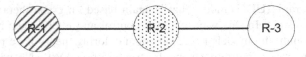

Fig. 5.6 A simple topology to illustrate graph coloring

Fig. 5.7 Graph coloring for
MR-WMN

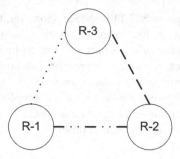

A different representation of graph coloring problem is necessary to capture multiple interfaces on each router. As depicted in Fig. 5.7 the colors are assigned to edges instead of vertices. In this example, we use different patterns instead of colors. Edge coloring assigns a color to each edge so that no two incident edges share the same color. This approach can always be transformed into the vertex version because edge coloring of a graph is just the vertex coloring of its line graph.

However, because the number of nonoverlapping channels on 802.11b/g network is limited to just 3, to enable an efficient channel assignment, it is necessary to introduce a weight associated with each edge on a graph. The weight indicates the importance of using different colors for corresponding vertices.

The theory on graph coloring is extensive; here we provide the fundamental formal definition of weighted coloring problem as given in literature [6]: "A proper coloring of a graph G is an assignment of a color to each vertex so that adjacent vertices receive distinct colors." Equivalently, it is a partition of the vertices into stable sets, where a set of vertices in G is stable (or independent) if no two are adjacent. This problem is known to be NP hard [9, 10].

A comprehensive overview of techniques used for channel assignment in cellular mobile telephony is provided in [11]. This reference is recommended to anyone who wishes to examine the detailed history and development of the problem. Mishra et al. in [12] use weighted graph coloring with the weight calculation based on a number of clients that are affected by the interference affecting an access point (AP) on a particular channel. Leith and Clifford [13] also use weighted graph coloring in the form of an interference graph. In their model each vertex represents a WLAN and edges represent interference between corresponding WLANs. Jain et al [14] use the same type of graph model as Mishra et al. but they use it in conjunction with other techniques to compute the optimum throughput that wireless network can support. Ramachandran et al. [15] extend the conflict graph model further into multiradio conflict graph (MCG). This model differs in a way that represents edges between the mesh radios as vertices instead of representing edges between the mesh routers as vertices as in the original conflict graph. In the literature use is also made of a unit disk graph (UDG) model that is again based on graph theory to reduce transmit power levels. In this way topology control and interference reduction is achieved. Although, the models based on graph coloring theory have proven their usefulness in modelling interference on infrastructure-based WLANs, we agree with the conclusion of [16] that graph coloring models do not adequately capture all the constraints of a multiradio WMN. In particular, the constraints such as: representations of partially overlapping channels and the interference when the source is external to the network.

5.3.2.2 Network Representation from Three Different Aspects

The aim of this subsection is to introduce the reader to different ways of representing a MR-WMN. The models described herein serve only as examples that illustrate several possibilities for representing WMN and are not necessarily the optimal representations.

Figure 5.8 has two parts: We depict five network models (rows A–E in Fig. 5.8 part 1 and 2) drawn from three aspects to demonstrate a variety of MR-WMN

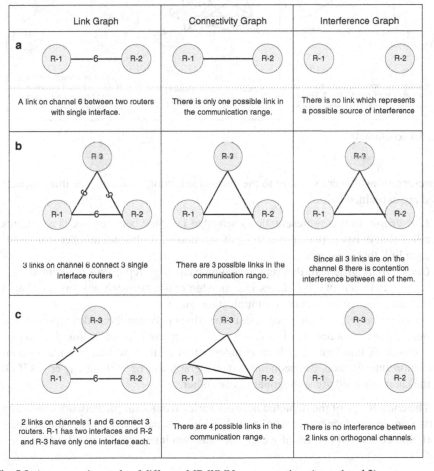

Fig. 5.8 A comparative study of different MR-WMN representations (parts 1 and 2)

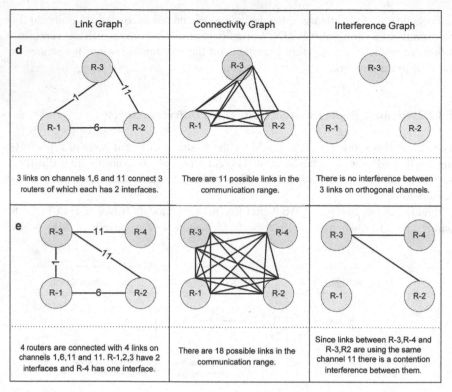

Link Graph	Connectivity Graph	Interference Graph
d		
3 links on channels 1,6 and 11 connect 3 routers of which each has 2 interfaces.	There are 11 possible links in the communication range.	There is no interference between 3 links on orthogonal channels.
e		
4 routers are connected with 4 links on channels 1,6,11 and 11. R-1,2,3 have 2 interfaces and R-4 has one interface.	There are 18 possible links in the communication range.	Since links between R-3,R-4 and R-3,R2 are using the same channel 11 there is a contention interference between them.

Fig. 5.8 (continued)

representations that are relevant to the channel selection problem, The three aspects used are explained below:

- *Link graph*. In this representation each edge i.e., line between nodes represents a link that exists between interfaces of two nodes. A channel number is used for each link that is assigned.
- *Connectivity graph*. In this representation each edge represents a potential connection between two interfaces i.e., an edge exists between any two interfaces that are within each others communication range.
- *Interference graph*. In this representation edges between the nodes represent the interference between two links. An edge is depicted for each link that can be a source of interference, which is independent of its type i.e., physical or contention interference. Consequently, such a graph can be without any edges if all the channels used are orthogonal to each other.

The complexity of the depicted network varies from a simple network consisting of two single interface routers (i.e., row A in Fig. 5.8; part 1) to a more complex network that contains 4 routers of which each has two interfaces (in row E of Fig. 5.8; part 2).

A key inference made from Fig. 5.8 (parts 1 and 2) is that the complexity of graph representation for a connectivity graph (column 2 in Fig. 5.8) increases significantly with each additional interface.

5.4 IEEE 802.11: Physical Layers and Their Limitations

The suitable assignment of channels in a MR-WMN is closely associated with the nature of the underlying physical layers of the 802.11 radio. Therefore in this section, we will study most significant aspects of 802.11 physical layers. Reasons for examining in details the 802.11 physical layers as well as identifying their limitations are enumerated below:

- The root causes of the interference related to the issues in 802.11a,b/g-based networks are in the restrictive features of their respective physical layers such as: spectrum availability and contention mechanism defined in medium access control (MAC) Layer.
- Interference is a cornerstone issue of channel assignment. As such, it is not feasible to design an effective channel assignment algorithm without taking into consideration the possible effects of interference on the underlying physical layers of the wireless network.
- An insight into the operation of the physical layers in a realistic scenario is necessary for the development, assessment and optimisation of any channel assignment algorithm. This process will facilitate the reduction in the number of steps necessary for the channel assessment as well as to alleviate its performance.

The discussion provided in this section will be limited only to the amendments of 802.11 given in 802.11a, 802.11b, 802.11g and the draft 2 proposal of 802.11n from February 2007 [17]. Further the amendments related to *radio resource measurements* part in 802.11k are covered by Sect. 5.4.2, which is relevant to 802.11n. As draft 2 proposal of 802.11n makes use of the 802.11k draft 4 as a reference for its spectrum measurement functions therefore the above stated amendments are discussed within the scope of Sect. 5.4.3.

802.11a and 802.11g as well as 802.11n use transmission techniques based on OFDM. Specifications for the individual components of OFDM for 802.11a are given in Table 5.1. The relatively higher data rates and robustness offered by such a modulation scheme has caused the 802.11g embedded devices to become increasingly popular within the period from 2003 to 2006. However, two major drawbacks related to the OFDM modulation, which as stated is used in 802.11g, are evident from the literature. These are:

1. Reduced range in comparison to 802.11b.
2. Drop in the transmission throughput when a mix of 802.11b- and 802.11g-based devices operate within the same network.

These drawbacks constitute the main reasons for an initial low penetration of 802.11g interfaces used as nodes in public APs or nodes in mesh networks.

Table 5.1 Specifications of modulation schemes in 802.11a

PHY mode	Pmin R (dBm)	Number of data bits per symbol	1,472 bytes transfer duration (μs)	SINRmin (dBm)	PHY Data rate (Mbits/s)	Bandwidth factor (is equal to PHY Data rate divided by 6 Mbps)
BPSK 1/2	-82	24	2,012	18	6	1
BPSK 3/4	-81	36	1,344	21	9	1.497
QPSK 1/2	-79	48	1,008	22	12	1.996
QPSK 3/4	-77	72	672	25	18	2.994
16QAM 1/2	-72	96	504	25	24	3.992
16QAM 3/4	-70	144	336	32	36	5.988
64QAM 2/3	-66	192	252	34	48	7.984
64QAM 3/4	-65	216	224	35	54	8.982

However, because of an increase in demand for higher network throughput rates the proportion of 802.11g-based interfaces for public AP has significantly increased in the last 2 years.

Over time, the deployment of 802.11 continued in the area of home networking and mobile computing platforms. The growth in the number of APs as well as inherent complexity of wireless networks has reflected the limitations of physical layer in 802.11.

5.4.1 ISM Bands Used for 802.11 Physical Layers

The ISM bands have been defined by the International Telecommunication Union Recommendation (ITU-R) in articles 5.138, 5.150, and 5.280 of the Radio Regulations. Individual countries' allocation of the bands designated in these sections may differ because of variations in national radio regulations. In USA use of the ISM bands is governed by Part 18 of the FCC rules, whereas Part 15 Subpart B contains the rules for unlicensed communication devices, including those that use the ISM frequencies. Thus, designers of equipment operating in the ISM bands in USA should be familiar with the relevant portions of Part 18 and Part 15 Subpart B of the FCC Rules.[4]

Three ISM bands used for IEEE 802.11 are:

- 2.400–2.500 GHz (centre frequency 2.450 GHz; used for 802.11b/g)
- 5.150–5.350 GHz (centre frequency 5.250 GHz; used for 802.11a)
- 5.725–5.875 GHz (centre frequency 5.800 GHz; used for 802.11a)

---

[4] In this chapter US standards will be used whenever not specified otherwise.

It is important to note that 802.11b/g uses one ISM band with a size of 100 MHz. On the other hand, 802.11a uses two bands with a total size of 350 MHz. Therefore, the number of nonoverlapping channels available for 802.11a is much higher than that for 802.11b/g.

As mentioned in Sect. 5.2, ISM bands are open for use by anyone, Furthermore they are extensively regulated (maximum power of transmitter is limited and depends on a regulatory domain). More information on ISM band and its use can be found on the ITU and FCC websites [18].

5.4.2 Effect of Interference on Link Throughput in 802.11-Based Networks

The main factor that affects network connectivity is the distance between the peer nodes. In addition, other factors that influence radio wave propagation are (1) obstacles in the path between a transmitter and receiver and (2) interference from other possible prominent radio sources. Interference causes a drop in signal strength, which triggers a change in the modulation type. This chain of events depicted in the Fig. 5.9, eventually leads to a reduction in the link throughput. To exemplify the significance of interference reduction: let us consider an extreme case when a channel suffers from a high noise and interference, which results in a 18 dBm SINR. If our algorithm selects an alternate channel with a SINR of 35 dBm this

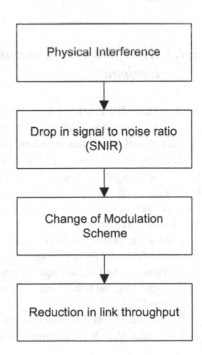

Fig. 5.9 Effect of interference on link throughput on 802.11 networks

Table 5.2 Interference factor (i-factor) between partially overlapping channels

Channel	1	2	3	4	5	6	7	8	9	10	11
SNR i-factor	0	0.22	0.60	0.72	0.77	1.00	0.96	0.77	0.66	0.39	0

will as per Table 5.1 result in an increase of available bandwidth of almost nine times! Furthermore, packet loss is proportional to interference. A high packet loss in turn can render a link unusable for QoS sensitive traffic such as voice over IP (VoIP). The interference caused by the use of partially overlapping channels can be analysed with the help of Signal to Noise Ratio (SNR) interference factor, which is termed as the i-factor and is provided by Mishra et al. in [19, 20]. The i-factor (provided in Table 5.2) can be used as a guidance for selecting partially overlapping channels once all nonoverlapping channels in a single interference domain are used up. The i-factor given in Table 5.2 is based on the drop in SNR of partially overlapping channels when the transmission occurs on channel 6. We can observe from Table 5.2 that the drop in SNR is decreased when partially overlapping channels further away from channel 6 are selected (e.g., channels 3 or 9). In this way the interference can be kept under control and the overall network throughput increased.

In the past year, wireless equipment based on draft 802.11n standard has gained an increase in popularity. However, these equipments use channel binding feature that enables the use of more than one nonoverlapping channel for a single link to increase the net throughput. The channel binding feature has thus resulted in a more pronounced interference problem.

5.4.3 IEEE 802.11k: Spectrum Awareness of 802.11-Based Equipment

At the outset IEEE 802.11-based networks were meant for use in a small area that serviced a limited number of APs. Because of this mindset, the incorporated radio resources and other measurement functions were inherent and very limited. However, a vast increase in the popularity of equipment based on these standards and its deployment in public hot spots such as airports and train stations have resulted in a higher spatial density. In turn, this has caused an increase in the level of mutual interference between co-located networks, which has motivated further work on extensions to the standard.

First such extensions were prompted by the European regulatory requirements for the 5-GHz frequency range used by 802.11a-based devices. IEEE task group 802.11h addressed these requirements in 2003. Focal point of the 802.11h amendment is the internal use of measurement data to attain dynamic frequency selection (DFS) and TPC.

To co-ordinate and accomplish these functions successfully, cellular system such as Universal Mobile Telecommunication System (UMTS) and Global System for

Mobile communications (GSM) specify the exchange of a variety of related information. As IEEE 802.11 networks have cellular like structure, which is similar to UMTS and GSM networks so an analogous functionality can be developed within the scope of 802.11.

An increase in demand for networks that cover metropolitan areas has resulted in an increase in density of APs and therefore an increase in interference. This is another important impetus for the introduction of new measurements procedures. With time, a need has arisen for the extension of the scope of information exchange between stations as well as a unified measurement mechanism for both frequency ranges (2.4 GHz, 802.11b/g and 5 GHz, 802.11a). This has led to the establishment of the IEEE Task Group 802.11k. At the moment this standard proposal is in the late draft 7.0 stage and the final approval is expected by the end of 2007. The main difference between 802.11h and 802.11k is that the former does not consider the use of measurement data on tasks such as TPC or dynamic frequency assignment; instead it focuses on the measurement and the transfer of measurement information between stations.

The significance of 802.11k in the development of channel assignment algorithms is twofold:

1. It facilitates in the development of an advanced TPC mechanism that reduces interference effects in MR-WMN.
2. It carries out a comprehensive link assessment and link selection process that is useful for the above point.

Although, some of these functions for the improvement of channel assignment mechanisms can be achieved with the use of previously discussed 802.11 amendments (see Sect. 5.4.3) only the 802.11k amendment offers an inherent protocol for radio resource measurements and the propagation of the results of such measurements through the wireless network.

The IEEE 802.11k amendment is specifically dedicated to the radio resource measurement. It introduces three additional information elements (1) AP channel report, (2) neighbor report and (3) receive channel power indication (RCPI). Also, nine additional request types are defined of which eight result in a report and one in a measurement pause. The AP channel report is periodically transmitted along with the beacon. The main objective of this transmission is to enable an easy identification of other APs, which in turn facilitates a fast handover.

In MR-WMN the ad-hoc mode of 802.11 is used and a fast handover facilitated by 802.11k does not represent an important factor (here we do not consider mobile clients and mesh nodes are considered to be static). On the other hand, beacon signals that are heard by all stations in the ad hoc mode and in particular the neighbour report can be used to expeditiously create a list of neighbouring stations. This is a possible way by which the number of steps and hence the resulting execution time of any measurement-based algorithm could be reduced. Similar to 802.11h, most of the measurement type requests in 802.11k apply generic request fields, which are described in details in the draft standard documents.

5.5 Algorithms for Channel Selection

It has been well established by now in the chapter that channel selection is a very challenging issue that influences the overall performance of the MR-WMN. To quickly comprehend as to why channel selection is challenging, we consider a simple approach to channel selection with the help of an example algorithm – called "first available channel." This approach simply selects the first channel from a pool of available channels; once all channels in the pool are exhausted new links cannot be created. However, this approach does not facilitate an efficient use of the spectrum and produces far from optimum results. In recent times, the research community has developed a significant number of algorithms that provide major improvements in spectrum usage. We commence this section with the taxonomy of existing solutions and then embark on a comprehensive review of the key algorithms for channel selection with specific attention on a mix of quantitative and qualitative attributes such as: performance, complexity, scalability and stability.

5.5.1 Taxonomy of Existing Algorithms

Many approaches to addressing the channel assignment problem are available in the contemporary literature. Historically they are almost all based on the developments made in cellular mobile telephony. To gain a further insight into this problem, we provide a brief classification that is categorically listed below and is with respect to:

1. Locality of channel assignment process that is based on:

   (a) Centralized solutions
   (b) Distributed solutions

2. Dynamics of assignment process, which is based on:

   (a) Fixed channel assignment
       – Specific channel can only be used in designated cells, different groups of channels may be assigned to adjacent cells, the same group can be assigned to the cells that are outside mutual interference range.
   (b) Dynamic channel assignment
       – All channels assignments are temporary, from time to time the situation is reassessed and channels are reassigned according to certain criteria.
   (c) Hybrid channel assignment
       – Available channels are divided in two groups, one group is used for fixed channel allocation and other is used for dynamic allocation.

3. Strategies deployed, which are:

   (a) Planning-based strategies
       – Everything is predefined in a master plan.

(b) Co-ordination-based strategies
- The channel selection is achieved through different co-ordination strategies between stakeholders.
(c) Measurement-based strategies
- Periodical measurements are undertaken to assess the available band. Based on the results channels are assigned.
(d) Hybrid strategies
- These strategies involve mechanisms such as co-ordinated measurements, planning augmented by periodic measurements and combined strategies.

5.5.2 Review of Algorithms for Channel Selection

Solutions that we are going to focus on in this chapter are dynamic measurements based. These comprise algorithms that not only take into account interference produced by a single network but also other co-located networks. Noise, that is generated by other transmissions in the band being used is also considered in these algorithms.

The work of [21] specifically targets the channel assignment problem on WMN. Authors have adopted their theoretical work in [22] and created a self-stabilizing distributed protocol and an algorithm for channel assignment.

The method of [21] assumes that the interference is symmetric and is based on an interference range of three hops. Their method results in improvements of only 20% compared to random channel assignment. In reality, most of the times interference will be asymmetric because neighbouring node interface may transmit on the same channel at different powers. In contrast, a better proposal would not assume symmetric interference and would not require a dedicated channel for frequency co-ordination, which is a significant advantage. Further the interference cost function in [21] has not been justified i.e., the cost function has not been based on an interference model. The other main limitation of their proposal, as well as the one by Raniwala and Chiueh [16] is the usage of a common channel on each node for the management of channel assignment. We believe that his approach should be avoided because it can be wasteful of bandwidth and imposes severe limitations on network capacity especially when nodes have only two interfaces. Furthermore, a strong source of interference on the frequency that is used for the co-ordination of channels can render the throughput of parts or the whole network unsatisfactory.

In [14], contrary to previous findings, authors state that the addition of new nodes can actually improve a per-node throughput because the richer connectivity provides increased opportunities for routing around interference "hotspots" in the network that offsets the increase in traffic load caused by the new nodes. They explain this by the fact that previous research has been done under the assumption that nodes always have data to send and are ready to transmit as fast as their wireless connection will allow. However, in realistic settings, sources tend to be bursty, so nodes on an average transmit at a slower rate than the speed of their wireless link.

In [14] authors use liner programming to model maximum achievable flow between the source and destination in absence of wireless interference and heuristics to obtain lower and upper bounds on the throughput. Also, they avoid making assumptions about the homogeneity of nodes with regards to radio range or other characteristics as well as regularity in communication patterns. The conclusion from the work in [14] was that neither multipath routing nor doubling the range of the radio increases cumulative throughput. On the other hand, by using two channels instead of one, the network may achieve the maximum possible throughput. Scenarios provided in [14] illustrate that the model they have developed could be a useful tool for analysis and capacity planning in wireless multihop networks. However, we believe that their model suffers from oversimplification. Although valid theoretical capacity bounds can be produced most of real world deployments are more complex − that involve neighbouring or co-located networks with unknown interference characteristic. Consequently in realistic circumstances we cannot simply obtain information that has to be fed into the proposed model and the capacity predicted by the model and in reality may vary significantly.

In [23] the authors have deployed heuristics that is based on interference measurements. Still, they do not define a threshold value range and a mechanism to keep a channel change under control. This may result in an infinite loop of channel changes that is caused by a slight variation in noise or by cyclical interference. Furthermore, it is not clear how much time (steps) the algorithm needs to achieve acceptable results as well if it can cope with dynamic environment.

Reference [16] considers a combined solution for channel assignment and routing issues in [24] and extend their previous proposal with the usage of a virtual control network instead of a dedicated interface-channel on each router. In other words this means that certain fraction of bandwidth is used on each channel for channel assignment and other management purposes rather than dedicating one exclusive channel. Their work contains careful analysis of all aspects of resource allocation problems relevant for 802.11-based WMN.

In [25] authors propose the usage of partially overlapping channels. Their interference model is theoretically based on a conflict graph and the interference data is acquired through the measurement of link pair interference. Reference [25] uses integer linear programming to obtain bounds of optimal solution and evaluate the proposed algorithm.

Reference [26] approach is based on the assumption that an interface can *dynamically* switch over from one channel to another. They present a distributed interface hybrid assignment strategy and their routing strategy selects routes that have low switching and diversity cost. However, a co-ordination protocol is required to assign the channels to fixed channels in the hybrid nodes i.e., nodes having an interface assigned with fixed channel and the other interfaces with switchable channels.

Reference [27] also propose a load aware based channel assignment. Although the work presented in their paper is of a good value but it assumes a centralized method for channel assignment, which also needs to keep track of the load in differ-

Table 5.3 A comparative study of some algorithms recently published in the literature

Related work attributes	[16, 24]	[28]	[29]	[25]	Algorithm that should be aimed for
Type of algorithm	Distributed/ centralized	Self-organizing	Self-organizing	Centralized and distributed	Self-organizing and distributed
Parameters	Interference + load	Interference	Interference, infrastructure mode	Interference	Interference + load
Dedicated channel for assignment	NO	NO (YES for previous work)	NO	NO	NO
Nonorthogonal channels used	NO	NO	NO	YES	YES
Transmit power control	NO	NO	NO	NO	Should be incorporated
Scalability	Addressed	Addressed	Not relevant	Partially addressed	Should be addressed
Stability	Not addressed	YES	NO	NO	Should be addressed
Capacity analysis	None specified	Non specified	None specified	Not specified	Should be specified

ent parts of the WMN. Their work is thus not scalable. However, they have shown a circular dependency between channel assignment, load on each link and routing.

In [15] the authors base interference estimate on the number of interfering radios on each channel supported by each router. An interfering radio is defined as a simultaneously operating radio that is visible to a router but external to the mesh. A visible radio is one whose packet(s) pass frame check sequence (FCS) checks and are therefore correctly received. However, this method is incomplete because it neglects interference caused by transmissions that are too weak for the signal to be decoded but still result with degradation of SNR on particular link. The other main drawback of last two proposals is the scalability because centralized algorithms are used. However, both proposals motivate further investigation because they indicate a 40% performance gains in comparison to static assignment.

Table 5.3 summarizes and compares the primary attributes associated with the key algorithms provided in the literature.

5.6 Thoughts for Practitioners

We list below some of the key thoughts that have been assimilated as a result of our work on this chapter:

1. A good channel assignment algorithm needs to effectively balance between the goals of maintaining connectivity and increasing aggregate bandwidth. The

problem definition will increase in complexity when we combine the constraints associated with routing and topology control along with the channel selection problem.

2. A key inference made from the use of graph theory for channel assignment is that the complexity of representation for a connectivity graph of a network system increases significantly with each additional radio interface.

3. 802.11k draft standard will be useful in the development of channel assignment algorithms as it will enable:

   - The development of an advanced TPC mechanism that reduces interference effects in MR-WMN.
   - A comprehensive link assessment and link selection process that is useful for the above point. This is because the 802.11k encompasses an inherent protocol for radio resource measurements and the propagation of the results of such measurements through the wireless network.

5.7 Directions for Future Research

This chapter has progressively built on the topic of channel assignment in MR-WMN by systematically explaining the concepts behind channel assignments. It has then explained the grand challenges posed by the channel assignment task in MR-WMN and some of the key works that have been proposed in the literature. We have contributed also to this research area through our work such as [30, 31] and as such we believe that the next possible steps to engage in would be:

1. Create an algorithm for channel assignment in MR-WMN that combines the approaches of topology control with Interference cost reduction. It is expected that this should lead to an increase in the overall channel capacity of the MR-WMN system as well as spectral efficiency. Some of the preliminary work done by us in this regard can be viewed at the following URL: http://www-staff.it.uts. edu.au/~debenham/prodan/

2. Create a mechanism that additionally takes into consideration the traffic load on the links in the system i.e., carries out channel assignment using topology control plus Interference cost reduction plus load distribution. As more factors are combined the creation of an efficient algorithm will be more complex.

3. The expected fourth generation (4G) of networks will encompass several heterogeneous wired and wireless networks offering seamless connectivity. In such a 4G environment the MR-WMN will be constituted of heterogeneous wireless routers using a mix of radio types. This will pose additional issues such as those associated with heterogeneous radio resource management, which could be another area of research.

5.8 Conclusions

MR-WMN need to have a suitable channel assignment approach so that an increase in the overall capacity gain is realized. However, the problem is very challenging especially because the 802.11-based radios have only a limited number of channels. As such, it is difficult to make the channel allocation without affecting the performance. To cater to a diverse reader audience, in this chapter simple terms ranging from channel, spectrum to more complex ones such as graph theory were first adequately explained with the help of simple illustrations. Further, to enable any interested reader to follow the chapter it was gradually approached for example by examining the ways in which 802.11 physical layer influences the channel interference and strategically providing recommendations for useful literature at several places within the text. Apart from explaining the emerging 802.11k standard and graph theory to alleviate the channel assignment problem, we also provided a critical review of the different approaches for channel assignment in MR-WMN. Overall this chapter draws out that even though a lot of effort has been put into addressing the channel assignment problem in MR-WMN but still this problem remains a fascinating area of research that needs to be explored further. By the time the readers have finished this chapter, they should have a clear grasp of how to successfully model a channel assignment problem with an emphasis on interference reduction.

5.9 Acronyms

AP	Access point
CCK	Complementary code keying
CSMA/CA	Carrier sense multiple access with collision avoidance
FCC	Federal communication commission
GSM	Global system for mobile communications
HR/DSSS	High rate direct sequence spread spectrum
IEEE	Institute of electrical and electronics engineers
ISM	Industrial, scientific and medical (radio bands)
LAN	Local area network
MAC	Medium access control
MR-WMN	Multiradio wireless mesh network
OFDM	Orthogonal frequency-division multiplexing
TPC	Transmit power control
UMTS	Universal mobile telecommunication system
WMN	Wireless mash network

5.10 Terminologies

1. *Channel selection*. The process of choosing appropriate channel for a link between two nodes. To minimize interference this process requires information about the usage of the band by existing links within the same interference domain.
2. *Communication range*. Is the range in which a reliable communication between two nodes is possible.
3. *Graph theory*. Is the area of mathematics that models the behaviour of interconnected structures.
4. *Interference range*. Is the range in which transmissions from one node can detrimentally affect the transmissions from other nodes on a same or partially overlapping channel. Interference range is always bigger than communication range. This is because even when signal degrades to a level that cannot be successfully decoded by a receiver it still causes degradation of other signals that are simultaneously transmitted on the same or partially overlapping channel.
5. *Multiradio Wireless Mesh Networks (MR-WMN)*. Is a mesh network that connects routers that contains more than one radio interface.
6. *Orthogonal channels or nonoverlapping channels*. The frequency range used by these channels is not overlapping; consequently there is no interference when these channels are used for simultaneous transmission within the same interference range.
7. *Partially overlapping channels*. The frequency range used by these channels is partially overlapping. When two or more links are created by using partially overlapping channels within the same interference range the interference occurs.
8. *Radio band*. The term *band* is used to describe spectrum range that includes more than one channel (e.g., 2–2.5 GHz – 802.11b/g band). Related terms: radio spectrum and radio channel.
9. *Radio channel*. A *radio channel* represents the radio spectrum within a specific range (e.g., 2.47–2.55 GHz), which is used to create a communications link between a transmitter and receiver. Related terms: radio spectrum and radio band.
10. *Radio spectrum*. The term *radio spectrum* is broadly used to describe the collection of electromagnetic wave frequencies within the range of approximately 3 Hz–300 GHz. Related terms: radio channel and radio band.

5.11 Questions

1. What is a channel and what as a partially overlapping channel?
2. Who regulates the radio spectrum in US?
3. What is a constructive and what destructive interference?

4. How many partially overlapping channels are in 802.11b/g (2.4 GHz) band (in US regulatory domain)?
5. Assuming that all depicted routers are inside a single interference domain, can you assign channels available in 2.4-GHz band (802.11b/g) so that each link is on a nonoverlapping channel? Explain your answer.

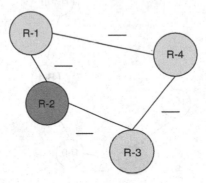

6. Suggest four channels in 2.4-GHz band for four links that are going to be used in a single interference range. Take into account that one of four links will be under high traffic load, one will be under low traffic load and remaining two will be under average traffic load. Show how you have obtained your results.
7. Draw a connectivity graph for the network described in Question 6. (Use illustrations in Sect. 5.3.2.2 for guidance.)

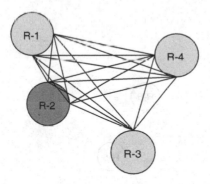

8. Draw an interference graph for the network described in Question 6. (Use illustrations in Sect. 5.3.2.2 for guidance.)
9. Figure given bellow depicts a MR-WMN with five routers of which three have two radio interfaces and two have a single interface each. The gray dotted circle around each router shows the interference domain of that specific router. How would you assign channels in 802.11b/g band to minimize the interference in this network?

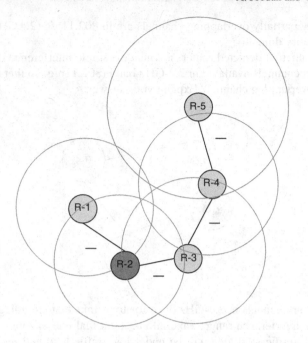

10. Figure given bellow depicts a MR-WMN with six routers; the gray dotted circle
    around each router shows the interference domain of that specific router. How
    would you assign channels in 802.11b/g band to minimize the interference in
    this network?

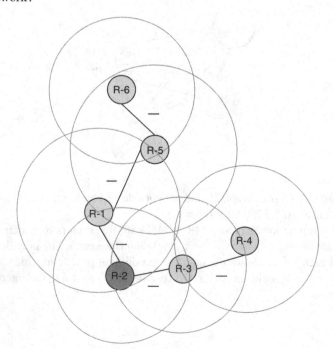

Acknowledgments This research is performed as part of an Australian Research Council Linkage Grant between Bell-Labs, Alcatel-Lucent and the University of Technology, Sydney (UTS), Australia.

References

1. J. E. Nuechterlein and P. J. Weiser, Digital Crossroads American Telecommunications Policy in the Internet Age. Cambridge, MA: MIT (2005).
2. L. Euler, The Seven Bridges of Königsberg Commentarii academiae scientiarum Petropolitanae **8**: 12 (1741).
3. R. Hekmat, Ad-hoc Networks: Fundamental Properties and Network Topologies. Dordrecht, The Netherlands: Springer (2006).
4. J. A. Bondy, and U. S. R. Murty, Graph Theory with Applications. London: Macmillan (1982).
5. S. Bornholdt and H. G. Schuster, Handbook of Graphs and Networks: From the Genome to the Internet. Weinheim: Wiley (2003).
6. R. Diestel, Graph Theory. Heidelberg, NY: Springer (2005).
7. B. H. Walke, S. Mangold, et al., IEEE 802 Wireless Systems: Protocols, Multi-Hop Mesh/Relaying, Performance and Spectrum Coexistence. New York, NY: Wiley (2006).
8. IEEE Task Group 802.11b, Higher-Speed Physical Layer Extension in the 2.4 GHz Band, vol. 97. New York, NY: The Institute of Electrical and Electronics Engineers (1999).
9. M. R. D Garey, S. Johnson, et al., Some simplified NP-complete problems. Sixth Annual ACM Symposium on Theory of Computing, Seattle, Washington, United States. New york, NY: ACM (1974).
10. M. R. Garey and D. S. Johnson, Computers and intractability: a guide to the theory of NP-completeness. San Francisco, CA: W. H. Freeman (1979).
11. I. Katzela and M. Naghshineh, Channel Assigment Schemas for Cellular Mobile Telecommunication Systems – A Comprehensive Survey. IEEE Personal Communications **3**: 10–31 (1996).
12. A. Mishra, S. Banerjeeb, et al., Weighted Coloring based Channel Assignment for WLANs. ACM SIGMOBILE Mobile Computing and Communications Review **9(3)**: 19–31 (2005).
13. D. J. Leith and P. Clifford, A self-managed distributed channel selection. Algorithm for WLANs. Proceedings of RAWNET, Boston, MA, USA (2006).
14. K. Jain, J. Padhye, et al., Impact of interference on multi-hop wireless network performance. ACM Annual International Conference on Mobile Computing and Networking (MOBICOM), San Diego, CA (2003).
15. K. N. Ramachandran, E. M. Belding, et al., Interference-Aware Channel Assignment in Multi-Radio Wireless Mesh Networks. INFOCOM 2006, Barcelona, Spain (2006).
16. A. Raniwala, and T.-c. Chiueh, Architecture and Algoriths for an IEEE 802.11-Based Multi-Channel Wireles Mesh Network. IEEE INFOCOM, Miami, USA (2005).
17. IEEE Task Group 802.11n Part 11: Wireless LAN Medium Access Control (MAC) and Physical Layer (PHY) Specifications: Amendment 11n: Enhancements for Higher Throughput. New York: The Institute of Electrical and Electronics Engineers (2007).
18. Federal Communication Commission (FCC), Home Page. Retrieved 15th November, 2007, from http://www.fcc.gov/(2007).
19. A. Mishra, E. Rozner, et al., Exploiting Partially Overlapping Channels in Wireless Networks: Turning a Peril into an Advantage. ACM/USENIX Internet Measurement Conference (2005).
20. A. Mishra, V. Shrivastava, et al., Partially overlapped channels not considered harmful. SIGMetrics/Performance, Saint Malo, France. New York, NY: ACM (2006).
21. B. -J. Ko, V. Misra, et al., Distributed Channel Assignment in Multi-Radio 802.11 Mesh Networks. New York, NY: Columbia University 2005.

22. B. -J. Ko and D. Rubenstein, A distributed, self-stabilizing protocol for placement of replicated resources in emerging networks. Eleventh IEEE International Conference on Network Protocols (ICNP), Atlanta, GA, USA. Washington, DC: IEEE (2003).

23. B. J. Leung and K. K. Kim, Frequency assignment for IEEE 802.11 wireless networks. Fifty-Eighth IEEE Vehicular Technology Conference (VTC 2003 – Fall) (2003).

24. A. Raniwala and T.-c. Chiueh, Architecting a High-Capacity Last-Mile Wireless Mesh Network. MobiCom, Philadelphia, USA. New york, NY: ACM (2004).

25. A. P. Subramanian, R. Krishnan, et al., Minimum interference channel assignment in multi-radio wireless mesh networks. Thirteenth International Conference on Network Protocols (ICNP 2005), Boston, USA (2005).

26. P. Kyasanur and N. Vaidya, Routing and interface assignment in multi-channel multi-interface wireless networks. IEEE Wireless Communications and Networking Conference, New Orleans, LA, USA (2005).

27. A. Raniwala, K. Gopalan, et al., Centralized channel assignment and routing algorithms for multi-channel wireless mesh networks. Mobile Computing and Communications Review 8(2): 15 (2004).

28. B. J. Ko, V. Misra, et al., Distributed Channel Assignment in Multi-Radio 802.11 Mesh Networks. WCNC Hong Kong, China. Washington, DC: IEEE (2007).

29. B. Kaufmann, F. Baccelli, et al., Self-Organization of Interfering 802.11 Wireless Access Networks, 22 (2005).

30. V. Mirchandani, A. Prodan, et al., A Method and Study of Topology Control based Self-Organization in Mesh Networks, IEEE AccessNets 2007, Ottawa (2007).

31. V. Mirchandani, A. Prodan et al., Impact of Topology Control on the Performance of a Self-Organization Scheme for Wireless Mesh Networks, Proc. ACM WICON, Austin, USA (2007).

Chapter 6
Routing, Interface Assignment and Related Cross-layer Issues in Multiradio Wireless Mesh Networks

Leonardo Badia, Marco Conti, Sajal K. Das, Luciano Lenzini, and Habiba Skalli

Abstract Many technological standards for Wireless Mesh Networks include the possibility to use several nonoverlapping channels for data transmission. This represents an opportunity that can be exploited by equipping the terminals with multiple network interfaces. This opens up an interesting challenge, namely, how to simultaneously use different frequencies, so as to limit collisions and therefore activate multiple simultaneous transmissions in the same geographic area. At the same time, this poses new issues; for example, network connectivity is reduced, because nodes that do not interfere are also unable to communicate with each other. Thus, more complex interface management techniques are required. Moreover, a paradigm shift from the classic routing schemes is needed. Usual approaches are not always satisfactory because they often use shortest-path heuristic and tend to concentrate transmissions to certain nodes. To efficiently exploit the presence of multiple channels instead, a proper routing algorithm should avoid congested links and possibly make use of an estimation of the actual network traffic. Therefore, cross-layer information exchange can be useful for an efficient functioning of the routing protocols. In this chapter, we will analyze all these issues and propose and identify possible solutions.

6.1 Introduction

Wireless mesh networks (WMNs) [1, 2] are a network technology currently under development to provide end users with broadband wireless connectivity. In such systems, each mobile terminal owned by an end user, called mesh client (MC), is linked through a single radio hop to a mesh router (MR), a fixed infrastructure node. All the MRs are, in turn, interconnected to each other in a multihop fashion so as to

L. Badia (✉)
IMT Lucca Institute for Advanced Studies, Piazza S. Ponziano 6, 55100 Lucca, Italy
e-mail: l.badia@imtlucca.it

S. Misra et al. (eds.), *Guide to Wireless Mesh Networks*, Computer Communications and Networks, DOI 10.1007/978-1-84800-909-7_6,
© Springer-Verlag London Limited 2009

form what is referred to as the network backbone. This kind of structure is easy to install because several low cost nodes can be added to improve the backbone connectivity. Moreover, MRs do not need to be battery-powered, because they can be easily placed in correspondence with a power outlet. Finally, the all-wireless structure does not require cable deployment, thus making WMNs appealing for connecting both vast rural regions and crowded urban areas where cable deployment is not cost-effective.

In general, to attach the WMN to the Internet, some special MRs, called mesh access points (MAPs), are equipped with wired connections and therefore can take the role of Internet gateways. Therefore, they usually have better computational capabilities than the other MRs, which work as simple relay nodes; for this reason, it is sensible to think of MAPs as the centers of the network management operations. On the other hand, this determines a higher cost of such nodes and therefore their number is reasonably limited. In most cases, just one or two MAPs are used; this will be also the case for the examples discussed throughout this chapter.

Because the communication between a MC and its reference MR is single-hop, most of the challenges of the WMN management are at the backbone level. This part of the network is similar to other kinds of wireless multihop networks, such as ad hoc and sensor networks. Differently from them, however, the main problems in the inter-MR communication do not relate to mobility and energy saving problems, which are avoided because of the assumptions made above. Instead, other major technical issues arise especially when the network size grows (scalability problem). Among them, one of the most challenging is represented by routing [3]. In fact, the performance of WMNs in this sense is, similar to any other multihop network, limited by wireless interference. The placement of additional relay nodes yet mitigates the problem, because it gives additional opportunities for traffic forwarding; however, the performance improvement is often limited and does not linearly scale with the number of nodes. Thus, the design of efficient routing algorithms plays a key role among WMN research topics.

Moreover, WMN solutions are often thought as utilizing existing standards, such as IEEE 802.11 [4], without any modification. On the one hand, this enables to use off-the-shelf network cards for the wireless mesh nodes, which keeps the infrastructure costs low. On the other hand, a straightforward adaptation of existing technologies, without taking into account the specific purposes of WMNs, will result in an inefficient management. In fact, these standards are commonly used in a different context; in particular, IEEE 802.11 is used almost exclusively in a single-hop fashion, whereas its collision avoidance mechanism is known to suffer from several problems in multihop scenarios, such as the decrease of network parallelism because of the exposed terminal problem [5].

In general, a compromise shall be sought between this inefficient usage and the design of entirely new protocols. A possible solution, in this sense, can be the idea of finding new applications of possibilities already envisioned by the protocol but scarcely used in practice. An example where this concept can be applied concerns the possibility of exploiting multiple portions of the available wireless spectrum. For example, the IEEE 802.11a/b/g specifications provide multiple channels, some of which can be regarded, with a good degree of approximation, as nonoverlapping (specifically, 3 channels for IEEE 802.11b/g and and 12 channels for IEEE 802.11a).

There are two possible approaches to deal with multiple channels. In the majority of the literature, it is assumed that they are perfectly nonoverlapping; in this chapter we will consider this case only. There is also an interesting line of research, discussed in more detail in the following, where partial overlap of the channel is taken into account with the aim to exploit it [6]. However, this approach requires to entirely reformulate the routing problem. The case of perfect nonoverlap is simpler, because it allows to regard the routing problem as a multicommodity allocation or a graph coloring issue. Notice that models for studying networks exploiting frequency diversity date back before the success of wireless networks, because they were already investigated, e.g., for optical fiber networks [7].

Although multiple channels can be introduced, and actually they are already available in existing standards, terminals are typically configured to operate on a single radio channel: in fact, in a single-hop scenario, this frequency diversity is mostly introduced to avoid collisions from different networks. In a WMN case, instead, this feature can be used to increase the number of transmissions that can be exchanged within a neighborhood. This imposes to differently tune the Network Interface Cards (NICs) of the involved MRs.

The opportunity given by multiple nonoverlapping channels is better exploited if more than one NIC is available at a single node. In this way, one can avoid, or at least mitigate, the need for dynamically tuning to a common frequency the interfaces of MRs that are meant to communicate with each other. As will be discussed in the following, fast frequency-switching transceivers are in fact not always feasible. Actually, the cost decrease for commodity hardware makes multi interface terminals economically sustainable, even though in general it is not possible, for many practical reasons to provide each node with a single NIC per every available channel. However, as shown in [8], the largest advantage in terms of network capacity, intended as traffic that can be transmitted over the network in a collision-free manner, is present already for a limited (though larger than one) number of NICs per node. The relative performance improvement when the number of interfaces approaches the number of available channels becomes marginal.

Thus, we will focus on multiradio, i.e., multichannel *and* multi-interface, WMNs. The investigations carried out in the following concern the strategy to determine the channels to which the NICs of every node shall be tuned, which can be regarded as a multiple allocation optimization problem, and how this affects routing strategies over the WMN.

There is a two-fold relationship between the routing and the interface assignment problems. First, when the routing algorithm is applied, two nodes i and j can communicate, and therefore it is possible to route traffic through a network link from i to j, only if they share a common channel assigned to at least one of their NICs. Conversely, to be realized efficiently, the interface assignment should take into account the routing pattern of the network. In fact, because the use of different channels decreases not only the mutual interference but also the network connectivity, it should leave the possibility of connecting the nodes along the main traffic routes and possibly decreasing the number of interfering links.

Classic routing protocols for multihop networks [9,10] may be easily extended to support multiple interfaces at each node. However, those protocols typically select shortest-hop routes, which may not be suitable for multichannel networks; as was noted in [11], routing metrics based on hop count only should be integrated by also taking into account the network load. Moreover, longer paths may be preferable if they allow to decrease interference and increase transmission parallelism. At the same time, more bandwidth should be given to nodes that support higher traffic, i.e., channels assigned to these links should be shared among a fewer number of nodes. More in general, the interface assignment strategy should be traffic-aware in the sense that it matches the distribution of traffic load in the mesh backbone.

For these reasons, in the following we will overview solutions presented in the literature and summarize basic criteria for routing and interface assignment in multi-radio WMNs, giving particular emphasis to the interaction between these two tightly related problems that can be efficiently managed with an adequate knowledge of the network traffic. In particular, we will discuss how to exploit the knowledge of the load on the links [12] and how to estimate it [13] and we give practical examples of application.

The rest of this chapter is organized as follows. In Sect. 6.2 we overview papers on routing and channel assignment in WMNs appeared in the literature. In Sect. 6.3 we give a comprehensive summary of different criteria that can be used to approach the problem. In Sect. 6.4 we formally state the problem and introduce definitions and notations. Section 6.5 describes a possible methodology to estimate the network load, which, as previously argued, is extremely useful to achieve a good cross-layer management of routing and interface assignment; additionally, it outlines an optimization framework for a routing-aware channel assignment problem, where load information is explicitly taken into account. Finally, the conclusions are drawn in Sect. 6.6.

6.2 Background

The problem of frequency selection in a multichannel networks inherits some approaches and methodologies, as well as the idea of using graph theory, from the problem of assigning channels in an optical network [7]. In this case however, the edges are fixed, because they correspond to a cabled connection between nodes. Thus, that topic resembles more closely the classic graph coloring problem. In the wireless case instead, the possibility of managing not only the frequency on which a connection is tuned to, but also the existence of the edge itself, requires an extended treatment. In this sense, another related problem is the frequency re-use planning in cellular networks, where graph representations have been also used [14].

An interesting line of research dealing with multichannel WMNs is based on the observation that most of the available channels are indeed partially overlapping. This, instead of being considered harmful, could be turned in an opportunity to achieve connectivity (though an imperfect one) in a less interference-prone way.

It is also possible to have a fully connected network and decrease interference while using a single NIC for all nodes.

Such an approach, investigated for example in [6] and [15], though very promising, implies to entirely reformulate the network management, and is therefore out of the scope of the present chapter, where we deal instead with adapting existing routing approaches to the multichannel case, and we consider different channels as perfectly separate in frequency.

Approaches for multiple orthogonal resource allocation mainly deal with time-division multiple access (TDMA), as for instance done by the earlier work reported in [16]. In fact, this paper proposes to introduce multiple time slots, with a special control slot where the users can rendezvous to negotiate the access in a distributed manner. However, this case can be easily extended, with few modification, to a frequency-division multiple access (FDMA) case. For example, [17] reports a description of the issues that need to be faced when dealing with multiradio multi-hop networks and proposes a similar strategy where a common control channel is used to coordinate a distributed assignment of multiple channels.

Because of the similarity between FDMA and TDMA multiplexing, some papers jointly investigate, together with routing, *both* channel assignment and packet scheduling over time [18–20]. In [18], the goal of finding a joint channel assignment, routing and scheduling technique that optimizes throughput of the MCs is studied. The problem is formulated as a linear programming (LP) framework. The approach used by this paper for tackling multichannel networks is similar to the one adopted in [21] where an analogous optimization framework is extended to the multichannel case. Under specific interference assumptions, necessary and sufficient conditions are described, under which collision free link schedule can be obtained. In particular, as done by most of the papers related to this topic, the protocol interference model is used, as introduced in [22]. This dictates to model interference through collisions, and can be equivalently mapped through a so-called *conflict graph*. Actually, such a model is not perfect, because it implies some approximations in modeling interference as pointed out, for example, in [23]. Nevertheless, it is quite simple and is, in fact, often used by those papers modeling channel assignment through LP frameworks. However, because the problem of achieving the optimal allocation of scheduling times over several frequencies is shown to be NP-hard, the final solution proposed by [18] is an efficient heuristic approach, which can be proved to be at most a given factor away from the optimum.

In [20] a similar problem of joint routing, channel assignment and scheduling is investigated, where the goal is again on throughput maximization. Interference is again modeled through a K-dimensional version of the protocol interference model. After that, the feasibility of a schedule is verified by means of a sufficient condition, that is considering whether the conflict graph can be properly colored, by using as many colors as TDMA slots so that conflicting edges are differently colored (i.e., they are active over different time instants).

Another similar optimization is also considered in [19]; to deal with the high complexity of the resulting problem, the solution is sought through Simulated Annealing [24], which is an evolutionary technique for LP problems offering a good trade-off between accuracy and computational complexity. The solution operates in two steps, i.e., the routing/channel assignment problem is split between two parts. First, routing is solved by means of a shortest-path strategy. Then, a simulated annealing algorithm tries to optimize the assignment of the NICs. Because this optimization technique needs a starting solution as input, channels are initially assigned randomly, provided that they satisfy interference constraints. Subsequently, the system evolves according to the simulated annealing procedure, which seeks to maximize the throughput.

An even simpler solution to overcome the NP-completeness of the problem is to propose efficient heuristic strategies. This methodology is adopted for example in [8, 25, 26]. In spite of their simplicity, these strategies can achieve good performance, especially in light of the fact that they do not need particularly complex computations. It is worth noting that, for the most, they employ the conflict graph model to represent interference, and therefore the proposed heuristic is related to graph coloring considerations.

All these approaches refer to a centralized solution, hence they assume the availability of a central controller (e.g., located in one of the MAPs) that takes care of solving the allocation problem and signalling the obtained solution to the other nodes. Instead, [12] proposes a decentralized maximization problem, where the interference constraints refer only to neighboring transmissions. An extended version, proposed in [15] by the same authors, investigates the case of partially nonorthogonal channels. This is done based on a technique in which a channel weighing matrix is calculated. An original aspect of this approach is that, even though interference is still based on the protocol model, or, equivalently, on conflict graphs, instead of simply preventing collision from arising at all, it is taken into account how they affect (i.e., degrade) the capacity of the links, which allows for a more tunable problem characterization.

6.3 Thoughts for Practitioners

In this section we review some practical criteria that have been proposed to determine interface assignment in multiradio WMNs. The technical contributions in this field are very heterogeneous for what concerns the depth of theoretical investigations. Thus, we try to discuss relevant points of interest that distinguish the existing proposals and we identify practical general criteria. The reported references can give further details on these topics.

6.3.1 Static vs. Dynamic Assignment

Interface assignment strategies can be classified according to the time-scale involved in the assignment, i.e., the rate of variability of the channel allocation.

Following [17], the schemes proposed in the literature can be divided into static and dynamic interface assignment. Hybrid schemes are also possible.

Within the static strategy, the interfaces are assigned with a constant value over time, or at least they are unchanged over a time period that is significantly larger than the packet scheduling time unit. The simplest possibility for a static assignment is the so-called common channel assignment (CCA), which was proposed in [25], actually more as a theoretical comparison scheme than a real policy. In CCA, the interfaces of each node are all assigned the same set of channels. For example, if each node has two radios, then the same two channels are used at every node. Hence, the connectivity of the network is the same as that of the single channel approach, possibly with redundant repetitions. Thus, there is still an advantage because multiple channels can be leveraged to increase throughput. However, the improvement achieved with respect to the single channel case is far below the highest potential gain of using multiple radios. Thus, varying channel assignment (VCA) strategies are usually proposed, where the variation is meant over space, not over time, as the assignment is still static, but allocates different sets of channels to different radio interfaces. VCA techniques are usually more efficient than CCA but have the potential risk of partitioning the network, and in general the length of routes between MRs increases.

In contrast, dynamic strategies allow all channels to be associated with any interface freely and continuously update the assignment that is potentially changed on a per packet basis. However, the challenge associated with this scheme is that whenever two nodes need to communicate with each other, a coordination scheme had to exist to ensure that they are on a common channel. For example, a common channel can be used as a rendezvous point to negotiate the allocation for the next transmission phase, as done in [27]. Another example is the slotted seeded channel hopping (SSCH) mechanism [28] in which each node switches channels synchronously in a pseudo-random sequence to allow all neighbors to meet periodically in the same channel.

The advantage of dynamic assignment [29], is the potential to exploit all channels with few interfaces. Their main problem relates to the demanding hardware requirements. In fact, real time services, which WMN are supposed to provide, have stringent delay requirements, which are therefore hardly met if the additional delay imposed by NIC switching time is introduced. Thus, these schemes have limited practicality unless expensive terminals are employed. Moreover, switching interfaces may result in a *deafness* problem, occurring when a node wants to communicate with another, which is tuned on another channel. Channel access issues arise, because the transmitter, being deaf, is unaware that the receiver may be busy in another transmission. This problem can be solved by introducing appropriate rendezvous on certain channels at certain time instants, and determines many challenges that are out of the scope of this chapter.

Finally, hybrid schemes apply a static scheme to some interfaces and a dynamic one for the rest. Examples of this kind are the link layer protocols described in [17] in which a VCA is used for the fixed interfaces. CCA may also be used for the fixed interfaces as is the case in the interference-aware channel assignment in [30]. In this case, there is certainty that any communication through some of the links can be established using the static part, but still the requirement of fast-switching NICs is present.

6.3.2 Centralized vs. Distributed Assignment

As any other resource allocation strategy, interface assignment schemes can be generally realized in centralized or distributed fashion. In the centralized schemes the channels are assigned by a central controller, usually located in one of the MAPs. In the distributed schemes, instead, each node assigns channels to its interfaces in a more loosely coordinated fashion, because no global network knowledge is available. Thus, the decision is based on neighborhood information. The complexity of this latter case is much lower, at the cost of lower efficiency. Especially, the effectiveness of distributed strategy is critical in relationship with routing awareness, which demands for network-wise knowledge.

In general, most of the techniques reviewed in this chapter are directly applicable within a centralized management. Extensions to distributed management are also possible, but they usually require information exchange to acquire some global knowledge at each node. Similar techniques to obtain a distributed implementation of routing and interface-assignment can be found for example in [8, 16, 29].

6.3.3 Heuristic vs. Optimization Strategies

As pointed out in Sect. 6.2, the joint routing and interface assignment problem can be investigated through a proper optimization framework, but the resulting complexity is very high. It is then possible to draw another classification of possible approaches, even though it does not relate to design aspects, but rather on practical methodologies to solve the problem. In fact, in the literature several papers investigate the problem through LP approaches [12, 18, 20, 21], but also many contributions proposing a heuristic approach [17, 25, 26, 29].

From a general point of view, these two choices are extreme points of a trade-off. LP solutions offer better accuracy, heuristics have lower complexity. Intermediate solutions are also possible, such as meta-heuristic techniques like Simulated Annealing, as proposed in [19]. However, we remark that these two possibilities are not perfectly separated. In fact, though LP approaches are usually limited to smaller WMNs and suffer from scalability problems, they can shed light on heuristic techniques in a more rigorous and appropriate manner. As a matter of fact, the aforementioned papers that give an LP formalization also investigate heuristic criteria to solve the problem inspired by the theoretical findings.

6.3.4 The Gateway Bottleneck

A practical criterion to assign channels to interfaces, useful especially for heuristic procedures, is to consider the MAPs at first, because during the execution of an algorithm the first nodes to receive an assignment can usually select the frequencies in a less constrained manner. In [31], where many inefficiencies possibly arising in WMNs are described, it was observed that the most congested nodes are likely to be the MAPs, where all the routes converge, a property referred to as *gateway bottleneck*. Also, the bottleneck is particularly limiting if a single gateway is present in the network; hence, it is suggested to always activate multiple MAPs (of course, this has beneficial effects not only in terms of network capacity, but also, e.g., in case of failure).

This implies that such nodes should be the ones where frequency diversity can be applied achieving the highest benefit. Especially if a single MAP is present, we could state a "rule of thumbs" of starting the channel assignment algorithm from it. Note also that in this case the property can be generalized, to some extent, by saying that the closer (in terms of number of hops) is a node to the MAP, the more critical can it be in terms of congestion. This is especially true for the node with the best connectivity to the gateway (e.g., in terms of highest rate, lowest interference, or both) among the neighbors of the gateway itself.

Actually, this strongly depends on the network topology. If the gateway has a single neighbor, the gateway bottleneck is simply translated to this node. On the other hand, if the network has a star topology, with all non-MAP nodes being neighbors of the gateway with relatively similar connectivity, there is no bottleneck whatsoever, or at least, no more than what dictated by the medium access control (MAC), because all multiple transmissions collide. However, in practical scenarios, the distance to the MAP in terms of number of hops can be a good heuristic weight to determine the priority in receiving a channel assignment. To some extent, this criterion is implicitly taken into account by certain existing heuristic algorithms [25,26].

6.4 Notation and Terminology

As done by many related contributions, we adopt in the following a graph-based representation of the WMN backbone. All terminals belonging to the backbone, i.e., all the MRs also including the MAPs, can be represented as *nodes* included in a set **N**. If two nodes can communicate, i.e., there exist conditions where they can exchange packets with sufficiently high success probability, we consider them as linked through a graph edge. This may require that all the other nodes in the backbone do not transmit, because the condition of successful transmission can be violated in the presence of interference from other nodes. For this reason, the existence of an edge is a necessary condition, but not a sufficient one, to have an error-free communication. In addition to the existence of an edge, also certain interference conditions must be verified, which may vary according to the interference model adopted. In this way, a notation is commonly achieved in many radio allocation problems, where the network is represented as a graph $G = (N, E)$, where the set $E \subseteq N \times N$ contains the network edges. Note that, from the physical point of view, the edges in E should be *directed*. This means that, given $i, j \in N$, $(i, j) \in E$ does not necessarily imply $(j, i) \in E$. Even though rarely taken into account, link asymmetry is very frequent in radio networks [32]. However, there are certain MAC protocols, most notably the IEEE 802.11 one, which explicitly assume the links to be bidirectional, e.g., for handshake exchange. In this case, it is implicitly assumed that nonsymmetric edges are discarded from E. This is actually a nontrivial assumption, as argued in [33], but we take it since it is both simple and also very common in the literature. In the following, we will therefore refer to this case and take edges as bidirectional. Most of the reasonings can however be easily extended to more general scenarios where directed links are present as well.

We observe that the terminology used throughout the literature concerning graph representation of the network is rather assorted: the existence of an edge from i to j is also sometimes referred to as "j is within communication range of i" or "node j can hear node i." Even though these descriptions are not rigorous from the propagation point of view, as the radio transmission involves more parameters than just distance, they are often adopted in the exposition and we sometimes will use them as well. Similarly, notice that "topology" is a term often used as a synonym of "graph," in particular channel assignment seen on graph representations is often referred to as "topology control" problem.

In channel assignment problems there is an additional requirement for network representation, i.e., to describe radio interfaces, and whether they are tuned on the same frequency, otherwise no communication can occur between them. Note that interference conditions are entirely orthogonal to this latter issue, i.e., to exchange packets, two nodes must at the same time meet the requirement of having a shared NIC allocation *and* interference free communication.

Usually, to depict frequency allocation, the graph representation is split in two parts. In both of them, the set of nodes N is the same, but they differ in the set of the edges. In the first one, called *physical topology* $G_P = (N, E_P)$, the set of the edges consider all possible connections among nodes, with the only requirement of radio propagation. However, when the channels are assigned to the radio interfaces, it could happen that some nodes do not share a channel where to communicate, even though they are linked through an edge in E_P (and therefore they can hear each other). To represent the network connectivity after the channel assignments have been determined, a *logical topology* $G_L = (N, E_L)$ is employed, where E_L is determined by imposing the additional condition that only nodes sharing a common channel can be linked through an edge. Actually, because there may be nodes sharing *more than one channel*, there can also be *multiple edges* in E_L linking the same pair of nodes. In this sense, E_L is not strictly speaking a subset of E_P because the channel graph may contain more than one element corresponding to the same edge in the physical topology. We also remark that the symmetry considerations previously made apply to both physical and logical topologies, because the property of sharing a channel assignment on a network interface is a symmetric property for any pair of nodes.

Moreover, we need a notation to specifically represent the channel assignment. If there are K orthogonal channels available, without loss of generality we can use the set of integers $\mathbf{K} = \{1, 2, \ldots, K\}$ to denote them. For all $i \in \mathbf{N}$, we denote with $v(i)$ the number of NICs owned by node i. The exact channel assignment is represented by an *interface allocation variable* denoted as y_i^q, where $i \in \mathbf{N}$ and $q \in \mathbf{K}$, which is a binary variable equal to 1 if node i has a NIC tuned on channel q and 0 otherwise. Note that $\sum_{r=1}^{K} y_i^q = v(i)$ for all nodes $i \in \mathbf{N}$. Similarly, if $i, j \in \mathbf{N}$ and $q \in \mathbf{K}$, we define a binary *channel edge variable* called x_{ij}^q and defined as equal to 1 if i can transmit to j using the q the channel, and 0 otherwise. If the link symmetry assumption holds, it is reflected in that $x_{ij}^q = x_{ji}^q$. These variables are connected through the relationship $x_{ij}^q = y_i^q \cdot y_j^q$.

An example of graph representation is given in Figs. 6.1 and 6.2, where the physical and the logical topologies, respectively, are shown for a sample network of six nodes with $K = 4$ channels. In this case, nodes a and f, which are shown to have wireline connection to the Internet, operate as MAPs, whereas the other nodes are ordinary MRs. For all nodes i, $v(i)$ is chosen equal to 2. In the logical topology (Fig. 6.2) the numbers written on the edges indicate the frequency on which they are established, and small numbers beside a node denote its NIC assignment.

First of all, the aforementioned difference between the two topologies can be observed. Some links of the physical topology can be absent in the logical topology, as is the case, e.g., for the edge (d, e). In Fig. 6.2, nodes d and e are not linked because they do not have a common interface assignment. On the other hand, all

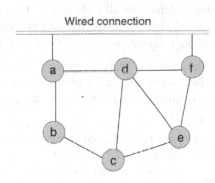

Fig. 6.1 1Physical topology of a sample network

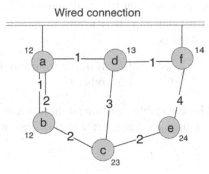

Fig. 6.2 Logical topology of a sample network

pairs of nodes in Fig. 6.1 are linked through one edge at most, whereas in Fig. 6.2 two edges connect nodes a and b because they share both of their interface assignment on channels 1 and 2.

By looking at Fig. 6.2, the interface allocation variables can be derived, for example $y_a^1 = y_a^2 = 1$, $y_a^3 = y_a^4 = 0$, or $y_e^2 = y_e^4 = 1$, $y_e^1 = y_e^3 = 0$. The channel edge variables are similarly determined, e.g., $x_{ab}^1 = x_{ab}^2 = x_{cd}^3 = 1$, $x_{ab}^3 = x_{ab}^4 = x_{de}^1 = 0$.

As discussed previously, in most of the investigations related to interface assignment, wireless interference is modeled through the so-called protocol model [22]. For our purposes this means that any edge $(i, j) \in \mathbf{E}_P$ is associated with a set $\mathbf{J}(i, j)$, called *conflicting link set*, containing all the edges $(x, y) \in \mathbf{E}_P$ whose activation on the same frequency than link (i, j) prevents a reliable transmission on it. For practical purposes, we adopt the convention of including also (i, j) in its own conflicting link set, i.e., $(i, j) \in \mathbf{J}(i, j)$, which simplifies the notation. The conflict relationship is mainly because of propagation phenomena; sometimes the conflicting link sets are defined based on simplified models, related for example to the distance between nodes. It is worth mentioning that this formulation is an abstraction useful for its conceptual simplicity, and for this reason will be used thereinafter. Yet, from the viewpoint of correctly modeling interference, more realistic descriptions, such as the so-called physical interference model [22] would be preferable. However, with some modifications, the reasonings presented in the following could be extended to alternative interference models as well. A detailed discussion about interference models is out of the scope of the present chapter. The interested reader can found overviews on this subject for example in [34, 35].

To instantiate the routing problem in the multichannel environment, we need also to define for all links $(i, j) \in \mathbf{E}_P$ a parameter $c_{ij}^{(P)}$ that describes their *physical capacity*, i.e, their nominal data rate (e.g., expressed in Mbps). For completeness, we can introduce a value $c_{ij}^{(P)} = 0$ if $(i, j) \notin \mathbf{E}_P$. According to whether edge (i, j) is reflected in the logical topology also, $c_{ij}^{(P)}$ will be mirrored into a *logical capacity* value. Because there are several channels, this latter value depends also on the channel q. Thus, for $i, j \in \mathbf{N}$ and $q \in \mathbf{K}$, we define $c_{ij}^{(q)}$ that can be larger than zero only if $x_{ij}^q = 1$.

Moreover, we denote with $\gamma^{(s,d)}$ the expected end-to-end traffic to be delivered from source s to destination d. Typically, in WMN either s or d will coincide with one of the MAPs. We also call $\lambda_{i,j}^q$ the amount of traffic (involving any pair source-destination) that passes through edge (i, j) over channel q. To put these quantities in relationship, it is useful to introduce a *binary routing variable* called $a_{i,j}^{(m,n),q}$ defined as

$$a_{i,j}^{(s,d),q} = \begin{cases} 1 & \text{if traffic from } s \text{ to } d \text{ is routed over } (i, j) \text{ on channel } q \\ 0 & \text{otherwise} \end{cases}. \qquad (6.1)$$

These variables will be put in relationship with each other in Sect. 6.5.3, where we use them to characterize traffic aware routing strategies.

6.5 Link Load Estimation and Traffic-aware Interface Assignment

The task of assigning channels to the available NICs can benefit from the exploitation of traffic information. In fact, because the purpose of utilizing multiple channels at the same time is to decrease interference and promote network parallelism, this should be done especially around the most congested links. In this section we discuss possible strategies to retrieve this knowledge and exploit it.

6.5.1 Link Load Estimation

There are different methods for deriving a rough estimate of the expected link traffic load. These methods depend on the routing strategy used (e.g., load balanced routing, multipath routing, shortest path routing, and so on). A possible approach is based on the concept of load criticality [13]. This method assumes perfect load balancing across all acceptable paths between each communicating pair of nodes. Let $P(s,d)$ denote the number of loop-free paths between a source-destination pair of nodes $(s,d) \in \mathbf{N} \times \mathbf{N}$, and let $P_l(s,d)$ be the number of them that pass through a given link $\ell \in \mathbf{E_P}$. Then the expected traffic load Φ_l on link ℓ is calculated as

$$\Phi_l = \sum_{(s,d) \in \mathbf{E_L}} \frac{P_l(s,d)}{P(s,d)} \cdot \gamma^{(s,d)}. \tag{6.2}$$

This equation implies that the initial expected traffic on a link is the sum of the loads from all acceptable paths, across all possible node pairs, that pass through the link. Because of the assumption of uniform multipath routing, the load that an acceptable path between a pair of nodes is expected to carry is equal to the expected load of the pair of nodes divided by the total number of acceptable paths between them.

Consider the logical topology as shown in Fig. 6.3 and assume that we have the three flows reported in Table 6.1.

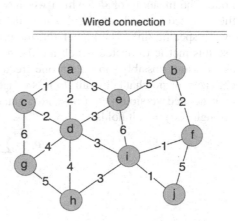

Wired connection

Fig. 6.3 Multichannel wireless mesh network

Table 6.1 Traffic profile with three flows

Source (s)	Destination (d)	$\gamma^{(s,d)}$ (Mbps)
a	g	0.9
i	a	1.2
b	j	0.5

Table 6.2 Possible flows between communicating nodes

(source, dest)	(a,g)	(i,a)	(b,j)
Possible paths	$a-c-g$	$i-e-a$	$b-f-j$
	$a-c-d-g$	$i-e-d-a$	$b-f-i-j$
	$a-d-g$	$i-d-a$	$b-e-i-j$
	$a-d-c-g$	$i-d-c-a$	$b-e-i-f-j$
	$a-d-h-g$	$i-d-e-a$	$b-e-d-i-j$
	$a-d-i-h-g$	$i-d-g-c-a$	
	$a-e-d-g$	$i-h-d-a$	
	$a-e-i-h-g$	$i-h-g-c-a$	
P(source, dest)	8	8	5

Because we have three different sources and destinations, we have

$$\Phi_\ell = \frac{P_\ell(a,g)}{P(a,g)} \cdot \gamma^{(a,g)} + \frac{P_\ell(i,a)}{P(i,a)} \cdot \gamma^{(i,a)} + \frac{P_\ell(b,j)}{P(b,j)} \cdot \gamma^{(b,j)}. \tag{6.3}$$

Furthermore, we calculate $P(s,d)$ for each flow. To this end, we need to determine all the possible source–destination paths, which can be achieved through a Route Discovery procedure [10]. Table 6.2 reports the results for the topology in Fig. 6.3. For practicality reasons, we have set an upper limit for the path length to 5 hops, e.g., by imposing a Time-To-Live to the Route Discovery broadcast packets.

From the above information, we can now calculate how many paths pass a specific link in the network topology. These values and the corresponding link traffic load Φ_ℓ calculated using (6.3) are shown in Table 6.3.

Based on these calculations, we can estimate the load between each neighboring node. The meaning of Φ_ℓ, which we have calculated throughout this example, is the expected traffic load of link ℓ, i.e., the amount of traffic expected to be carried over a specific link. The higher Φ_ℓ, the more critical the link. The idea is now to use this metric to decide which are the most congested points in the network, so as to assign possibly more than one frequency to heavily loaded links and fewer channels, or no channel at all, to less congested edges. Also, as Φ_ℓ can be seen as an estimated version, i.e., a measurement, of the the amount of traffic that passes through $(i,j) = \ell$, it holds

$$\Phi_\ell \approx \sum_{q=1}^{k} \lambda_\ell^q. \tag{6.4}$$

Table 6.3 Possible flows between communicating nodes

ℓ	$P_\ell(a,g)$	$P_\ell(i,a)$	$P_\ell(b,j)$	Φ_ℓ (Mbps)
a–c	2	3	0	0.675
c–g	2	2	0	0.525
c–d	2	1	0	0.375
d–g	2	1	0	0.375
a–d	4	3	0	0.9
g–h	0	1	0	0.15
d–h	1	1	0	0.2625
a–e	2	2	0	0.525
d–e	1	2	1	0.5125
d–i	1	3	1	0.6625
h–i	2	2	0	0.525
e–i	1	2	2	0.6125
b–e	0	0	3	0.3
b–f	0	0	2	0.2
f–i	0	0	2	0.2
i–j	0	0	2	0.2
f–j	0	0	2	0.2

Thus, if the variables λ_ℓ^q are available, they can be used in place of Φ_ℓ which depends on some a priori assumptions such as the perfect load balancing among the edges.

Moreover, several related issues open up. First of all, the strategy to weigh the different paths considers all of them as identical. Actually, there may be conditions that make a path less likely to be used for routing traffic, e.g., if it is very long. On the other hand, it is not true either that shortest hops are to be preferred. As discussed in [11], simple hop count may not be the most appropriate metric to decide on the best routes toward the destination. Thus, in general the determination of quantities $P(s,d)$ is a possible interesting subject for further research.

At the same time, the Φ_ℓ metric can be used only as a rough estimate of the load. Importantly, because channel assignment may affect how $\mathbf{E}_P$ is reflected to $\mathbf{E}_L$, there may be the case that some links are turned off by the absence of a common channel between the involved nodes. In this case, it is not possible to route traffic over them, and therefore the expected traffic load should be recomputed. Thus, also the study of these interactions and possible proposals about how to use similar metrics to infer where congestion is likely to arise are a possible challenging topic to investigate further.

6.5.2 Link Capacity Estimation

The link capacity, or the portion of channel bandwidth available to a link, is determined by the number of all physical links in transmission range of its transmitter or its receiver, i.e., in its conflicting link set, that are also assigned to the same channel. Obviously, the exact short-term instantaneous bandwidth available to each link is dynamic and continuously changing depending on several propagation and interference phenomena [13]. The goal here is to derive an approximation of the long-term bandwidth share available. Thus, the capacity $b_{ij}^{(q)}$ assigned to link (i,j) on channel q can be obtained using the following equation:

$$b_{ij}^{(q)} = \frac{\lambda_{ij}^q}{\sum\limits_{(x,y) \in J(i,j)} \lambda_{xy}^q} \cdot c_{ij}^{(q)}.$$ (6.5)

Note that if $v(i) = v$, constant for all the nodes,

$$\sum_{q=1}^{K} b_{ij}^{(q)} \approx \frac{\Phi_{ij} v \cdot c_{ij}^{(P)}}{\sum\limits_{(x,y) \in J(i,j)} \Phi_{xy}}.$$ (6.6)

In other words, the capacity share available to a link is approximately proportional to its expected load.

6.5.3 Traffic-Aware Joint Interface Assignment and Routing

Giving the preliminaries defined in Sect. 6.4 and the results reported previously, we may specify relationships among the variables that can be used, for example, in an LP context as done by [12]. We stress the important aspect that a comprehensive framework includes channel assignment (represented by variables y_i^q and x_{ij}^q), routing variables $a_{i,j}^{(m,n),q}$, and finally traffic information (variables $\gamma^{(s,d)}$). Thus, it is appropriate to refer to the resulting model as a traffic-aware joint interface assignment and routing. We focus on the model only, whereas the solution techniques are out of the scope of the present analysis. Only, we remark here that the model is rather general and can be solved in a plethora of ways, including exact and approximate, centralized and distributed ones.

The variables of the model are related as per the following relationship, which can be seen as LP constraints. The aggregate traffic on a given link depends on the routing variables and the traffic requirements, so that

$$\lambda_{i,j}^q = \sum_{(s,d) \in \mathbf{N} \times \mathbf{N}} a_{i,j}^{(s,d),q} \gamma^{(s,d)}.$$ (6.7)

The effective capacity $c_{ij}^{(q)}$ of link (i,j) on any channel q cannot exceed the nominal capacity $c_{ij}^{(P)}$ and it is zero if i and j do not share channel assignment q. Thus,

$$c_{ij}^{(q)} = x_{ij}^q c_{ij}^{(P)}.$$ (6.8)

Moreover, the aggregate traffic $\lambda_{i,j}^q$ must be less than $c_{ij}^{(q)}$. Actually, in [12] it is proposed to strengthen this constraint by including a parameter $\Lambda \leq 1$. The motivation is that perfect capacity sharing among all interfering links is not true in practice. Thus, this constraint may be ineffective because it overestimates the effective capacity. Obviously, this is just an artifice and other solutions to cope with this problem are possible as well. Then, we impose

$$\lambda_{i,j}^q \leq \Lambda c_{ij}^{(q)}. \tag{6.9}$$

Finally, we impose a constraint describing conservation of the flows, i.e.,

$$\sum_{\substack{j \in N \\ (i,j) \in E_P}} \sum_{q=1}^{K} a_{i,j}^{(s,d),q} \gamma^{(s,d)} - \sum_{\substack{j \in N \\ (i,j) \in E_P}} \sum_{q=1}^{K} a_{j,i}^{(s,d),q} \gamma^{(s,d)} = \begin{cases} \gamma^{(s,d)} & \text{if } s = i \\ -\gamma^{(s,d)} & \text{if } d = i \\ 0 & \text{otherwise} \end{cases} \tag{6.10}$$

At this point, several metrics can be chosen as the metric to optimize. For example, following again [12], we can choose to minimize the ratio between load and available capacity share on the most congested link. This implies to optimize the utilization of the most congested link and results in the following objective:

$$\min_{(i,j) \in E_P} \max_{x_{ij}^q = 1} \frac{\lambda_{i,j}^q}{b_{ij}^{(q)}}. \tag{6.11}$$

This somehow determines a performance bound in terms of capacity, which is independent of the absolute values of load requirements $\gamma^{(s,d)}$. In fact, they can be rescaled until constraint (6.9) is violated. Therefore, the most congested link gives the capacity bottleneck for the throughput of the whole network. Of course, other objectives are possible as well, for example also introducing fairness considerations. Finally, once the objective function has been identified, the problem can be approached by both LP optimization frameworks and heuristic techniques, and both in a centralized and a distributed manner. The choice of the specific technique to use mostly relates to general design issues such as the computational capability of the terminals.

6.6 Directions for Future Research

Even though many algorithms have been proposed in this context, the design of efficient techniques for interface assignment and routing in multiradio WMN is still an open issue. In the previous sections, we have identified certain possible enhancements to the usual routing and channel assignment metrics. However, the research community need also to face the issue of implementing these techniques within an optimization framework.

In this context, two related problems appear to be of primary importance. First of all, *scalability* is known as the main challenge not only, e.g., for routing problems, but also for any resource allocation issue in WMNs. Because the impact of WMN is expected to be very pervasive, and it is often assumed that at least hundreds of nodes can be part of the network, we must acknowledge a difficulty in identifying practical algorithms for large networks. This involves the trade-off between exact solutions, whose computational complexity may explode as the number of variables (nodes times interfaces) can be extremely high, and heuristic techniques, which can often manage WMNs with many nodes but are very difficult to validate, because it is hard to tell how far from optimality they are. Moreover, another problem, which still relates to scalability issues, is to identify where the source of the computational capabilities is located, i.e., how to *coordinate* the mesh routers to achieve an efficient allocation. In this sense, another trade-off is involved, namely, centralized vs. distributed management. Centralized solutions can work only if the MAP is powerful enough and the number of nodes is not high, so that global awareness about nodes and channels is possible. Otherwise, distributed solutions should be sought. However, these techniques do not always achieve the same performance than centralized management.

For these reasons, it is key that new research on the topics of routing and interface assignment in multiradio WMNs involves a significant effort to determine efficient optimization techniques with low computational complexity, and also distributed implementations that approach the performance of centralized solutións. Moreover, we recognize the study of *clustered* networks [36] as a possible application of these principles. Aggregating terminals in small clusters that are easy to manage allow a dramatic reduction of the computational complexity. If the network partitioning is performend efficiently, the solution found is still close to the optimal. Finally, clustered managements of WMNs can be seen as an intermediate solution between the fully distributed (but also inefficient) and the fully centralized (with acute computational problems) approaches.

6.7 Conclusions

In this chapter, after having classified existing proposals according to their diverse characteristics, we have highlighted the motivations that suggest the benefit of using traffic aware channel assignment. This point has been further explored by presenting examples on how link load and capacity can be estimated, and this knowledge can be exploited.

We emphasize the importance of the interactions between interface assignment and routing for the capacity performance of multichannel WMNs. Routing and interface assignment can benefit from simple information passing, where the two layers are still separated but cooperating. Moreover, if the terminal capabilities allow for it, one can also think of merging together the related strategies with a cross-layer approach.

To sum up, from a general viewpoint there are strong expectations about multira-dio WMNs providing end users with high network capacity. However, routing and interface assignment, require a careful, and possibly joint, investigation because of their tight interdependencies. Traffic aware algorithms, which offer the opportunity to turn this relationship to an advantage, appear as very promising to make this goal easier to reach.

6.8 Terminologies

1. *Wireless mesh network (WMN)*. It is a communication network, where clients are connected via radio to routers that are in turn interconnected via multi-hop wireless links. Its structure is entirely wireless, thus making WMNs espe-cially applicable where cable deployment is difficult or too expensive. Because the wireless medium is intrinsically broadcast, the radio nodes belonging to the WMN need special procedures to work in harmony with each other and enable dedicated communications.
2. *Mesh router (MR)*. It is a wireless element of a WMN that does not gener-ate traffic but only serves to relay the traffic of the clients and convey it to a gateway (or vice versa). Actually, the structure of a WMN comprises multiple MRs that are interconnected with each other, so as to create a multihop wireless backbone. As communications over the backbone are limited by wireless inter-ference, special techniques can be used to decrease the mutual interference of MRs, such as making them operate on different frequencies.
3. *Mesh access point (MAP)*. It is a special Mesh Router that is also connected to other external networks, e.g., the Internet, typically through a cabled con-nection. It can be therefore considered as a gateway for the network. However, because of the wireless structure, it also becomes a critical point for the rout-ing, because of the so-called *gateway bottleneck* phenomenon. Indeed, the usual congestion caused by the convergence of the routes at the gateway is compli-cated by the fact that, as for any other node, interference can block some of the communications. Thus, its role in the WMN has to be carefully planned.
4. *Network interface card (NIC)*. Also called network adapter, it is the hardware component that enables the communication over the network. It involves both PHY (physical) and MAC layer capability. In particular, we are concerned in this chapter with NICs providing access over a wireless channel. Thus, a node can be supplied with multiple NICs to enable simultaneous communications on different channels, which is a way to avoid wireless interference.
5. *Topology*. Multihop networks are often represented as a graph, where the ver-tices are the MRs and the edges are the communication links among them. In this context, "topology" is often used as a synonym of "graph." However,

when multiple frequencies are introduced, different graph representations (and therefore, different topologies) need to be considered, where the set of vertices, i.e., the MRs, is always the same but the set of edges changes.

6. *Physical topology*. The *physical topology* corresponds to a graph representation of the multichannel WMN where an edge is drawn between two nodes if it is theoretically possible for them to communicate. This requires, of course, that the nodes communicate on the same frequency and wireless interference is absent. The physical topology corresponds to a graph representing the potential connectivity before any channel assignment procedure.

7. *Logical topology*. The *logical topology* depicts the connectivity of the WMN after a specific interface assignment procedure, so that any edge of the logical topology is kept only if the transmitter and the receiver actually share an NIC tuned on the same channel. If all the NICs are tuned on the same channel, the logical topology is equal to the physical topology. However, in general, the logical topology is different, for even multiple links are present between two nodes if more NICs tuned on the same channels can connect them. Alternatively, a link of the physical topology may be absent in the logical topology if there is no pair of NICs at both nodes with the same channel assignment.

8. *Wireless interference models*. According to the most common classifications, wireless interference models fall under two main classes: *protocol* and *physical* interference models. Protocol interference models describe interference as a binary relationship, i.e., two links either interfere or do not interfere with each other. Physical interference models take a more detailed approach with considerations taken from the physical layer. The most common version of the physical model corresponds to evaluate the Signal-to-Interference Ratio at the receiver, and check whether this is above a given threshold describing correct reception. Note that this also allows nonbinary evaluation of interference.

9. *Conflicting link set*. In the protocol interference models, each link $e = (i, j)$ is associated with its conflicting link set $\mathbf{J}(e) = \mathbf{J}(i, j)$, containing all the links whose simultaneous activation with e is forbidden (the protocol interference model describes interference as a binary relationship). In other words, if a transmission is taking place on any link belonging to $\mathbf{J}(e)$, e has to either stay silent or use another frequency, and vice versa. Otherwise, interference will destroy the communication.

10. *Load criticality*. A useful criterion to allocate channels is to exploit frequency diversity to alleviate network congestion. This can be achieved by allocating more different channels to *critical* links of the network. To this end, a possible approach requires at first to estimate the expected load Φ_ℓ for any edge ℓ of the physical topology. To this end, it is possible to use a simple a priori assumption such as uniform distribution of the end-to-end traffic over all possible paths, or perform measurements of the per-hop load. After this evaluation, channels may be assigned to fairly subdivide the expected load over all links, e.g., by minimizing the load on the more critical edge.

6.9 Questions

1. Determine the logical topology for the physical topology shown in the picture below.

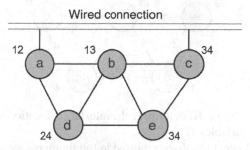

2. Consider the physical topology reported in the figure below. Channel assignment has been performed for all nodes but node b, which has two NICs. How can these two interfaces tuned so that every edge of the physical topology corresponds with at least one edge in the logical topology?

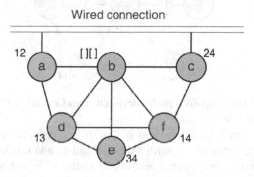

3. Discuss pros and cons of the dynamic channel assignment approach.
4. What is the "gateway bottleneck" and what does it imply, both in terms of limitations and practical approaches?
5. Consider a 7-node physical topology $G_P = (N, E_P)$, i.e., where $|N| = 7$. Assume all nodes have three NICs and the network is fully connected, that is, there is an edge between any two nodes in N. Further, assume all links are symmetric and bi-directional. Determine:

(a) The number of edges $|E_P|$ in the physical topology.
(b) The number of edges $|E_L|$ in the logical topology that results from CCA, i.e., the same channel for all NICs even belonging to the same node.
(c) The number of edges $|E_L|$ in the logical topology that results from a channel assignment procedure imposing the same triplet of different channels (say, (1,2,3)) for the three NICs belonging to any node.
(d) The number of edges $|E_L|$ in the logical topology that results from a channel assignment procedure where five nodes have their NICs set to (1,2,3) and two nodes have their NICs set to (1,2,4).

6. Consider the logical topology reported in the figure below.

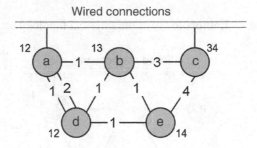

For every $i, j \in \mathbf{N}, q \in \mathbf{K}$, determine the interface allocation variables y_i^q, and the channel edge variables x_{ij}^q.

7. Consider the logical topology reported in the figure below.

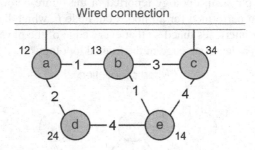

Determine all the loop-free paths between a and e, called $P(a, e)$, and between c and d, called $P(c, d)$.

8. Consider the same logical topology of Question 7. Assume two flows are present in the network: from a to e, with expected end-to-end traffic $\gamma^{(a,e)} = 1.8$ Mbps, and from c to d, with expected end-to-end traffic $\gamma^{(c,d)} = 1.5$ Mbps.
   According to the load criticality method with uniform traffic repartition over all paths (see Sect. 6.5.1), determine the expected load on each of the links below.

9. Consider a pair of nodes i, j whose conflicting set $\mathbf{J}(i, j)$ includes, beyond (i, j), the following edges of the physical topology: $e_1, e_2, e_3, e_4, e_5, e_6$. In the logical topology e_1, e_2, e_3 are tuned on channel 1, e_4, e_5 are tuned on channel 2, and e_6 is tuned on both. Assume that $c_{xy}^{(P)} = 10$ Mbps for any x, y.
   Traffic is 2.0 Mbps between i and j, and as reported below on edges e_k.

Index k	1	2	3	4	5	6
Load of e_k on channel 1	3.0	1.2	0.8	0	0	1.0
Load of e_k on channel 2	0	0	0	2.4	1.1	2.0

Assuming fair bandwidth share, determine $b_{ij}^{(q)}$ for $q = 1, 2$ in the following cases:

(a) Nodes i and j share one NIC assignment on channel 1.
(b) Nodes i and j share one NIC assignment on channel 2.
(c) Nodes i and j share two NIC assignments on both channels 1 and 2, and the traffic is equally split between the resulting two links in the logical topology.

10. Consider the same setup of Question 9 (point (c)) but now assume we want to take the objective of optimal utilization into account, as per (6.11). Assume link (i, j) is the most critical of the network. How should its traffic be split between channels 1 and 2?

References

1. I.F. Akyildiz, X. Wang, and W. Wang, Wireless mesh networks: a survey. Comput. Networks (Elsevier) 47(4): 445–487 (2005).
2. R. Bruno, M. Conti, and E. Gregori, Mesh networks: commodity multihop ad hoc networks. IEEE Commun. 43(3): 123–131 (2005).
3. X. Hong, K. Xu, and M. Gerla, Scalable routing protocols for mobile ad hoc networks. IEEE Network 16(4): 11–21 (2002).
4. Wireless LAN medium access control (MAC) and physical layer (PHY) specification", IEEE Std. 802.11, (1997).
5. S. Basagni, M. Conti, S. Giordano, and I. Stojmenovíc, Eds., Mobile Ad Hoc Networking. IEEE/Wiley, New York, NY (2004).
6. A. Mishra, V. Shrivastava, S. Banerjee, and W. Arbaugh, Partially overlapped channels not considered harmful. Proc. ACM SIGMetrics Perform., 63–74 (2006).
7. D. Banerjee and B. Mukherjee, A practical approach for routing and wavelength assignment in largewavelength-routed optical networks. IEEE J. Select. Areas Commun. 14(5): 903–908 (1996).
8. P. Kyasanur and N.H. Vaidya, Capacity of multichannel wireless networks: impact of number of channels and interfaces. Proc. ACM MobiCom, 43–57 (2005).
9. D.B. Johnson, D.A. Maltz, and J. Broch, DSR: The dynamic source routing protocol for multihop wireless ad hoc networks. In: Perkins C.E. (Ed.) Ad Hoc Networking. Addison-Wesley, London, Chap. 5, 139–172 (2001).
10. C. Perkins, E. Belding-Royer, and S.R. Das Ad Hoc On-Demand Distance Vector (AODV) Routing. IETF RFC 3561 (2003).
11. R. Draves, J. Padhye, and B. Zill Routing in multi-radio, multi-hop wireless mesh networks. Proc. ACM MobiCom, 114–128 (2004).
12. A.H. Mohsenian Rad and V.W.S. Wong Joint optimal channel allocation, interface assignment, and MAC design for multi-channel wireless mesh networks. Proc. IEEE INFOCOM, 1469–1477 (2007).
13. A. Raniwala, K. Gopalan, and T. Chiueh Centralized channel assignment and routing algorithms for multi-channel wireless mesh networks. ACM Mobile Comput. Commun. Rev. (MC^2R) 8(2): 50–65 (2004).
14. I. Katzela and M. Naghshineh Channel assignment schemes for cellular mobile telecommunicationsystems: a comprehensive survey. IEEE Pers. Commun. 3(3): 10–31 (1996).
15. A.H. Mohsenian Rad and V.W.S. Wong, Partially overlapped channel assignment for multichannel wireless mesh networks. Proc. IEEE ICC, 3770–3775 (2007).
16. I. Cidon and M. Sidi, Distributed assignment algorithms for multihop packet radionetworks. IEEE Trans. Comput. 38(10): 1353–1361, (1989).
17. P. Kyasanur and N.H. Vaidya, Routing and interface assignment in multi-channel multi-interface wireless networks. Proc. IEEE WCNC 4: 2051–2056 (2005).

18. M. Alicherry, R. Bhatia, and L.E. Li, Joint channel assignment and routing for throughput optimization in multiradio wireless mesh networks. IEEE J. Select. Areas Commun. 24(11): 1960–1971 (2006).
19. Y.Y. Chen, S.C. Liu, and C. Chen, Channel assignment and routing for multi-channel wireless mesh networks using simulated annealing. Proc. of IEEE Globecom, 1–5 (2006).
20. X. Meng, K. Tan, and Q. Zhang, Joint routing and channel assignment in multi-radio wireless mesh networks. Proc. IEEE ICC 8: 3596–3601 (2006).
21. M. Kodialam and T. Nandagopal, Characterizing the capacity region in multi-radio multi-channel wireless mesh networks. Proc. ACM MobiCom, 73–87 (2005).
22. P. Gupta and P.R. Kumar, The capacity of wireless networks. IEEE Trans. Inform. Theory 46(8): 388–404 (2000).
23. G. Brar, D. Blough, and P. Santi, Computationally efficient scheduling with the physical interference model for throughput improvement in wireless mesh networks. Proc. ACM MobiCom, 2–13 (2006).
24. S. Kirkpatrick, C.D. Gelatt, and M.P. Vecchi, Optimization by simulated annealing. Science 220: 671–680 (1983).
25. M.K. Marina and S.R. Das A topology control approach for utilizing multiple channels in multi-radio wireless mesh networks. Proc. Int. Conf. Broadband Netw., 381–390 (2005).
26. H. Skalli, S. Ghosh, S.K. Das, L. Lenzini, and M. Conti Channel assignment strategies for multi-radio wireless mesh networks: issues and solutions. IEEE Commun. Mag. 45(11): 86–95 (2007).
27. J. So and N.H. Vaidya Multi-channel MAC for ad hoc networks: handling multi-channel hidden terminals using a single transceiver. Proc. ACM MobiHoc, 222–233 (2004).
28. P. Bahl, R. Chandra, and J. Dunagan, SSCH: Slotted seeded channel hopping for capacity improvement in IEEE 802.11 ad-hoc wireless networks. Proc. ACM Mobicom, 216–230 (2004).
29. A. Raniwala and T. Chiueh, Architecture and algorithms for an IEEE 802.11-based multi-channel wireless mesh network. Proc. IEEE INFOCOM, 2223–2234 (2005).
30. K. Ramachandran, K. Almeroth, E. Belding-Royer, and M. Buddhikot, Interference aware channel assignment in multi-radio wireless mesh networks. Proc. IEEE INFOCOM, 1–12 (2006).
31. J. Jun and M.L. Sichitiu, The nominal capacity of wireless mesh networks. IEEE Wireless Commun. 10(5): 8–14 (2003).
32. D. Kotz, C. Newport, R.S. Gray, J. Liu, Y. Yuan, and C. Elliott, Experimental evaluation of wireless simulation assumptions. Proc. ACM MSWiM, 78–82 (2004).
33. M.K. Marina and S.R. Das, Routing performance in the presence of unidirectional links in multihop wireless networks. Proc. ACM MobiHoc, 12–23 (2002).
34. L. Badia, A. Erta, L. Lenzini, and M. Zorzi, A general interference-aware framework for joint routing and link scheduling in wireless mesh networks. IEEE Netw. 22(1): 32–38 (2008).
35. A. Iyer, C. Rosenberg, and A. Karnik, What is the right model for wireless channel interference?. Proc. Qshine, No. 2, invited paper (2006).
36. L. Badia, N. Bui, M. Miozzo, M. Rossi, and M. Zorzi, On the exploitation of user aggregation strategies in heterogeneous wireless networks. Proc. of the 11th IEEE Int. Workshop CAMAD, 8–15 (2007).

Chapter 7
Wireless Mesh Network Routing Under Uncertain Demands

Jonathan Wellons, Liang Dai, Bin Chang, and Yuan Xue

Abstract Traffic routing plays a critical role in determining the performance of a wireless mesh network. Recent research results usually fall into two ends of the spectrum. On one end are the heuristic routing algorithms, which are highly adaptive to the dynamic environments of wireless networks yet lack the analytical properties of how well the network performs globally. On the other end are the optimal routing algorithms that are derived from the optimization problem formulation of mesh network routing. They can usually claim analytical properties such as resource use optimality and throughput fairness. However, traffic demand is usually implicitly assumed as static and known a priori in these problem formulations. In contrast, recent studies of wireless network traces show that the traffic demand, even being aggregated at access points, is highly dynamic and hard to estimate. Thus, to apply the optimization-based routing solution in practice, one must take into account the dynamic and uncertain nature of wireless traffic demand. There are two basic approaches to address the traffic uncertainty in optimal mesh network routing (1) predictive routing that infers the traffic demand with maximum possibility based in its history and optimizes the routing strategy based on the predicted traffic demand and (2) oblivious routing that considers all the possible traffic demands and selects the routing strategy where the worst-case network performance could be optimized. This chapter provides an overview of the optimal routing strategies for wireless mesh networks with a focus on the above two strategies that explicitly consider the traffic uncertainty. It also identifies the key factors that affect the performance of each routing strategy and provides guidelines towards the strategy selection in mesh network routing under uncertain traffic demands.

J. Wellons (✉)
Department of Computer Science, Vanderbilt University, VU Station B 351824, Nashville, TN 37235, USA
e-mail: jonathan.wellons@vanderbilt.edu

S. Misra et al. (eds.), *Guide to Wireless Mesh Networks*, Computer Communications and Networks, DOI 10.1007/978-1-84800-909-7_7,

7.1 Introduction

Wireless mesh networks [1, 2], which now offer a rapid and inexpensive solution to last-mile broadband Internet access, are attracting ever greater attention and widespread deployment. A wireless mesh network is composed of local access points and wireless mesh routers, which form an organic backbone structure that forwards traffic between mobile clients and the Internet.

Traffic routing plays a critical role in determining the performance of a wireless mesh network. Thus it has attracted extensive recent research. The key challenges come from the scarce wireless channel resource, high dynamic link quality, and the uncertain traffic demands. The proposed approaches address these challenges in different ways. On one end of the spectrum are the heuristic algorithms [3–6]. Although many of them are adaptive to the dynamic environments of wireless networks, these algorithms lack the analytical properties of how well the network performs globally (e.g., whether the scarce channel resource is shared in an optimal and fair fashion). On the other end of the spectrum, there are theoretical studies that formulate mesh network routing as optimization problems [7, 8]. The routing algorithms derived from these optimization formulations can usually claim analytical properties such as resource use optimality and throughput fairness. In these optimization frameworks, traffic demand is usually implicitly assumed as static and known a priori. Recent studies of wireless network traces [9], however, show that the traffic demand, even being aggregated at access points, is highly dynamic and hard to estimate. Such observations have significantly challenged the practicability of the existing optimization-based routing solutions in wireless mesh networks.

There are two basic approaches to address the traffic uncertainty in optimal mesh network routing:

- *Predictive routing* [10,11], which infers the traffic demand with maximum probability based in its history and optimizes the routing strategy based on the predicted traffic demand. Underlying predictive routing is the assumption that past behavior is a good indicator of the future.
- *Oblivious routing* [12], which makes no assumption on traffic demand and considers all the possible traffic demands and selects the routing strategy where the worst-case network performance is optimized.

This chapter provides an overview of optimal routing algorithms for wireless mesh networks: optimal routing under fixed demand, predictive routing, and oblivious routing based primarily on several recent works [10–12]. It focuses on the latter two routing strategies and shows how they explicitly consider the traffic uncertainty in their problem formulation and algorithm design. In this chapter, we also identify the key factors that affect the performance of each routing strategy and provide guidelines towards the strategy selection in mesh network routing under uncertain traffic demands.

The remainder of this chapter is organized as follows. Section 7.2 presents the network and system model. Section 7.3 reviews the background knowledge, i.e., the optimal routing algorithm under fixed demand. Section 7.4 presents the predictive mesh network routing strategy. Section 7.5 presents the oblivious mesh network routing formulation and algorithm. Section 7.6 evaluates and compares the performance of different routing strategies. Sections 7.7 and 7.8 provide thoughts for practitioners and the directions for future research. Finally, Section 7.9 concludes the chapter.

7.2 Model

7.2.1 Network and Interference Model

In a multihop wireless mesh network, local access points aggregate and forward traffic for the mobile clients that are associated with them. They communicate with each other and with the stationary wireless routers to form a multihop *backbone* network, which forwards the user traffic to the Internet gateways. Figure 7.1 shows an example of wireless mesh network. We use $w \in W$ to denote the set of gateways in the network and $s \in S$ to denote the set of local access points that generate traffic in the network. Local access points, gateways and mesh routers are collectively called mesh nodes and denoted by the set V.

In a wireless network, packet transmissions are subject to location-dependent interference. Here we consider the *protocol model* presented in [13]. We assume that all mesh nodes have the uniform transmission range denoted by R_T. Usually the interference range is larger than its transmission range, which is denoted as

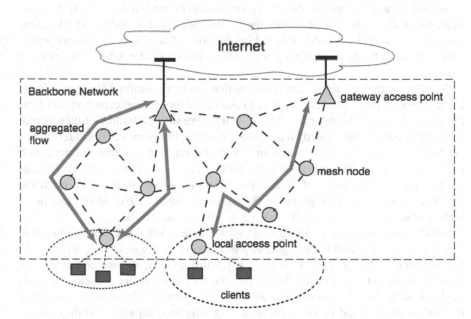

Fig. 7.1 Illustration of wireless mesh network

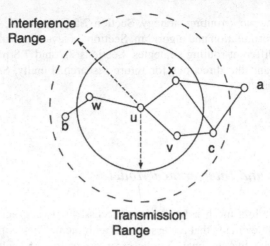

Fig. 7.2 Transmission and interference network

$R_I = (1 + \Delta)R_T$, where $\Delta \geq 0$ is a constant. For simplicity, in this chapter we assume that each node is equipped with one radio interface, which operates on the same wireless channel as others. Let $r(u, v)$ be the distance between two nodes u and v (u, $v \in V$). In the protocol model, packet transmission from node u to v is successful, if and only if (1) the distance between these two nodes $r(u, v)$ satisfies $r(u, v) \leq R_T$ and (2) any other node $x \in V$ within the interference range of the receiving node v, i.e., $r(x, v) \leq R_I$, is not transmitting. If node u can transmit to v directly, they form an edge $e = (u, v)$. As an example shown in Fig. 7.2, nodes w, x, v are within the transmission range of node u, thus they can transmit the node u directly. At the same time, nodes w, v, x, b, c are all within the interference range of node u, which means the signal from node u could be heard by any node of w, v, x, b, c, and vice versa. Thus they must be silenced, if they are not the intended sender, when u is receiving a packet.

We assume that the maximum data rate that can be transmitted along an edge is the same for all edges, and denote it as c (also called the channel capacity). Let E be the set of all edges. We say two edges e, e' interfere with each other, if they cannot transmit simultaneously based on the protocol model. Further we define *interference set $I(e)$* of an edge e as the union of e with the set of edges that interfere with e. Figure 7.3 is an illustration of the interference set of edge (u, v). The circles are the interference ranges of node u and v, and the union of these two circles is the interference range of edge (u, v). So the interference set $I(u, v)$ of edge (u, v) includes (u, v), (a, b), (v, b), (v, a), (a, u), (x, u), and (x, y).

Finally, we introduce a virtual node w^* to represent the Internet, as shown in Fig. 7.4 w^* is connected to each gateway with a virtual edge $e^* = (w^*, w)$, $w \in W$. Further, let $E' = E \cup \{e^*\}$ and $V' = V \cup \{w^*\}$. For simplicity, we assume that the link capacity in the Internet is much larger than the wireless channel capacity, and thus the bottleneck always appears in the wireless mesh network. Under this assumption, the virtual edges could be regarded as having unlimited capacity and they do not interfere with any of the wireless transmissions.

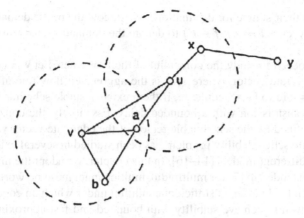

Fig. 7.3 Interference set

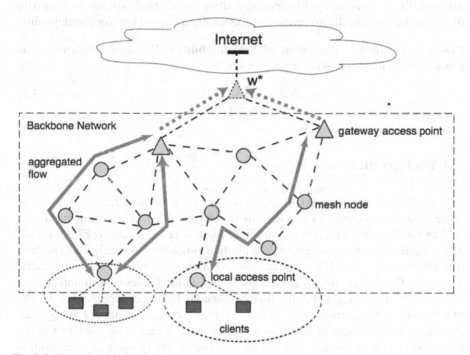

Fig. 7.4 Illustration of virtual node

7.2.2 Traffic Model and Schedulability

This chapter studies the routing strategies for wireless mesh *backbone* networks. Thus it only considers the aggregated traffic between the local access points and the Internet gateways. Here we call the aggregated traffic in (or out) a local access point a *flow* and denote it as $f \in F$, where F is the set of all aggregated flows. All flows will take w^* as their source (or destination). We denote the traffic demand of flow f as d_f and use vector $\boldsymbol{d} = (d_f, f \in F)$ to denote the demand vector consisting of all flow demands.

Now we proceed to study the constraint of the flow rates. Let $\boldsymbol{y} = (y(e), e \in E)$ denote the edge rate vector, where $y(e)$ is the aggregated flow rate along e. Edge rate vector $\boldsymbol{y}$ is said to be schedulable, if there exists a stable schedule that ensures every packet transmission with a bounded delay. Essentially, the constraint of the flow rates is defined by the schedulable region of the edge rate vector $\boldsymbol{y}$.

The edge rate schedulability problem has been studied in several existing works, which lead to different models [14–16]. In this chapter, we adopt the model in [15], which is also extended in [7] for multiradio, multichannel mesh network. In particular, Kumar et al. [15] present a sufficient condition under which an edge scheduling algorithm is given to achieve stability with bounded and fast approximation of an ideal schedule. Alicherry et al. [7] present a scheme that can adjusts the flow routes and scale the flow rates to yield a feasible routing and channel assignment. Based on these results, we have the following claim as a sufficient condition for schedulability.

Claim 7.1 (Sufficient Condition of Schedulability). *The edge rate vector* $\boldsymbol{y}$ *is schedulable if the following condition is satisfied*:

$$\forall e \in E, \sum_{e' \in I(e)} y(e') \leq c. \tag{7.1}$$

7.3 Background

This section provides the background of optimal mesh network routing, introduces its problem formulations, and reviews its algorithm under fixed traffic demand.

The existing works on optimal multihop wireless network routing [7, 8, 14] usually formulate it as a throughput optimization problem that maximizes the flow throughput, while satisfying the fairness constraints. In this formulation, traffic demand is fixed and reflected as the flow weight in the fairness constraints. Recall that $f \in F$ is the aggregated traffic flow between the local access points and the virtual gateway (i.e., Internet) and $\boldsymbol{d} = (d_f, f \in F)$ is the demand vector consisting of all flow demands. Consider the fairness constraint that, for each flow f, its throughput being routed is in proportion to its demand d_f. The goal of throughput maximization routing is to maximize λ (so-called *scaling factor*) where at least $\lambda \cdot d_f$ amount of throughput can be routed for flow f.

To balance the traffic load, flow f could be routed over multiple paths, let $\mathcal{P}_f$ be the set of unicast paths that could route flow f, and $x_f(P)$ be the rate of flow f over path $P \in \mathcal{P}_f$. Obviously the aggregated flow rate y_e along edge $e \in E$ is given by $y_e = \sum_f$, where f is taken over all paths $\in \mathcal{P}_f$ which contain the edge e. Based on the sufficient condition of schedulability in Claim 7.1 (7.1), we have that

$$\sum_{e' \in I_e} \sum_{f:P \in \mathcal{P}_f \& e' \in P} x_f(P) \leq c. \tag{7.2}$$

To simplify the above equation, we define $A_{eP} = |I_e \cap P|$ as the number of wireless links path P passes in the interference set I_e. The throughput optimization routing with fairness constraint is then formulated as the following linear programming (LP) problem:

$$P_T : \text{maximize } \lambda \tag{7.3}$$

$$\text{subject to } \sum_{P \in \mathcal{P}_f} x_f(P) \geq \lambda \cdot d_f, \forall f \in F \tag{7.4}$$

$$\sum_{f \in F} \sum_{P \in \mathcal{P}_f} x_f(P) A_{eP} \leq c, \forall e \in E \tag{7.5}$$

$$\lambda \geq 0, x_f(P) \geq 0, \forall f \in F, \forall P \in \mathcal{P}_f. \tag{7.6}$$

In this problem, the optimization objective is to maximize λ, such that at least $\lambda \cdot d_f$ units of data can be routed for each aggregated flow f with demand d_f. Inequality (7.4) enforces fairness by requiring that the comparative ratio of traffic routed for different flows satisfies the comparative ratio of their demands. Inequality (7.5) enforces the capacity constraint by requiring the traffic aggregation of all flows passing wireless link $e \in E$ satisfy the sufficient condition of schedulability. This problem formulation follows the classical maximum concurrent flow problem.

Although the above throughput maximization routing problem formulation is widely used in designing optimal mesh network routing strategies under known demands, it is not suitable to study the routing performance under dynamic and uncertain traffic demand. Here we consider a formulation based on another routing performance metric – network *congestion* (or *use*). In the Internet, *link use* is commonly used for traffic engineering [17], whose objective is to minimize the use at the most congested link under given traffic demand. However, link use cannot be straightforwardly applied to multihop wireless networks, such as mesh backbone network, as a metric of routing performance because of the location-dependent interference. In what follows, we define the *network congestion based on the use of the interference set* as the routing performance metric and outline the relation between the formulation of the throughput optimization problem and the congestion minimization problem.

Let $x'_f(P)$ be the rate of flow f on path P under traffic demand d_f. It is obvious that $\sum_{P \in \mathcal{P}_f} x'_f(P) = d_f$. The traffic being routed within the interference set I_e is then given by $\sum_{f \in F} \sum_{P \in \mathcal{P}_f} x'_f(P) \cdot A_{eP}$. Formally, the *congestion* of an interference set I_e is defined as its *use* (i.e., the ratio between its load and the channel capacity) and denote it as θ_e

$$\theta_e = \frac{\sum_{f \in F} P \in \mathcal{P}_f x'_f(P) A_{eP}}{c}. \tag{7.7}$$

Further, the *network congestion* is defined as the maximum congestion among all the interference sets, i.e.,

$$\theta = \max_{e \in E} \theta_e. \tag{7.8}$$

The network congestion minimization routing problem is then formulated as follows:

$$P_C : \text{minimize } \theta \tag{7.9}$$

$$\text{subject to } \sum_{P \in \mathcal{P}_f} x'_f(P) \geq d_f, \forall f \in F \tag{7.10}$$

$$\sum_{f \in F} \sum_{P \in \mathcal{P}_f} x'_f(P) A_{eP} \leq c \cdot \theta, \forall e \in E \tag{7.11}$$

$$\theta \geq 0, x'_f(P) \geq 0, \forall f \in F, \forall P \in \mathcal{P}_f. \tag{7.12}$$

To reveal the relation between P_T and P_C, we let $\theta = 1/\lambda$ and $x'_f(P) = x_f(P)/\lambda$. Problem P_C is then transformed to

$$P'_C : \text{minimize } \frac{1}{\lambda} \tag{7.13}$$

$$\text{subject to } \frac{1}{\lambda} \sum_{P \in \mathcal{P}_f} x_f(P) \geq d_f, \forall f \in F \tag{7.14}$$

$$\frac{1}{\lambda} \sum_{f \in F} \sum_{P \in \mathcal{P}_f} x'_f(P) A_{eP} \leq c \cdot \theta, \forall e \in E \tag{7.15}$$

$$\lambda \geq 0, x'_f(P) \geq 0, \forall f \in F, \forall P \in \mathcal{P}_f, \tag{7.16}$$

which is obviously equivalent to the throughput optimization problem P_T.

If the demand vector d is known, both problem P_T and P_C could be solved by a LP-solver such as [18, 19]. To reduce the complexity for practical use, the work of [10] also presents a fully polynomial time approximation algorithm for problem $\{P_T\}$, which finds a ε-approximate solution. The key to a fast approximation algorithm lies on the dual of this problem, which is formulated as follows. First, we assign a price $\mu \mu_e$ to each set S_e for $e \in E$. The objective is to minimize the aggregated price for all interference sets. As the constraint, Inequality (7.18) requires that the price $\sum_{e \in E} A_{eP} \mu_e$ of any path $P \in \mathcal{P}_f$ for flow f must be at least μ_f, the price of flow f. Further, Inequality (7.19) requires that the weighted flow price μ_f over its demand d_f must be at least 1.

$$D_T : \text{mininmize } \sum_{e \in E} c \cdot \mu_e \tag{7.17}$$

$$\text{subject to } \sum_{e \in E} A_e P \mu_e \geq \mu_f, \forall f \in F, \forall P \in \mathcal{P}_f \tag{7.18}$$

$$\sum_{f \in F} \mu_f d_f \geq 1. \tag{7.19}$$

Table 7.1 Routing algorithm under fixed demand

Mesh Network Routing Under Fixed Demand
1   $\forall e \in E, \mu_e \leftarrow \beta/c$
2   $x_f(P) \leftarrow 0, \forall P \in \mathcal{P}_f, \forall f \in F$
3   while $\sum_{e \in E} c \cdot \mu_e < 1$
4   for $\forall f \in F$ do
5       $d'_f \leftarrow d_f$
6   while $\sum_{e \in E} c \cdot \mu_e < 1$ and $d'_f > 0$ do
7         $P \leftarrow$ lowest priced path in $\mathcal{P}_f$ using μ_e
8         $\delta \leftarrow \min\{d'_f, \min_{e \in P} A_{eP}\}$
9         $d'_f \leftarrow d'_f - \delta$
10         $x_f(P) \leftarrow x_f(P) + \delta$
11         $\forall e$ s.t. $A_{eP} \neq 0, \mu_e \leftarrow \mu_e(1 + \epsilon \delta A_{eP})$
12   end while
13   end for
14   end for

Based on the above dual problem D_T, the fast approximation algorithm is presented in Table 7.1. The properties of this algorithm are shown as follows.

Property 7.1. If $\beta = (|E|/(1 - \varepsilon))^{-1/\varepsilon}$, then the final flow generated by FMR is at least $(1-3\varepsilon)$ times the optimal value of P. The running time is $O(\frac{1}{\varepsilon^2}[\log|E|$ $(2|D||T_{\text{fmr}}||F|\log|F| + |E| + \log U)]) \cdot T_{\text{mp}}$, where U is the length of the longest path in G and T_{mp} is the running time to find the shortest path.

7.4 Predictive Mesh Network Routing

The predictive mesh network routing is based on a two-tier framework as shown in Fig. 7.5, which integrates traffic modeling and routing optimization.

- *Traffic modeling* derives the traffic model of a wireless mesh network. The model should be dependable at characterizing the long term traffic demand, yet agile at containing the uncertain traffic dynamics in the short term. The traffic modeling component needs to produce traffic demand estimations as inputs to the network optimization component.
- *Routing optimization* determines the routing strategy that distributes the traffic along different routes so that minimum congestion will be incurred even under dynamic traffic. To achieve this goal, the routing optimization decision should effectively take into account the traffic demand estimation results from the traffic modeling component.

7.4.1 Traffic Prediction

First we study the dynamic behavior of aggregated traffic at access points. Our goal is to (1) develop a reliable prediction method that is able to estimate the aggregated

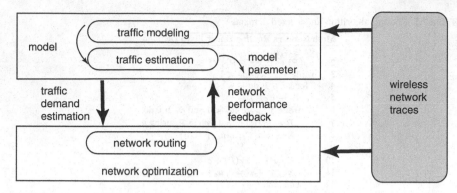

Fig. 7.5 Integrated framework of traffic modeling and network optimization

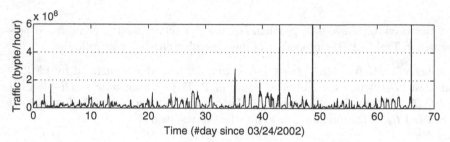

Fig. 7.6 Incoming traffic time series of ResBldg97AP3 (March 24, 11 P.M., 2002–June 9, 10 P.M., 2002)

traffic demand of an access point based on its historical data and (2) develop a statistical model to characterize the prediction errors. The predicted traffic demand will serve as the input of predictive mesh network routing algorithms, which will be presented Sect. 7.4.2.

To develop such a traffic demand model, we study the traces collected at the campus wireless LAN network of Dartmouth College in Spring 2002 [20]. By analyzing the *snmp* log from each access point, we derive the dynamic behavior of aggregated traffic demand. To illustrate our analysis procedure, we choose one of the access points (ResBldg97AP3) as an example. The time series of its incoming traffic is plotted in Fig. 7.6. From the figure, we can easily observe that (1) the traffic demand is nonstationary over large time scales because of the diurnal and weekly working cycles; (2) compared with the traffic behavior in the backbone Internet [17], the traffic at an access point is significantly bursty because of the insufficient level of multiplexing. The above observations are consistent with the findings in [9].

The first step of our analysis is to identify and remove the daily and weekly cyclic patterns in the time series. This requires us to calculate the weekly/daily cyclic average. Formally, let us denote $x(t)$ as the *raw traffic series*. We estimate the moving average of this series based on the same time of the same day of the week, i.e.,

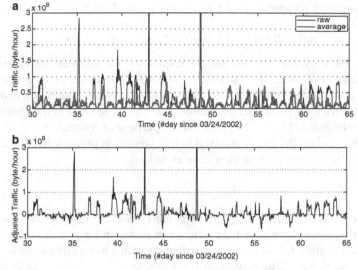

Fig. 7.7 Traffic series in 5 weeks

$$\bar{x}(t) = \sum_{i=1}^{W} x(t - 24 \times 7 \times i)/W, \qquad (7.20)$$

where W is the size of moving window. To eliminate the effect of bursty traffic, we also filter out the spike traffic during the above averaging procedure. Figure 7.7a plots the raw traffic as well as its moving average with $W = 5$. By removing the cyclic effect from the raw data, we derive the *adjusted traffic series* $y(t)$ as follows.

$$y(t) = x(t) - \bar{x}(t). \qquad (7.21)$$

The adjusted series of the one shown in Fig. 7.7a is given in Fig. 7.7b. This adjusted traffic exhibits short-term (a few hours) traffic correlations. We model the adjusted traffic series with an autoregressive process as follows.[1]

$$y(t) = \beta_1 y(t-1) + \beta_2 y(t-2) + \cdots + \beta_K y(t-K) + c, \qquad (7.22)$$

where K is the process order. To apply this model for prediction, we estimate the parameters of this process. Given N observations $y_1, y_2, \ldots, y_n$, the parameters $\beta_1, \ldots, \beta_k$ are estimated via least squares by minimizing

$$\sum_{t=K+1}^{N} [y(t) - \beta_1 y(t-1) - \cdots - \beta_K y(t-K)]^2. \qquad (7.23)$$

---

[1] Ideally, $y(t)$ should have zero mean. In some cases, $y(t)$ has a small mean value, which needs to be removed before fitting an autoregressive process.

Based on these parameters, we further derive the adjusted traffic prediction $\hat{y}(t)$ as follows:

$$\hat{y}(t) = \beta_1 y(t-1) + \beta_2 y(t-2) + \cdots + \beta_K y(t-K). \tag{7.24}$$

Figure 7.8 illustrates the estimation results for the adjusted traffic series in Fig. 7.7b, where $K = 2$, $\beta_1 = 0.531$, $\beta_2 = 0.469$. The figure plots the predicted series for the adjusted traffic as well as its raw data. In this figure, the number of observations used for parameter estimation is $N = 60$. The fitted traffic series is also plotted for the interval [720, 790] hour for the purpose of comparison.

We now consider the errors involved in this prediction process. In particular, we define the adjusted traffic prediction error as follows.

$$\varepsilon_y(t) = y(t) - \hat{y}(t). \tag{7.25}$$

Based on this definition, Fig. 7.9a plots the cumulative distribution function of the prediction error of the adjusted traffic series shown in Fig. 7.8. It is obvious that the error distribution follows normal distribution with a mean close to zero.

Finally, we define traffic prediction $\hat{x}(t)$ follows:

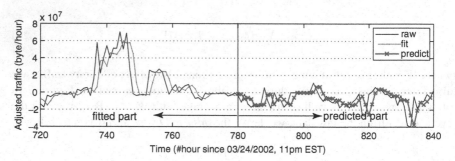

Fig. 7.8 Adjusted traffic and its prediction

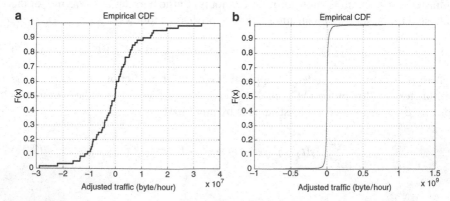

Fig. 7.9 Cumulative density function of prediction error (**a**) Prediction error for adjusted traffic and (**b**) Prediction error for entire series

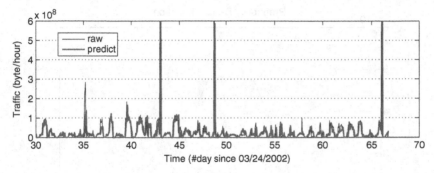

Fig. 7.10 Raw traffic vs. predicted traffic

$$\hat{x}(t) = [\bar{x}(t) + \hat{y}]^{+}. \tag{7.26}$$

Figure 7.10 plots the predicted traffic series $\hat{x}(t)$ in comparison with the raw traffic. We can see the predicted traffic closely matches the real (raw) traffic. The cumulative distribution function of the prediction error $\varepsilon_x(t)$, which is defined as $\varepsilon_x(t) = x(t) - \hat{x}(t)$, is plotted in Fig. 7.9b. It clearly shows that this distribution also follows normal distribution with a near-zero mean.

Thus we could consider the traffic demand at time t as a random variable $X(t)$ that follows normal distribution with mean $\hat{x}(t)$ and the same variance as $\varepsilon_x(t)$. Figure 7.11 shows an example distribution of the predicted traffic demand of the 976th hour.

To summarize, the presented prediction method provides two prediction models: mean value and statistical distribution. These two traffic prediction models will serve as the inputs for predictive routing algorithms, which are presented in Sect. 7.4.2.

7.4.2 Predictive Routing Optimization

There are two predictive routing algorithms [11] – one takes the mean value of the predicted traffic demand as input, the other takes the statistical distribution of the predicted traffic demand as input.

The mean-value predictive routing algorithm is a natural integration of the optimal routing algorithm under fixed demand (Sect. 7.3), where the traffic demand d_f at time t takes the mean value of the predicted traffic demand $\hat{x}(t)$. In what follows, we will focus on the statistical-distribution predictive routing.

First, we model the traffic demand of an aggregated flow $f \in F$ using a random variable D_f, which follows the following discrete probability distribution

$$Pr(D_f = d_f^i) = q_f^i, \tag{7.27}$$

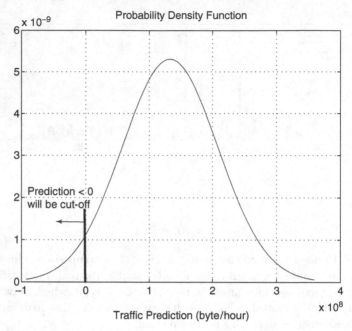

Fig. 7.11 Predicted traffic distribution

where $D_f = \{d_f^1, d_f^2, \ldots, d_f^m\}$ is the set of values for D_f with nonzero probabilities. Let $d = (d_f, d_f \in D_f, f \in F)$ be a sample traffic demand vector, D be the corresponding random variable, and $\mathcal{D}$ be the sample space. We further assume that the demand from different access points are independent from each other. Thus the distribution of D is given by the joint distribution of these random variables as follows.

$$Pr(D = d) = Pr(D_f = d_f^i, f \in F) = \Pi_{f \in F} q_f^i. \tag{7.28}$$

Let us consider a traffic routing solution $(x_f(P), P \in \mathcal{P}_f, f \in F)$ that satisfies the capacity constraint (Inequality (7.5)). It is obvious that λ is a function of d:

$$\lambda(d) = \min_{f \in F} \left\{ \frac{x_f}{d_f} \right\}, \tag{7.29}$$

where $x_f = \sum_{P \subset \mathcal{P}_f} x_f(P)$. Further let us consider the optimal routing solution under demand vector d. Such a solution could be easily derived based on Algorithm I shown in Table 7.1. We denote the optimal value of λ as $\lambda^*(d)$. We further define the *performance ratio* ω of routing solution $(x_f(P), P \in P_f, f \in F)$ as follows

$$\omega(d) = \frac{\lambda(d)}{\lambda^*(d)}.$$

Obviously, the performance ratio is also a random variable under uncertain demand. We denote it as Ω. Ω is a function of random variable D. Now we extend the wireless mesh network routing problem to handle such uncertain demand. Our goal is to maximize the expected value of Ω, which is given as follows

$$E(\Omega) = Pr(D = d) \times \frac{\lambda(d)}{\lambda^*(d)}. \tag{7.30}$$

We abbreviate $Pr(D = d)$ as $p(d)$. It is obvious that $\sum_{d \in D} p(d) = 1$. Formally, we formulate the throughput optimization routing problem for wireless mesh backbone network under uncertain traffic demand as follows.

$$P_U : \text{maximize} \sum_{d \in D} p(d) \frac{\lambda(d)}{\lambda^*(d)} \tag{7.31}$$

subject to $\forall d \in D$, where $d - (d_f, f \in F)$

$$\sum_{P \in P_f} x_f(P) \geq \lambda(d) \cdot d_f, \forall f \in F \tag{7.32}$$

$$\sum_{f \in F} \sum_{P \in P_f} x_f(P) A_{eP} \leq c, \forall e \in E \tag{7.33}$$

$$\lambda \geq 0, x_f(P) \geq 0, \forall f \in F, \forall P \in P_f. \tag{7.34}$$

Similar to problem P_T, the constraints of P_U come from the fairness requirement and the wireless mesh network capacity. In particular, Inequality (7.32) enforces fairness for all demand $d \in D$, and Inequality (7.33) enforces capacity constraint as Inequality (7.5) in problem P_T.

Now we consider the dual problem D_U of P_U. Similar to D_T, the objective of D_U is to minimize the aggregated price for all interference sets. However, in Inequality (7.37), for each sample demand vector d, the aggregated price of all flows weighted by their demand needs to be larger than its probability.

$$D_U : \text{maximize} \sum_{e \in E} c \cdot \mu_e \tag{7.35}$$

subject to $\sum_{e \in E} A_{eP} \mu_e \geq \mu_f, \forall f \in F, \forall P \in P_f \tag{7.36}$

$$\sum_{f \in F} \mu_f d_f \geq \frac{p(d)}{\lambda^*(d)}, \forall d \in \mathcal{D}, \quad \text{where } d = (d_f, f \in F). \tag{7.37}$$

Now we present an approximation algorithm for P_U in Table 7.2. Note that because the channel capacity c will not affect the final result of the algorithm, we simply omit it here. In the work of [11], we are able to prove the following properties with this algorithm.

Property 7.1. If $\beta = (|E|/(1-\varepsilon))^{-1/\varepsilon}$, then the final flow generated by the above algorithm is at least $(1-3\varepsilon)$ times the optimal value of P_U. The running time is

$$O\left(\frac{1}{\varepsilon^2} [\log |E| (2|D| |T_{\text{fmr}}| |F| \log |F| + |E| + \log U)]\right) \cdot T_{\text{mp}},$$

Table 7.2 Routing algorithm under uncertain demand

Mesh Network Routing Under Uncertain Demand
1 $\forall e \in E, \mu_e \leftarrow \beta$
2 $x_f(P) \leftarrow 0, \forall P \in \mathcal{P}_f, \forall f \in F$
3 **loop**
4   **for** $\forall f \in F$ **do**
5        $\bar{P} \leftarrow$ lowest priced path in $\mathcal{P}_f$ using μ_e
6        $\mu_f \leftarrow \sum_{e \in E} A_{e\bar{P}} \mu_e$
7   **end for**
8   **for** $\forall d \in \mathcal{D}$ **do**
9        $\mu_d \leftarrow \sum_{f \in F} \mu_f d_f \frac{\lambda^*(d)}{p(d)}$
10   **end for**
11      $\mu^{\min} \leftarrow \min_{d \in \mathcal{D}} \mu_d$
12      $d^{\min} \leftarrow \arg\min_{d \in \mathcal{D}} \mu^{\min}$
13   **if** $\mu^{\min} \geq 1$
14   **return**
15   **for** $\forall f \in F$ **do**
16        $d'_f \leftarrow d_f^{\min}$
17   **while** $d'_f > 0$ **do**
18          $P \leftarrow$ lowest priced path in $\mathcal{P}_f$ using μ_e
19          $\delta \leftarrow \min\{d'_f, \min_{e \in P} \frac{1}{A_{eP}}\}$
20          $d'_f \leftarrow d'_f - \delta$
21          $x_f(P) \leftarrow x_f(P) + \delta$
22          $\forall e$ s.t. $A_{eP} \neq 0, \mu_e \leftarrow \mu_e(1 + \epsilon\delta A_{eP} \times \frac{\lambda^*(d^{\min})}{p(d^{\min})})$
23   **end while**
24   **end for**
25 **end loop**

where U is the length of the longest path in G, T_{mp} is the running time to find the shortest path, and T_{fmr} is the running time of the optimal routing algorithm under a fixed demand.

7.5 Oblivious Mesh Network Routing

In contrast to the predictive routing that establishes traffic models based on time-series analysis and optimizes towards the traffic demands with maximum possibility, oblivious routing makes no assumptions on the traffic model, rather it considers all traffic demand possibilities and optimizes towards the worst-case scenario. To formally study the oblivious routing strategy, we need a performance metric that could characterize the worst-case congestion under all possible traffic demand.

7.5.1 Routing

First, let us examine the formal description of *routing*, which specifies how traffic in each flow is distributed across the network. In the previous formulation (P_C), a routing is characterized through the traffic load distribution along different paths (i.e., $x'_f(P)$). This description of a routing depends on the traffic demand of each flow. When we have to consider all possible traffic demands, it becomes infeasible. In fact, a routing strategy could be modeled independently of the traffic demand, which is the core of the oblivious routing problem formulation.

Formally, we define a *routing* by the fraction of each flow that is routed along each edge $e \in E'$. We use $\phi_f(e)$ to denote the fraction of demand of flow f that is routed on the edge $e \in E'$. Thus, a routing could be specified by the set $\phi = \{\phi_f(e), f \in F, e \in E'\}$. Recall that the demand of flow $f \in F$ is denoted by d_f. Therefore, the amount of traffic demand of f that needs to be routed over e in routing ϕ, denoted by $y'_f(e)$, is given as follows:

$$y'_f(e) = d_f \cdot \phi_f(e). \tag{7.38}$$

Thus the *congestion* θ_e *of an interference set* $I(e)$ is given by

$$\theta_e = \sum_{e' \in I(e)} \sum_{f \in F} \frac{y'_f(e')}{c} \sum_{e' \in I(e)} \sum_{f \in F} \frac{d_f \cdot \phi_f(e')}{c}. \tag{7.39}$$

We further use $\theta(\phi, d) = \max_{e \in E} \theta_e(\phi, d)$ to denote the network congestion under a certain routing ϕ and traffic demand vector d.

7.5.2 Oblivious Performance Ratio

Now we proceed to study the performance metric that could characterize a "good" routing solution under all possible traffic demands. We start with the *optimal routing* $\phi^{opt}(d)$ for a certain demand vector d, which would give the minimum congestion under this demand, i.e.,

$$\theta^{opt}(d) = \min_\phi \theta(\phi, d). \tag{7.40}$$

Now we define the *performance ratio* $\gamma(\phi, d)$ of a given routing ϕ on a given demand vector d as the ratio between the network congestion under the routing ϕ and the minimum congestion under the optimal routing, i.e.,

$$\gamma(\phi, d) = \frac{\theta(\phi, d)}{\theta^{opt}(d)}. \tag{7.41}$$

The performance ratio γ measures how far ϕ is from being optimal on the demand d. Now we extend the definition of performance ratio to handle uncertain traffic demand. Let D be a set of traffic demand vectors. Then the performance ratio of a routing ϕ on D is defined as the worst-case performance ratio for all demands in D, i.e.,

$$\gamma(\phi, D) = \max_{d \in D} \gamma(\phi, d) \tag{7.42}$$

A routing ϕ^{opt} is optimal for the traffic demand set D if and only if

$$\phi^{\text{opt}} = \arg\min_{\phi} \gamma(\phi, D) \qquad (7.43)$$

meaning ϕ^{opt} minimizes the performance ratio under the worst-case scenario. When the set D includes all possible demand vectors d, we refer to the performance ratio as the *oblivious performance ratio*. The oblivious performance ratio is the worst performance ratio a routing obtains with respect to all possible demand vectors. To study the optimal routing strategy under uncertain traffic demand, we are interested in the *optimal oblivious routing* problem, which finds the routing that minimizes the *oblivious performance ratio*. We call this minimum value the *optimal oblivious performance ratio*.

It is worth noting that the performance ratio γ is invariant to scaling. Thus to simplify the problem, we only consider traffic demand vectors D that satisfy $\theta^{\text{opt}}(d) = 1$, instead of considering all possible traffic vectors. In this case,

$$\gamma(\phi, D) = \max_{d \in D} \theta(\phi, d). \qquad (7.44)$$

Thus the goal of oblivious routing is given by

$$\min_{\phi} \max_{\theta^{\text{opt}}(d)=1} \theta(\phi, d). \qquad (7.45)$$

7.5.3 Flow Conservation

Traffic into and out of a mesh node must be conserved. In P_C, a *path representation* of the routing is being used $(x'_f(P))$, which implicitly formulates the flow conservation. Here, because we use an *edge representation* of the routing $(\phi_f(e))$, the flow conservation has to be explicitly formulated. In particular, for the node $v \in V'$ that only relays for flow f (i.e., neither source nor destination), we have the following relations:

$$\forall f \in F, \sum_{e=(u,v)} y'_f(e) - \sum_{e=(u,v)} y'_f(e) = 0$$

if v is a relay of f $\qquad (7.46)$

$$\forall f \in F, \sum_{e=(u,v)} y'_f(e) - \sum_{e=(u,v)} y'_f(e) = -d_f$$

if v is the source node of f. $\qquad (7.47)$

Summarizing the above discussions, the *oblivious mesh network routing* problem is formulated as follows

$$\boldsymbol{P_O}: \tag{7.48}$$

$$\text{minimize } \theta \tag{7.49}$$

$$\text{subject to } \sum_{e' \in I(e)} \sum_{f \in F} \frac{y'_f(e')}{c} \leq \theta, \forall e \in E \tag{7.50}$$

$$\sum_{e=(u,v)} y'_f(e) - \sum_{e=(u,v)} y'_f(e) = 0$$

$$\forall f \in F, \forall v \in V', \text{ if } v \text{ is a relay of } f$$

$$\sum_{e=(u,v)} y'_f(e) - \sum_{e=(u,v)} y'_f(e) = -d_f$$

$$\forall f \in F, \forall v \in V', \text{ if } v \text{ is the source node of } f$$

$$\forall f \in F, \forall e \in E, y'_f(e) = d_f \cdot \phi_f(e) \geq 0$$

$$\theta \geq 0, \forall d \text{ with } \theta^{\text{opt}}(d) = 1.$$

Different from $\boldsymbol{P_C}$, the oblivious mesh routing problem $\boldsymbol{P_O}$ cannot be solved directly, because it is taken over all demand vectors, and $\theta^{\text{opt}}(d)$ is an embedded maximization in the minimization problem.

Here we use a similar method as in [21], which provides a LP formulation of the oblivious routing problem. The key insight is to look at the dual problem of the slave LPs of the original oblivious routing problem. Given a routing $\phi_f(e)$, the constraints (7.50) can be tested by solving, for each interference set $I(e)$, the following "slave LP," and testing if the objective is $\leq \theta$ or not.

$$\max \sum_{e' \in I(e)} \sum_{f \subset F} \frac{d_f \cdot \phi_f(e')}{c} \tag{7.51}$$

$$\text{subject to } \phi_f(e) \text{ is a routing};$$

$$\text{constraints } (51), (52) \tag{7.52}$$

$$\sum_{e' \in I(e)} \sum_{f \in F} y_f(e') \leq c.$$

In the dual formulation, we first introduce interference set weights $\pi_e(e')$ for every pair of interference sets e and e'. Each π variable can be thought of as a dual multiplier on the capacity constraint. There are three essential properties shown in Theorem 7.1.

Theorem 7.1. *A routing ϕ has oblivious ratio $\leq \theta$ if and only if there exist weights $\pi_e(e')$, for every pair of interference set $I(e)$, $I(e')$, e, $e' \in E$, such that*

P1 $\forall e, e' \in E, \sum_{e' \in E} c \cdot \pi_e(e') \leq \theta;$

P2 $\forall$ paths $h, \forall f \in F, \forall e \in E$

$$\sum_{e' \in I(e)} \phi f(e') \leq c \cdot \sum_{\alpha \in E} \pi_e(a) |I(a) \cap h|$$

P3 $\forall$ interference sets $I_e, I'_e, \pi_e(e') \geq 0.$

Note that the number of paths between any two nodes grows exponentially with the size of the network (in P2). To retain polynomial solvability, we introduce a variable $\zeta_e(u, v)$ for each edge e and node pair u and v, which is the length of the shortest path from u to v based on interference set weights $\pi_e(e')$. The introduction of these variables allows us to replace the exponential number of constraints in P2 with a polynomial number of constraints. Summarizing the above discussions, the LP formulation of problem P_O is given as follows

$$P_{LP} : \text{minimize } \theta$$

$$\phi \text{ is a routing} \tag{7.53}$$

$$\forall e, e' \in E : \sum_{e' \in E} c \cdot \pi_e(e') \leq \theta \tag{7.54}$$

$$\forall e \in E, \forall f : u \to v \in F : \sum_{e' \in I(e)} \phi_f(e')/c \leq \zeta_e(u, v) \tag{7.55}$$

$$\forall u \in V, \forall \alpha = (v, w) \in E, \sum_{\alpha' \in I(\alpha)} \pi_e(\alpha') + \zeta_e(u, v) - \zeta_e(u, w) \geq 0$$

$$\forall u \in V, \zeta_e(u, u) = 0$$
$$\forall u, v \in V, \zeta_e(u, v) \geq 0 \tag{7.56}$$
$$\forall e, e' \in E, \pi_e(e') \geq 0$$

In the above formulation, (7.54) can be explained by property P1. Property P2 and the shortest interference set paths account for (7.55), and finally property P3 appears at (7.56). The problem P_{LP} is a single polynomial-size LP instance, which can be solved with any LP solver. Our choice of LP solver was *lp solve* [19] an open source Mixed Integer Linear Programming (MILP) solver.

7.6 Simulation Study

In this section, we simulate the predictive and oblivious routing strategies over a variety of mesh network setups. Our goal is to evaluate and compare their performance and identify the key factors that impact the performance. Two other routing strategies, namely oracle routing and shortest-path routing, are used as the baseline strategies for comparison. We describe the routing strategies that are evaluated in the simulation study as follows.

- *Oracle Routing (OR)*. In this strategy, the traffic demand is known a priori. It runs every hour based on the up-to-date traffic demand and returns the optimal set of routes. As a result, no other routing strategies can outperform *OR*, and we used it to provide a performance upper bound.
- *Shortest-Path Routing (SPR)*. This strategy is agnostic of traffic demand, and returns a fixed routing solution purely based on the shortest distance (number of hops) from each mesh node to the gateway. Many mesh network routing

heuristics resemble the shortest-path routing strategy. We evaluate this strategy to quantitatively contrast the advantage of routing strategies that explicitly consider traffic uncertainty.

- *Predictive Routing (PR)*. This strategy attempts to adjust to changing the traffic demand. Future demand is estimated based on the historical data every hour based on the traffic prediction method presented in Sect. 7.4.
- *Oblivious Routing (OBR)*. This strategy is oblivious to the traffic demands. It considers all possible traffic demands that may be imposed on the network and finds a routing that optimizes the worst-case congestion using the algorithm presented in Sect. 7.5.

Notice that *SPR* and *OBR* will compute the traffic routes only once and use them during the entire simulation time, whereas the *OR* and *PR* need to compute and update the routes every hour.

To realistically simulate the traffic demand at each LAP, we employ the traces collected in the campus wireless LAN network. The network traces used in this work are collected in Spring 2002 at Dartmouth College and provided by CRAWDAD [20].

By analyzing the *snmp* log trace at each access point, we are able to derive its 1,108-h incoming and outgoing traffic volume beginning 12:00 A.M., March 25, 2002 EST. We select the access points from the Dartmouth campus wireless LAN and assign their traffic traces to the LAPs in our simulation. The traffic assignment is given in Table 7.3 in one of the random topologies as shown in Fig. 7.12.

We experiment with the above routing strategies along the time range [108, 1108], a 1,000-h period excerpted from the trace.[2] Note that all the simulation results presented in this section use 108 as the zero point.

We start by presenting the congestion achieved by all strategies during the entire 1,000-h simulation period. As seen in Fig. 7.13, *OR* constantly achieves the minimum worst-case congestion among others, because of its unrealistic capability to know the actual traffic demand. We note that the burstiness of θ applies to all strategies including *OR*. This observation comes from the burstiness of the traffic load in the *snmp* log trace, which is caused by the insufficient level of traffic multiplexing at wireless local access points.

To filter out the noise caused by traffic burstiness, in Fig. 7.14a, we normalize θ achieved by other strategies by the same value of *OR*. Because *OR* always achieves the minimum θ among others, this ratio will end up at least 1. Also we take a close-up look during the hour range [190, 290]. Here, *PR, SPR*, and *OBR* achieve less

Table 7.3 Traffic assignment from trace file

AP	31AP3	34AP5	55AP4	57AP2	62AP3	62AP4	82AP4	94AP1	94AP3	94AP8
Node ID	22	18	57	5	55	20	53	3	56	27

[2] Note that the beginning of the trace [0,107] is used as training data, thus it is not included in the simulation result.

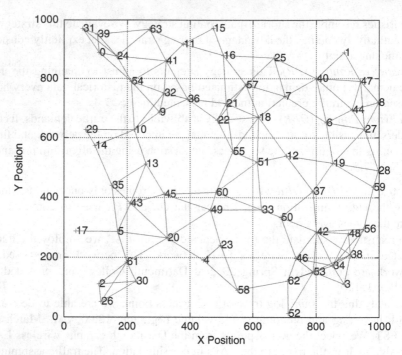

Fig. 7.12 Mesh network topology

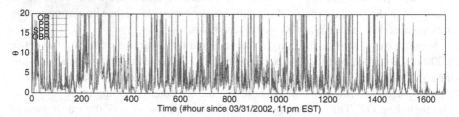

Fig. 7.13 Overview of all strategies

than two times the optimal congestion in most cases. The above observations get clearer when we sort out the normalized congestion ratio for the three strategies in Fig. 7.14b. It is clear that both *PR* and *OBR* that integrate the traffic prediction with the optimal routing outperform the *SPR* strategy that is agnostic about the traffic demand. Further, *PR* achieves lower congestion than *OBR* for many time points because of more comprehensive representation of the traffic demand estimation. However, in other cases (less than 10% of the time), the worst-case congestion of *PR* is substantially higher than *OBR*. This problem can be mostly attributed to the fundamental inaccuracy of traffic prediction that is highly sensitive to the traffic's erratic behavior.

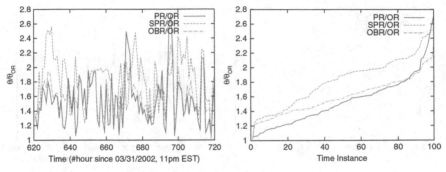

Fig. 7.14 (**a**) Congestion ratio (θ/θ_{OR}) and (**b**) sorted view

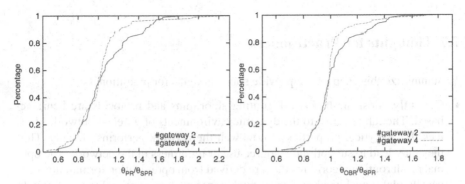

Fig. 7.15 (**a**) CDF of congestion ratio (θ_{PR}/θ_{SPR}) and (**b**) CDF of congestion ratio

We also investigate the performance of *PR* and *OBR* in a representative random topology with Internet gateways near the perimeter. For a more complete picture, we also investigate cases with two gateways and with four gateways. Each of these topologies has a total of 64 nodes, including ten access points receiving traffic from mobile clients, the gateways, and the remaining nodes forwarding traffic on behalf of the Internet and the access points. The points are distributed at random over a simulation square 1,000 m on the side, with an interference range of 155 m. For simplicity, the transmission range is equal to the interference range.

In both the four-gateway and the two-gateway scenarios, we run *PR, OBR,* and *SPR* using the demand data from the Dartmouth trace. Figure 7.15 plots the congestion ratios of *PR* over *SPR* and *OBR* over *SPR*. In both pictures, *OBR* and *PR* outperform *SPR* in more than 50% of the cases. During the time when they are inferior to *SPR*, the worst-case ratio is bounded by 2. Also when we increase the number of gateways from 2 to 4, both ratios decrease. Obviously, *SPR* takes advantage of this topology change, because of the fact that more gateways will diversify the shortest paths from access points to nearest gateways, and also shrink the lengths of the paths. A direct comparison between Predictive Routing and Oblivious Routing is shown in Fig. 7.16.

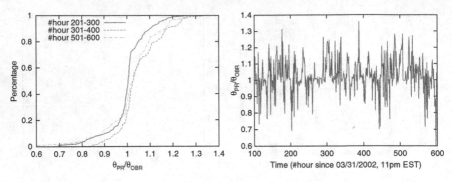

Fig. 7.16 (**a**) Congestion ratio (θ_{PR}/θ_{OBR}) and (**b**) CDF of congestion ratio

7.7 Thoughts for Practitioners

To summarize this chapter, we provide some thoughts for practitioners:

- Currently, most mesh network routing algorithms and protocols are heuristic-based. Though adaptive to the dynamic environments of wireless networks, they lack the analytical properties of how well the network performs globally. Thus they may lead to suboptimal resource use or unfairness in the network. The optimal mesh routing algorithms that are derived from optimization formulations can usually claim analytical properties such as resource use optimality and throughput fairness. However, they usually have strong assumptions on static and known traffic demand, which have been shown to be unrealistic by studies of wireless network traces [9]. Thus there is a critical need to investigate the optimal mesh network routing strategies that can accommodate traffic uncertainty.

- Predictive routing and oblivious routing are two optimal routing strategies that address traffic uncertainty in mesh network routing. Their designs, however, are based on different principles. (1) *Predictive routing* infers the traffic demand with maximum possibility based in its history and optimizes the routing strategy based on the predicted traffic demand. Underlying predictive routing is the assumption that past behavior is a good indicator of the future. (2) *Oblivious routing*, which makes no assumption on traffic demand and considers all the possible traffic demands. In particular, oblivious routing selects the routing strategy where the worst-case network performance is optimized. For a given mesh network, it is important to know which routing strategy would provide a better performance.

- Through the simulation study, we find predictive routing performs better under consistent traffic demand compared to highly variable demand. Furthermore, oblivious routing, being a stateless routing, is unaffected by the traffic behavior. The performance of both algorithms is sensitive to demand and topology, suggesting that the optimal choice for deployment should be based on local parameters.

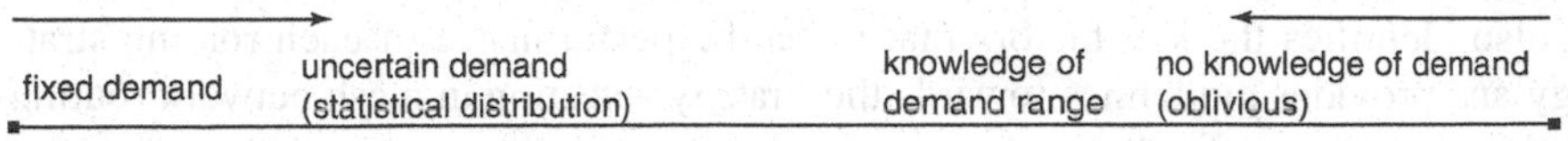

| fixed demand | uncertain demand (statistical distribution) | | knowledge of demand range | no knowledge of demand (oblivious) |

Fig. 7.17 Research space for route optimization

7.8 Directions for Future Research

This chapter studies optimal routing strategies for wireless mesh networks with attention to traffic demand uncertainty over time and provable robustness. Two approaches are reviewed and discussed in this chapter. Here we outline several possible directions for future research.

- *Traffic modeling and estimation.* The predictive routing strategy is sensitive to traffic dynamics and the prediction accuracy. To obtain a higher prediction accuracy, the future research needs to develop appropriate traffic models that can be integrated with network optimization formulations. The key problem involved is how to parameterize the traffic models to represent its structure with a small number of parameter values that can be estimated from the data. Based on the traffic model, traffic estimation needs to develop reliable estimation methods that determine the values of the parameters that provide robust and high accurate traffic estimate.
- To incorporate traffic uncertainty and dynamics, and integrate different traffic models, future research should explore the full spectrum of research outlined in Fig. 7.17 from two directions. One side of the spectrum starts with the fixed-demand network optimization, where the traffic demand is known as a fixed single-value scalar; then it extends to handle the scenarios where the traffic demand is represented using a random variable with statistical distribution. The other side of the spectrum starts with the oblivious optimization problem where the traffic demand is completely unknown, where it can be refined to handle the cases where the range of the traffic demand is known.

7.9 Conclusions

This chapter studies the optimal routing strategies for wireless mesh networks. Different from existing works that implicitly assume traffic demand as static and known a priori, this chapter considers the traffic demand uncertainty. It presents two approaches to address the traffic uncertainty in optimal mesh network routing (1) predictive routing that infers the traffic demand with maximum possibility based in its history and optimizes the routing strategy based on the predicted traffic demand and (2) oblivious routing that considers all the possible traffic demands and selects the routing strategy where the worst-case network performance could be optimized.

It also identifies the key factors that affect the performance of each routing strategy and provides guidelines towards the strategy selection in mesh network routing under uncertain traffic demands.

7.10 Terminologies

Some of the most important terms in this chapter are defined below.

1. *Interference set.* The Interference Set of an edge e is the set of other edges with which it cannot simultaneously transmit on the same channel. One or both of the endpoints for an edge in the interference set of e are within the interference range of one of e's endpoints.
2. *Oblivious routing.* A routing algorithm that is independent of the actual demands on a network.
3. *Predictive routing.* A routing algorithm that attempts to extrapolate future routing demands and plans the network flows accordingly.
4. *Demand erraticity/elasticity.* The degree to which traffic demand changes over time. Also known as "burstiness."

7.11 Questions

1. Explain the factors that must be taken into account when deploying a wireless mesh network.
2. What is the intuition behind the differing strengths of oblivious and predictive routing?
3. What affect does traffic aggregation at network endpoints have on the burstiness of the traffic?
4. Name some of the key factors that make traffic more predictable. Name some that make it less predictable.
5. Using Fig. 7.12, how many neighbors would node 55 have if the transmission range was 100 m? How many if it was 200 m? In general, with randomly distributed nodes, what is the asymptotic relationship between the degree and the transmission range? Your answer should be a simple polynomial, such as "linear."
6. Suppose the traffic demands at nodes A, B, and C in a wireless mesh network are given in the following table.

Time	A	B	C
1	100	200	1
2	200	200	3
3	300	100	8
4	400	400	1

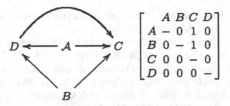

Fig. 7.18 Sample network and demand matrix for question 7

(a) Which of nodes A, B, and C would you say has the most erratic demand? Which is least erratic? Formalize your intuition.

(b) During which of the three time intervals does the traffic demand change the most? Justify your answer.

7. Consider the graph topology shown in Fig. 7.18 with the corresponding demand matrix. Assume the rows are the traffic sources and the columns are the destinations. There are many ways to route the traffic. Calculate the shortest path routing and also the routing that minimizes the network congestion.

References

1. Seattle wireless. http://www.seattlewireless.net
2. Mit roofnet. http://www.pdos.lcs.mit.edu/roofnet/
3. R. Draves, J. Padhye, and B. Zill, Routing in multi-radio, multi-hop wireless mesh networks. Proc. ACM Mobicom (2004).
4. S. Biswas and R. Morris, Exor: Opportunistic multi-hop routing for wireless networks. Proc. ACM SIGCOMM (2005).
5. A. Raniwala and T. Chiueh, Architecture and algorithms for an IEEE 802.11-based multi-channel wireless mesh network. Proc. IEEE INFOCOM (2005).
6. H. Wu, F. Yang, K. Tan, J. Chen, Q. Zhang, and Z. Zhang: Distributed Channel Assignment and Routing in Multi-radio Multi-channel Multi-hop Wireless Networks. IEEE JSAC, Special issue on multi-hop wireless mesh networks **24**(11), 1972–1983 (2006).
7. M. Alicherry, R. Bhatia, and L. Li, Joint channel assignment and routing for throughput optimization in multi-radio wireless mesh networks. Proc. ACM MobiCom (2005).
8. J. Tang, G. Xue, and W. Zhang, Maximum throughput and fair bandwidth allocation in multi-channel wireless mesh networks. Proc. IEEE INFOCOM (2006).
9. X. Meng, S.H.Y. Wong, Y. Yuan, and S. Lu, Characterizing flows in large wireless data networks. Proc. ACM MobiCom (2004).
10. L. Dai, Y. Xue, B. Chang, and Y. Cui, Throughput optimization routing under uncertain demand for wireless mesh networks. Proc. IEEE MASS (2007).
11. L. Dai, Y. Xue, B. Chang, Y. Cao, and Y. Cui, Integrating traffic estimation and routing optimization for multi-radio multi-channel wireless mesh networks. Proc. IEEE INFOCOM (2008).
12. J. Wellons, L. Dai, Y. Xue, and Y. Cui, Predictive or Oblivious: A Comparative Study of Routing Strategies for Wireless Mesh Networks Under Uncertain Demand. Proc. IEEE SECON (2008).
13. P. Gupta and P. Kumar, The capacity of wireless networks. IEEE Trans. Inform. Theory 388–404 (2000).

14. K. Jain, J. Padhye, V. Padmanabhan, and L. Qiu, Impact on interference on multi-hop wireless network performance. Proc. Mobicom (2003).
15. V.S.A. Kumar, M.V. Marathe, S. Parthasarathy, and A. Srinivasan, Algorithmic aspects of capacity in wireless networks. Proc. ACM SIGMETRICS, 133–144 (2005).
16. Y. Xue, B. Li, and K. Nahrstedt, Optimal resource allocation in wireless ad hoc networks: A price-based approach. IEEE Trans. Mobile Comput. 5(4), 347–364 (2006).
17. H. Wang, H. Xie, L. Qiu, Y.R. Yang, Y. Zhang, and A. Greenberg, Cope: traffic engineering in dynamic networks. Proc. ACM SIGCOMM (2006).
18. Ilog cplex mathematical programming optimizers. http://www.ilog.com/products/cplex
19. The lp solve mixed integer linear programming (milp) solver. http://lpsolve.sourceforge.net/5.5/
20. A community resource for archiving wireless data at dartmouth. http://crawdad.cs.dartmouth.edu/
21. D. Applegate and E. Cohen, Making intra-domain routing robust to changing and uncertain traffic demands: Understanding fundamental tradeoffs. Proc. ACM SIGCOMM, 313–324 (2003).

Chapter 8
Routing Metrics for Wireless Mesh Networks

Georgios Parissidis, Merkourios Karaliopoulos, Rainer Baumann, Thrasyvoulos Spyropoulos, and Bernhard Plattner

Abstract Routing in wireless mesh networks has been an active area of research for many years, with many proposed routing protocols selecting shortest paths that minimize the path hop count. Whereas minimum hop count is the most popular metric in wired networks, in wireless networks interference and energy-related considerations give rise to more complex trade-offs. Therefore, a variety of routing metrics has been proposed especially for wireless mesh networks providing routing algorithms with high flexibility in the selection of best path as a compromise among throughput, end-to-end delay, and energy consumption. In this paper, we present a detailed survey and taxonomy of routing metrics. These metrics may have broadly different optimization objectives (e.g., optimize application performance, maximize battery lifetime, maximize network throughput), different methods to collect the required information to produce metric values, and different ways to derive the end-to-end route quality out of the individual link quality metrics. The presentation of the metrics is highly comparative, with emphasis on their strengths and weaknesses and their application to various types of network scenarios. We also discuss the main implications for practitioners and identify open issues for further research in the area.

8.1 Introduction

Routing in wireless mesh networks has been a highly popular research topic during the last decade. Whereas many routing function objectives are the same as in wired networks and the Internet, wireless mesh networks add several new dimensions that make the routing problem less straightforward and more interesting at the same

M. Karaliopoulos (✉)
Computer Engineering and Networks Laboratory (TIK), ETH Zurich, Gloriastrasse 35, 8092 Zurich, Switzerland
e-mail: karaliopoulos@tik.ee.ethz.ch

S. Misra et al. (eds.), *Guide to Wireless Mesh Networks*, Computer Communications and Networks, DOI 10.1007/978-1-84800-909-7_8,

time. As a result, although experience and wisdom gained by wired networks have guided the first steps in the wireless domain, in many cases there was need for novel approaches and solutions.

In the Internet, network nodes are quite static; changes in connectivity may happen but are not frequent. As a result, routing protocols for wired networks proactively maintain routes from all nodes to every other node, by propagating the occasional topology update as soon as it occurs. However, the topology of wireless mesh networks changes much more dynamically than in wired networks. This is primarily because of node mobility on the one hand, e.g., in mobile ad hoc networks (MANETs) or hybrid networks with both mobile and static nodes, and the impairments of wireless links because of propagation phenomena, on the other hand. Wireless networks end up often being only intermittently connected, so that the use of proactive routing protocols and the overhead related to route maintenance become less attractive.

We summarize here the additional challenges related to wireless mesh networks:

Node mobility. Wireless mesh nodes may move. As a result, links may break and network topology may change frequently; in graph-theoretic terms, the connectivity graph varies more quickly with time. This makes route maintenance much more complex than in wired networks.

Wireless propagation phenomena. In the wireless environment, node transmissions are physically broadcast and subject to radio propagation dynamics, such as shadowing and multipath fading. Even in the case of links between static nodes, the received signal varies considerably over time, giving rise to "grey zones." In grey zones, the overall link quality does not allow data traffic transmission. Nevertheless, occasional control traffic data transmission may still succeed yielding a false view of the network connectivity and resulting in frequent route failures and re-establishments.

Energy constraints. In many cases, energy preservation and elongation of battery lifetime may become the primary objectives for network operation. Advances in battery technology are significantly slower than those in nanotechnology and electronics. Thus, the available power will continue to be a performance bottleneck for handheld, low-end devices and sensors, in scenarios where nodes move and operate for long periods without access to the electricity grid.

Lack of centralized control. One of the most attractive features of wireless mesh networks is self-organization. Various functions, such as medium access control and routing, are carried out in a fully distributed manner with minimal human intervention. They are not subject to any centralized network management processes of the kind practiced in wired networks. However, the drawback is that most decisions are made by individual nodes having primarily knowledge about their local environment only. This leaves little margin for network optimizations that require global knowledge about the network state. More critically, the network operation itself assumes the cooperation of all nodes, rendering the network more vulnerable to node misbehavior practices.

The need to think differently when it comes to wireless mesh networks is also reflected in the large variety of routing metrics that have been proposed along with routing protocols, to enable efficient data delivery in the wireless context. This does not mean that routing metrics in wired networks are not in abundance. Besides minimum hop count (shortest path first), which is the alma matter of metrics, the literature is quite rich in other metrics that have either been more "intelligent" in pursuing minimum delivery delay or have prioritized other aspects of network performance [1]. Load sharing and balancing, fault-tolerance, low jitter, and high throughput rank high on the list of goals that have determined the costs of links and paths in the network and have driven the routing decisions. Whereas these objectives remain relevant in wireless mesh networks, there are additional concerns that may complement or overshadow traditional objectives.

This chapter identifies different categories of routing metrics proposed for wireless mesh networks and describes the rationale of each category. Some metrics are treated in more detail, either because they were the first to introduce a new approach or because they are being considered, themselves or their variations, in standardization procedures. Our description is deliberately comparative, pointing to the similarities and differences amongst the different categories and the relative advantages of each metric. Wherever appropriate, we draw references to studies that have already made such comparisons between the metrics discussed.

8.2 Background

It is possible to group routing metrics into broader categories according to a number of criteria. The optimization goal, the way required information for the metric computation is collected, and the way the route (path)[1] metric is related to individual link metrics, have been selected as a nonexhaustive list of attributes for systematically characterizing and classifying them in this chapter.

8.2.1 Optimization Objectives

A routing metric is essentially a value assigned to each route, and used by the routing algorithm to select one, or more, out of a subset of routes discovered by the routing protocol. These values generally reflect the cost of using a particular route with respect to some optimization objective, and could take into account both application and network performance indicators. More specifically, the objective of the routing algorithm and thus the routing metric may be to:

- *Minimize delay*. This is often the default objective of the routing function. The network path over which data can be delivered with minimum delay is selected. If queuing delays, link capacity, and interference are not taken into account, then delay minimization ends up being equivalent to hop-count minimization.

---

[1] The terms *path* and *route* are used interchangeably throughout this chapter.

- *Maximize probability of data delivery.* For non real-time applications, the main requirement is to achieve a low data loss rate along the network route, even at the expense of increased delay. This is equivalent to minimizing the probability of data loss between network end-points.
- *Maximize path throughput.* In that case, the primary aim is the selection of an end-to-end path consisting of links with high capacity.
- *Maximize network throughput.* Contrary to the first three objectives, which are user application-oriented, network throughput is a system objective. The objective may be formulated explicitly as the maximization of data flow in the whole network or, more implicitly, through the minimization of interference or retransmissions.
- *Minimize energy consumption.* Energy consumption is rarely an issue in wired networks. However, it becomes a major concern in sensor and MANETs, where the battery lifetime constrains the autonomy of network nodes.
- *Equally distribute traffic load.* This objective is more general. Here, the aim is to ensure that no node or link is disproportionately used and could be achieved, for example, by minimizing the difference between the maximum and minimum traffic load over the network links. Load balancing may have an indirect effect on other objectives such as battery lifetime and per node throughput.

It is worthwhile noting here that the first three objectives in the above list are concerned with individual application performance, namely they optimize the performance for a given source-destination pair, whereas the last three are "system-oriented" objectives focusing on the performance of the network as a whole. Furthermore, routing metrics may consider more than one of the aforementioned objectives. In that case, the multidimensional metric combines different cost values, weighting them appropriately to account for the relative prioritization of the respective objectives.

8.2.2 Link and Path Metrics

The ultimate decision to be made by routing will be about the selected path(s); therefore, the final metric value that will be the subject of comparison will relate to the whole path. However, the path metric needs to be somehow derived as a function of the individual metric values estimated for each link in the path. The actual function to be used varies and highly depends on the actual metric in question. The most widely used functions are:

- *Summation.* The link metric values are added to yield the path metric. Examples of additive metrics are the delay or number of retransmissions experienced over a link.
- *Multiplication.* Values estimated over individual links are multiplied to get the overall path metric. The probability of successful delivery is an example of a multiplicative metric.

- *Statistical measures (minimum, maximum, average).* The path metric coincides with the minimum, average, or the maximum of values encountered over the path links. Example of the first case is the path throughput, which is dictated by the minimum link throughput (bottleneck link) over all hops included in a network path.

8.2.3 Metric Computation Method

There are also various ways in which network nodes acquire the information they need for the computation of the routing metric:

- *Reuse of locally available information.* Information required by the metric is available locally at the node, usually as result of the routing protocol operation. Such information may include the number of node interfaces, number of neighbor nodes (connectivity degree), length of input and output queues.
- *Passive monitoring.* Information for the metric is gathered by observing the traffic coming in and going out of a node. No active measurements are required. In combination with other information, passive measurements can yield, for example, an estimate for the available capacity.
- *Active probing.* Special packets (probes) are generated for measuring the properties of a link/path. This method incurs the highest overhead on the network, which is directly dependent on the frequency of measurements.
- *Piggyback probing.* This method also involves active measurements. However, these measurements are now carried out by including probing information into regular traffic or routing protocol packets. With piggyback probing, no additional packets are generated for metric computation purposes, thus reducing the overhead for the network. Piggyback probing is a common method to measure delay.

Raw information about a link, acquired from passive or active measurements, usually requires some processing before it can be used to construct efficient and stable link metrics. Measured network parameters (e.g., delay or link loss ratio) are often subject to high variation. It is usually desired that short-term variations do not influence the value of a metric. Otherwise, rapid oscillations of the metric value could, depending on the actual metric context, result in the phenomenon of *self-interference*, quite early observed in Internet applications [2]: once a link is recognized as good, it is chosen by the routing protocol and starts getting used till it is overloaded and assigned with a worse metric value. As traffic starts being routed around that link, its metric value increases again and the effects starts anew.

Therefore, metric measurements are subject to some filtering over time. Different metrics apply different types of filtering including:

- *Fixed history interval.* An average is computed over a fixed number of previous measurement samples.
- *Dynamic history window.* An average is computed over a number of previous measurement samples, which varies depending on the current transmission rate.

- *Exponential weighting moving average (EWMA)*. Measurement samples are weighted so that the impact of past samples on the current value of the metric decays exponentially with the sample age. Every time a new sample d_{sample} is obtained, the value of the metric is updated as: $d_{avg} = \alpha \cdot d_{avg} + (1 - \alpha) \cdot d_{sample}$ with $\alpha \in [0, 1]$ being the weighting factor and d_{avg} the current metric value.

8.3 Routing Metrics

In this section we describe routing metrics proposed for wireless mesh networks. Firstly, we discuss topology-based metrics and demonstrate the performance disadvantage of the hop count metric in wireless mesh networks. We then argue in favor of more elaborate metrics that can address the additional challenges of those networks and present the main metrics proposed up-to-date in literature. The presentation groups metrics in four categories, namely (1) signal strength-based, (2) active probing-based, (3) mobility-aware, and (4) energy-aware.

8.3.1 Topology-Based Metrics

The main advantage of topology-based routing metrics is their simplicity. Examples of relevant topological information are the number of neighbors of each node, and the number of hops and/or paths towards a particular destination. The metrics almost always take into account connectivity information that becomes available locally by the routing protocol, without requiring additional passive or active measurements.

In general, the topology definition in wireless networks is less straightforward than in wired networks. First of all, links are physically broadcast. The link definition between two nodes is a *soft* definition; a link is said to exist as long as the one node is within the transmission range of the other, which is a function of the sender node transmit power, the reception sensitivity of the receiving node and the propagation environment. In fact, varying the transmit power of nodes lies at the heart of the topology control function, an important tool for engineering wireless mesh networks.

Another complication in wireless mesh networks is related to the link asymmetry. Although node X may receive successfully packets from node Y, it may well be that node Y cannot receive packets of node X. The reason is different interference levels at the neighborhood of the two nodes. This asymmetry has to be taken into account when making routing decisions, in particular for bidirectional traffic (e.g., TCP traffic).

Although topology-based metrics do not take into account several variables that have an impact on the network and application performance, their simplicity makes them highly popular. In fact, one of them, the hop-count metric, is by far the most popular metric in wired networks and, as such, one of the first considered in wireless mesh networks as well.

8.3.1.1 Hop Count

The concept of the hop count metric is simple: every link (hop) counts equally as one unit, independent of the quality or other characteristics of the link. The ease of implementation has made hop count the most widely used metric in wired networks; it is implicitly or explicitly the default metric in many popular wireless mesh network routing protocols, such as OLSR [3], DSR [4], DSDV [5], and AODV [6].

The rationale for minimizing the hop metric is straightforward. Fewer hops on the data path imply smaller delay (higher throughput) and reduced waste of network resources, whether these involve network links or buffers or computational power. The implicit assumption is the existence of error-free links, which is almost always the case with wired networks.

On the contrary, links in wireless mesh networks cannot be assumed error-free. The wireless radio propagation environment, external and network-internal interference, and, when relevant, node mobility result in intermittent connectivity among the network nodes. Minimum hop count tends to select more distant nodes. Depending on the flexibility in setting the transmit power, a node has two options:

- The node may increase the transmit power to achieve a target probability of successful delivery despite the large distance to the receiver. The result of the minimum-hop count in this case is increased power consumption, which may be a concern for low-end, battery-powered devices.
- On the other hand, when the transmit power is fixed, the probability of data loss over the more distant link increases (on average). The risk of retransmissions is higher, implying additional energy consumption at the node, more interference at the network, and, eventually, increased delay. We illustrate this scenario with a simple example below.

Example 8.1. (a) Assume that the probability of packet loss between the node S–D in Fig. 8.1 is p_1 in both directions, $S \to D$ and $D \to S$. Likewise, the probability of loss over both hops of path SHD is p_2, again in both directions. A packet transmission is considered successful when the data packet is correctly received in the forward direction and an ACK packet is correctly received in the reverse direction, as in the unicast 802.11x transmission mode. What would be the minimum value of loss p_1, under which the minimum-hop path SD would result in larger delay than the two-hop path SHD? Assume, for simplicity, that the number of retransmissions at the link-layer is infinite.

Answer:

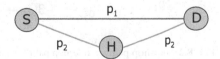

Fig. 8.1 One-hop path vs. two-hop path

Given that the propagation delay is small (in the order of μs) compared to the transmission delay (in the order of ms), the overall end-to-end delay of the packet is directly proportional to the total number of hop transmissions (including retransmissions) along the path.

The number of transmissions over a hop with symmetric packet probability loss, i.e., $p_f = p_r = p$, is a Geometric RV with parameter $(1 - p)^2$; the mean expected number of transmissions, assuming infinite retransmissions, equals $1/(1 - p)^2$.

The end-to-end normalized delay over paths SD and SHD are:

$$D_{SD} = \frac{1}{(1 - p_1)^2} \text{ and } D_{SHD} = \frac{2}{(1 - p_2)^2}.$$

Therefore,

$$D_{SD} \geq D_{SHD} \Rightarrow \frac{1}{(1 - p_1)^2} \geq \frac{2}{(1 - p_2)^2} \Rightarrow \frac{(1 - p_2)}{(1 - p_1)} \geq \sqrt{2} \Rightarrow p_1 \geq \frac{\sqrt{2} - 1 + p_2}{\sqrt{2}}$$

and the minimum required value for p_1 to get smaller delay over the two-hop path is

$$p_{1,min} = p_1|_{p_2=0} = \frac{\sqrt{2} - 1}{\sqrt{2}} = 0.29.$$

(b) Repeat the same calculation for the S–D pair in Fig. 8.2.
With the same rationale, the end-to-end normalized delays over paths S–2–4–6–D and S–1–3–5–7–D are:

$$D_{S246D} = \frac{4}{(1 - p_1)^2} \text{ and } D_{S1357D} = \frac{5}{(1 - p_2)^2}.$$

Therefore,

$$D_{S246D} \geq D_{S1357D} \Rightarrow \frac{4}{(1 - p_1)^2} \geq \frac{5}{(1 - p_2)^2} \Rightarrow \frac{(1 - p_2)}{(1 - p_1)} \geq \frac{\sqrt{5}}{2} = 1.12$$

$$\Rightarrow p_1 \geq \frac{0.12 + p_2}{1.12}.$$

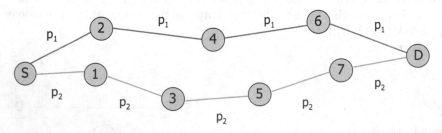

Fig. 8.2 Four-hop path vs. five-hop path

and the minimum required value for p_1 to get smaller delay over the two-hop path is

$$p_{1,\min} = p_{1|p_2=0} \approx 0.1.$$

Therefore, when there is no flexibility in increasing the transmit power or when this is not appealing because of energy constraints, there are scenarios where the rationale of the minimum-hop metric, as known from the wired networks, is cancelled. As the example suggests, it is more likely that the decisions made using the minimum hop count metric differ from the optimum ones along paths with many hops, because there the loss probability values over the minimum hop links do not have to be as high as in paths with one or two hops. In any case, the message coming out of the example is that knowledge of the dynamically changing loss probabilities over the network links could support wiser routing decisions in wireless mesh networks.

This remark was made quite early by researchers in the field. What took more time and experimentation was the method to obtain this information from the network. Using the signal strength measurements to infer these loss probabilities was historically the first attempt in this direction.

8.3.2 Signal Strength-Based Metrics

Signal strength has been used as link quality metric in several routing schemes for wireless mesh networks. The hypothesis is: because a packet is successfully received when the signal strength exceeds some threshold value, the signal strength could be viewed as a good indicator of the link quality. Nowadays, commodity wireless network adapters provide an average received signal strength value for every successfully received packet.

Signal strength values have been used in routing in two different ways:

- As control parameters for excluding routes with "bad" quality link from the route selection process
- As conventional routing metrics, where some function of the signal strength is considered in the link(path) cost function

8.3.2.1 Signal Strength as Control Parameter for Eliminating Routes

In [7] signal strength is measured passively upon packet reception. A preemptive region around a source is introduced and a path is considered likely to break when the power of the received packet becomes lower than a predefined preemptive threshold $P_{\text{threshold}}$. The threshold is defined as:

$$P_{\text{threshold}} = \frac{P_O}{r_{\text{preemptive}}^4},$$

where P_0 is a constant for each transmitter/receiver pair that depends on the antenna gain and height and $r_{\text{preemptive}}$ is the radius of the node's preemptive region. Receiver nodes generate a protocol specific warning message towards the source as soon as the reported signal strength of a received packet drops below $P_{\text{threshold}}$. Then, the source will search for a higher quality path to route its packets. Generally, more stable average values can be generated by having a number of message exchange rounds. The main disadvantage of the proposed preemptive routing mechanism is its assumption that links are symmetric. Because this does not often hold in reality (for example, see [8]), the proposed mechanism may suffer from instabilities. Nevertheless, the evaluation of the proposed method with simulations shows significant improvement in the performance of the AODV and DSR routing protocols, because the proposed modifications result in reduced number of broken paths.

In a similar approach, the signal stability-based adaptive routing (SSAR) in [9] uses periodic link-layer beacons to get estimates of the link quality and forward route discovery packets only via routes involving stable links with good signal strength. The difference with the preemptive routing is that the decision to eliminate routes is not taken by the source but is rather distributed, with each node dropping packets from links with weak signal rather than issuing warning messages.

8.3.2.2 Signal Strength as Routing Metric

Punnoose et al. [10] convert the signal strength into a link quality factor, which is then used to assign weights to the links. For a route consisting of M hops, the link quality factor L of the route is estimated as

$$L = \prod_{s=1}^{M} \left(1 - Q((P_{\text{pred}_i} - P_{\text{th}})/\sigma)\right),$$

where $Q(\cdot)$ is the Q-function[2], P_{pred_i} is the theoretically predicted power received by the ith node from the $(i-1)$th node, P_{th} is the receiving threshold, and σ is the variance of signal variations, which are assumed to be normally distributed.

The link quality factor is the product of probabilities computed for each hop that at a certain time in the future the signal level will be above the receiving threshold. The theoretically predicted power is calculated as follows: using linear position extrapolation based on the input data from GPS positioning and velocity information, estimates for the positions of all nodes one second in the future are calculated. These positions, along with some propagation model are used to obtain P_{pred_i}, whereas the default values for the variance of signal is $\sigma = 6\,\text{dB}$ and for the receiving threshold $P_{\text{th}} = 60\,\text{dBm}$.

---

[2] The Q-function is defined as $Q(z) = \int_z^\infty \frac{1}{\sqrt{2\pi}} e^{-\frac{y^2}{2}} \, dy$.

8.3.2.3 Correlation of Signal Strength with Probability of Successful Packet Delivery

Although correlation of signal strength and loss is assumed by the above metrics, the actual existence of such correlation is addressed in two studies.

In [11], the focus is on the packet delivery performance in sensor networks. It is reported that high signal strength implies low packet loss, however low signal strength does not necessarily imply high packet loss.

A similar observation is made in [8]. Link-level measurements in a wireless mesh network (Roofnet) demonstrate that although the signal strength values do affect the delivery probability, one cannot expect to use them as a predictive tool. This is clearly shown in Fig. 8.3, which plots the link delivery probabilities at different rates vs. the average S/N (minimum signal-to-interference ratio for successful reception). Although the specification of the wireless card used in Roofnet suggests that the range of signal strength values for which the packet error rate would be between 10 and 90% (intermediate loss rates) is only 3 dB wide, the actual measured range of intermediate loss rates is much broader. Experiments using a hardware channel emulator demonstrate that an essential cause of intermediate loss rates is multipath fading because of reflections in the radio environment.

The results of both studies are aligned regarding the impact of the signal strength upon the delivery probability, but also the difficulty to get a mapping function

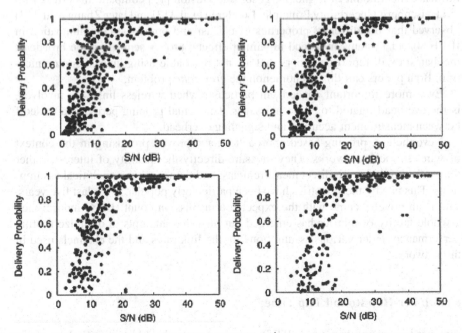

Fig. 8.3 Delivery probability at 1, 2, 5.5, and 11 Mbit s^{-1} vs. the averages S/N. Each data point represents an individual sender-receiver pair. Figure is adapted with permission from [8]

between the two quantities. The signal strength, at least the values reported by most commercial 802.11 cards, represent coarse average values of the received signal strength and do not capture channel fading effects. Therefore, the signal strength does not lend itself to reliable estimates of the probability of packet loss over the network links, which, as discussed in Sect. 8.3.1.1 could drive more intelligent routing decisions. An alternative method to obtain these probabilities is via active probe measurements.

8.3.3 Active Probing-Based Metrics

Inferring the probabilities of data loss in the network links via the signal strength values is one possibility; as discussed in Sect. 8.3.2, the results were not very promising. The alternative approach is to carry out active measurements and use probe packets to directly estimate those probabilities.

Probing introduces various challenges. One concern with it is that it should be treated as normal traffic in the network, e.g., the packet sizes of probes should be equal to the actual traffic data so that what probes measure is as close to the target as possible. Likewise, probe packets should not be prioritized or treated preferentially in the network. On the other hand, if the probing packets are interlaced with the regular traffic (so-called intrusive or in-band measurement), the probes themselves influence the amount of traffic. Ferguson and Huston [12] compare this effect with the Heisenberg Uncertainty Principle. Lundgren et al. [13] and later Zhang et al. [14] observed that the different properties of unicast and broadcast communication in IEEE 802.11 systems may lead to similar effects: probes sent using the broadcast mechanism will report neighbors that are not reachable using unicast communication. Both papers call this phenomenon the *grey-zone* problem.

Even more important concern, in particular when wireless links are involved, is the overhead related to probe messages. The actual probing period is a tradeoff between measurement accuracy and signaling overhead.

Nevertheless, probing-based approaches have proved promising in the context of wireless mesh networks. They measure directly the quantity of interest, rather than inferring it from indirect measurements, and do not rely on analytical assumptions. This is why these metrics have been particularly popular in the last five years. The main novelty came with the expected transmission count (ETX) metric; then a whole family of metrics has emerged out of it that attempts to optimize routing performance under various assumptions for the link rates and the channels used in the network.

8.3.3.1 Per-Hop Round Trip Time

The per-hop round-trip time (RTT) metric reflects the bidirectional delay on a link [15]. To measure the RTT, a probe carrying a timestamp is sent periodically to each neighboring node. Then each neighbor node returns the probe immediately. This probe response enables the sending node to calculate the RTT value. The path RTT metric is simply the addition of the link RTTs estimated over all links in the route. The RTT metric is a load-dependent metric, because it comprises queuing, channel contention, as well as 802.11 MAC retransmission delays. Besides the probe-related overhead, the disadvantage of RTT is that it can lead to route instability (phenomenon of self-interference).

8.3.3.2 Per-Hop Packet Pair Delay

The Per-Hop Packet Pair (PktPair) delay involves the periodic transmission of two probe packets back-to-back, one small and one large, from each node. The neighbor node then measures the interprobe arrival delay and reports it back to the sender. This technique is designed to overcome the problem of distortion of RTT measurements because of queuing delays. The PktPair metric is less susceptible to self-interference than the RTT metric, but it is not completely immune, as probe packets in multihop scenario contend for the wireless channel with data packets. To understand this, consider three nodes A, B, and C in a chain where A sends data to C via B. Data packets sent to node B contend with probe packets of B destined to C. This increases the PktPair metric between B and C and consequently increases the metric along the path from A to C. Performance evaluation on an indoor wireless testbed showed that RTT performed 3–6 times worse than the minimum hop count, Packet Pair or ETX metrics in terms of TCP throughput [16]. As RTT is more sensitive to load, it performs worse than PktPair.

Both the RTT and PktPair metrics measure delay directly, hence they are load-dependent and prone to the self-interference phenomenon. Moreover, the measurement overhead they introduce is $O(n^2)$, where n is the number of nodes. On the contrary, the metrics presented below are load-independent and the overhead they introduce is $O(n)$.

8.3.3.3 Expected Transmission Count

ETX is one of the first routing metrics based on active probing measurements specifically designed for MANETs. Starting with the observation that minimum hop count is not optimal for wireless networks, De Couto et al. [17] proposed a metric that centers on bidirectional loss ratios. ETX estimates the number of transmissions (including retransmissions) required to send a packet over a link. Minimizing the number of transmissions does not only optimize the overall throughput, it does also minimize the total consumed energy if we assume constant transmission power levels, as well as the resulting interference in the network [18]. Let d_f be the expected forward delivery ratio and d_r be the reverse delivery ratio, i.e., the probability that the acknowledgement packet is transmitted successfully. Then, the probability that a packet arrives and is acknowledged correctly is $d_f \cdot d_r$. Assuming that each attempt to transmit a packet is statistically independent from the precedent attempt, each transmission attempt can be considered a Bernoulli trial and the number of attempts till the packet is successfully received a Geometric variable, $\text{Geom}(d_f \cdot d_r)$; therefore, the expected number of transmissions is:

$$\text{ETX} = \frac{1}{d_f \cdot d_r}.$$

The delivery ratios are measured using link-layer broadcast probes, which are not acknowledged at the 802.11 MAC layer. Each node broadcasts a probe packet every second including in its probes the number of probes received from each neighboring node over the last w seconds ($w = 10$ in [17]). Each neighbor of a sender node A can then calculates the d_r value to A each time it receives a probe from node B, as the ratio of the reported count over the maximum possible count w. The whole process is summarized in Fig. 8.4.

Node B reports with the latest broadcast probe the number of probes x received over the previous time window w. Node A estimates the probability that a data packet will be successfully transmitted to B in a single attempt. It also counts the number of probes y received from node B over the same time and gets the ETX value for the link. The ETX along a path is defined as the sum of the metric values of the links forming the path.

The main advantages of the ETX metric are its independence from link load and its account for asymmetric links. In other words, ETX does not try to route around congested links and therefore it is immune to the phenomenon of self-interference. Measurements conducted on a static test-bed network show that ETX achieves up to two times higher throughput than minimal hop-count for long links. ETX is one of the few non hop-count metrics that has been implemented in practice in MANETs, e.g., as part of the OLSR protocol daemon (OLSRD) over multiple platforms [19].

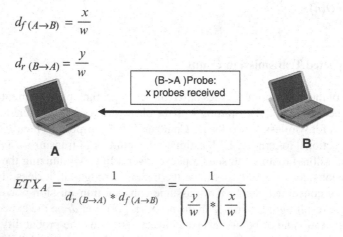

$$d_{f\,(A \to B)} = \frac{x}{w}$$

$$d_{r\,(B \to A)} = \frac{y}{w}$$

(B->A)Probe:
x probes received

A   B

$$ETX_A = \frac{1}{d_{r\,(B \to A)} * d_{f\,(A \to B)}} = \frac{1}{\left(\dfrac{y}{w}\right) * \left(\dfrac{x}{w}\right)}$$

Fig. 8.4 ETX metric estimation for node A

The essential disadvantage of the ETX metric, as already mentioned earlier, is the overhead injected in the network in the form of probe packets. Furthermore, because broadcast packets are small and are sent at the lowest possible rate, the estimated packet loss may not be equal to the actual packet loss of larger data packets sent at higher rates. Moreover, it does not directly account for the link transmission rate; two links with different transmission rates, hence different transmission delays, may have the same packet loss rate. Finally, ETX is only relevant for radio interfaces that perform retransmissions.

8.3.3.4 Expected Transmission Time, Medium Time Metric, and Weighted Cumulative Expected Transmission Time

Draves et al. in [20] observe that ETX does not perform optimally under certain circumstances. For example, ETX prefers heavily congested links to unloaded links, if the link-layer loss rate of congested links is smaller than on the unloaded links. They address this proposing the expected transmission time (ETT) metric incorporating the throughput into its calculation. If S is the size of the probing packet and B the measured capacity of a link, then the link ETT is defined as follows:

$$\text{ETT} = \text{ETX} \times \frac{S}{B}.$$

A similar metric, called medium time metric (MTM), was independently proposed by Awerbuch et al. in [21]. The metric estimate for link l and packet p is a function of the link transmission rate, $\text{rate}(l)$, and the packet size, $\text{size}(p)$, and is given by

$$\tau(l,p) = \frac{\text{overhead}(l) + \frac{\text{size}(p)}{\text{rate}(l)}}{\text{reliability}(l)},$$

where the overhead(l) term accounts for the per-packet overhead of the link that includes control frames, back-off, and fixed headers, and the reliability(l) term equals to the fraction of packets successfully delivered over the link. It is straightforward to see that there is an one-to-one correspondence between the terms: $\text{size}(p) \leftrightarrow S$, $\text{rate}(l) \leftrightarrow B$, and $\text{ETX} \leftrightarrow 1/\text{reliability}(l)$, as used in the equations describing the MTM and ETT metrics, respectively. The only difference between the two metrics lies in the explicit account for MAC-related overheads in the MTM metric; although, it seems that subsequent definitions of the ETT metric have also accounted for this term [22].

Draves et al. propose to use packet-pairing techniques (see Sect. 8.3.3.2) to measure the transmission rate on each link at the expense of additional measurement overhead. On the contrary, Awerbuch et al. recommend the use of interlayer communication, so that the routing layer can have access to relevant information and statistics maintained by the physical and MAC layer. This would require some standard interface that, at least for the moment, is not available on most wireless network adapter cards.

Draves et al. go one step further in their work to suggest computing the path metric as something more than just the sum of the metric values of the individual links in this path. Pure summation of link metrics does not take into account the fact that concatenated links interfere with each other, if they use the same channel. As many wireless technologies, including 802.11a/b/g, provide multiple nonoverlapping channels, they propose an adaptation of the ETT metric accounting for the use of multiple channels, namely the weighted cumulative ETT (WCETT).

Let k be the total number of channels of a network; the sum of transmission times over all hops on channel j is defined as:

$$X_j = \sum_{i \text{ uses channel } j} \text{ETT}_i \quad 1 \leq j \leq k.$$

As the total path throughput will be dominated by the bottleneck channel, which has the largest X_j, they propose to use a weighted average between the maximum value and the sum of all ETTs. This results in the formula:

$$\text{WCETT} = (1 - \beta) \sum_{i=1}^{n} \text{ETT}_i + \beta \max_{1 \leq j \leq k} X_j$$

with $0 \leq \beta \leq 1$ being a tunable parameter. The authors describe different interpretation possibilities for this parameter. In their static test-bed implementation they showed that WCETT outperformed ETX by a factor of two and minimal hop count by a factor of four, when two different IEEE 802.11 radio cards per station were used. The main disadvantage of the WCETT metric is that it is not immediately clear if there is an algorithm that can compute the path with the lowest weight in polynomial or less time.

8.3.3.5 Metric of Interference and Channel Switching

The metric of interference and channel switching (MIC) [23] improves WCETT by addressing the problem of intraflow and interflow interference. The MIC metric of a path p is defined as follows:

$$\text{MIC}(p) = \frac{1}{N \times \min(\text{ETT})} \sum_{\text{link } l \in p} \text{IRU}_l + \sum_{\text{node } i \in p} \text{CSC}_i,$$

where N is the total number of nodes in the network and $\min(\text{ETT})$ is the smallest ETT in the network, which can be estimated based on the lowest transmission rate of the wireless cards. The two components of MIC, interference-aware resource usage (IRU) and channel switching cost (CSC) are defined as:

$$\text{IRU}_l = \text{ETT}_l \times N_l,$$

$$\text{CSC}_i = \begin{cases} w_1 & \text{if } \text{CH}(\text{prev}(i)) \neq \text{CH}(i) \\ w_2 & \text{if } \text{CH}(\text{prev}(i)) = \text{CH}(i) \end{cases} \quad 0 \leq w_1 \leq w_2,$$

where N_l is the set of neighbors that the transmission on link l interferes with, $CH(i)$ represents the channel assigned for node i's transmission and $prev(i)$ represents the previous hop of node i along the path p.

The IRU_l component copes for the interflow interference and corresponds to the aggregate channel time consumed (or the amount of capacity resource consumed) on a link l. In other words, this component includes the expected transmission time for an intended sender as well as the time neighbor nodes have to defer in CSMA/CA MAC protocols and favors a path that consumes less channel time of its neighboring nodes. The CSC component represents the intraflow interference, favoring paths with more diversified channel assignments and penalizing paths with consecutive links using the same channel.

The MIC metric provides better performance because it considers intra/interflow interference and channel diversity. The disadvantage of the metric is the high over-head needed to estimate the per path $MIC(p)$ value. Each node should be aware of the total number of nodes in the network; in large networks, this may become a very expensive operation.

8.3.3.6 Multichannel Routing Metric

Kyasanur and Vaidya [24] extend WCETT in a different direction than MIC does; they take into account the cost of changing channels. Let $InterfaceUsage(i)$ be the fraction of time a switchable interface spends on transmitting on channel i and let $p_s(j)$ be the probability that the used interface is on a different channel when we want to send a packet on channel j. If we assume that the total of the current interface idle time can potentially be used on channel j, we can estimate $p_s(j)$ as

$$p_s(j) = \sum_{\forall i \neq j} InterfaceUsage(i).$$

Let *SwitchingDelay* denote the switching latency of an interface. This value can be measured offline. Then, the cost of using channel j is measured as

$$SC(c_i) = p_s(j) \times SwitchingDelay.$$

To prevent frequent channel switching of the chosen paths, a switching cost is included into the ETT metric, so that the resulting multichannel routing (MCR) metric becomes:

$$MCR = (1 - \beta) \sum_{i=1}^{n} (ETT_i + SC(c_i)) + \beta \max_{1 \leq j \leq k} X_j.$$

Simulation results evaluating the MCR metric have shown that network capacity can be improved by using multiple channels, even if only two interfaces per node are available.

8.3.3.7 Modified ETX and Effective Number of Transmissions

Most of the ETX derivatives described so far in Sects. 8.3.3.4–8.3.3.7 expand the applicability of ETX into various directions not well captured by the original definition of the metric, such as the use of multiple channels that may interfere with each other, and the variation of link transmission rates and packet sizes. Nevertheless, all of them maintain at their core the estimator of successful delivery and expected number of transmissions, as it was coined in the original ETX proposal.

On the contrary, Koksal and Balakrishnan [25] focus exactly on the accuracy of the loss estimator function. The starting point is that, under certain conditions such as links with low average loss rate but high variability, the estimation capacity of the mean statistic is poor. They propose two alternative statistics for the estimation of required number of transmissions over a link.

- Modified ETX (mETX), is defined as $mETX = \exp(\mu + (1/2)\sigma^2)$ with μ being the estimated average packet loss ratio of a link and σ^2 the variance of this value. Like ETX, mETX is additive over concatenated links.
- Effective number of transmissions (ENT) is defined as $ENT = \exp(\mu + 2\delta\sigma^2)$. The parameter δ acts as an additional degree of freedom with respect to mETX; for $\delta = 1/4$, ENT coincides with mETX. Its value depends on the number of subsequent retransmissions, which will cause the link layer protocol to give up a transmission attempt.

Empirical observations of a wireless mesh network suggest that mETX and ENT rate could achieve a 50% reduction in the average packet loss, when compared with ETX.

Measurement-based approaches have two major disadvantages. The first one is the data overhead they impose on the network. The second one has to do with the achievable accuracy and reliability of the measurements. This clearly does not scale for small to moderate error rates, even when measurements are carried out via broadcast packets. These considerations motivated Parissidis et al. [26] to take a different approach. They propose a simple yet accurate interference-aware routing metric based on the estimation of the successful transmission probability on a link in the presence of interference from other nodes in the network.

Compared to probe-based approaches, the advantage of their derivation is that all metric inputs can be available (or estimated) locally at each node, avoiding all measurement-related pitfalls. Performance evaluation in a large set of experiments in the presence of intraflow and interflow interference, shows that their interference-aware routing metric performs at least as good as ETX and minimum hop count across a large set of experiments, because it directly accounts for interference, the primary cause of performance degradation in wireless multihop networks.

8.3.4 Mobility-Aware Metrics

The metrics that are based on active measurements with probe packets, such as the ETX and its derivatives described in Sects. 8.3.3.3–8.3.3.7, outperform the hop count metric in static networks. The situation appears to be reversed in mobile scenarios. As nodes move around, links may come up and down altering the optimal routes in the network. Metrics relying on measurements need some time to update their estimate of the link quality and this may result in significant performance degradation, in particular when routes change multiple times within the duration of the data transfer. On the contrary, the minimum-hop count metric can use the new links almost as quickly as they become available [16].

Mobility-aware metrics aim at the selection of routes with higher expected lifetime to minimize the routing overhead related to route changes and their impact on throughput. The metrics largely use signal strength measurements and their rate of variation to infer the stability of links and routes. The path average degree of association stability, as proposed in the context of associativity-based routing (ABR) in [27], and the affinity metric defined in [28] and reused by the route-lifetime assessment-based routing (RABR) protocol in [29], are example metrics of this category.

8.3.4.1 Link Associativity Ticks and Path Average Degree of Association Stability

Mobile nodes transmit link-layer beacons at fixed time intervals (default value: 1 s) and measure the received number of probs (*associativity ticks*) from their neighbors. These values serve as indicators of the actual stability of the link. Low values of associativity ticks imply mobile nodes in high mobility state, whereas high associativity ticks, beyond some threshold value A_{thr}, are obtained when a mobile node is more stable. The underlying assumption of the metric is that nodes alternate between periods of transition/migration and idleness.

The average degree of association stability over route R, A_{ave}^R, is estimated as a function of the associativity ticks over all links along the route

$$A_{ave}^R = \frac{1}{n} \sum_{l \in R} \mathbf{1}_{A_l \geq A_{thr}},$$

where $\mathbf{1}$ is the logical indicator function and n is the number of links in route R. In ABR, the routes considered for selection are only those with relay load lower than some threshold. The selected route is simply the one with the highest average degree of association stability. In case two routes feature the same average degree of association stability, the route with the minimum hop count is selected.

8.3.4.2 Link Affinity and Path Stability

The link affinity is an estimator of the link lifetime. The affinity of a link l_e is related to the received power over that link P_e, its rate of change, and a threshold P_{THR}, determining whether the link is broken or not. Each node samples periodically, every interval dt, the strength of the signal received over l_e. Defining the signal strength change rate as $\Delta P = (P_e(\text{current}) - P_e(\text{previous}))/dt$ and the average rate of signal strength change as ΔP_e^{ave}, the link affinity is determined by:

$$a_e = \begin{cases} \text{high,} & \text{if } \Delta P_{\text{ave}} > 0 \\ (P_{\text{THR}} - P_e)/\Delta P_e^{\text{ave}}, & \text{if } \Delta P_{\text{ave}} < 0 \end{cases}.$$

The affinity between two nodes A and B is then given by:

$$\eta_{\text{AB}} = \min\ [a_{\text{AB}}, a_{\text{BA}}].$$

The route stability is then given by the minimum of the affinities of all links lying in the route

$$\eta_R = \min_{l \in R} n_l.$$

The route is selected as long as the estimated value for its stability exceeds the required time to transfer data, whose estimate equals the time required to transmit data over the link capacity C. A correction factor f accounts for the imprecision of the metric, so that the check performed for the route is:

$$D_{\text{AB}}/C < f \cdot \eta_{\text{AB}}.$$

If the inequality holds, the route R is selected. Otherwise, the next available route, if it exists.

In both aforementioned approaches, the link metrics are piggybacked on the route discovery packets that propagate from the source towards the destination. The decision upon the route selection is taken at the receiver.

8.3.4.3 Mobility-Model Driven Metrics

Mcdonald and Znati [30] propose another routing metric, which defines a probabilistic measure of the availability of links that are subject to link failures caused by node mobility. They base their considerations on a random walk mobility model. Each node mobility pattern is characterized by three values that describe the statistical distribution of the mean and variance of the speed of a node as well as an average interval time. Together with an estimated communication radius, Mcdonald and Znati derive a sophisticated function, which estimates the expected availability of a link.

Various other metrics were proposed, based on other mobility models. Among them are the metrics described by Gerharz et al. [31] and Jiang et al. [32] that estimate the average residual lifetime of a link. However, the weak link in all these studies is the assumption that all nodes have similar mobility characteristics. In mesh networks, this obviously is not the case.

8.3.5 Energy-Aware Metrics

In contrast with wired networks, energy consumption may represent an essential constraint in wireless mesh networks. Sensors as well as small and battery-operated wireless devices have restricted battery lifetime and are most vulnerable to the energy constraints. Energy-related objectives are often at odds with performance related objectives. For example, choosing paths so that the overall delay(throughput) is minimized may result in overuse of certain nodes in the network and premature exhaustion of their battery. Therefore, energy concerns have to be properly reflected in the definition of routing metrics.

The total energy consumed when sending and receiving a packet is influenced by various factors such as the wireless radio propagation environment, interference from simultaneous transmissions, MAC protocol operation, and routing algorithm. Unsuccessful reception because of interference (external, inter-flow or intra-flow interference) results in retransmissions and higher energy consumption. The essential objectives of routing metrics targeting at minimizing energy consumption are then (1) to minimize overall energy consumption and (2) to maximize the time until the first node runs out of energy.

8.3.5.1 Minimal Total Power Routing

One of the first proposals in energy-aware routing is to minimize the per packet consumed energy. The rationale of the metric, called minimal total power routing (MTPR) metric in [33], is that this way the overall energy consumption is minimized. Singh et al. [34] formalize this idea as follows: let $e_{i,j}$ denote the energy consumed for transferring a packet from node i to the neighboring node j. Then, if the packet has to traverse the path p, including nodes $n_1, \ldots, n_k$, the total energy E required for the packet transfer is

$$E = \sum_{i=1}^{k-1} e_{n_i, n_{i+1}}.$$

Out of the full set P of possible paths, the route p' that minimizes total energy is selected

$$p' = \{p \in P | E^p < E^q, \quad \forall q \in P\}.$$

Interestingly, when considering lightly loaded paths and good links, the MTPR metric tends to yield the same route with the minimum hop-count metric. In those cases, both the overall delay and the energy consumption are proportional to the hop count of the path; hence minimizing the one is equivalent to minimizing the other. The situation changes when we consider error-prone or high-contention links, where more than one transmission attempts are required to get the packet through. Then, the MTPR may select a different route resulting in higher hop count; this is similar to what the ETX metric and its derivatives do, as discussed in Sect. 8.3.3.

A disadvantage of this packet-oriented metric is that it does not directly take into account the nodes' remaining battery lifetimes. It is quite probable that seeking for routes that minimize the per-packet energy consumption, one might end up with nodes that forward traffic from multiple concurrent flows and consume their battery power much faster than other nodes.

8.3.5.2 Minimum Battery Cost Routing

To address the aforementioned problem and balance the energy consumption over all nodes in a network, the battery capacity of a node is taken into consideration in the routing metric definition. The *"minimum battery cost routing"* (MBCR) [35] is based on the remaining battery capacity of the node. The ratio of battery capacity R_{brc} is defined as

$$R_{brc} = \frac{E_i}{E_{max}} = \frac{\text{Battery remaining capacity}}{\text{Battery full capacity}}.$$

Under the assumption that all nodes have the same battery full capacity, a cost value $f_i(E_i)$ is assigned to each node n_i based on its residual battery capacity E_i

$$f_i(E_i) = \frac{1}{E_i}.$$

Then the total available battery cost along a path p is the sum of the battery costs of all nodes along the route

$$R_{brc}^p = \sum_{n_i \in p} f_i(E_i).$$

Out of the full set P of possible paths, the one selected, p', features minimum total battery cost, hence maximum total residual battery capacity

$$p' = \left\{ p \in P | R_{brc}^p < R_{brc}^q, \quad \forall q \in P \right\}.$$

The apparent disadvantage of MBCR is that the selected route may well feature individual nodes with small remaining battery capacity. In Fig. 8.5, for example, path 1 will be selected even though the individual battery value for node 3 is very high ($f_3 = 90$). To address this problem, a classification of nodes in three categories based on the cost value $f_i(E_i)$ is proposed in [36]. The first category consists of nodes with less than 10% of their initial battery capacity. The routing algorithm in this case avoids paths with nodes of this category, as long as there is an alternative path. The second category includes nodes that their remaining energy is between 10 and 20% of their initial energy. This signifies that the nodes are running out of energy and the routing algorithm should also avoid them if possible. Otherwise, a node is not treated specially. Referring to the example illustrated in Fig. 8.5, path 2 would have been selected. Simulation evaluation showed an increase of nodes' lifetimes of up to 65% under low-traffic and up to 25% under heavy-traffic scenarios.

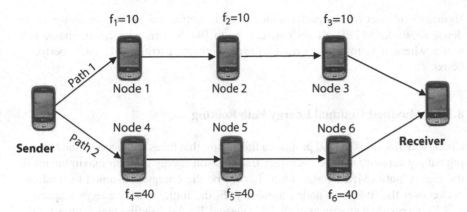

Fig. 8.5 Route selection based on energy cost value. Route 1 is selected over route 2, although it involved one node (node 3) with very low residual battery capacity

8.3.5.3 Min–Max Battery Cost Routing

The min–max battery cost routing (MMBCR) metric [34] addresses more explicitly the shortcoming of the original MCBR metric in avoiding nodes with very low residual battery capacity along paths with high overall battery capacity. The idea is to select a path, which minimizes the maximum power required at any node in a network. In agreement with the formulation in Sect. 8.3.5.2, with MMCBR the chosen path p' must fulfill

$$p' = \min_{p' \in P} \max_{n_i \in p'} f_i(E_i)$$

Simulation results show significant reduction of energy consumption by using shortest-cost routing as opposed to shortest-path routing. However, there is no guarantee that the MMBCR minimizes the total energy consumed over the path, making clear a trade-off between individual node and overall system energy optimization.

8.3.5.4 Conditional Max–Min Battery Capacity Routing

Toh in [37] merges MTPR and MMBCR into one single hybrid routing metric called conditional max–min battery capacity routing (CMMBCR) metric. Firstly, CMM-BCR searches paths using MTPR, with the restriction that all nodes need to have a remaining percentage battery capacity that exceeds a threshold value γ. If there is no such path, MMBCR is used.

The threshold γ effectively operates as a tuning knob that can shape the behavior of the metric towards the one or the other metric; when $\gamma = 0$, the CMMBCR degenerates to the MTPR metric, whereas for $\gamma = 100$ CMMBCR behaves like the MMBCR metric.

Kim et al. [38] compare MTPR, MMBCR, and CMMBCR. Their first finding was that overhearing the transmissions of some neighboring nodes does have a significant impact on the performance of each metric and all behave similarly. In dense networks MTPR allows connections to live longer, whereas in sparse networks where it is more important to avoid network partition MMBCR performs better.

8.3.5.5 Maximal Residual Energy Path Routing

Chang and Tassiulas in [39] propose a link metric that takes into account the remaining battery capacity and the necessary transmission energy for their maximum residual energy path (MREP) algorithm. Let $e_{i,j}$ be the energy consumed to send one packet over the link from node i to node j, E_j the initial battery energy (capacity), and $\underline{E}_j$ the residual energy at node j. Chang and Tassiulas define two metrics for the $i \to j$ link: The remaining energy $d_{i,j}$ of a node, defined as

$$d_{i,j} = \frac{1}{\underline{E}_j - e_{i,j}},$$

and the inverse of the residual capacity of a node in terms of packets that can be delivered with the remaining energy:

$$d_{i,j} = \frac{e_{i,j}}{\underline{E}_j}.$$

Performance evaluation with simulations in scenarios with highly mobility, both metrics came quite close to a theoretically predicted average node lifetime (theoretical values are calculated using linear programming). Refining their work in [40], they propose a more general formula:

$$d_{i,j} = e_{i,j}^{x_i} \underline{E}_i^{-x_2} E_i^{x_3},$$

where x_1, x_2, and x_3 are nonnegative weighting parameters. Simulation evaluation reveal that with reasonable setting of the parameters, the theoretical maximal lifetime, the worst-case lifetime, and the transfer reliability can be well approximated.

8.3.5.6 Power- and Interference-Based Metric

Michail and Ephremides in [41] study the problem of energy-efficient routing in a more concrete context, namely that of connection-oriented traffic. Every node avails one or more radio interfaces and can make use of a set of k frequency channels to communicate with its neighbors. The authors incorporate interference by considering that each transmission *blocks* certain hops (sender–destination node pairs) and seek to minimize both the transmission power required for the communication of nodes i and j and the number of blocked hops (links)

$$c_{i,j} = \begin{cases} \frac{P_{i,j}}{P_{\max}} + \frac{|B_{i,j}|}{|B|}, & \text{if } \sum_{k=1}^{m} f^{(i,j)}(k) > 0 \\ \infty \end{cases},$$

where $c_{i,j}$ is the cost of a link from node i to j, $P_{i,j}$ is the power needed for successful transmission, $P_{\max}$ is the maximum transmission power, $|B_{i,j}|$ is the number of blocked links from i to j, and $|B|$ is the overall number of links in the network. The metric gets a finite value as long as there is at least one frequency channel available for communication between the two nodes.

Their metric is called power- and interference-based metric (PIM). Its performance evaluation is carried out with simulation but is limited to a comparison with another metric considered in the same paper, the minimum power metric MPM, which only considers the energy consumed in each transmission $(c_{i,j} = P_{i,j})$. The results show that PIM outperforms MPM in terms of energy consumption, while achieving better fairness in terms of energy expenditure per node.

8.3.6 Routing Metrics in Standardization Arena

Standardization work with respect to Wireless Multihop Networks is carried out primarily within the IEEE, as part of the standardization work on various aspects of the 802.11x family of protocols. The respective working group is the 802.11s, which, as the time of writing, is working on an IEEE 802.11s standard specification [42]. The standard addresses routing recommending the use of the Airtime Link metric as the default routing metric. The metric is a measure of the amount of consumed channel resources for transmitting a frame over a particular link. The airtime cost c_a for each link is calculated according to the following formula:

$$c_a = \left[O_{ca} + O_p + \frac{B_t}{r} \right] \cdot \frac{1}{1 - e_{fr}},$$

where O_{ca} is the channel access overhead, O_p the MAC protocol overhead, B_t the number of bits of a constant test frame depending on the IEEE 802.11 transmission technology, r the transmission bit rate in Mbit s^{-1} on the current conditions with frame error rate e_{fr}. Interestingly, the airtime metric definition points directly to the MTM and ETT metrics described in Sect. 8.3.3.4.

The Airtime Link Metric parameters for the two main IEEE 802.11 physical layers are listed in Table 8.1.

Standardization work within the internet engineering task force (IETF) has mainly focused on the MANETs; the homonymous working group (WG) has issued various RFCs on routing protocols such as AODV (RFC 3561), OLSR (RFC 3626), and DSR (RFC 4728). Work on network performance metrics is carried out within the IPPM WG. Although the work is quite general and does not focus on routing, there are several RFCs addressing practical aspects of measurements that have direct

Table 8.1 Airtime link metric constants

Parameter	802.11a	802.11b
Channel access overhead: O_{ca}	75 µs	335 µs
Protocol overhead: O_p	110 µs	364 µs
Number of bits in test frame: B_t	8,224	8,224

application to the area of routing metrics as well. Examples are the RFC 2680 on one-way packet loss, and RFC 2681 on round-trip delay.

8.4 Taxonomy

In Table 8.2, the routing metrics presented in Sect. 8.8.3 are classified according to the criteria selected in Sect. 8.8.2, namely the optimization objective, method used to acquire the needed information to compute the metric, and function used to compute the metric along a path.

8.5 Thoughts for Practitioners

The presentation of routing metrics and their inline discussion lend themselves to several conclusions that could be of interest to wireless mesh networking practitioners:

- *There is no "one size fits all" solution for routing in wireless mesh networking.* This is no surprise, because the principle applies to many different areas of network engineering. There is a great variety of protocols, which have been proposed with different applications and priorities in mind. For example, energy-aware metrics are more appropriate for sensor networks or low-end, battery-powered devices that must operate without access to electricity grid for large intervals of time. Metrics that rely on active probing appear to have superior performance well in static wireless mesh networks. On the contrary, in high-mobility scenarios, mobility-aware metrics may result in selection of better routes.
- *Simplicity does not always pay off.* Shortest-path routing has seen enormous success in wired networks, primarily because of its simplicity. The combined dynamics of the wireless radio propagation, interference, node mobility, and, where relevant, energy constraints result in error-prone links and highly dynamic network topology, making the routing task much more challenging. The minimum hop count metric is not adequate in these cases, if optimum performance is sought after.
- *There are multiple tradeoffs amongst routing metrics, even when their objectives are identical.* The result is high flexibility at network configuration level and the

Table 8.2 Taxonomy of routing metrics

Metrics	Optimization objectives	Metric computation method	Path metric function
Hop count	Minimize delay	Use of locally available information	Summation
Signal Strength based			
Preemptive routing [7]	Maximize expected route lifetime	Use of locally available information	Not defined, routing algorithm decision
SSAR [9]			
Link quality factor [10]			
Active Probing			
Per hop RTT [16]	Minimize delay	Active probing	Summation
Per hop PktPair [16]	Maximize probability of data delivery		
ETX [17]			
ETT [20]			
MTM [21]			
WCETT [20]			
MCR [24]			
Modified ETX [25]			
ENT [25]			
MIC [23][1]			
Interference-aware			
Interference-aware [26]	Minimize delay	Use of locally available information	Summation
	Maximize probability of data delivery		
Mobility aware			
ABR [27]	Maximize expected route lifetime	Active probing	Not defined, routing algorithm decision
Link affinity metric [28]		Metrics piggybacked to route discovery packets	
Energy-aware			
MTPR [33]	Minimize energy consumption	Use of locally available information[2]	Summation
MBCR [35]			
CMMBCR [37]			
MREP [39]			
PIM [41]			
MMBCR [34][3]			
Standardization			
AirTime [42]	Minimize delay	Active probing	Not defined, routing algorithm decision
		Use of locally available information	

[1] Equally distribute traffic load
[2] The routing algorithm disseminates the information needed to calculate the metric
[3] Order statistics (min–max)

possibility to tailor the routing behavior to the requirements and constraints of a particular network scenario. For example, link loss estimation accuracy can be traded with control data overhead; and network/system battery lifetime can be compromised with optimum application performance.

- *There is a large margin for improvement of current network card hardware.* What seems to be lacking at the moment is a clean interface between the network/routing layer and the lower layers of the radio interface, which would allow to take advantage of all the information and state maintained at the lower radio layers. Proposed routing metrics based on signal strength appear to be hard constrained by the limited monitoring and reporting capabilities of the network cards.

8.6 Directions for Future Research

There are several issues with respect to routing metrics for wireless mesh networks that could benefit from further research. We discuss two of them, which in our opinion are the most important ones.

Multiple access interference has always been one of the main concerns when building wireless networks. Whereas its impact is quite well understood and addressed in infrastructure-based cellular networks, its characteristics and impact in wireless mesh networks are less straightforward. There is consensus in the research community that the level of interference should be an input for routing protocols. Most routing metrics that have been proposed to overcome the inefficiencies of minimum hop count routing rely on active probing methods to drive routing decisions. The main disadvantages of these approaches are that they impose additional overhead and they suffer from inaccuracy and responsiveness to network node mobility. However, none of these metrics capture the impact of interference explicitly. In fact, it is not even clear how to best measure interference [22]. Interference-aware routing can avoid the pitfalls of the measurement-based approaches, and is an open area of research that could result in improvements in path metric computation and consequently in route selection.

Currently the 802.11x suite of standards does not provide much information to higher layers. The only channel quality measure reported from commodity wireless adapters is the "Received Signal Strength Indicator" (RSSI) value, which is also vendor-dependent. However, standardization efforts within IEEE 802.11 are preparing standards (802.11k [43] for wireless LANs and 802.11s [42] for wireless mesh networks), which will enable higher layers to obtain detailed channel condition information from the PHY and the MAC layers and provide additional flexibility with respect to transmit power control. These standards will include signal strength measurements and neighbor reports containing information on neighboring nodes as well as link quality metrics such as the Airtime metric. The use of this information to develop more sophisticated and efficient routing metrics is expected to be an area for future research.

8.7 Conclusions

This chapter presented an overview of routing metrics specifically designed for wireless mesh networks. Whereas in wired networks the hop count metric remains the most attractive solution, in wireless mesh networks interference, link asymmetry, mobility, and energy-related considerations represent additional challenges that require more elaborate solutions. The proposed routing metrics address partially the aforementioned issues considering different optimization objectives and applying various techniques such as signal strength measurements, active probing, energy consumption monitoring, and prediction of link breakage because of node mobility. Further improvement over what is achievable today is expected through better understanding the impact of interference and the exploitation of MAC and PHY layer information that will be available from commodity wireless adapters in the future.

8.8 Terminologies

1. *Routing metric.* A routing metric is a value assigned by a routing algorithm and used to determine whether one route performs better than another.
2. *Hop count.* Hop count is the number of wireless links (hops) traversed by a packet from its source to its destination.
3. *Minimum hop count.* The minimum number of "hop" among all available paths between a source and a destination (shortest route).
4. *Link metric.* A value quantifying the quality of a link. This value is used by the routing algorithm to determine a route between a source and a destination.
5. *Path cost function.* A function to derive the path metric from the individual link metric values estimated for each link in the path.
6. *Path metric.* The cost of a path estimated out of the metrics of the path links by use of the *path cost function*.
7. *Active probe measurement.* A method of measuring the properties of a link/path, whereby special packets (probes) are generated and sent periodically to infer properties of a link.
8. *ETX.* Popular active probe measurement metric predicting the number of retransmissions required to deliver a packet all the way to its destination.
9. *Self-interference.* Once a link is recognized as good, it is chosen by the routing protocol and starts being used till it gets overloaded and is assigned with a worse metric value. As traffic starts being routed around this link, its metric value increases again and the effects starts anew. The phenomenon of self-interference results in route oscillations.
10. *Airtime link metric.* A measure of the amount of consumed channel resources for transmitting a frame over a particular link. Recommended metric within the forthcoming 802.11s standard.

8.9 Questions

1. Mention possible relations between link metrics and the respective path metrics. Give one example of path metric that results from the summation and another that results from the multiplication of link metrics.
2. Which are the four ways used generally by (wireless multihop) network nodes to obtain the information they require for the computation of the actual routing metric value? List them in increasing order of control traffic overhead they generate.
3. What is the self-interference phenomenon when we refer to routing metrics?
4. Explain why the minimum hop count routing metric does not always yield minimum delay paths in wireless multihop networks.
5. Why is the ETX metric less prone to the self-interference phenomenon than the RTT metric and the PktPair Delay metric?
6. What are the disadvantages of the ETX metric?
7. What drawbacks of ETX does each of its derivatives address (ETT, WCETT, mETX).
8. Describe the main trade-off introduced by the MTPR and the MMBCR metric. How does the CMMBCR metric combine properties of the two metrics?
9. Why active probe measurements perform less satisfactorily in scenarios with high mobility?
10. What is the recommended routing metric in the IEEE 802.11s forthcoming standard? Which one(s) of the reviewed routing metrics does it resemble?

References

1. G. Apostolopoulos, R. Guerin, S. Kamat, and S.K. Tripathi. Quality of service based routing: A performance perspective. In Proceedings ACM Sigcomm, Vancouver, BC, Canada, October (1998).
2. A. Khanna and J. Zinky. The revised arpanet routing metric. ACM Sigcomm Review, 19(4), 45–56, (1989).
3. T. Clausen and P. Jacquet. Optimized link state routing protocol (OLSR). IETF RFC 3626, October (2003).
4. D. Johnson and D. Maltz. Dynamic source routing in ad hoc wireless networks. Mobile Computing, Kluwer, Dordrecht, The Netherlands, (1996).
5. C. Perkins and P. Bhagwat. Highly dynamic destination-sequenced distance-vector routing (dsdv) for mobile computers. Proceedings ACM Sigcomm, London, UK, October (1994).
6. C. E. Perkins and E. M. Royer. Ad hoc on-demand distance vector routing. Proceedings of Second IEEE Workshop on Mobile Computing Systems and Applications, New Orleans, LA, 90–100, February (1999).
7. T. Goff, N. Abu-Aahazaleh, D. Phatak, and R. Kahvecioglu. Preemptive routing in ad hoc networks. Proceedings ACM Mobicom, Orlando, FL, USA, July (2001).
8. D. Aguayo, J. Bicket, S. Biswas, G. Judd, R. Morris. Link-level measurements from an 802.11b mesh network. Proceedings ACM Sigcomm, Portland, OR, USA, August (2004).
9. R. Dube, C. Rais, K. Wang, and S. Tripathi. Signal stability based adaptive routing (SSA) for ad hoc mobile networks. IEEE Personal Communications, 4(1), 36–45, (1997).

10. R. Punnose, P. Nitkin, J. Borch, and D. Stancil. Optimizing wireless network protocols using real time predictive propagation modeling. Proceedings IEEE Rawcon, Denver, CO, USA, August (1999).

11. J. Zhao and R. Govindan. Understanding packet delivery performance in dense wireless sensor networks. Proceedings ACM SenSys, Los Angeles, CA, USA, November (2003).

12. P. Ferguson and G. Huston. Quality of Service – Delivering QoS on the Internet in Corporate Networks, Wiley, New York, NY, (1998).

13. H. Lundgren, E. Nordström, and C. Tschudin. Coping with communication gray zones in IEEE 802.11b based ad hoc networks. Proceedings IEEE WoWMoM, New York, NY, USA, September (2002).

14. H. Zhang, A. Arora, and P. Sinha. Learn on the fly: Data-driven link estimation and routing in sensor network backbones. Proceedings IEEE Infocom, Barcelona Spain, April (2006).

15. A. Adya, P. Bahl, J. Padhye, A. Wolman, and L. Zhou. A multi-radio unification protocol for IEEE 802.11 wireless networks. Proceedings BroadNets, San Jose, CA, USA, October (2004).

16. R. Draves, J. Padhye, and B. Zill. Comparison of routing metrics for static multi-hop wireless networks. Proceedings ACM Sigcomm, New York, NY, USA, 133–144, August (2004).

17. D. De Couto, D. Aguayo, J. Bicket, and R. Morris. A high throughput path metric for multi-hop wireless routing. Proceedings ACM Mobicom, San Diego, CA, USA, September (20030.

18. C. E. Koksal and H. Balakrishnan. Quality aware routing in time-varying wireless networks. IEEE Journal on Selected Areas of Communication Special Issue on Multi-Hop Wireless Mesh Networks, 24(11), 1984–1994, (2006).

19. T. Lopatic. OLSRD link quality extensions. Internet page: http://www.olsr.org/docs/READMELink-Quality.html, December (2004).

20. R. Draves, J. Padhye, and B. Zill. Routing in multi-radio, multi-hop wireless mesh networks. Proceedings ACM Mobicom, New York, NY, USA, 114–128, (2004).

21. B. Awerbuch, D. Holmer, and R. Rubens. The Medium Time Metric: High throughput route selection in multi-rate ad hoc wireless networks. Springer Mobile Networks and Applications, 11(2), 253–266, (2006).

22. V. Bahl. A crash course in mesh networking. Proceedings ACM Sigcomm 2006 Tutorial, Pisa, Italy, September (2006).

23. Y. Yang, J. Wang, and R. Kravets. Designing routing metrics for mesh networks, Proceedings IEEE WiMesh, Santa Clara, CA, USA, (2005).

24. P. Kyasanur and N. H. Vaidya. Routing and link-layer protocols for multi-channel multi-interface ad hoc wireless networks. ACM Sigmobile Mobile Computing and Communications Review, 10(1), 31–43, (2006).

25. C. E. Koksal and H. Balakrishnan. Quality aware routing in time-varying wireless networks. Journal on Selected Areas of Communication Special Issue on Multi-Hop Wireless Mesh Networks, 24(11), 1984–1994, (2006).

26. G. Parissidis, M. Karaliopoulos, M. May, T. Spyropoulos, and B. Plattner. Interference in wireless multihop networks: A model and its experimental evaluation. Proceedings IEEE WoWMoM, Newport Beach, CA, USA, June (2008).

27. C. -K. Toh. Associativity-based routing for ad hoc mobile networks. IEEE Personal Communications, 4(2), 103–139, (1997).

28. K. Paul, S. Bandyipadhyay, A. Mukherjee, and D. Saha. Communication-aware mobile hosts in ad-hoc wireless network. Proceedings IEEE ICPWC, Jaipur, India, February (1999).

29. A. Agarwal, A. Ajuja, J. P. singh, and R. Shorey, Route-lifetime based routing (RABR) protocol for mobile ad-hoc networks. Proceedings IEEE ICC, New Orleans, LA, USA, June (2000).

30. A. Mcdonald and T. Znati. A path availability model for wireless ad-hoc networks. Proceedings IEEE Wireless Communications and Networking Conference (WCNC), New Orlean, LA, USA, September (1999).

31. M. Gerharz, C. de Waal, M. Frank, and P. Martini. Link stability in mobile wireless ad hoc networks. Proceedings IEEE LCN, 30–39, Tampa, FL, USA, November 2002.

32. S. Jiang, D. He, and J. Rao. A prediction-based link availability estimation for mobile ad hoc networks. Proceedings IEEE Infocom, Anchorage, AL, USA, 1745–1752, April (2001).

33. K. Scott and N. Bambos. Routing and channel assignment for low power transmission in PCS. Proceedings IEEE ICUPC, Cambridge, MA, USA, vol. 2, 498–502, October (1996).
34. S. Singh, M. Woo, and C. Raghavendra. Power-aware routing in mobile ad hoc networks. Proceedings Fourth Annual International Conference on Mobile Computing and Networking, Dallas, TX, USA, 181–190, October (1998).
35. J.-P. Sheu, C.-T. Hu, and C.-M. Chao. The Handbook of Ad Hoc Wireless Networks, Chapter Energy-Conserving Grid Routing Protocol in Mobile Ad Hoc Networks, CRC, Boca Raton, FL, (2003).
36. N. Gupta and S. R. Das. Energy-aware on-demand routing for mobile ad hoc networks. Proceedings International Workshop on Distributed Computing, Mobile and Wireless Computing, London, UK, vol. 2571, 164–173, January (2002).
37. C.-K. Toh. Maximum battery life routing to support ubiquitous mobile computing in wireless ad hoc networks. IEEE Communications Magazine, 39(6), 138–147, (2001).
38. D. Kim, J. Garcia-Luna-Aceves, K. Obraczka, J. Cano, and P. Manzoni. Performance analysis of power-aware route selection protocols in mobile ad hoc networks. Proceedings Networks, Atlanta, GA, USA, August (2002).
39. J. Chang and L. Tassiulas. Maximum lifetime routing in wireless sensor networks. Annual Allerton Conference on Communication, Control, and Computing, Monticello, IL, USA, September (1999).
40. J. Chang and L. Tassiulas. Energy conserving routing in wireless ad-hoc networks. Proceedings IEEE Infocom, Tel Aviv, Israel, March (2000).
41. A. Michail and A. Ephremides. Energy-efficient routing for connection-oriented traffic in wireless ad-hoc networks. Mobile Networks and Applications, 8(5), 517–533, (2003).
42. IEEE, Draft amendment: ESS mesh networking, IEEE P802.11s Draft 1.00, November (2006).
43. IEEE, Draft amendment, IEEE P802.11k/D9.0, Sep (2007).

Chapter 9
Reliable Transport in Multihop Wireless Mesh Networks

Simon Heimlicher and Bernhard Plattner

Abstract Traditional wireless access networks require a large number of stationary access points. Multihop wireless mesh networks use relaying among nodes to cover the same area with a smaller number of access points. However, communication among mobile nodes is challenging since mobility and radio signal propagation impairments lead to frequent disruptions of end-to-end paths, which violates fundamental assumptions of widely used network protocols such as TCP. In this chapter, we explore measures to improve the network performance and the user experience in multihop wireless mesh networks.

9.1 Introduction

In this chapter, we look at a feature pertaining to wireless communication among mobile nodes: *intermittent connectivity*. Intermittent connectivity refers to the phenomenon that end-to-end routes are frequently disrupted temporarily or permanently. We focus our discussion on *multihop wireless mesh networks*. In this class of mesh networks, data is relayed among mesh nodes such as laptops, handheld devices, or cellular phones, and access points connected to the Internet. This increases the coverage as compared to single-hop mesh networks, where every node needs to be connected to an access point directly. Note that intermittent connectivity is a serious issue in any multihop wireless network, including sensor and ad hoc networks.

The most prominent causes for intermittent connectivity are node mobility and radio signal propagation impairments. Further causes are temporary or permanent node failures, which may occur because of power saving efforts or complete loss of

S. Heimlicher (✉)
Computer Engineering and Networks Laboratory (TIK), ETH Zurich, Gloriastrasse 35, 8092 Zurich, Switzerland
e-mail: heimlicher@tik.ee.ethz.ch

S. Misra et al. (eds.), *Guide to Wireless Mesh Networks*, Computer Communications and Networks, DOI 10.1007/978-1-84800-909-7_9,
© Springer-Verlag London Limited 2009

power of battery-driven devices. In Definition 9.1, we define a first set of terms that will help us to further discuss this issue.

Definition 9.1 (Intermittent Connectivity).

- *Disruption.* A period of disruption between a pair of nodes is a period during which no end-to-end connectivity is available among these nodes.
- *Disruptive network.* A network that must be assumed to sometimes be disrupted is called a disruptive network.
- *Disconnected network.* A disruptive network that in general lacks end-to-end connectivity is referred to as a disconnected network.
- *Intermittently connected network.* A disruptive network that in general provides connectivity but where routes are frequently disrupted is termed an intermittently connected network.

9.1.1 Impact of Intermittent Connectivity

The majority of mobile communication today is mobile-to-fixed, i.e., communication between a mobile node and a fixed cellular base station or a wireless access point, but mobile-to-mobile communication is increasing in popularity. Multihop wireless mesh networks are currently being deployed in cities around the world. In such networks, nodes in range of an access point relay data on behalf of nodes that are further away, extending the coverage of each access point. Upcoming mobile devices such as portable music players, pocket organizers, and cellular phones that are equipped with wireless network interfaces are predestined to participate in multihop wireless mesh networks.

Definition 9.2 (Scenario).

- *Multihop wireless mesh network.* A multihop wireless mesh network comprises a set of gateways, i.e., access points that are connected to the Internet, and a set of mesh nodes in need of Internet access. These nodes relay data on behalf of each other to increase the coverage of the access points.

In the early days of wireless networking, it was conjectured that all wired and wireless networks would converge at the network layer. In simulation experiments of routing protocols for mobile wireless networks [1–3] it appeared as though end-to-end connectivity was merely a matter of a sufficiently high node density and a decent routing protocol.

More recently, real-world measurement studies of wireless link capacity [4, 5] revealed that wireless links are frequently disrupted even among stationary nodes, invalidating fundamental assumptions of simulation studies. Depending on the scenario, node density may be so sparse in real-world wireless networks [6], that end-to-end connectivity becomes infeasible [7]. Even if an end-to-end path exists for some time, signal propagation impairments, interference, and node mobility incur frequent network partitioning.

Under intermittent connectivity, the performance of popular applications that depend on reliable end-to-end communication, such as web browsing, e-mail, or instant messaging, suffer. The most widely used network protocol family is the TCP/IP [8, 9] protocol suite, which has served as the core of the Internet for three decades. When the TCP/IP protocol suite was conceived, all networking was wired and this is reflected in the design of these protocols. Wired networks are in general connected over reliable links and disruptions are not an issue; rather it is assumed that an end-to-end path exists between source and destination for the whole duration of the communication.

The original objective of the TCP/IP protocol suite was to interconnect heterogeneous, wired networks. In the TCP/IP protocol stack, providing a service to the applications between the two end points of a connection is the task of the transport layer. The heterogeneity of the networks to be connected required a transport layer implementation that made minimal assumptions about intermediate network elements and resulted in the end-to-end transport layer of the current Internet architecture.[1]

An end-to-end transport protocol is implemented only at the end points, i.e., at the source and the destination node, of a connection. This allows its use over almost any intermediate network that is capable of forwarding data. However, the performance of end-to-end transport protocols in intermittently connected networks is limited because such protocols transfer data only when a connected path is available.

9.1.2 Transport Principles

In this chapter, we *revisit* the decision to implement the transport layer end-to-end. There are several alternatives to end-to-end transport that may provide better performance under intermittent connectivity. One approach is called *custody transfer* and is investigated in the context of delay- and disruption-tolerant networking [11]. In this context, it is assumed that source and destination nodes are rarely, if ever, connected and the only way of communication is by asynchronous multihop forwarding. Custody transfer is implemented in the bundle protocol [12], which operates on self-contained *bundles* of packets and moves these entities from one node to the next in a store-and-forward manner. Because the bundle protocol is aimed specifically at disconnected networks, it operates without an end-to-end control loop and lacks the end-to-end semantics provided by reliable transport protocols such as TCP.

For multihop wireless mesh networks, a more suitable alternative is a distributed implementation of the transport layer along the network nodes that lie on the data transport path; we refer to this approach as *hop-by-hop transport*. The performance of hop-by-hop transport is superior to the end-to-end alternative in many aspects but it comes at the price of higher protocol complexity and additional memory and processing requirements. Hop-by-hop transport is not a new paradigm at all. In wired networks, hop-by-hop transport was considered for instance as a means to reduce

---

[1] In [10], the safe operating area of TCP's congestion control algorithm is discussed in terms of packet error rates and bandwidth × delay product.

the buffer size requirements in high-speed network routers [13, 14] or to improve the efficiency of the network as a whole [15, 16].

In multihop wireless mesh networks, hop-by-hop transport may transfer data along partial paths and deliver the same amount of data with less packet transmissions and lower latency. In general, a reduced number of link transmissions results in lower energy consumption and interference and also has a favorable impact on the capacity of the network. Furthermore, implementing flow and congestion control on a per-hop basis allows for finer-grained adjustment of the transmission rate, retransmission timeout and other transmission parameters and provides higher use. These factors make hop-by-hop transport solutions attractive in stationary and mobile wireless environments, and we will study this approach in detail in Sect. 9.4. In the following definition, we recap the three data transport principles we have introduced so far.

Definition 9.3 (Transport Principles).

- *End-to-end transport.* A protocol implementing end-to-end transport only runs at the end points of the connection. This means that no transport layer protocol is required at intermediate nodes, allowing end-to-end transport to run over any intermediate network elements that implement the network layer.
- *Hop-by-hop transport.* A protocol implementing hop-by-hop transport runs at all nodes that lie on the network path, allowing data transfer to be controlled at each link separately. In particular, lost data can be retransmitted by the node before the link where the loss occurred. In addition, hop-by-hop transport uses end-to-end acknowledgment to provide reliability.
- *Custody transfer.* Custody transfer is a way of transferring a bundle of packets from a source to a destination node in a network where no end-to-end connectivity can be assumed. Under this paradigm, the responsibility to deliver the bundle always rests with the node that carries the bundle and moves to the next node upon successful transfer. Note that custody transfer on purpose does not provide end-to-end reliability and cannot be used as a direct replacement of reliable transport protocols such as TCP.

The rest of this chapter is organized as follows. In Sect. 9.2, we review prior work on transport layer issues in mobile wireless networks. In Sect. 9.3, we discuss in detail hop-by-hop transport. In Sect. 9.4, we give recommendations to practitioners, and in Sect. 9.5, we outline future work. Section 9.6 concludes the chapter. Important terms are summarized in Sect. 9.7; review questions and references are provided in Sects. 9.8 and 9.9, respectively.

9.2 Background

In this section, we first provide an overview of prior work aiming to improve the performance of TCP/IP in wireless networks. We then introduce two alternative transport schemes aimed at disruptive networks.

9.2.1 Prior Work

The design of the TCP/IP protocol suite reflects the characteristics of the scenario it was created for: a network comprising only wired links and fixed hosts running on mains power. In the past 30 years, many new scenarios have come up that differ substantially from TCP/IP's original target environment.

As a consequence, a lot of effort was put into analyzing and improving the performance of the TCP/IP suite in wireless communication scenarios. There is broad consensus that TCP in particular is not suitable for wireless networks of an intermittently connected nature. TCP was found to:

- Under-use links because of its congestion control algorithm [17]
- Frequently incur route failures when probing for available bandwidth [18]
- Share communication bandwidth unevenly among short and long (in terms of hop count) connections [19]

Over the past decade, countless patches were proposed to enhance TCP performance over wireless links. For a survey of these efforts, please refer to [17, 20]. Three approaches have been investigated in detail. The straightforward approach is (1) to improve the reliability of the link layer [21] and its interaction with TCP [22]. As shown by Vaidya et al. [23], the sender can only distinguish the causes of packet loss with assistance from intermediate nodes or the destination. Enhancing the sender to react appropriately to the cause of the packet loss (2) was studied in [24–26]. However, even if the sender responds appropriately to losses, in wireless environments, the end-to-end loss recovery mechanism employed by TCP is overly expensive in terms of time, bandwidth, and energy, which is particularly harmful in energy-constrained mobile networks. A further approach is to split the connection at the last stationary node, i.e., the cellular base station or wireless access point, and then running a separate control loop on the last hop [27–30]. This approach, however, is limited in applicability. A more general sibling of the split-connection approach is hop-by-hop transport, and will be discussed in detail in Sect. 9.3.

9.2.2 Alternatives to End-to-End Transport

In the presence of frequent disruptions, the transport layer needs to be tolerant to extended periods lacking end-to-end connectivity. In this section we discuss prior work on two alternatives to end-to-end transport: hop-by-hop transport and custody transfer, defined in Sect. 9.1.2 in Definition 9.3.

The major difference between hop-by-hop transport and custody transfer is their target scenario. Hop-by-hop transport aims to use both connected and disconnected periods and provide the same semantics as end-to-end transport. Custody transfer is based on the assumption that no end-to-end path may ever exist, and providing end-to-end reliability is not attempted.

9.2.3 Hop-by-Hop Transport

Even though hop-by-hop transport is similar to link layer retransmission schemes in that it also recovers from packet losses locally, hop-by-hop transport runs above the routing protocol and is able to also retransmit lost packets over a new next hop after a route change. The major advantages of hop-by-hop vs. end-to-end transport are (1) hop-by-hop transport is able to forward data towards the destination without an end-to-end route as long as some parts of the path are available and (2) hop-by-hop transport recovers from packet loss locally.

Hop-by-hop has been considered as an alternative to end-to-end transport for more than three decades. Gitman [31] evaluates packet delay and channel use under hop-by-hop and end-to-end retransmission in an analytical model of an early wireless network with an uncorrelated packet loss process in store-and-forward packet-switching communication networks. In his model, he focuses on the packet radio network (PRN), which was an early wireless network in the 1970s. He uses variable link failure probability to model congestion that is higher in the vicinity of super nodes called stations. He observes that hop-by-hop retransmission leads to lower delay and higher channel use for packets that traverse many hops or if the channel's packet loss rate is high.

In [32], DeSimone et al. investigate the interaction between link layer and transport layer retransmissions analytically and using simulation under the assumption of uncorrelated packet loss. Their results clearly show that the link layer retransmissions are much more effective against high link packet loss rates. They observe a negative impact of link layer retransmissions on the throughput of the end-to-end transport protocol if packet loss is lower than a certain threshold.

Note that [31] and [32] only consider uncorrelated packet loss. Intermittent connectivity is characterized by extensive periods of disruption, which cannot be captured without correlation in the loss process. In Sect. 9.3, we will consider a model of intermittent connectivity based on a correlated loss process.

One reason that hop-by-hop transport is not supported by the TCP/IP suite are scalability concerns relating to its higher complexity. As hop-by-hop transport requires per-flow state management in intermediate nodes, it appears to be limited in scalability. However, for instance in [16], Kortebi et al. suggest that per-flow fair queuing is feasible even in large-scale high-speed networks because the number of active flows needing scheduling is counted in hundreds. In [13], a reliable hop-by-hop flow control protocol aimed at ATM virtual circuits is proposed that is claimed to be resilient to errors and allow efficient buffer sharing. In [14], an analytical model and simulation experiments are used to compare end-to-end and hop-by-hop transport in a high-speed wired network. It is found that the hop-by-hop scheme adapts to changes in the traffic intensity more quickly and thus uses resources at the bottleneck more effectively than the end-to-end scheme. In [15], hop-by-hop flow control is evaluated against end-to-end flow control in the context of high-speed wired networks. The authors find that the hop-by-hop scheme achieves lower packet loss, requires smaller buffers throughout the network, and provides lower latency.

However, to the best of our knowledge, not too many concrete hop-by-hop transport protocols for wireless environments are available. Kopparty et al. [19] report that in mobile networks, long (multihop) TCP connections suffer more severely from route failures due to mobility than short connections. They propose a protocol called Split TCP, which splits TCP connections crossing many hops at some of the intermediate nodes. If a node is a so-called proxy, it buffers TCP packets and acknowledges them to the previous proxy by sending a local acknowledgment packet. The buffered packets are then forwarded as usual and buffered again at the next proxy. End-to-end reliability is provided by end-to-end acknowledgment and retransmission. In simulation experiments, the authors find that Split TCP provides performance improvements between 10% and 50%. In [33], Sundararaj et al. provide an analytical treatment of the concatenation of multiple TCP connections based on the well-known analytical model by Padhye et al. [34].

Split TCP was one of the first protocols to propose hop-by-hop transport, but it still relies on the basic TCP mechanisms. CAT [18] is a more recent hop-by-hop protocol that is targeted at multihop wireless scenarios. CAT employs a rate-based flow and contention control mechanism that evaluates the level of contention to calculate the sending rate at every intermediate hop. This particular feature appears to be of tremendous value in wireless scenarios as it reduces the number of packets lost because of bandwidth probing dramatically. CAT is reported to deliver similar performance improvements as Split TCP.

In [35], Yi et al. present an optimization framework for congestion control algorithms in multihop networks with the constraint imposed by typical wireless MAC protocols. This theoretical work suggests that distributed congestion control is feasible in wireless multihop networks. However, the generality of the derivation is limited by the assumptions that there is no packet loss and link capacities are time-invariant.

Real-world deployments of wireless mesh networks [36–38] and testbeds [39] exist, but only few measurement studies of alternative transport layer protocols are available. In [40], a hop-by-hop congestion control protocol is evaluated in a wireless mesh network testbed and the authors find an improved performance as compared to both TCP and the ad hoc transport protocol (ATP) [41].

9.2.4 Custody Transfer

Communication in a disconnected network is investigated in the context of *delay-tolerant networking* (DTN). The DTN Research Group (DTNRG) [11] develops the bundle protocol [12]. This delay- and disruption-tolerant overlay transport protocol is based on the concept of custody transfer. Custody transfer is a principle of data transfer that is similar to hop-by-hop transport but does away with the end-to-end control loop. This allows custody transfer to operate even in a network where no end-to-end path exists between source and destination. One could see custody transfer as an unreliable version of hop-by-hop transport. In wireless mesh networks, we assume that nodes are connected most of the time and we do not discuss further disconnected communication approaches.

In Sect. 9.3, we will have a closer look at hop-by-hop transport using simulation experiments and analytical modeling.

9.3 Hop-by-Hop Transport

In this section, we look at hop-by-hop transport as one way of providing reliable communication in intermittently connected environments. We first estimate upper bounds on the throughput of hop-by-hop and end-to-end transport, then we present a simulation case study comparing a simple hop-by-hop transport protocol against TCP in a mobile wireless multihop network. Finally, we present numerical results from an analytical model of hop-by-hop and end-to-end transport over an intermittently connected multihop path.

9.3.1 Theoretical Bounds

The consideration of hop-by-hop transport may be due to multiple reasons. In wired networks, hop-by-hop transport provides finer-grained control over the flow of data between two nodes and it allows lower buffer capacities. Under intermittent connectivity, we are primarily interested in the capability of hop-by-hop transport to bring data closer to the destination even during disruptions, and to recover from packet loss on a per-hop basis.

For a chain topology, it is straightforward to show analytically that the upper bound on the achievable throughput is higher with hop-by-hop transport. Consider a source–destination pair communicating over a chain of H wireless links (hops). Assume that every link is only available a fraction p of the time, as a result of the wireless environment, node mobility, and the routing protocol. An upper bound for the achievable link throughput can be obtained under the following simplifying assumptions.

With end-to-end transport control, data can only be transferred when *all* links are available contemporaneously. During those periods, the throughput, denoted by T^e, is bounded by the link capacity C, i.e.,

$$T^e \leq \begin{cases} C, & \text{all links available;} \\ 0, & \text{otherwise.} \end{cases}$$

Given the availability p, we can write:

$$T^e \leq p^H \cdot C.$$

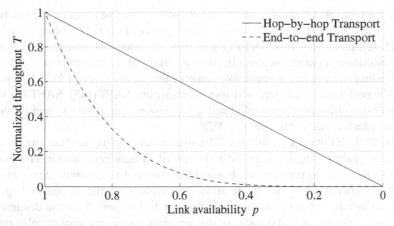

Fig. 9.1 Normalized throughput of hop-by-hop vs. end-to-end transport

In contrast, hop-by-hop transport is capable of utilizing individual links given that the node at the sending end of each link has data to send. Thus, the throughput of hop-by-hop transport, T^h, is bounded by the fraction of links that are available, denoted by h, over the path length, H, i.e.,

$$T^h \leq \begin{cases} C, & \text{all links available;} \\ \dfrac{h}{H} \cdot C, & h \text{ out of } H \text{ links available;} \\ 0, & \text{otherwise.} \end{cases}$$

With p, in this case the upper bound is given by

$$T^h \leq p \cdot C.$$

In Fig. 9.1, we plot the normalized upper bound of the throughput for availability ranging from $p = 1$ (full availability) down to $p = 0$ (no availability). As expected, hop-by-hop transport degrades much more gracefully than the end-to-end alternative as the availability decreases.

To determine, to what extent this higher upper bound of hop-by-hop transport can be leveraged to provide better performance, we use simulation experiments comparing concrete hop-by-hop and end-to-end transport protocols, respectively, in a multihop mobile network scenario.

9.3.2 Simulation Experiments

We investigate the performance of hop-by-hop in comparison to end-to-end transport in a simulation study. We use a model of a multihop wireless mesh network where each node acts both as a router and an end system. We focus on the behavior of the hop-by-hop and the end-to-end transport protocol over the mobile leg of the path. Therefore we do not model any gateways. The network scenario is a sparse mobile multihop network; as a result, there are frequent route disruptions and end-to-end connectivity is intermittent. We compare the performance of TCP NewReno [42] to the performance of store-and-forward transport (SAFT) [43]. SAFT is a simple hop-by-hop transport protocol aimed at wireless mobile networks and provides a service interface identical to that of TCP.

In brief, SAFT works as follows. The source node splits the data stream into so-called segments. Segments are the end-to-end data management unit. The source node is responsible to provide end-to-end reliability and retransmits segments that are not acknowledged by the destination. Segments are typically rather large to reduce the number of acknowledgments that need to be sent from the destination to the source. For the actual transfer of the segments, the source further splits every segment into smaller chunks we refer to as fragments, as shown in Fig. 9.2.

Fragments are the per-hop data management unit and are typically only a few packets long. They are forwarded through the network in a store-and-forward manner and acknowledged by each intermediate node to its precursor. Every intermediate node attempts to forward a fragment to the next node until either the node acknowledges the fragment or the maximal number of attempts is exhausted. After that, fragments are still kept in memory as long as buffer space permits. The rationale behind caching fragments that have been acknowledged by the next hop already is that these fragments may still get lost later on down the path and memory is usually cheaper than communication. In the event that a fragment is lost at a further point, the source node initiates a retransmission of the corresponding segment. Now, the odds are high that the path of the retransmission shares some or all nodes with the previously used path. Upon receipt of a packet belonging to a cached fragment, the receiving node immediately acknowledges the whole fragment to the sending node and starts transferring the fragment to its successor. Essentially, only the first packet of the retransmitted fragment is sent again if the receiver already has the fragment in its cache, saving a lot of bandwidth and reducing interference and power consumption.

SAFT performs flow and congestion control both on an end-to-end basis and a per-hop basis, limiting the number of outstanding segments and fragments using a sliding window algorithm. The estimates of the optimal sending rate and fragment retransmission timeout are determined per link, incorporating the current level of

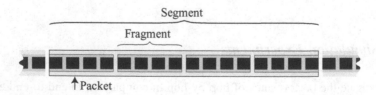

Fig. 9.2 SAFT data management units

contention at the MAC layer. Multiple connections that use the same next hop share these estimated values to reduce the need for bandwidth probing.

In our simulation study [43], 30 nodes move according to the random waypoint mobility model within an area of $1,000\,m \times 3,000\,m$. Because the wireless range is only $250\,m$, disruptions occur with high likelihood. Routes are established on-demand via the AODV [44] routing protocol. In our scenario, we run a simple messaging application that has the following connection pattern. At the beginning of each run, ten source–destination pairs are chosen at random among all nodes. The designated source nodes begin transmitting ten messages of $100\,kB$ each within the first $100\,s$. The total amount of data to be sent through the network is thus $10,000\,kB$.

In Fig. 9.3, we plot the data transfer progress over time for ten connections from a random simulation run. The bold curve indicates the total amount of data transferred as a percentage of $10,000\,kB$. With TCP, the complete transfer takes about 1 h; SAFT delivers the same amount of data within one quarter of the time. The plotted scenario illustrates that the progress of the data transfer with SAFT is considerably steadier than with TCP. The plot of the TCP progress shows periods where the destination nodes do not receive any data. This alternation between transfer and idle periods is also visible in the hop-by-hop protocol's plot, but the idle periods are shorter because SAFT is capable of transferring data along partial paths.

Our analysis of the trace files revealed that TCP uses mostly single-hop connections, and often only one connection is transmitting while the rest of the connections do not transfer any data at all. SAFT exhibits better sharing characteristics, using a mix of single- and multihop routes simultaneously. Additionally, we found that SAFT starts transmitting data much earlier than TCP. The late start of TCP is partly due to the three-way handshake it uses to initiate connections. And if the handshake succeeds, TCP is still prone to disruptions as it frequently induces pseudo route failures, i.e., route failures that are not due to node mobility but rather because of packet loss induced by TCP's bandwidth probing.

Overall, the extensive set of simulation experiments we report about in [43], suggests that a hop-by-hop protocol such as SAFT is able to make better use of the communication opportunities than an end-to-end protocol like TCP. However,

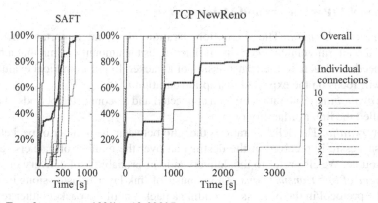

Fig. 9.3 Transfer progress. $100\% \equiv 10,000\,kB$

the performance of the transport protocol is dependent on many variables (transport layer configuration, routing protocol, mobility), so that the simulation results can only be viewed as a case study. To enable more positive statements about the relative performance of the two transport alternatives under intermittent connectivity in general, we now turn our focus on performance evaluation based on analytical modeling.

9.3.3 Analytical Performance Evaluation

In this section, we present numerical results derived from an analytical model of an intermittently connected network. We consider a chain of H hops and evaluate the performance of hop-by-hop and end-to-end transport and a link layer retransmission scheme. We evaluate the probability of successful delivery to the destination p_D, (henceforth referred to as delivery ratio) and the expected number of link transmissions $E[X]$. As in the previous simulation study, we focus our modeling efforts on the mobile leg of the path in a multihop wireless mesh network and do not consider the gateways in our analysis.

The goal of this comparison is to investigate the effectiveness of the different retransmission mechanisms employed by the three protocols under study. The analytical model we use to gain the numerical results we study below is described in [45]. The model is implemented as a discrete-time Markov chain [46]. This means that time is slotted; one time unit could be seen as one packet transmission at the link layer. Under this model, every link is either in the *up* state, corresponding to an operational link with zero packet loss, or in the *down* state, corresponding to a disrupted link with 100% packet loss. The distribution of the disruption period duration is geometric. The parameters of the model are the number of hops H, the mean duration of disruptions $E[D]$, and the link availability p. The number of hops is set to $H = 5$ for this evaluation, corresponding to a multihop network. The end-to-end round-trip time, including queuing delays and processing overhead, can be assumed to be around 20 time units. In the definition below, we summarize the parameters and metrics we use throughout our analysis.

Definition 9.4 (Parameters and Metrics).

- *Disruption duration.* The disruption duration, D, is an important metric to describe intermittent connectivity. It determines the amount of time that a protocol needs to devote to the transmission of a packet. In our numerical evaluation, we will focus on the expected disruption duration $E[D]$.
- *Link availability.* The ratio between connected and disconnected periods of a link is called the link availability or p.
- *Delivery ratio.* The delivery ratio is the ratio between the number of packets that are successfully received at the destination over the number of packets sent by the source. We also refer to the delivery ratio as probability of delivery or p_D.
- *Number of link transmissions.* The number of link layer transmissions incurred by the protocol in the process of sending a packet until the packet either reaches

the destination or it is discarded after too many failed attempts is referred to as the number of link transmissions, X. We usually consider its expected value, $E[X]$.

9.3.3.1 Protocols Under Study

The *end-to-end transport protocol* performs retransmissions from the source; the number of attempts made by the source is limited by the transmission limit L. The *hop-by-hop transport protocol* initiates retransmissions from the node before the link where the packet loss occurred, up to L times per hop. The *link layer protocol* operates exactly the same as the hop-by-hop transport protocol, but employs a different timing regime.

Under all three protocols, retransmissions are spaced over time according to an exponential back-off algorithm. In the case of the end-to-end and the hop-by-hop transport protocol, retransmissions are initiated in an interval that begins at twice the round-trip time, i.e., 40 time units, and is doubled after every attempt. The link layer retransmission scheme begins at one time unit and also doubles the interval for every subsequent attempt.

9.3.3.2 Hop-by-Hop vs. End-to-End Transport

We evaluate the response of the three schemes to a wide range of mean disruption durations $E[D]$, ranging from $E[D]_{\min} = 1$ to $E[D]_{\max} = 100,000$ time units. We consider the link availability values $p = 99\%$ and $p = 70\%$, corresponding to link loss rates $q = 1\%$ and $q = 30\%$. The number of transmission attempts is limited by the transmission limit $L = 7$ for all protocols.

For the following discussion, we introduce a measure that we call *transmission period*, T_X. The transmission period denotes the maximal period of time along which a packet is retransmitted before it is discarded. Intuitively, if any link is in the down state for longer than this period, the protocol will probably fail to deliver the packet. The transmission period is determined by the maximum number of transmissions L and by the retransmission timeout algorithm of the protocol. In the case of the hop-by-hop and end-to-end transport protocol, the transmission period is

$$T_X^e = T_X^h := \sum_{k=1}^{L-1} 40 \times 2^{k-1} = 2,520 \text{ time slots,}$$

corresponding to approximately 126 round-trip times. For the link layer scheme, we have

$$T_X^l := \sum_{k=1}^{L-1} 1 \times 2^{k-1} = 63 \text{ time slots or roughly three round-trip times.}$$

9.3.3.3 Link Availability $p = 0.99$

In Figs. 9.4 and 9.5, we plot the delivery ratio and the expected number of link transmissions, respectively, for a link availability $p = 0.99$. As shown in Fig. 9.4, for longer disruptions, the delivery ratio of all schemes approaches a performance floor. This is due to the following. For all disruption periods that are longer than the transmission period of any retransmission scheme, the scheme necessarily fails to deliver that packet. As the mean disruption duration on the x axis increases, the likelihood of this condition increases, and at very long disruption durations, the delivery ratio of all schemes approaches the performance floor. The floor is well-defined and equal to $p_{D,min} = p^H = 0.9510$, i.e., it is the same as the probability of successful transmission without retransmissions.

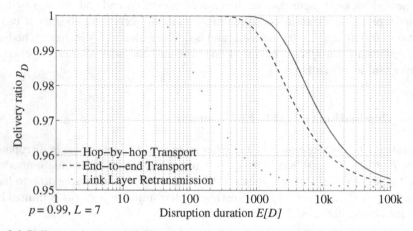

Fig. 9.4 Delivery ratio p_D; $p = 0.99$

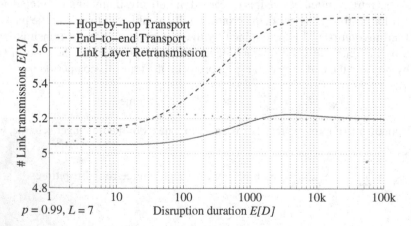

Fig. 9.5 Expected number of link transmissions $E[X]$; $p = 0.99$

Even at the very high link availability of 99% we use in this first set of plots, the difference of the response to extensive disruption periods among the link layer scheme on the one hand and the two transport layer schemes on the other hand becomes apparent. The link layer scheme retransmits lost packets almost back-to-back, and as a result its transmission period only covers a short time span. The transport layer schemes space their transmission attempts over a long period and succeed with higher probability for short and medium disruption durations.

The expected number of link transmissions depicted in Fig. 9.5, it appears to look similar for the three schemes. One slight but crucial difference is that with the end-to-end scheme, $E[X^e]$ is still increasing at the far end of the disruption duration, whereas with the other two schemes that employ per-hop retransmission, $E[X^h]$ and $E[X^l]$ decrease noticeably at the point where their delivery ratio, p_D, approaches the floor.

We learn from this comparison that the transmission period plays a crucial role in determining the capability of a scheme to operate under long disruptions. Furthermore, we see a small discrepancy between hop-by-hop and end-to-end transport in the number of link transmissions. To better show these differences, we consider a lower link availability in the next pair of plots.

9.3.3.4 Link Availability $p = 0.7$

In Figs. 9.6 and 9.7, we plot the delivery ratio and the expected number of link transmissions, respectively, but this time for a rather low link availability of $p = 0.7$.

In Fig. 9.6 of the delivery ratio, we see similar trends as in the previous set of plots with $p = 0.99$ in Fig. 9.4. Again, the delivery ratio of the link layer protocol is much more sensitive to the disruption duration and thus becomes significantly inferior at medium and long disruption durations to the transport layer schemes,

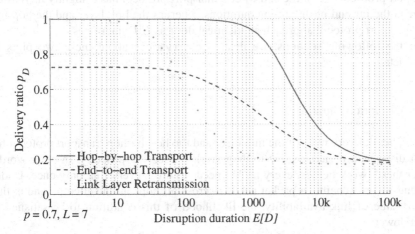

Fig. 9.6 Delivery ratio p_D; $p = 0.7$

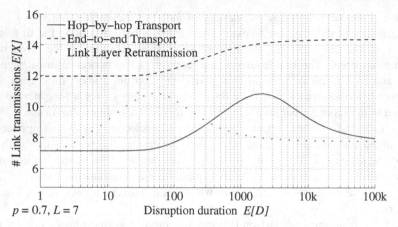

Fig. 9.7 Expected number of link transmissions $E[X]$; $p = 0.7$

and for adequately high values of $E[D]$, all three schemes reach an identical performance floor.

In contrast to the previous plots with $p = 0.99$, now the end-to-end scheme is limited by an upper bound; even for the shortest expected disruption duration, i.e., $E[D] = 1$. The link layer and the hop-by-hop transport scheme still achieve a maximal delivery ratio very close to unity. As a result, there is a crossover of the delivery ratio values of the link layer and the end-to-end transport protocol.

Regarding the number of link transmissions spent per packet plotted in Fig. 9.5, both the link layer and the hop-by-hop transport protocol spend about the same number of transmissions; the maximum value of the transport scheme is located at a higher value of $E[D]$. For all values of $E[D]$, the expected number of link transmissions is considerably lower with these two protocols than with the end-to-end transport protocol. Also, the end-to-end transport protocol has a slightly increasing curve at the far end of disruption durations, whereas the link layer and the hop-by-hop transport protocol approach an identical floor as $E[D]$ increases.

In the following sections, we discuss observations about these two plots in more detail.

Short Disruption Durations

The higher delivery ratio of the link layer and the hop-by-hop transport protocol for short disruption periods is due to their work conserving property. In other words, under these two schemes, every packet needs to cross every link only once. Under the end-to-end scheme, a packet needs to cross all hops in one attempt, and at this lower value of link availability, the likelihood of this condition to be satisfied is much lower.

Medium and Long Disruption Durations

However, if $E[D]$ is longer than the transmission period of the link layer scheme, the link layer scheme collapses because a necessary condition for successful delivery is that no link be disrupted longer than the transmission period. Because the transport layer schemes cover a longer period with their transmission attempts, they are more likely to be still transmitting when a disrupted link has switched to the up state.

Performance Floor of Delivery Ratio

The performance floor that all three protocols approach as the mean disruption duration increases is a result of the strong positive correlation in the loss process for large values of $E[D]$. For large values of $E[D]$, the link availability p, which is also the probability of a link being up, determines success or failure of a transmission, independent of at which point in time this transmission occurs. The limited number of transmission attempts are likely to all occur during a period where the links do not switch their state; thus either the first transmission succeeds, or none of the retransmissions will succeed either. Therefore, at high values of $E[D]$, the delivery ratio is the same with all three protocols and given by $p_{D,\min} = p^H = 0.1681$.

Crossover of the Link Layer and the End-to-End Transport Protocol

The crossover between the curves of the two schemes for $p = 0.3$ at $E[D] = 75$ highlights the two physical properties governing the relative performance of these schemes. The first is the work conservation of the link layer protocol, which crosses each link only once; this property is not shared by the end-to-end transport protocol and thus the link layer scheme enjoys a relative advantage. The second is the relative length of the transmission period and the disruption period. Having the transmission period longer than the disruption period is an advantage because if the transmission period is too short, packet retransmissions are likely to fail. As the link layer scheme sends retransmitted packets in closely located time slots, it is susceptible to lengthy disruption periods. The end-to-end transport scheme spaces retransmissions over time because it requires a longer retransmission timeout and thus has a relative advantage in this range.

Dominance of the Hop-by-Hop Transport Protocol over the Link Layer and the End-to-End Transport Protocol

The hop-by-hop transport protocol dominates the end-to-end transport protocol both in delivery ratio and expected number of transmissions over the whole range of mean disruption durations and for both values of p we consider. An explanation for this dominance may be that hop-by-hop transport unites the two advantages of the link layer and the end-to-end transport scheme discussed in the previous paragraph. Note that hop-by-hop transport also dominates the link layer scheme for all disruption durations where the correlation in the loss process is positive, which is the case for all medium and long-term disruptions (cf. [45]).

To summarize, in the above study of an analytical model of hop-by-hop vs. end-to-end transport vs. link layer retransmission, we found the following:

1. The *transmission period* of a protocol determines to a large extent its capability to survive lengthy periods of disruption. Thus, any link layer scheme is not suitable to provide resilience against disruptions.
2. The retransmission mechanism (end-to-end or hop-by-hop) determines the efficiency of retransmissions. Thus, end-to-end retransmission as employed by the end-to-end transport scheme is not efficient if disruptions occur frequently.
3. Hop-by-hop transport dominates both alternatives in terms of delivery ratio and expected number of link transmissions.

In Sect. 9.4, we draw conclusions for practitioners from these results.

9.4 Thoughts for Practitioners

Handling intermittent connectivity in a real-world scenario requires careful analysis of the methods that are available. In particular, the options depend on the set of protocols at disposal. In the following, we briefly review available methods to handle intermittent connectivity and then look at three different scenarios and give considerations as to which methods appear most beneficial.

9.4.1 Countermeasures

It is crucial to understand that most mechanisms aiming to alleviate packet loss are not effective against extensive periods of disruption if they are applied on an end-to-end basis. Forward error correction (FEC) or automated repeat request (ARQ), for example, may improve the performance during connected periods but have no effect during disruptions. Note that there are also alternatives to end-to-end transport such as custody transfer [12], but those are not applicable in wireless multihop mesh networks because they lack the end-to-end semantics required by most applications mesh networks are used for.

As seen in Sect. 9.3.3, hop-by-hop transport is capable of handling intermittent connectivity better than end-to-end transport or a link layer-based approach. Because there are no established standard hop-by-hop transport protocols available, hop-by-hop protocols are only applicable if at least the gateways are extensible with a custom transport protocol.

Hop-by-hop transport requires cross-layer information exchange between routing and transport protocol and one might consider merging multiple layers. However, the drawbacks of merged layers are manifold. For one, the flexibility to exchange layers separately is no longer given. Second, the fact that broad functionality is provided by a single entity makes it more difficult to verify this entity's proper operation. A unified scheduling/routing/transport layer, for instance, might benefit the performance of a wireless mesh network, but it should not be underestimated, how many different issues need to be tackled by such a "super layer." In most cases, it appears to be more appropriate to stick to the layering principle and use cross-layer information only if it provides considerable performance gains.

9.4.2 Application Scenarios

In the following subsections, we consider three scenarios in more detail. These scenarios differ in the degree to which gateways and end systems can be adapted.

9.4.2.1 Commodity Gateways

In this environment, we assume that commodity gateways are deployed. Under these circumstances, no changes can be made to the protocols running at the transport layer and the only optimization potential lies at the lower layers, such as the routing or link layer protocol.

9.4.2.2 Customized Gateways, Commodity End Systems

In a wireless mesh network based on customized gateways, there is the possibility of deploying a protocol stack tailored to its purpose. This requires that the connection be split into three parts: Two parts that run the transport protocol used by the end points and a third part running customized software within the mesh network. In a recent study of a hop-by-hop transport protocol that runs among the gateways in such a scenario [40], the authors report a considerable performance improvement over standard TCP.

Deploying a unified link/transport layer approach in this context may also be feasible, depending on the wireless network interface and the accompanying driver software.

9.4.2.3 Peer-to-Peer Mesh Network

If commodity laptops, personal computers, cellular phones, etc. are used as both gateways and end systems, they need to be equipped with dedicated mesh network routing software. Thus, also a customized transport layer implementation could be run. The possibilities are wider than in the above scenario where only the gateways run custom protocols. In this scenario, the whole part of the path within the mesh network including the client host may run a custom transport protocol.

Overall, multihop wireless mesh networks are amenable to optimization at all layers up to and including the transport layer. Hop-by-hop transport protocols are a promising way to leverage the power of customizable gateways and even end systems. However, some questions remain open and we give an overview of future work in Sect. 9.5.

9.5 Directions for Future Research

In the broad context of handling intermittent connectivity, there are still many open questions. First, there is no metric that captures the impact of disruptions on the network performance. Packet loss rate or mean disruption duration are metrics that together may give some indication, but they do not directly represent the performance degradation that is caused by disruptions. Ways to measure the degree of intermittent connectivity might be based on the performance impact or on their statistical properties.

More realistic modeling and analysis methods would allow to study in more detail and with higher degree of realism the impact of different kinds of disruptions on a given network scenario.

Only little real-world measurement data of intermittently connected networks exists. There are quite a few measurement studies of disconnected networks, but they are not immediately applicable to study intermittently connected scenarios.

9.6 Conclusions

In this chapter, we studied the phenomenon of intermittent connectivity and measures to reduce its impact in the context of multihop wireless mesh networks. We found that popular techniques to handle packet loss, such as link layer or end-to-end retransmission, are not applicable in the case of intermittent connectivity and we studied hop-by-hop transport as a promising alternative. Hop-by-hop transport is more efficient than end-to-end transport and is capable of moving data towards the destination even during periods where no end-to-end connectivity is available. Using simulation experiments and analytical modeling, we found that hop-by-hop transport dominates the end-to-end alternative in a broad set of scenarios.

Our practical consideration of three different scenarios shows that the main reason that hop-by-hop transport is not in wider use today is the lack of a standard hop-by-hop transport protocol implementation.

As directions for future work, we see many open questions in the area of intermittently connected networks in terms of evaluation methodology and in terms of real-world implementations of concrete protocols.

9.7 Terminologies

- *Multihop wireless mesh network.* A multihop wireless mesh network comprises a set of gateways, i.e., access points that are connected to the Internet, and a set of mesh nodes in need of Internet access. These nodes relay data on behalf of each other to increase the coverage of the access points.
- *Disruption.* A period of disruption between a pair of nodes is a period during which no end-to-end connectivity is available among these nodes.
- *Disruptive network.* A network that must be assumed to sometimes be disrupted is called a *disruptive network*.
- *Disconnected network.* A disruptive network that in general lacks end-to-end connectivity is referred to as a *disconnected network*.
- *Intermittently connected network.* A disruptive network that in general provides connectivity but where routes are frequently disrupted, is termed an *intermittently connected network*.
- *End-to-end transport.* A protocol implementing end-to-end transport only runs at the end points of the connection. This means that no transport layer protocol is required at intermediate nodes, allowing end-to-end transport to run over any intermediate network elements that implement the network layer.
- *Hop-by-hop transport.* A protocol implementing hop-by-hop transport runs at all nodes that lie on the network path, allowing data transfer to be controlled at each link separately. In particular, lost data can be retransmitted by the node before the link where the loss occurred. In addition, hop-by-hop transport uses end-to-end acknowledgment to provide reliability.
- *Custody transfer.* Custody transfer is a way of transferring a bundle of packets from a source to a destination node in a network where no end-to-end connectivity can be assumed. Under this paradigm, the responsibility to deliver the bundle always rests with the node that carries the bundle and moves to the next node upon successful transfer. Note that custody transfer on purpose does not provide end-to-end reliability and cannot be used as a direct replacement of reliable transport protocols such as TCP.
- *Disruption duration.* The *disruption duration* is an important metric to describe intermittent connectivity. It determines the amount of time that a protocol needs to devote to the transmission of a packet.
- *Link availability.* The ratio between connected and disconnected periods of a link is called the *link availability*.

9.8 Questions

Unless stated otherwise, we consider the scenario of a mobile multihop wireless mesh network as defined in Sect. 9.11, Def. 9.2.

1. Give two reasons to deploy hop-by-hop transport in the given scenario.
2. Come up with two reasons not to deploy hop-by-hop transport.
3. What is the advantage of end-to-end transport? Give two advantages.
4. What are two reasons not to deploy end-to-end transport?
5. Why can custody transfer not be used in the given scenario?
6. What is the major difference of custody transfer to hop-by-hop transport?
7. Why is the duration of disruption periods crucial in determining if a link-layer retransmission scheme improves the performance?
8. What is the reason why a routing protocol, in general, is not suitable to provide disruption-tolerance?
9. Does it make sense to run a hop-by-hop transport protocol over a link layer that performs retransmissions?
10. What is the minimal amount of cross-layer information that is required by hop-by-hop transport?

References

1. C.E. Perkins and P. Bhagwat, Highly dynamic destination-sequenced distance-vector routing (DSDV) for mobile computers. In: Conference on Communications Architectures, Protocols and Applications. SIGCOMM'94, ACM, pp. 234–244. ACM, New York, NY (1994).
2. D.B. Johnson and D.A. Maltz, Dynamic source routing in ad hoc wireless networks. In: Imielinski, T. and Korth, H. (Eds.) Mobile Computing, vol. 353. Kluwer, Dordrecht (1996). URL citeseer.ist.psu.edu/johnson96dynamic.html
3. C.E. Perkins and E.M. Royer, Ad-hoc on-demand distance vector routing. In: Workshop on Mobile Computer Systems and Applications. WMCSA'99. IEEE, p. 90. IEEE Computer Society, Washington, DC, (1999).
4. D. Aguayo, J. Bicket, S. Biswas, G. Judd, R. Morris, Link-level measurements from an 802.11b mesh network. In: Conference on Applications, Technologies, Architectures, and Protocols for Computer Communications. SIGCOMM '04. ACM, pp. 121–132. ACM, New York, NY (2004).
5. V. Lenders, J. Wagner, and M. May, Analyzing the impact of mobility in ad hoc networks. In: ACM/Sigmobile Workshop on Multi-hop Ad Hoc Networks: from Theory to Reality (REAL-MAN), Florence, Italy (2006).
6. P. Juang, H. Oki, Y. Wang, M. Martonosi, L.S. Peh, and D. Rubenstein, Energy-efficient computing for wildlife tracking: design tradeoffs and early experiences with ZebraNet. In: Proceedings of the Tenth International Conference on Architectural Support for Programming Languages and Operating Systems (ASPLOS-X), San Jose, CA, USA (2002).
7. J.C. Kuo and W. Liao, Hop count distribution of multihop paths in wireless networks with arbitrary node density: Modeling and its applications. IEEE Transactions on Vehicular Technology **56**, 2321–2331 (2007).
8. J. Postel: Internet Protocol. IETF RFC 791 (1981).
9. J. Postel: Transmission Control Protocol. RFC 793 (Standard) (1981). URL http://www.ietf.org/rfc/rfc793.txt. Updated by RFC 3168.

10. V. Jacobson and M. Karels, Congestion avoidance and control. In: Proceedings of the ACM Symposium on Communications Architectures and Protocols (SIGCOMM'88), pp. 314–329 (1988). URL citeseer.ist.psu.edu/654992.html

11. Delay Tolerant Networking Research Group. http://www.dtnrg.org/

12. K. Scott and S. Burleigh, Bundle Protocol Specification. draft-irtf-dtnrg-bundle-spec-10.txt (2007).

13. C.M. Ozveren, R. Simcoe, and G. Varghese, Reliable and efficient hop-by-hop flow control. IEEE Journal on Selected Areas in Communications 13(4), 642–650 (1995).

14. P.P. Mishra, H. Kanakia, and S.K. Tripathi, On hop-by-hop rate-based congestion control. IEEE/ACM Transactions on Networking 4(2), 224–239 (1996).

15. D. King, K. Walker, and D. Platt, The price we pay for using TCP. International Conference on Networks 2004 (ICON 2004), vol. 1, pp. 9–13. IEEE, Washington, DC (2004).

16. A. Kortebi, L. Muscariello, S., and Oueslati, J. Roberts, Evaluating the number of active flows in a scheduler realizing fair statistical bandwidth sharing. In: International Conference on Measurement and Modeling of Computer Systems. SIGMETRICS'05. ACM, pp. 217–228. ACM, New York, NY (2005).

17. H. Elaarag, Improving TCP performance over mobile networks. ACM Computer Surveys 34(3), 357–374 (2002).

18. K.Y. Lee, S.S. Joo, and J. dong Ryoo, CAT: Contention aware transport protocol for IEEE 802.11 MANETs. Vehicular Technology Conference 2006, VTC 2006, vol. 2, pp. 523–527. IEEE, Washington, DC (2006).

19. S. Kopparty, S. Krishnamurthy, M. Faloutsos, S. Tripathi, Split TCP for mobile ad hoc networks. In: Proceedings of the IEEE Global Communications Conference (GLOBECOM 2002) (2002).

20. A.A. Hanbali, E. Altman, and P. Nain, A survey of tcp over ad hoc networks. IEEE Communications Surveys and Tutorials 7(3), 22–36 (2005).

21. E. Ayanoglu, S. Paul, T.F. LaPorta, K.K. Sabnani, and R.D. Gitlin, AIRMAIL: A link-layer protocol for wireless networks. Wireless Networks 1(1), 47–60 (1995).

22. H. Balakrishnan, S. Seshan, and R.H. Katz, Improving reliable transport and handoff performance in cellular wireless networks. ACM Wireless Networks 1(4), 469–481 (1995).

23. S. Biaz and N.H. Vaidya, Distinguishing congestion losses from wireless transmission losses: A negative result. In: International Conference On Computer Communications and Networks (ICCCN 1998), October 12–15, 1998, Lafayette, Lousiana, USA, pp. 722–731. IEEE, Washington, DC (1998).

24. K. Chandran, S. Raghunathan, S. Venkatesan, and R. Prakash, A feedback based scheme for improving tcp performance in ad-hoc wireless networks. In: Proceedings of the International Conference on Distributed Computing Systems (ICDCS'98) (1998).

25. G. Holland and N.H. Vaidya, Analysis of TCP performance over mobile ad hoc networks. In: International Conference on Mobile Computing and Networking. MobiCom'99. ACM, New York, NY (1999).

26. M. Zhang, B. Karp, S. Floyd, and L. Peterson, RR-TCP: A reordering-robust TCP with DSACK. In: International Conference on Networking Protocols (ICNP 2003), Los Alamitos, CA, USA, vol. 00. IEEE Computer Society, Washington, DC (2003).

27. A. Bakre and B. Badrinath, I-TCP: Indirect TCP for mobile hosts. In: Proceedings of the 15th International Conference on Distributed Computing Systems (ICDCS'95). IEEE Computer Society, Los Alamitos, CA (1995).

28. R. Caceres and L. Iftode, Improving the performance of reliable transport protocols in mobile computing environments. IEEE Journal on Selected Areas in Communications 13(5), 850–857 (1995).

29. R. Yavatkar and N. Bhagawat, Improving end-to-end performance of TCP over mobile internetworks. In: Mobile'94 Workshop on Mobile Computing Systems and Applications. ACM, New York, NY (1994).

30. K. Brown and S. Singh, M-TCP: TCP for mobile cellular networks. In: Proceedings of ACM SIGCOMM'97 27(5), 19–43 (1997).

31. I. Gitman, Comparison of hop-by-hop and end-to-end acknowledgment schemes in computer communication networks. IEEE Transactions on Communications [legacy, pre-1988] **24**(11), 1258–1262 (1976).

32. A. DeSimone, M.C. Chuah, and O.C. Yue, Throughput performance of transport-layer protocols over wireless LANs. In: Global Telecommunications Conference, 1993. GLOBECOM'93, Houston, TX, USA, pp. 542–549. IEEE, Washington, DC (1993).

33. A. Sundararaj and D. Duchamp, Analytical Characterization of the Throughput of a Split TCP Connection. Technical Report 2003–04, Department of Computer Science, Stevens Institute of Technology (2003).

34. J. Padhye, V. Firoiu, D.F. Towsley, and J.F. Kurose, Modeling TCP Reno Performance: A Simple Model and Its Empirical Validation. IEEE/ACM Transactions on Networking **8**(2), 133–145 (2000).

35. Y. Yi and S. Shakkottai, Hop-by-hop congestion control over a wireless multi-hop network. IEEE/ACM Transactions on Networking **15**(1), 133–144 (2007).

36. J. Camp, J. Robinson, C. Steger, and E. Knightly, Measurement driven deployment of a two-tier urban mesh access network. In: MobiSys'06: Proceedings of the Fourth International Conference on Mobile Systems, Applications and Services, pp. 96–109. ACM, New York, NY (2006).

37. J. Eriksson, S. Agarwal, P. Bahl, and J. Padhye, Feasibility study of mesh networks for all-wireless offices. In: MobiSys'06: Proceedings of the Fourth International Conference on Mobile Systems, Applications and Services, pp. 69–82. ACM, New York, NY (2006).

38. K. Xu, S. Bae, S. Lee, and M. Gerla, TCP behavior across multihop wireless networks and the wired internet. In: International Workshop on Wireless Mobile Multimedia. WOWMOM'02. ACM, pp. 41–48. ACM, New York, NY (2002).

39. A. Zimmermann, M. Gunes, M. Wenig, U. Meis, and J. Ritzerfeld, How to study wireless mesh networks: A hybrid testbed approach. In: AINA'07: Proceedings of the 21st International Conference on Advanced Networking and Applications, pp. 853–860. IEEE Computer Society, Washington, DC (2007).

40. A. Raniwala, S. Sharma, P. De, R. Krishnan, and T. cker Chiueh, Evaluation of a stateful transport protocol for multi-channel wireless mesh networks. In: Fifteenth IEEE International Workshop on Quality of Service, Evanston, IL, USA, 2007, pp. 74–82 (2007).

41. K. Sundaresan, V. Anantharaman, H.Y. Hsieh, and R. Sivakumar, ATP: A reliable transport protocol for ad-hoc networks. In: International Symposium on Mobile Ad hoc Networking and Computing. MobiHoc'03. ACM, New York, NY (2003).

42. S. Floyd, T. Henderson, and A. Gurtov, The NewReno Modification to TCP's Fast Recovery Algorithm. RFC 3782 (Proposed Standard) (2004). URL http://www.ietf.org/rfc/rfc3782.txt

43. S. Heimlicher, R. Baumann, M. May, B. Plattner, The transport layer revisited. In: Proc. of IEEE COMSWARE 2007. Bangalore, India (2007).

44. C. Perkins, E. Belding-Royer, and S. Das, Ad hoc On-Demand Distance Vector (AODV) Routing. RFC 3561 (Experimental) (2003). URL http://www.ietf.org/rfc/rfc3561.txt

45. S. Heimlicher, M. Karaliopoulos, H. Levy, and M. May, End-to-end vs. hop-by-hop transport under intermittent connectivity (invited paper). In: Proc. of ACM/ICST Autonomics 2007. Rome, Italy (2007).

46. I. Iosif and A.V.S. Gikhman, The Theory of Stochastic Processes II, Reprint of the First Ed, 2004 edn, 1975. Series: Classics in Mathematics. Springer, Berlin (2004).

Chapter 10
Transport Protocols for Wireless Mesh Networks

Ka Lun Eddie Law

Abstract Transmission control protocol (TCP) provides reliable connection-oriented services between any two end systems on the Internet. With TCP congestion control algorithm, multiple TCP connections can share network and link resources simultaneously. These TCP congestion control mechanisms have been operating effectively in wired networks. However, performance of TCP connections degrades rapidly in wireless and lossy networks. To sustain the throughput performance of TCP connections in wireless networks, design modifications may be required accordingly in the TCP flow control algorithm, and potentially, in association with other protocols in other layers for proper adaptations. In this chapter, we explain the limitations of the latest TCP congestion control algorithm, and then review some popular designs for TCP connections to operate effectively in wireless mesh network infrastructure.

10.1 Introduction

Recommended by the Internet engineering task force (IETF), transmission control protocol (TCP) [1–5] offers connection-oriented, full-duplex, point-to-point communication services for data transfers on the Internet. TCP operates at the transport layer (the Layer 4) in the OSI seven-layer network model. In wired networks, TCP has been working properly with millions of connections, and handling packet loss events effectively, which usually happens at routers suffering with packet congestion conditions. The proliferation of mobile computing devices, and the maturation of wireless technologies, e.g., 802.11 wireless local area networks (WLANs), will lead to the creations of wireless mesh network (WMN) infrastructure in the near

K.L. Eddie Law (✉)
Department of Electrical and Computer Engineering, Ryerson University, 350 Victoria Street, Toronto, ON, Canada M5B 2K3
e-mail: eddie@ee.ryerson.ca

S. Misra et al. (eds.), *Guide to Wireless Mesh Networks*, Computer Communications and Networks, DOI 10.1007/978-1-84800-909-7_10,

future. Unfortunately, TCP congestion control algorithms are not designed to deal with, for example, the random packet loss events in wireless medium. In the following, we briefly review the latest recommended and commonly used TCP congestion control algorithms, and the basic design concept of WMNs. To delivering information successfully across multihop WMNs, modifications of TCP congestion control algorithms, and, in some cases, in collaborations with changes in other layers' protocols may be necessary.

10.2 Background on Transmission Control Protocol

TCP is usually used in a client–server computer communication model [6]. Typically, a computing device initiates a server process through a series of application programming interface (API), commonly called socket, system calls to its operating system. Then, a client device can establish a full-duplex communication connection to a server through another socket system call to its operating system. A socket system call creates sending and receiving buffers locally for data communications. Packets can then be sent across the Internet from a sending buffer at one side to the receiving buffer at the other side.

TCP is an end-to-end transport layer protocol, and packets are always called segments at the transport layer. Packet routing operations are network layer's functions, which should be independent of Layer 4's operations. In wired networks, packet loss events usually happen at regions of congestion where network routers do not have sufficient resources to handle all arriving packets. As a result, some packets have to be dropped. These packet losses happen usually during packet routing and forwarding operations, because network routers cannot handle these arriving packets. With this concept in mind, TCP has been designed for providing reliable information communications. Reliability is typically achieved through retransmissions of lost packets. Because client and server nodes are not explicitly notified by networks regarding these segment loss events, TCP congestion control algorithm is needed and its initial design goal is to tackle the congestion issues in wired networks. The first version of TCP congestion control algorithm was known as Tahoe [7], which was created in late 1980s by Van Jacobson.

10.2.1 Congestion Control Algorithms

Through the years, many different versions of congestion control algorithms have been created and the latest popular one is known as NewReno [3, 5]. NewReno is the de facto standard and recommended by the IETF. Cubic [8] is the default implementation in the latest Linux kernel operating system. Its design goal is mainly to serve large delay-bandwidth product connections. However, Cubic also carries some doubtful design problems [9], and it has not been recommended by IETF. In the following, the designs of TCP NewReno congestion control mechanisms are explained in detail, thus enabling us to quantify modifications to operate in wireless environment. The congestion control algorithm in NewReno consists of four operating mechanisms. They are Slow-Start, Congestion Avoidance, Fast Retransmit, and Fast Recovery [3, 5]. The amount of segments sent for a connection is controlled by a sliding window at the sender. And the size of this sliding window is parametrically controlled by a local variable called congestion window, *cwnd*. Furthermore, the value of *cwnd* is governed by the operating principle of TCP, the additive-increment and multiplicative-decrement (AIMD) mechanism. The designs of these mechanisms were started to deal with congestion events on the Internet, that is, in wired networks. In the following, these control mechanisms are outlined.

10.2.1.1 Slow-Start

When a TCP connection has just been established for file transfer, it is in the Slow-Start mode. A sender usually sets *cwnd* to one-segment size, which is equal to the negotiated sender's maximum segment size (SMSS). This implies only one segment can be sent out when a connection has just been initiated. Upon correctly receiving the segment, the receiver replies with an acknowledgment (ACK) packet. The acknowledgment number indicates all packets with sequence numbers ahead of this number have been received properly. Hopefully, this ACK packet can reach the source node; otherwise, the unacknowledged segments at sender are considered lost and retransmitted upon retransmission timeout (RTO) timer events occur. If this ACK packet reaches the source properly, the lower bound of the sliding window is shifted forward, and the *cwnd* parameter is incremented by one more segment size. That is

$$cwnd = cwnd + \text{SMSS}.$$

10.2.1.2 Congestion Avoidance

At the sender, when the *cwnd* reaches a preset value, which is commonly called slow-start threshold *(ssthresh)*, then the TCP connection leaves Slow-Start and enters congestion avoidance phase. In this phase, the goal is to increase *cwnd* approximately by one-segment size per round-trip time (RTT). That is, upon receiving an ACK segment, the sender updates the *cwnd* parameter with

$$cwnd = cwnd + \frac{\text{SMSS}^2}{cwnd}.$$

10.2.1.3 Fast Retransmit

The size of a sliding window becomes large when *cwnd* keeps increasing. It may reach a stage that too many segments have been sent and some network links may not be able to handle these packets. Consequently, some packets have to be dropped, and burst segment losses commonly happen in wired networks. The packet loss events can be observable at sender upon receiving multiple returning ACK packets with identical acknowledgment numbers. In the specification [3], a packet is defined as lost if there are three duplicated ACK packets returning. If this event happens, then the connection enters fast retransmit phase. This packet should be retransmitted immediately without waiting for the timeout event, and the connection then enters Fast Recovery.

10.2.1.4 Fast Recovery

Because of the occurrence of a packet loss event, the value of *ssthresh* should be changed in this phase. It is set to the larger value of either one-half of the currently unacknowledged outstanding packet size, or two-segment size. The amount of currently unacknowledged outstanding segments is called flight-size in the TCP specification [3]. In Fast Recovery, the *cwnd* is then set to *ssthresh* plus three-segment size. This value enables the sender to detect another packet loss event again. The operations are

$$ssthresh = \max\left\{\frac{\text{flight–size}}{2}, 2 \cdot \text{SMSS}\right\},$$
$$cwnd = ssthresh + 3 \cdot \text{SMSS}.$$

Furthermore, *cwnd* is incremented by one segment size upon receiving each packet with duplicate ACK again, i.e., $cwnd = cwnd + \text{SMSS}$. For NewReno [5], only the packet that acknowledges all previously unacknowledged segments can reset *cwnd* to *ssthresh*, and the TCP connection re-enters the Congestion Avoidance phase. Otherwise, the connection stays in Fast Recovery mode.

If, in case, a timeout event occurs, then both the *ssthresh* and *cwnd* parameters are required to be changed. They are set to

$$ssthresh = \max\left\{\frac{\text{flight–size}}{2}, 2 \cdot \text{SMSS}\right\},$$
$$cwnd = \text{SMSS},$$

and then enters the Slow-Start phase. Although the NewReno congestion control algorithm has been successful in dealing with millions of connections on the Internet, its performance degrades rapidly in wireless networks. For multiple-hop wireless networks, a sender of a TCP connection may have difficulty in transmitting large-size data file.

10.3 Wireless Mesh Networks and Transport Protocols

Disregard the similarities to the mobile ad hoc networks (MANETs), WMNs [10] are designed based on the concept of traditional wired networking infrastructures, which are interconnected with network routers. An example of a WMN is shown

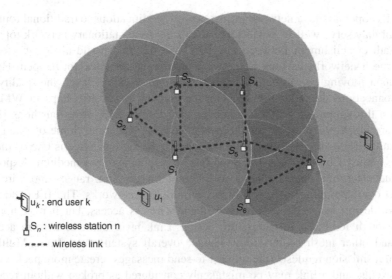

Fig. 10.1 Wireless mesh architecture

in Fig. 10.1. Large-scale wired networks, such as the Internet, consist of different types of routers for traffic relay in network cores, and access purposes at network edges. On the other hand, WMNs are semi-stationary networks. Although wireless mesh routers (or base station) should be stationary in a well-planned wireless infrastructure, additions and deletions of mesh routers are relatively easy in WMNs based on measured traffic intensity and demand of services. In addition, a wireless mesh router or base station can serve both the relaying and accessing purposes. In Fig. 10.1, the base station s_1 can serve an end-user u_1, and relay traffic for node s_2, if desirable. Unfortunately for wireless networks, the signal strength and transmission quality of a wireless connection between mesh routers may change because of various conditions. For example, the changes may happen because of weather, radio signals, or as worse as a new building constructed between two wireless base stations. After all, WMNs offer more advantages because of dynamic self-organization and self-configuration. As long as isolations do not occur among mesh routers, the nodes in networks can automatically establish different ad hoc network architecture and maintain the mesh connectivity.

In a WMN, the number of wireless hops between a source and a destination is a variable. Routing protocol developments in mesh networks are in general established based on numerous developed algorithms in MANETs. Route setup may depend on quality of a wireless link. As a result, a path may change comparatively often in wireless networks. Suppose that an end-user u_1 wants to reach u_2 in Fig. 10.1. A short path can have four hops, e.g., $u_1–s_1–s_5–s_7–u_2$. But a long path can have eight hops, e.g., $u_1–s_1–s_2–s_3–s_4–s_5–s_6–s_7–u_2$. Both paths are feasible as long as they meet the path quality requirements. But for TCP, the number of wireless hops is a significant parameter to the path quality.

WMNs operate like a network infrastructure. Modifications to traditional routing protocol may serve well in WMNs because of the semi-stationary network topologies. Path establishment between any two base stations should always be feasible in mesh networks as long as network partition issue does not happen. However, upon moving consideration from network to transport layer, the quality of TCP connections degrades rapidly with the number of wireless hops in WMNs. The data throughput of a TCP connection can drop to 23% after four hops [11]. There are numerous reasons for the performance degradation. Some of them are listed in the following. For example, the random packet loss issue is one of them, and it is caused by the high bit error rate (BER) in wireless medium. Exposed and hidden terminal problems due to different transmission ranges and wireless channel overlapping are other problems in multihop networks. The IEEE 802.11 protocol design has been operating well for wireless access, but it may not be good enough for mesh infrastructure design. Link layer contention from access users and other mesh routers may degrade overall system throughput. Multiple frame transmission requests (the request-to-send messages) create more packet collision events, and a link may be mistakenly considered as broken without receiving acknowledging replies (the clear-to-send messages). A broken link event may be sufficient to terminate a regular TCP connection. For WMNs, there are many other design issues required thorough investigations that include MAC designs, route recomputation due to broken path, network partitioning, multipath routing problems, energy efficiency, and user mobility. Indeed, successful file transmission relies on the transmission reliability provided by the TCP. But the current TCP congestion control algorithm itself is not designed to operate in wireless medium.

10.3.1 Flight-Size and Loss Probabilities

As aforementioned, the size of a segment is bounded above by SMSS in TCP. Packet with one SMSS is usually considered as one basic unit in TCP designs for performance analysis. If the BER in a wireless network is high, then the size of a segment actually has impact on the throughput performance. The value of SMSS is negotiated and fixed between sender and receiver after a TCP connection has been started. The number of segments sent is then controlled by the *cwnd* parameter. For example, if the BER of a lossy medium is ξ and the size of a packet is SMSS $= m$ bits, then the probability that a packet may be corrupted during transmission is $\gamma_m = 1 - (1 - \xi)^m$.

Suppose the source has a large file to send. Assuming that it is a persistent TCP session, and it goes through a wireless forward path with BER ξ. To consider a better networking scenario and simplify the analysis, we assume that the ACK packets can always reach the sender after a RTT with a perfect reverse path. Furthermore, an ACK packet only shifts forward the bound of sliding window by one segment, and triggers the transmission of another segment. However, the data packets may drop because of random errors. Suppose that a packet loss event has just occurred at this particular instant, and the number of unacknowledged packets, i.e., flight-size,[1] is equal to n packets. Without considering the additive-increment of *cwnd*, suppose that k denotes the number of RTTs, and it may go to infinity in this analysis. It indicates that ultimately the flight-size diminishes nonlinearly, and it is equal to

$$n(1 - \gamma_m)^k \xrightarrow{k \to \infty} 0, \tag{10.1}$$

for $0 < \gamma_m < 1$. If k is large, the connection has a flight-size of zero, which is independent of the value of n. Also, if γ_m is large, then the connection also has a flight-size of zero even if k is small. This implies that BER has a significant impact on the performance of TCP connections in wireless media. Upon considering the additive-increment operation during Slow Start, that is, each successfully delivered data packet induces a perfectly returned ACK and increases the *cwnd* by extra SMSS bytes. Then the flight-size in networks should be

$$\sum_{z=0}^{k} n(1 - \gamma_m)^z = n \left[1 - (1 - \gamma_m)^{k+1} \right] / \gamma_m \tag{10.2}$$

after k round-trips. With the additional segment increments per RTT (exponential growth rate), the flight-size is given in (10.2). Even in this kind of perfect networking condition assumptions, i.e., only forward path is suffering with BER, ultimately the flight-size still goes to zero for large k in (10.2). Based on this analytical result, if flight-size is small for a TCP connection in wireless network, the *cwnd* is small, and the size of sliding window is then also small. In other words, a sender with current TCP specification can not complete a file transfer if it is a large file for the wireless network with BER to handle.

This indicates that the random BER can restrain the growth of *cwnd* parameter. With small sliding window size, data rate is also slow. This states that fundamentally we need to improve the designs of TCP congestion control algorithms for WMNs. Subsequently, there have been a few proposals to modify the TCP algorithms as well as some other layers' protocols to improve the performance of TCP, for example, in MANETs [12–14]. At the end, only partial system improvements can be achievable [15]. Actually, some of these designs also fit well to improve the TCP performance in WMNs. In the following, we classify them into three categories: transport layer flow control, cross-layer, and split connections. They will be discussed below.

10.3.2 Transport Layer Flow Control

Without resorting to modifications to other layers' protocols is always the most desirable method to improve TCP throughput in wireless networks. If this is achievable, then we may not even need to change anything in existing network devices, such as routers, apart from updating the transport layer TCP stacks in end users'

---

[1] The flight-size is defined in terms of bytes. But for analysis purpose, it is measured in terms of the number of outstanding unacknowledged packets in this section.

computing devices. This may also imply a family of new control algorithm may be designed to fit the new needs. As discussed before, the TCP Tahoe, Vegas, Reno, NewReno algorithms have been designed to resolve congestion in wired network. This is also why these algorithms are called congestion control algorithms. In this regard, novel TCP algorithmic designs for wireless networks should be called "flow control algorithm," instead of "congestion control algorithm." Again, the advantage of having new end-to-end TCP flow control design is that no changes are required in mesh routers.

In TCP NewReno congestion control algorithms, the Fast Retransmit, and Fast Recovery mechanisms are designed to deal with bursty packet loss events, but not the random packet loss events. Therefore, some proposals have been made that attempt to solve the problem of random packet losses. One example is TCP Westwood+ [16]. Its design is based on TCP Vegas and its primitive goal is to work on a mixture of wired and wireless links. The TCP Westwood+ tries to find an available bandwidth estimate of a TCP connection without discriminating the differences between congestion and random loss events. An estimate for setting $cwnd$ is obtained by filtering a stream of returning ACK packets. If there is a packet loss event upon receiving three duplicate ACKs, then both the $cwnd$ and $ssthresh$ are set to a bandwidth estimate b_k times the recorded minimum round-trip time ($\text{RTT}_{\min}$). That is,

$$cwnd = ssthresh = \max\{b_k \cdot \text{RTT}_{\min}, 2 \cdot \text{SMSS}\}.$$

A sample of available bandwidth $b_k = d_k / \text{RTT}_k$ is computed for the kth RTT, where d_k is the amount of data acknowledged during the last round-trip time, RTT_k. The amount d_k is determined by a proper counting procedure that also considers delayed ACKs and duplicate ACKs. A duplicate ACK counts for one delivered segment, a delayed ACK [1] counts for two segments, a cumulative ACK counts for one segment or for the number of segments exceeding those already accounted for by previous duplicate acknowledgements. Furthermore, TCP Westwood+ has another choice in setting the estimate by using the exponentially weighted moving average mechanism of bandwidth samples b_k. This design can avoid the rapid fluctuation of the bandwidth estimate. The estimate is $\widehat{b}_k$, which is

$$\widehat{b}_k = \alpha \cdot \widehat{b}_{k-1} + (1 - \alpha) \cdot b_k,$$

where α is set to 0.9.

Unfortunately, Westwood+ only works in a single-hop wireless network, for example, a satellite network, because it relies heavily on returning ACK messages for calculating estimate. Therefore, it does not work in multihop WMNs. As of today, many proposed solutions can improve the TCP performance on multihop wireless networks albeit each of them can only solve a small part of the whole problem. Integration of these different solutions may be potentially improve the overall TCP performance in wireless networks. The designs that may possibly work in transport layer are as follows.

10.3.2.1 Selective Acknowledgment Option

Different types of TCP options can be appended at the back of a TCP protocol header [1]. One of them has been found useful to work in wireless environment and it is the selective acknowledgment (SACK) option [2, 4]. This option allows a receiver to inform the sender that up to at most four blocks of noncontinuous segments that have been received properly. Upon receiving this information, a sender does not retransmit data segments that have already been received properly at the destination. The SACK option significantly reduces the amount of data traffic in networks, and it is crucial to improve the system performance. Furthermore, the SACK option can be used by TCP connections running any types of flow control algorithms. However, the goodput of SACK option still degrades rapidly with increased erasure rate [17]. Erasure implies packets are received but corrupted because of the BER in air medium.

10.3.2.2 State Frozen

As stated in a TCP recommended standard document [18], the number of consecutive timeout retransmissions is bounded at a sender. If a connection cannot be re-established with these retransmission attempts, this connection has to be ungracefully terminated. This implies a TCP connection cannot be sustainable by retransmitting lost segments. A long disconnection simply terminates a TCP connection by default. Although a "keep alive" message can also sustain the connectivity of a TCP connection, it still requires a successful transmission of the message. In wireless network, broken link or path happens frequently. Therefore, suspending a TCP state is one of the simplest methods to let a connection re-establish when a receiver can receive packets again. The term "standby mode" is used commonly regarding this mechanism. Also, this mechanism can be added to any other algorithmic designs.

10.3.2.3 Delayed ACKs

Bit errors occur at the physical layer and cause random packet losses in wireless links. Similarly, packet loss also happens when too many packets sent across the broadcast medium along a wireless link because of the frame collision and corruption problems. As a result, delayed ACK is another optional feature in TCP specification [1]. Normally, a delayed ACK acknowledges the receptions of two consecutively sequenced packets. According to [19], the delayed ACK system should be made adaptively and dynamically. For adaptation, an ACK should be sent immediately if out-of-order (OOO) arrival has just occurred [1], and the receiver should keep an estimate of the next segment arrival after receiving a segment without sending an ACK message. An ACK timeout is introduced and should always larger than the next segment arrival estimate. If an ACK timeout occurs, then an ACK should be sent immediately. From [19], the number of packets should be acknowledged by a delayed ACK should be dynamically varying. An ACK may acknowledge up to four packets if the traffic flow is perceived to be in good condition. The delayed ACK design reduces the amount of traffic in networks with the goal of diminishing packet loss because of collisions. Delayed ACK designs can also be added to other designs easily.

10.3.2.4 Retransmission Timeout

For each segment sent, there is an associated RTO timer, which is stored locally in the sender. In [20], a heuristic approach is defined to distinguish between route failures and network congestion. When two timeouts expire in sequence, which implies there are no ACK packets returning when the second RTO expires, the sender can conclude that a route failure may have occurred, and the unacknowledged packet should be retransmitted. In the design, instead of using the "exponential" backoff algorithm to increase RTO in regular TCP, the RTO remains fixed until the route is re-established and the retransmitted packet is acknowledged. TCP selective and delayed acknowledgment options can be used in conjunction with fixed RTO, performance enhancements can be observable. Because it is different from the regular TCP specification, the proposed design works well in WMNs, but it should not work with users in wired networks. Moreover, the adoption of two consecutive timeouts is a heuristic design, and needs more research investigation.

10.3.2.5 Congestion Avoidance

Slow Congestion Avoidance (SCA) [21] or fractional increment of $cwnd$ [22] attempt to limit the growth rate of $cwnd$ with smaller than one segment per RTT. Typically, it can be an one-segment increment for multiple RTTs. In [22], it explicitly states the next segment should be sent when cwnd reaches one-segment size. However, a lower window bound should also be introduced to avoid a long idle period. Similar to delayed ACK mechanism, lower amount of traffic flow is sent than in regular TCP connection, and hence the packet loss rate because of collision may also be reduced.

10.3.2.6 Forward Error Correction

Another trivial design but algorithmic complicated design is to introduce forward error correction (FEC) mechanism in TCP segment. Bit error in wireless link creates a concept of erasure channel. In regular TCP, a packet with corrupted bytes is dropped at the receiver given that it can still arrive at destination. But with FEC, a number of packets with small error bits can be corrected at the receiver. Another design found in [23] is to send extra packets based on simplified Reed–Solomon codes. The missing packets can then be reconstructed at the destination when certain percentages of packets arrive at the receiver safely. The benefit of this design is that the impact of returning ACK messages lessens, and the effect of packet losses can be reduced. There are drawbacks, which include requiring extra computation times and resources. If, in case, FEC field is added to a segment, then modification in network card firmware is required to avoid dropping erratic packets before executing correction attempts.

10.3.2.7 TCP-DOOR

TCP detection of out-of-order and response (DOOR) is an end-to-end approach [24]. TCP-DOOR interprets OOO deliveries as indications of route failures in wireless networks. Because of the random packet losses, OOO events may happen more often in networks with wireless links. The detection of OOO events can be accomplished through either a sender-based or a receiver-based mechanism. Because the ACK numbers in returning messages should have the nondecreasing property, a sender can detect an OOO event if a later arriving ACK message has a smaller value. However, when there are returning duplicate ACK messages, the sender may require extra information to detect an out-of-order event. Hence, a new one byte option field, the ACK duplication sequence number (ADSN), is introduced to notify the sender. The receiver increments the ADSN when it decides to send out duplicate ACK packets to notify the sender. On the other hand, to notify the receiver, the sender should increment another new two-byte TCP option field, the TCP packet sequence number (TPSN). The TPSN is incremented and transmitted with each TCP packet, including the retransmitted packets, to enable the receiver to detect an OOO event. Upon observing an OOO event, the receiver should notify the sender by setting a new specific option bit (OOO bit) in the ACK packet header. These two mechanisms enable the detections of route failures in both directions.

Upon detecting an OOO event in TCP-DOOR, the sender should temporarily disable the congestion control algorithm for a predefined T_1 duration, and freezes the TCP states. The sender also continues to detect if three duplicate ACKs have been arrived in the last T_2 time duration. If this event has happened, the TCP instantly returns to the states it has frozen instead of moving into the Fast Retransmit and Fast Recovery stages.

The TCP-DOOR design has been identified to work better in low BER multipath wireless networks. Similar to TCP Westwood+, its operations heavily rely on the successful receptions of ACK packets. In fact, any mechanisms that rely on the arrivals of returning ACK packets have to work in low BER wireless network environments. Consequently, the TCP-DOOR mechanism may not work well in the multihop environment in WMN. As of today, algorithmic modifications made in transport layer have only made small impacts on the performance outcome. Cross-layered designs discussed below may have noticeable overall system improvements.

10.3.3 Cross-Layered Designs

The popular TCP NewReno congestion control algorithm can handle bursty traffic effectively in wired networking environment with relatively low packet loss rate. The aforementioned flow control modifications in transport layer help improve the TCP performance in dealing with, for example, the random packet losses in wireless networks. But in WMNs, there are other types of erroneous scenarios, such as the failures of wireless links. Wireless link failures happen in link layer, which then affect the path setup between a sender and receiver. In order words, this is the route failure in network layer as it affects the routing path in the WMNs. Indeed, false link failure can also happen if there are too many frames sent on a wireless link. Collisions of multiple consecutive request-to-send (RTS) messages may occur and the upstream mesh router may mistakenly conclude it a broken link. Other problems in link and network layers include the capture effect, end-user handoff operations, etc. This is why cross-layered flow control designs play important roles in solving erroneous events in other layers besides the transport layer.

10.3.3.1 Explicit Link Failure Notification

Explicit link failure notification (ELFN) technique [11] is a popular design that many newer designs rely on. The ELFN is a cross-layered design between TCP and the routing protocol. Route failures may happen often in wireless networks. If it happens, the routing protocol should inform the sender with a route failure message. In the design, this message is piggybacked with an ELFN message. Functionally, this ELFN message is equivalent to the "host unreachable" Internet Control Message Protocol (ICMP) control message. The piggybacked information contains the TCP sequence number, addresses, and port numbers of both the sender and receiver. Upon receiving the ELFN message, the source node disables its retransmission timers, and enters the "standby" mode. During the standby period, the sender regularly sends probe messages to check if a route has been restored. Actually, the first data segment queued for the TCP connection is the probe message. If an acknowledgment of the probe packet has been returned, then the sender departs the standby mode, resumes its retransmission timers, and continues the normal operations.

The interval among sending probe packets has been evaluated. Its value may likely link to different network topologies. It should depend on the RTT between the source and destination nodes. As stated in [11], the preferred probe interval should depend on the values of RTO and *cwnd* as they may have impacts on performance upon route restoration. With the experiments, the TCP connection, upon restoring the states before entering the standby mode, performs better than initializing *cwnd* to 1 segment size and/or RTO to 6 s, which is the default value of RTO used in NewReno. Because probe packets are sent periodically to detect if routes have been re-established, the ELFN design has not been performing well in cases of high network traffic scenarios. The reason is that the ELFN makes TCP behave more aggressive. If there are many active connections, false broken links may also happen with ELFN design. Besides, further investigations are required to harmonize the routing protocols with the ELFN design.

10.3.3.2 TCP-Feedback

Similar to ELFN, TCP-Feedback (TCP-F) [25] is a feedback-based approach to handle route failures. This approach enables a TCP sender to distinguish between losses because of route failure, or because of network congestion or frame collision. TCP-F also works with network layer. When a routing agent detects the disruption of a route, it explicitly sends a route failure notification (RFN) packet to the sender. Upon receiving the RFN packet, the source enters a "snooze" state, and state information of the connection is frozen. That is, it stops sending packets and freezes all its state parameters including the timer values and *cwnd* size. The sender remains in this snooze state until it is notified regarding a route restoration through a route re-establishment notification (RRN) packet. When the source node is notified, it departs the snooze state, restores the state information, and resumes transmissions based on the saved *cwnd* and RTO values. Potentially, the design may be blocked indefinitely if the RRN has been lost. To avoid staying forever in the snooze state, the TCP sender starts a route failure timer upon receiving the RFN message. When this timer expires, the regular congestion control algorithm should be invoked normally. The control messages are explicit in this design without generating large amount of messages as in the case of ELFN design. Certainly, the appropriate value of the route failure timer for WMN requires more analytical and experimental works.

10.3.3.3 TCP/RCWE

The TCP with Restricted Congestion Window Enlargement (TCP/RCWE) [26] works with the ELFN mechanism. This is a cross-layered design that relies on ELFN to handle broken link issues. The design focus of TCP/RCWE is mainly on solving random packet losses by observing the value of RTO in the transport layer. If the RTO increases, then *cwnd* is not increased. If the RTO decreases or keeps unchanged, then modifications of *cwnd* follow the basic TCP operational rules. TCP/RCWE is found to have small sliding window size, which naturally leads to better goodput with fewer packet losses.

10.3.3.4 ATCP

Although the design of ad hoc TCP (ATCP) [27] focuses on improving TCP performance in MANETs, it should work well in WMNs and use feedbacks from the network layer. Apart from route failure issue, ATCP also attempts to solve the problem of high BER. In the design, a layer known as ATCP is inserted between the TCP and IP layers. The ATCP listens to the state information in network layer provided by the explicit congestion notification (ECN) messages [28] and the "destination unreachable" control messages in ICMP. The ATCP may assign a TCP sender into one of the following three states: persist state, congestion control state, or retransmit state.

For example, upon receiving a "destination unreachable" message, the TCP sender should enter a persist state. In this state, the state information is frozen and no packets can be sent unless a new route has been established and found by, similar to ELFN, probing the network regularly. The ECN in network layer is used to explicitly notify the sender about network congestion along the path being used. If an ECN message has been received, then TCP congestion control algorithm should be invoked without needing to wait for a timeout event. If detecting a packet loss event, i.e., receiving three duplicate ACK messages, the ATCP layer stops forwarding the third duplicate ACK message, retransmits the lost packet immediately, and sets the connection into persist state. The ATCP sets a TCP connection to the normal congestion control state only if it receives the next ACK message. The design of ATCP offers interoperability among TCP sources or destinations that do not implement ATCP.

The design of ATCP is simple and robust. Apart from route failure problems, ATCP also works with high BER links, network congestion, and packet reordering issues. Unfortunately, the probing mechanism used to detect route reestablishment suffers similar problems as discussed in the ELFN design.

10.3.3.5 TCP-BuS

TCP Buffering capability and Sequence information (TCP-BuS) [29] also uses feedback from the network layer to detect route failure events. The design introduces a concept of buffering capability in the intermediate mesh routers. The TCP-BuS works with a source-initiated on-demand associativity-based routing (ABR) protocol. The routing protocol has the capability of associating saved segments regarding different TCP connections. There are five enhancement features in TCP-BuS, which are:

- *Explicit notification.* There are two types of control messages to notify a source node regarding route failure and route re-establishment. The explicit route disconnection notification (ERDN) is sent when a route failure occurs. When a network node detects the failure of route, it sends the ERDN message to the sender, which should stop sending data packets. This specific network node, known as Pivoting Node (PN), should attempt to re-establish the route locally using a localized query (LQ) message. If a new route can be established, the PN node sends the explicit route successful notification (ERSN) to the TCP sender, which can then resumes data transmission.

- *Extending timeout values.* In route reconstruction (RRC) phase, packets are buffered along the path from the source to the Pivoting Node until a new partial route is established. These buffered packets are delivered after the route is reconstructed. To avoid triggering too many timeout events while waiting for the new route at sender during the RRC phase, the values of RTO timers for buffered packets are doubled.
- *Selective retransmission request.* Because the durations of RTO timers are doubled, it may take too long for the lost packets, not saved along the path, to get retransmitted by the source. Hence, the receiver should use selective retransmission request to indicate the sender to retransmit the lost packets selectively.
- *Avoiding unnecessary requests for fast retransmission.* When the route is restored, the receiver notifies the sender about the lost packets along the path from PN to the destination. On receiving this notification, the source retransmits the lost packets. However, packets buffered along the path from sender to PN may arrive earlier at destination than sender's retransmitted packets. These duplicate packets induce the receiver to generate multiple duplicate ACK messages, which in turn causes the sender to move into Fast Retransmit operating phase. Therefore, TCP-BuS works with the ABR routing mechanism to suppress these duplicate ACK messages.
- *Reliable retransmission of the control message.* The routing control messages, ERDN and ERSN, must be sent reliably to guarantee the correctness of TCP-BuS operations. The reliability is achievable through overhearing the wireless communication channel after transmitting the control messages. If a node sent a control message without overhearing this message relayed during a timeout, it concludes that the control message has been lost and retransmits this control message.

The buffering and control message retransmission techniques from TCP-BuS present a feasible solution for route failures in WMNs. It is comparatively easy to adopt the PN model in WMNs because the PN are basically carrying out network layer operations only. However, the mappings of ABR designs and the buffering algorithms are not straightforward. The performance measurements in [29] demonstrated that TCP-BuS outperformed regular TCP and TCP-F algorithmic designs.

10.3.3.6 LRED and Adaptive Pacing

A TCP connection typically grows its *cwnd* larger than the optimal value in wireless multihop channel. The larger *cwnd* may overload the network and often lead to packet losses because of collisions. Therefore, link random early detection (LRED) [30] is recommended to fine-tune the link layer packet dropping probability with a goal to sustain a TCP *cwnd*, hopefully, around its optimal value. The LRED increases the packet dropping probability when the link layer contention level, measured by the moving average of retransmissions, exceeds a minimum threshold. When the retransmission count is larger than the threshold, the dropping/marking probability is set to the minimum of computed dropping probability or a preset upper bound value. Furthermore, LRED can work with ECN-enabled TCP. Adaptive pacing [30] is also proposed to better coordinate channel access along the packet forwarding path. It typically solves the exposed terminal problem by adding one-packet transmission time to the backoff interval. Improvements on TCP NewReno performance have been demonstrated through simulations.

10.3.3.7 Loss-Tolerant TCP

The design of loss-tolerant TCP (LT-TCP) [17] is to work in high loss regimes in ECN environments by adding adaptive SMSS (not only *cwnd*) and packet-level FEC mechanisms [23]. LT-TCP carries out congestion response only if it receives ECN messages. Given a congestion window, a number of segments can be sent. In the algorithm, the per-window loss fraction samples are averaged using an exponentially weighted moving average with parameter 0.5. LT-TCP focuses on erasure loss and it has a few main building blocks, which are:

- *Proactive FEC.* The number of FEC packets [23] per window used (a proactive FEC) is a function of the erasure estimate. The SMSS is adjusted to allow one or more FEC packets per window.
- *Adaptive MSS and granulation.* For FEC, the *cwnd* have to send at least certain number of packets, subject to the limits of introduced minimum SMSS and maximum SMSS. As the window increases (in bytes), the SMSS is increased in steps of prefixed sizes, provided it does not decrease the window granulation. If necessary, SMSS is adjusted to accommodate the proactive FEC. When ECN arrives, both the window and SMSS are halved (subject to minimum SMSS constraint).
- *Reactive FEC.* Because proactive FEC may be insufficient because of the variance in loss patterns, the sender transmits certain number of reactive FEC packets, and this number of packets depends on (1) the currently estimated loss rate, (2) the number of proactive FEC packets sent for this block, and (3) the number of holes left to be filled to completely decode this block.

As reported, the simulation result of LT-TCP does not degrade rapidly as observed with SACK option. The drop in performance is graceful because of its resilience at higher error rates. LT-TCP is not sensitive to RTT and is robust to burstiness because of the FEC design. Hence, it performs better compared to SACK, especially as the average error rate increases. However, LT-TCP may possibly only present partial solution for TCP to work in multihop WMNs.

Introducing coding theory in TCP, such as FEC, to combat random packet loss is comparatively newer design mechanism for TCP flow control algorithms. Besides, network coding can reduce certain amount of traffic flow using exclusive OR (XOR) operations distributed in networks. There may be more research results in this area in future.

10.3.4 Split Connections

Splitting a long multihop TCP connection into multiple shorter TCP connections in a WMN is the most trivial and also the most reliable method in forwarding a data file. Certainly, the drawback is that it may take a much longer time to transfer a file, especially if the size of a file is large. Almost requiring no changes to TCP specification, Split TCP proposal is one of the simplest and straightforward designs [31]. The initial design goal of Split CP is to improve the system throughput and resolve issue of unfairness. The details of Split TCP are described below.

10.3.4.1 Split TCP

Even without considering link failure events and user mobility issues, the data throughput of a TCP connection simply drop drastically after sending data through merely a few hops of wireless communication links [12]. The most trivial method is simply to partition a long multihop wireless path into multiple shorter TCP subpaths. Hence, Split TCP splits a long wireless route in wireless networks into multiple shorter localized segment paths for easier connection management. The interfacing node between two localized segments is called a proxy. In a mesh router, its routing agent determines if the node should act as a proxy according to the inter-proxy distance parameter. If it is a proxy, then it intercepts TCP packets, buffers them, and acknowledges their receptions to the previous proxy or the source node by sending a local acknowledgment (LACK) message. Also, the proxy is responsible for delivering the packets at an appropriate rate to the next local segment. Upon receiving a LACK message from the next proxy or the destination node, the proxy purges acknowledged packets from its buffer. To ensure the traffic reliability between the source and destination, an ACK message should be sent from the destination to the source, as in the regular TCP specification. Split TCP partitions the transport layer functionalities into end-to-end reliability, local congestion control, and overall system integrity. All nodes in Split TCP should have two TCP connection control parameters. At the sender, there should be two transmission windows, the congestion window and the end-to-end window. The congestion window is a sub-window of the end-to-end window. The congestion window changes according to the arrival rate of LACK messages from the next proxy. The end-to-end window changes according to the rate of arrival of the end-to-end ACK messages from the destination. Therefore, each proxy should have a *cwnd* that controls the transmission rate among proxies.

The recommended inter-proxy distance is of between three and five hops. Split TCP impacts favorably on both throughput and fairness. This result matches well the multihop performance reported in [11]. Indeed, Split TCP is the simplest way to improve overall system throughput. Certainly, one hop Split TCP offers the highest throughput and reliability performances. But it is also the most expensive implementation because of the cost of buffers and network overhead at all mesh routers.

10.4 Thoughts for Practitioners

Currently, there are no commercial WMNs. Most of the existing wireless networks are 802.11 WLANs, which operate in hot-spot locations for serving access purposes. Research in TCP flow control algorithm for WMNs is still at its early stage. But practitioners can experience the performance of different experimentally designed TCP congestion control algorithms with Linux operating systems.

Since the version 2.6.13, Linux supports pluggable congestion control algorithms. Readers can obtain a list of congestion control algorithms that are available in the kernel upon typing

```
> sysctl net.ipv4.tcp_available_congestion control
```

Unfortunately, there are no TCP designs implemented for WMNs. If readers are interested, readers are encouraged to implement different versions of TCP for testing WMNs.

10.5 Directions for Future Research

Designs of transport layer protocols for WMNs are still at the early research stage. Many proposed designs in this article have been validated through simulations only. And in reality, there are no commercial WMNs. This is far from a mature technology. Indeed, there are many research directions that are still open for future research. Examples include:

- TCP flow control algorithm for wireless networks,
- Cross-layered designs, for example, MAC+TCP,
- Performance evaluation,
- Testbed experiments.

10.6 Conclusions

Although the latest popular TCP NewReno congestion control algorithm offers one of the best scalable performances on the Internet, it does not serve well in WMNs. In this chapter, we review the basic algorithmic mechanisms in NewReno. Furthermore, TCP is expected to have problems in sending large-size file, or traversing through a large number of wireless hops. Fundamentally, TCP belongs to transport layer and it would be the best if all changes in TCP can be made in transport layer. Some modifications made in transport layer work to improve TCP performance in wireless networks, but none of them at the current stage can offer a complete solution. Indeed, some performance degradations occur because of network, link, and physical layers. Combining cross-layered solutions with TCP flow control algorithm should be the desirable methods.

10.7 Terminologies

1. *Transmission control protocol.* The commonly used transport layer protocol for reliable information transfer across the Internet.
2. *Congestion control algorithm.* The control algorithm deployed in a TCP sender to control traffic delivery to avoid causing excessive network congestion events in networks.
3. *Additive-increment multiplicative-decrement.* The basic operating principle of the TCP congestion control algorithm for offering fairness of sharing resources among multiple connections.
4. *NewReno.* The latest recommended TCP congestion control algorithm from the Internet engineering task force (IETF).
5. *Cubic.* The default TCP congestion control algorithm deployed in the latest Linux operating system.
6. *Flight-size.* The amount of segments sent but not acknowledged by the recipient.
7. *Wireless mesh networks.* Wireless mesh networks are composed of wireless routers or access points that facilitate the connectivity and intercommunication of wireless clients through multihop wireless paths.
8. *Cross-layered design.* A general accepted approach to design protocols and algorithms for wireless networks.
9. *Explicit link failure notification.* Explicit link failure notification (ELFN) technique is a cross-layered design for TCP to operate in wireless networks. Many newer designs have been designed based on ELFN.
10. *Split connection.* A simple way to break one TCP connection into multiple TCP connections in multihop wireless networks. An intermediate proxy terminates a TCP connection and then starts a new TCP to the next one.

10.8 Questions

1. What is the design goal of the slow start (SS) process in TCP? Please describe in detail the operations of the SS in TCP NewReno.
2. In the latest TCP flow control RFC, the starting value of a host's congestion window (cwnd) for the SS is set to be either 1 SMSS or 2 SMSSs. Before this RFC, the starting cwnd was set to 1 SMSS only. Can you try to explain the reason behind the increase from 1 SMSS to 2 SMSSs?
3. A protocol designer develops a new mechanism known as the fast start (FS) at the receiver side. The operation of the FS is to split an acknowledgement into multiple acknowledgements within one round trip time (RTT). Can you interpret the goal of this design?
4. There is a sending host transmits an infinitely large file using a TCP connection to a receiving host in a wired local area network. Suppose this is the only connection in the network. Given that the RTT is D, packet loss probability is p,

what is the throughput, T, of the connection in steady state? Suppose all packets are in fixed-sizes, which are 1 SMSSs. [Hint: Ignore the Slow Start process, one packet loss always happens but recovered through the Fast Retransmit/Fast Recovery, and the connection always runs in Congestion Avoidance process.]

5. What are wireless mesh networks?
6. Can you outline a few mechanisms that can be used in the transport layer for combating random packet loss in wireless networks?
7. Can you outline the five operating features of the TCP buffering capability and sequence information (TCP-BuS)?
8. What are the three main operating features of loss-tolerant TCP (LT-TCP)?
9. Outline the operations of explicit link failure notification (ELFN)?
10. What is the command to change the congestion control algorithm in Linux operating system?

References

1. J. Poster, *Transmission Control Protocol*, STD 7, RFC 793, Internet Engineering Task Force (IETF), Sept. (1981).
2. M. Mathis, J. Mahdavi, S. Floyd, and A. Romanow, *TCP Selective Acknowledgment Options*, RFC 2018, IETF, Oct. (1996).
3. M. Allman, V. Paxson, and W. Stevens, *TCP Congestion Control*, RFC 2581, IETF, Apr. (1999).
4. S. Floyd, J. Mahdavi, M. Mathis, and M. Podolsky, *An Extension to the Selective Acknowledgment (SACK) Option for TCP*, RFC 2883, IETF, Jul. (2000).
5. S. Floyd, T. Henderson, and A. Gurtov, *The NewReno Modifications to TCP's Fast Recovery Algorithm*, RFC 3782, IETF, Apr. (2004).
6. J. F. Kurose and K. W. Ross, *Computer Networking: A Top-Town Approach Featuring the Internet*, 2003, Addison Wesley, Reading, MA.
7. V. Jacobson, Congestion avoidance and control, *Computer Communication Review*, 18(4), 314–329, (1988).
8. L. Xu and I. Rhee, CUBIC: A new TCP-friendly high-speed TCP variant, *Proceedings of Workshop on Protocols for Fast Long Distance Networks (PFLDnet) 2005*, 3–4 Feb. 2005, Lyon, France.
9. D. J. Leith, R. N. Shorten, and G. McCullagh, Experimental evaluation of cubic TCP, *Proceedings of Protocols for Fast Long Distance Networks (PFLDnet) 2007*, 7–9 Feb. 2007, Los Angeles, CA, USA.
10. I. F. Akyildiz and X. Wang, A survey on wireless mesh networks, *IEEE Communications Magazine*, 43(9), 23–30, (2005).
11. G. Holland and N. Vaidya, Analysis of TCP performance over mobile ad hoc networks, *Wireless Networks*, 8, 275–288, (2002).
12. H. Elaarag, Improving TCP Performance over Mobile Networks, *ACM Computing Surveys*, 34(3), (2002).
13. X. Chen, H. Zhai, J. Wang, and Y. Fang, TCP performance over mobile ad hoc networks, *Canadian Journal of Electrical and Computer Engineering*, 29(1/2), 129–134, (2004).
14. A. Al Hanbali, E. Altman, and P. Nain, A survey of TCP over ad hoc networks, *IEEE Communications Surveys*, 7(3), 22–36, Third Quarter (2005).
15. S. Charoenpanyasak, B. Paillassa, and F. Jaddi, Experimental study on TCP enchancement interest in ad hoc networks, *Third International Conference on Wireless and Mobile Communications (ICWMC'07)*, 2007, Guadeloupe.

16. L. A. Grieco and S. Mascolo, Performance evaluation and comparison of Westwood+, NewReno, and Vegas TCP congestion control, *ACM Computer Communication Review*, 34(2), 25–38, (2004).
17. V. Subramanian, K. K. Ramakrishnan, S. Kalyanaraman, and L. Ji, Impact of interference and capture effects in 802.11 wireless networks on TCP, *Second International Workshop on Wireless Traffic Measurements and Modeling (WitMeMo'06)*, Aug. 2006, Boston.
18. R. Braden, Editor, *Requirements for Internet Hosts – Communication Layers*, STD 3, RFC 1122, IETF, Oct. (1989).
19. R. de Oliveira and T. Braun, A dynamic adaptive acknowledgment strategy for TCP over multihop wireless networks, *IEEE Infocom'05*, 3, 1863–1874, (2005).
20. T. Dyer and R. Boppana, A comparison of TCP performance over three routing protocols for mobile ad hoc networks, *ACM MobiHoc'01*, (2001), Long Beach.
21. S. Papanastasiou and M. Ould-Khaoua, TCP Congestion Window Evolution and Spatial Reuse in MANETs, *Journal of Wireless Communications and Mobile Computing*, 4(6), 669–682, (2004).
22. K. Nahm, A. Helmy, and C.-C. J. Kuo, TCP over multihop 802.11 networks: Issues and performance enhancement, *ACM MobiHoc'05*, 277–287, May 25–27, (2005).
23. L. Baldantoni, H. Lundqvist, and G. Karlsson, Adaptive end-to-end FEC for improving TCP performance over wireless links, *IEEE International Conference on Communications (ICC'04)*, 7, 4023–2027, Jun. (2004).
24. F. Wang and Y. Zhang, Improving TCP performance over mobile ad hoc networks with out-of-order detection and response, *ACM MobiHoc'02*, June (2002), Lausanne.
25. K. Chandran, S. Ragbunathan, S. Venkatesan, and R. Prakash, A feedback-based scheme for improving TCP performance in ad hoc wireless networks, *International Conference on Distributed Computing Systems (ICDCS98)*, May (1998), Amsterdam.
26. M. Gunes and D. Vlahovic, The performance of the TCP/RCWE enhancement for ad-hoc networks, *IEEE International Symposium on Computers and Communication*, 43–48, (2002).
27. J. Liu and S. Singh, ATCP: TCP for mobile ad hoc networks, *IEEE Journal on Selected Areas in Communications*, 19(7), 1300–1315, (2001).
28. K. Ramakrishnan, S. Floyd, and D. Black, *The Addition of Explicit Congestion Notification (ECN) to IP*, RFC 3168, IETF, Sept. (2001).
29. D. Kim, C. Toh, and Y. Choi, TCP-BuS: Improving TCP performance in wireless ad hoc networks, *Journal of Communications and Networks*, 3(2), June (2001).
30. Z. Fu, P. Zerfos, H. Luo, S. Lu, L. Zhang, and M. Gerla, The impact of multihop wireless channel in TCP throughput and loss, *IEEE Infocom'03*, 3, 1744–1753, (2003).
31. S. Kopparty, S. V. Krishnamurthy, M. Faloutsos, and S. K. Tripathi, Split TCP for mobile ad hoc networks, *IEEE Globecom'02*, Nov. (2002), Taipei.

Chapter 11
Congestion Control in Wireless Mesh Networks

Bahareh Sadeghi

Abstract In this chapter, layer-2 hop-by-hop congestion control mechanisms for wireless mesh networks are studied. First, background on development of congestion control as a research area is presented. Then, congestion is defined and the causes of congestion in a mesh network are identified. The steps and challenges for development of a congestion control protocol are explored by dividing congestion control mechanism into three main functions: congestion detection, signaling, and resolution. Stability and fairness in the context of rate control are studied, and global stability in the presence of delay is identified as the main challenge faced by hop-by-bop congestion control algorithms in multihop networks. The congestion control framework in the IEEE 802.11 draft standard on mesh networking is then presented. The chapter concludes with thoughts for practitioners and areas for future research in this field.

11.1 Introduction

Mesh networks have become increasingly ubiquitous as a natural extension to the Wireless Local Area Networks (WLAN). Many WLAN-based proprietary mesh solutions have been developed for different usage scenarios, ranging from home networking to metro and emergency services, both in industry, e.g., [1,2], as well as in academia [3,4]. To enable interoperability among the different existing solutions by different developers and vendors, there has been an effort to define a mesh networking standard in the IEEE standardization body. This effort has led to formation of a new task group, Mesh Networking Task Group (TGs), in IEEE 802.11 Working Group. The charter of this task group is to develop an amendment to IEEE 802.11 MAC that includes enhancements for multihop communication among nodes.

B. Sadeghi (✉)
Intel Corporation, 2111 NE 25th Avenue, JF3-206, Hillsboro, OR 97124, USA
e-mail: bahareh.sadeghi@intel.com

S. Misra et al. (eds.), *Guide to Wireless Mesh Networks*, Computer Communications
and Networks, DOI 10.1007/978-1-84800-909-7_11,

The IEEE 802.11 MAC protocol has been designed for single-hop networks where all nodes are in close proximity and within radio range of one another. IEEE 802.11 MAC is based on Carrier Sense Multiple Access with Collision Avoidance (CSMA/CA) [5]. In a single-hop network, the network throughput is equal to the sum of the throughput of all nodes. Hence, in an IEEE 802.11-based network the network throughput is maximized if all nodes seek to transmit as many packets as possible following the specified CSMA/CA medium access rules. In a multihop network, however, only packets that reach their final destination count toward the throughput of the network; packets that are lost midway before reaching their final destination result in lost throughput because the resources consumed to transmit them midway through the network are wasted. Packet loss in the network may happen because of channel error or queue overflow, i.e., congestion, in the network.

A congestion control function in the network is responsible for preventing the occurrence of congestion, as well as for alleviating the impact of congestion on network throughout if it occurs. Although these goals can be achieved by imposing limitation on the rate of access of the source nodes to the network resources, the congestion control scheme should ensure that a fair distribution of resources among nodes is maintained.

In this chapter we study layer-2 congestion and congestion control mechanisms. The remainder of the chapter is organized as follows. In Sect. 11.2, we present the background and related work on congestion control in wireless mesh networks. Congestion is defined in Sect. 11.3. We then study the impact of congestion on system performance in Sect. 11.4. Section 11.5 discusses the congestion control mechanisms and their functional building blocks, followed by a discussion of the impact of congestion control on stability and fairness in Sect. 11.6. We present the congestion control framework that is part of the IEEE 802.11s draft standard in Sect. 11.7 and then provide some thoughts for practitioners in Sect. 11.8. Directions for future research are presented in Sect. 11.9, and the chapter is concluded in Sect. 11.10.

11.2 Background

The theory of congestion control is based on the seminal work of Frank Kelly [6], who showed how an *optimal* network-wide rate allocation can be achieved among nodes that individually control their rates. Definition of *optimal* in this context depends on the fairness objective, such as max–min fair or proportional fair rate allocations. Based on the model developed by Kelly, congestion control takes the form of a distributed optimization algorithm. Although many of the congestion control algorithms can be modeled by using Kelly's framework, his model makes the assumption that at a given link, all flows observe the same price.

Protocols for congestion control were first developed for the Internet and led to numerous publications on the subject during 1990s. These publications covered a wide range of solutions and optimizations on TCP [7–12], complementing the work undertaken in IETF that resulted in publication of TCP [13]. In addition to the end-to-end solutions like TCP, hop-by-hop congestion control for the Internet was also well studied [14–16].

With growing popularity of wireless data networks in the late 1990s, many researchers examined the performance of TCP over wireless [17, 36]. In wireless links, as opposed to the traditional wired networks, the packet loss is not only due to congestion, but also due to channel errors and lossy handovers. Because TCP has a unique reaction to all packet losses regardless of their cause, its end-to-end performance is severely degraded over wireless links.

The extended number of back-to-back wireless links in the mobile ad hoc and sensor networks further impacts TCP performance. The interactions between TCP and the MAC may result in severe unfairness and flow starvation in such networks. Careful adjustment of TCP parameters, as well as MAC behaviors such as explicit link layer acknowledgments or a reduced congestion window size, are required to improve the performance [18, 19].

The fact that characteristics specific to wireless links impact the overall network performance has changed the traditional view of network links. Links are no longer viewed as pipes with a fixed capacity. Rather, the characteristics and variations of the wireless link are now being incorporated in what is called cross-layer design. In cross-layer design, MAC algorithms and higher-layer protocols take into account the underlying PHY dynamics to improve the overall performance. Cross-layer design can achieve significant improvements in performance by optimizing the control over multiple protocol layers [20].

Cross-layer design in development of congestion control algorithms has been recently well explored [21,22] and has led to many solutions proposing cross design of PHY modulation schemes, power control, MAC algorithms and scheduling, TCP optimizations, and routing protocols.

Ensuring stability and fairness in multihop wireless networks in the presence of congestion control mechanisms has been identified as a challenge and has gained recent attention [23–25]. The research in this area has made a case for the feasibility of hop-by-hop congestion control schemes and has proved the stability of such solutions in absence of delay. It is, however, known that hop-by-hop schemes can result in spatial spread of congestion in the network, and stability is only achieved under specific conditions.

11.3 Definition of Congestion

Congestion occurs in a mesh network as a result of MAC queue buildup at an intermediate node on the path of a multihop flow. A forwarding node experiences congestion when the arrival rate of packets to the node is greater than its forwarding rate. Many factors may contribute to congestion, including network topology, number of flows, the traffic characteristics of the flows and their routes, as well as channel capacity and the available transmission rate at the physical layer. From the perspective of a single node, however, there are two major local causes for queue buildup: *neighborhood congestion* and *high packet arrival rate*.

11.3.1 Neighborhood Congestion

The rate at which a node empties the packets in its queue depends, not only on the physical rate that the packets are transmitted at, but also on the frequency that a node successfully accesses the channel for packet transmission. In IEEE 802.11 MAC, nodes follow a CSMA/CA-based MAC for accessing the channel, where the nodes within radio range of one another contend for access to the channel. As the number of active nodes (nodes with nonempty queues contending for the media) within transmission range of one another increases, the probability of successful access to the channel decreases [26]. Hence, in a congested neighborhood, where the available resource is less than what is required by the active nodes, the probability of queue buildup at the nodes increases.

Note that for a successful packet transmission in a multihop network, not only must the transmitter successfully grab the channel, but the receiver should also be available and sense a free media. Figure 11.1 shows an example of a node that has a low opportunity for transmission (Node B) because the immediate receiver of its flow (Node C) is located in a congested neighborhood and, because of hidden terminal problem [27], the receiver is not clear to receive packets.[1]

It should also be noted that when a node has a low chance of successful transmission because of the media being busy (and not because of its immediate receiver being busy), the node itself is equivalently less available as a receiver for the nodes that have packets destined to it. Hence, one might argue that local congestion in the neighborhood of a congested node would contribute less to further queue buildup at the congested node compared to local congestion in its receiver's neighborhood.

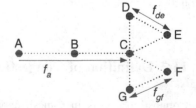

Fig. 11.1 Example network topology where Node B suffers packet buildup because of congestion in neighborhood of Node C

---

[1] In the figures, nodes that are within radio range of one another are connected with a dotted line.

11.3.2 High Packet Arrival Rate

Regardless of how fast a node can successfully transmit packets, if the number of incoming packets is higher than the exit rate of the packets from the queue, congestion will occur in the node. The packet arrival rate at an intermediate node depends on the traffic model of the flows, as well as on the available routes. If a node is on the route of many flows, the probability of it suffering from congestion increases. This is usually the case for the nodes that are closer to the gateways in a multihop mesh network and thus on the route of multiple ingress and egress flows.

11.4 Impact of Congestion on System Performance

Congestion results in packet loss if its duration lasts long enough to result in overflow of the queue in a node. Loss of packets on their way to destination in an intermediate node on a multihop path results in waste of resources used to transfer packets from their original source to the congested node.

Example 11.1. Consider an IEEE 802.11 system, in which all nodes within transmission range of one another are given equal share of the channel access time; thus the network is long-term-throughput fair under the scenario that the PHY transmission rate of all nodes is equal. Figure 11.2 illustrates an example of congestion in such a system where all the transmitters are within radio range of one another. In this figure, nodes A, B, and C are within transmission range of one another and contend for channel access. Node D is the receiver of two flows originating from nodes A and B, and is within transmission range of only node C and not nodes A and B. The queue of Node C builds up because its share of the channel is equal to either node A or B, and hence can only forward half of the packets that enter its queue to the final destination of Node D.

Example 11.2. In the example scenario captured in Fig. 11.3, we consider three different cases: only f_b is active, only f_a is active, and both flows are active. Let L_A and L_B denote the offered load to nodes A and B, respectively. Also, we show the capacity of the two links with C_{AB} and C_{BC}.

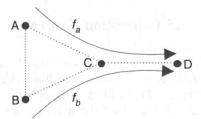

Fig. 11.2 Network scenario of Example 1

Fig. 11.3 Network scenario of Example 2

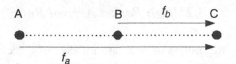

1. In the first case, where only f_b is active, $L_A = 0$. Based on the definition of congestion presented in Sect. 11.3, queue buildup and congestion can still occur in node B if

$$L_B > C_{BC}. \tag{11.1}$$

   In this case, although there is a queue buildup at node B, there is no wasted throughput because the packets that might get dropped because of queue overflow have not been transmitted over the air.
2. In this case, $L_B = 0$ and $L_A > 0$. The condition for occurrence of congestion at node B is

$$\min(C_{AB}, L_A) > C_{BC}. \tag{11.2}$$

   Note that the amount of incoming traffic to node B is bound by the capacity of the link between nodes A and B. In this case, if congestion occurs, the wasted capacity in the network is $\min(C_{AB}, L_A) - C_{BC}$.
3. In the last scenario, both flows f_a and f_b are active. It is clear that if $C_{BC} \leq C_{AB}$, Node B suffers congestion as it needs to transmit traffic of flow f_b as well as to forward the packets belonging to flow f_a. The congestion will not occur if:

$$C_{BC} \geq \min(C_{AB}, L_A) + L_B. \tag{11.3}$$

In the case of congestion, the wasted capacity in the network is equal to the traffic of f_a that is dropped at node B, and depends on the order that the traffic of each of the flows is served at node B. Assuming that node B serves traffic from both flows at a rate equal to their relative arrival rate, the wasted throughput because of congestion is

$$\min(C_{AB}, L_A) - C_{BC}\frac{\min(C_{AB}, L_A)}{\min(C_{AB}, L_A) + L_B}. \tag{11.4}$$

The example above illustrated how the congestion in the network results in waste of throughput. In Sect. 11.5, we discuss how a congestion control scheme can release the wasted resources, to be used by other flows in the network, by adjusting the MAC transmission rate of flows at their source.

11.5 Congestion Control

Congestion control is the mechanism implemented in the network to prevent, reduce, or resolve congestion. The most well-known congestion control mechanism is TCP and its variants, which are implemented in the application layer and adaptively adjust the transmission rate of the nodes to the available end-to-end resources in the network. Although end-to-end congestion control schemes like TCP are the preferred solution for the Internet, in a wireless mesh network a hop-by-hop congestion control mechanism has a better performance [24]. A mesh network does not have the scalability issues associated with the Internet, and, in comparison to an end-to-end solution, a layer-2 hop-by-hop mechanism is capable of reacting more quickly to the system dynamics and congestion. Additionally, a MAC-based function not only is effective regardless of the traffic type, but also does not inherit the problems associated with degraded performance of TCP over wireless [28].

Any congestion control scheme consists of three main functions for detection, signaling and resolution of the congestion. In this section we study each of these three functions.

11.5.1 Congestion Detection

Efficiency of a congestion control scheme is dependent on how early and how accurately the congestion is detected in the network. Given the definition of congestion provided in Sect. 11.3, it is clear that the congestion can be detected by monitoring the MAC queue buildup locally at each node. However, a queue buildup can be the effect of transient system dynamics and, hence, temporary. A congestion control scheme should be able to recognize and react accordingly to transient behaviors, as opposed to congestion, by choosing the measurement window appropriately. Too large of measurement window will result in slow reaction to congestion, and too small of a measurement window may result in performance oscillation and instability. Additionally, a careful choice for the timescale in which a node reacts to the congestion control is required to ensure stability (see Sect. 11.6).

Although the congestion can be simply detected at a node by monitoring the queue buildup, additional measurements performed at a node on the local and over-the-air traffic provide more information regarding the congestion in the neighborhood that can potentially be used to improve the efficiency of the congestion control scheme. Additionally, further monitoring of the packets, identifying the sources, the destinations, and possibly the flows the packets belong to, provide valuable information that could potentially be used to efficiently remove congestion. Such information, however, is costly to achieve and in most cases would require deep packet inspection that would add to the implementation complexity and the memory requirements.

11.5.2 Congestion Signaling

Existence of congestion, as well as the congestion status, is communicated through congestion signaling to the nodes that are involved in creation of congestion. Nodes that receive this message then need to take action based on the information received.

As discussed in Sect. 11.3, from the point of view of a congested node, the queue buildup can be attributed to two different causes: the high rate of incoming traffic, and the congested neighborhood. Hence, there are two approaches that a congestion control scheme may take to reduce and/or remove the congestion:

1. Notifying the neighborhood of presence of congestion and asking all the nodes in the neighborhood to reduce their rate of accessing the channel so that there is a higher opportunity for the congested node to successfully access the channel and transmit the packets accumulated in its queue.
2. Notifying the upstream nodes to reduce the amount of forwarding traffic transmitted to the congested node to an amount that the node can manage to successfully forward without a queue buildup.

Clearly, there is also a third approach that applies both these options simultaneously.

Without going through the graph theory analysis of how either approach impacts the congestion in the network, we use two examples to illustrate how either of these approaches might impact the network in undesirable ways.

Example 11.3. Figure 11.4 shows an example network in which node A is congested because of being on the route of multiple flows. Node A broadcasts a congestion control signaling message notifying the neighborhood of the presence of congestion. Node B, which is in transmission range of node A, receives this message and accordingly reduces its MAC transmission rate (see Sect. 11.5.3.2). However, node B itself is a forwarding node on the path of flow f_b, which does not interfere with any of the flows traversing Node A. Reduction of the forwarding rate of flow f_b at Node B can lead to congestion at node B, which itself will trigger congestion signaling and spread of the congestion notification. This can lead to a chain reaction and spread of congestion throughout the network. Note that in this example scenario, if the source nodes for flows traversing through node A reduce their transmission rates

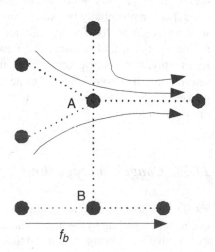

Fig. 11.4 Network scenario of Example 3

Fig. 11.5 Network scenario
of Example 4

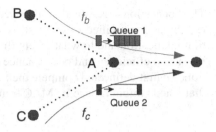

to what can be successfully forwarded by node A, other parts of the network need
not be affected. It is important to notice that rate reduction at the source nodes of the
congested flows simply means avoidance of transmission of packets that will otherwise end up being dropped because of congestion, thus, avoiding waste of resources
used to transmit them half way through the network to node A.

Example 11.4. Figure 11.5 illustrates the example where node B and node C, the
upstream nodes of the congested node A, are notified of the congestion. Assume
that the MAC protocol used in the network is IEEE 802.11e, where different traffic
types are divided into four categories and queued separately [5]. Further, assume
that the congestion at node A is related to only one traffic type, say the traffic type
of the flow traversing through node B, and node C does not have such a traffic
source. In such a scenario, the amount of information node A communicates with
its upstream nodes has a direct impact on performance of the congestion control
scheme in relation to throughput and fairness. In the ideal case, only node B should
be notified of congestion and asked to reduce the rate of its transmission to node A.

These examples show that, as discussed in Sect. 11.5.1, the type and the granularity of information that is included in the congestion control signaling, how this
information is used by the recipient of the signaling message, and the choice of
recipients of the congestion control signaling can all have a significant impact on
efficiency of the congestion control scheme.

There is a tradeoff between the amount of information included in the signaling
messages and the overhead caused by the congestion control scheme. Frequency
of the signaling is another factor contributing to the overhead. A congestion control
protocol might require nodes to periodically signal their congestion status or only do
so when a change in status occurs. When an aperiodic signaling scheme is in place,
use of a timer is required to ensure that nodes do not act based on stale information.

11.5.3 Congestion Resolution

The nodes that are notified of presence of congestion in the network should react to
the signaling information received in a manner that leads to congestion resolution.

The congestion is resolved as a consequence of the reduction of MAC transmission rate by nodes notified of congestion. How quickly the nodes react to the congestion notification and to what extent they reduce their transmission rate have a direct impact on efficiency and performance of the congestion control scheme. In this section, we first define and compute the *target rate* and then describe the different ways that a node might reduce its MAC transmission rate to meet its target rate.

11.5.3.1 Target Rate

The *target rate* is the maximum rate of transmission that will not cause congestion in the downstream nodes.

Computation of the target rate can be done by either the congested node and communicated to the upstream nodes using congestion signaling, or alternatively, a node receiving congestion signaling can compute the target rate based on the information available locally, as well as information received in the congestion signaling. The target rate depends not only on the forwarding rate of the intermediate nodes but also on the network topology and system policies, including fairness requirements in the network.

Figure 11.6 illustrates an example of how a congested node might compute the target rate of multiple flows traversing through it. Assume node A in Fig. 11.6 experiences congestion because of being on the route of f_l flows to destination node B. Note that because node A has $n - 1$ neighbors, then $f_l \leq n$, as node A can itself be the source of a flow going through link l to Node B. Moreover, let C_l denote the average capacity of node A over link l between nodes A and B. The fairness characteristics of the network specify the share of each flow from C_l. Here for simplicity we assume that all the flows are entitled to equal shares of this available capacity [29]. If, however, one flow has a smaller demand than what its fair share of the throughput of the congested link is, then the extra capacity can equally be shared among the other flows.

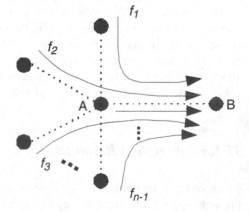

Fig. 11.6 Example network scenario where node A is on the route of n flows to destination node B and hence experiences congestion

The problem becomes more complicated when there are multiple flows originating from the same nodes with different priorities that require different treatments. In such a scenario, the target rate for each of the flows depends on the priority of the flow both within a node and among all the nodes, the offered load at the source, network-wide fairness policies, as well as the capacity available for each particular traffic type downstream.

An efficient congestion control scheme should be able to react accordingly to the dynamics of the system; as the network returns to a noncongested state, the excess capacity that becomes available should be used by the source of a previously congested flow to increase its number of transmitted packets. In the presence of the system dynamics, the following information is required by a node to calculate the target rate:

- The available capacity that is a function of channel idle time measured by the node, the PHY transmission rate of each link, and the contention overhead in the neighborhood.
- The available forwarding rate at the downstream node.
- The offered load on the upstream link.

Note that the available capacity is shared by all the links within transmission range of one another.

Assuming that the offered load of the upstream link, L, and the forwarding capacity of the downstream link, F, are learnt via signaling, and that the resources are equally shared among all flows and all the flows belong to the same priority class, then the target increase rate (packets per measurement window), R, can be estimated as

$$R = \min\left(L, \min\left(\frac{1}{n}\frac{T_{idle}C}{P+CH}, F\right)\right), \qquad (11.5)$$

where C is the average transmission rate, T_{idle} represents the average channel idle time per measurement window, the number of average active nodes (links) is shown by n, H is the average overhead per packet in time units, and p is the average packet size [30].

11.5.3.2 Local Rate Control

When a node receives a congestion control signaling message, it is required to adjust its MAC transmission rate such that it is limited to the target rate. There are multiple ways the MAC transmission rate can be controlled:

- The rate of incoming packets to MAC can be controlled by higher layers, e.g., by application layer functions such as TCP.
- A rate controller, e.g., a leaky bucket [31], implemented at the MAC layer can be used. A rate controller can control the rate such that it is limited to the target rate specified by the congestion control mechanism. However, implementation of such a controller adds complexity and cost.

- Control of the MAC transmission rate by modification of the MAC parameters is an alternative solution [32] that has a lower complexity and is also compatible with existing standards. Next we investigate this solution.

IEEE 802.11 MAC specifies different parameters that control the access of nodes to the channel without allowing for differentiation among the nodes or the traffic types. In IEEE 802.11e, however, four Access Categories (AC) are defined based on the QoS requirements of the applications and allow for differentiation and prioritizing of different traffic types. IEEE 802.11e MAC provides service differentiation by defining additional MAC parameters, including, Arbitration Interframe Spacing (AIFSN), the minimum and maximum contention window sizes, CW_{min} and CW_{max}, and Transmission Opportunity (TXOP). These parameters, which are locally stored at the nodes, can further be used to modify the MAC transmission rate. Although achieving an accurate transmission rate in a real system using the MAC parameters is very challenging, the channel access rate and transmission rate of nodes can be efficiently increased and decreased by modification of these parameters [30].

Figure 11.7 illustrates the IEEE 802.11e MAC parameters [5]. TXOP is the duration that a node is allowed to possess the channel after successfully accessing the channel and AIFSN is the minimum duration that a node should remain idle before contending for the channel after a busy period. AIFSN, CW_{min}, and CW_{max} control the channel access probability for different nodes. Although modification of each of these parameters can potentially change the transmission rate of the nodes, in [33] it is shown that modification of AIFSN has a more direct impact on the transmission rate compared to the other parameters. It is, however, important to note that the relative difference among the AIFSN values for different traffic categories and nodes is used to provide service differentiation and prioritization. Adjustment of these values, if not performed with utmost care, can have a negative impact on the performance of the network and cancel the service differentiation provided by IEEE 802.11e MAC. The modification of these parameters should be performed such that the relative priority among different traffic types and access categories is maintained

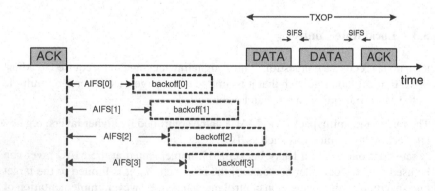

Fig. 11.7 Illustration of MAC parameters in IEEE 802.11e

not only within each node but also among all the nodes in the network; this is more of a challenge in a distributed network with many nodes independently and dynamically adjusting their MAC parameters.

11.6 Stability and Fairness

In the previous sections we discussed how presence of congestion in a network impacts the network throughput. We also established that to remove the congestion a MAC transmission rate control should be imposed on the flows that suffer packet drop at the congested nodes. With such a rate control, the congestion control protocol succeeds in saving network resources that otherwise were to be used to transmit packets mid-way through the network without ever reaching their final destination. It is clear that if these released resources are not used by other nodes in the network, there will be no change in the overall network throughput, as the number of packets that are delivered to the final destination nodes remains unchanged. However, efficient use of these freed-up resources results in an increase in network throughput.

In this section, we study ways that the available resources are allocated in the network and the impact of rate allocation on stability and fairness. To do so we consider both the analytical approach and the practical aspects to be considered in developing a congestion control protocol.

11.6.1 Analytical Approach

The congestion control can be viewed as a global rate allocation problem where the network capacity is fully used, i.e., there is no packet loss because of congestion in the network, and a notion of fairness among the nodes is satisfied. Such a rate allocation problem can be formulated as a utilization maximization problem that can be solved by distributed optimization algorithms [6].

Let S denote the set of flows in the network, each with transmission rate of $x_s, s \in S$. Satisfaction of each source transmitting at rate x_s is captured by the utility function $U_s(x_s)$, which is assumed to be concave, nondecreasing and twice continuously differentiable. Let the set of available links in the network be L, and the capacity of each link $l, l \in L$ be shown by C_l. Further, let r_{ls} be equal to 1 if link l is on the path of flow s and equal to 0 otherwise. The congestion control problem can then be formulated as follows

$$\max \sum_{s \in S} U_s(x_s)$$
$$\text{subject to } \sum_{s \in S} \leq C_l \quad \text{for all } l \in L. \tag{11.6}$$
$$\text{and } x_s \geq 0, \quad s \in S$$

The optimization problem of (11.6) is shown to be equivalent to the problem of satisfying fairness objectives if the utility function $U_s(x_s)$ is appropriately chosen. A general form of the utility function is given as

$$U_s(x_s) = w_s \frac{x_s^{1-\beta}}{1-\beta}, \quad \beta > 0, \tag{11.7}$$

where w_s, $s \in S$ denotes the weight [34]. Using the utility function of (11.7) in problem formulation of (11.6) results in different objectives as a function of β

$$\beta \rightarrow \begin{cases} 0 & \text{maximized weighted throughput} \\ 1 & \text{weighted proportional fairness} \\ 2 & \text{minimized weighted potential delay} \\ \infty & \text{max} - \text{min fairness} \end{cases} \tag{11.8}$$

The maximization problem presented in (11.6) results in a unique globally optimized solution when C_l is assumed constant, e.g., in wired networks. However, the solution is more complex for wireless networks where C_l is variable and dependent on the PHY parameters and MAC scheduling. There are two ways of addressing the variable nature of capacity of wireless links; one way is to simplify the constraint set of (11.6) by specifying a rate region from the set of feasible rates available for each link. The solution to the rate allocation problem would then be computed based on this limited rate region. The other solution is to formulate C_l as a function of underlying PHY parameters; this approach results in a solution that is a joint optimization of rates as well as underlying PHY and MAC resources (these resources include power allocations, modulation schemes, MAC transmission opportunities, etc). For example, [21] defines $C_l(t)$ as a function of power allocation on all the links at time t.

The stability region of a rate allocation scheme is the set of offered loads under which the network is stable, i.e., the queue lengths across the network remain finite. It is shown that the convergence of a congestion control algorithm and its fairness characteristics are tightly related to its stability region.

11.6.2 Practical Aspects

A congestion control protocol may or may not take an active role in specifying how the released resources should be used within the network [30]. In Sect. 11.5.3.1, we provided an example of how the extra capacity that becomes available as the congestion is removed can be allocated to different sources by explicit indication of the target rate. A congestion control protocol might use a similar approach for actively allocating freed up resources in the network. However, an explicit allocation of resources among the nodes has several drawbacks that make it almost impossible to achieve in practice.

- The signaling and notification of the rates will not be limited to signaling because of congestion, as in principle there is no difference between the two cases where resources become available as a result of congestion control or termination of a flow. Hence, the problem is changed to defining a generic resource allocation scheme.
- For a protocol to fairly and efficiently allocate resources, it should be aware of the topology, as well as offered loads, flow routes, and link capacities. Gaining this information in a distributed network introduces an increased overhead. Such a scheme would suffer from high complexity and high sensitivity to system dynamics.
- Providing fairness is extremely challenging. The resource allocation scheme should ensure fairness by allocation of resources based on a preferred fairness model. However, the characteristics of wireless channels, the known shortcomings of the MAC in resource allocation in a mesh network, as well as interaction with the higher layers, would make achieving the fair rates extremely difficult.

A practical approach for handling the released resources in the network is *not* to take an active role in allocating them and instead allow the nodes (except those that are rate-controlled by congestion control protocol) to contend for the resources based on the rules of the underlying MAC. Hence, the congestion control protocol inherits the fairness characteristics of the underling MAC, and there is not any additional complexity and overhead associated with the allocation of resources. The stability of such congestion control schemes is studied in [25, 30] where it is proved that in a network consisting of a single contention region, i.e., a region in which all links mutually contend, a passive contention control scheme leads to stability. It is also shown that in networks with instantaneous feedback, even in the presence of multiple overlapping contention regions, a passive congestion control scheme is self-stabilizing. In general though, the congestion can appear elsewhere in the network if the rate limitation function is not performed for a long enough duration to allow for the changes in the rates to be propagated and rate adjustments to be completed throughout the network.

11.7 Congestion Control in IEEE 802.11s Draft Standard

A new standard for mesh networking is being developed in IEEE 802.11 Task Group S. This standard extends IEEE 802.11e MAC with enhancements for support of multihop communication among the nodes [35]. To improve the efficiency of the multihop communication, a congestion control signaling framework is introduced. Although the standard is not finalized, in this section we provide an overview of the congestion control framework currently present in the draft standard.

The congestion control framework in IEEE 802.11s draft standard provides an extensible solution for support of different congestion control schemes. Note that although the standard and the nodes may support multiple congestion control schemes, the standard allows for only one active congestion control protocol in a mesh network at any given time. The defined framework specifies a simple default signaling protocol and allows use of different congestion control schemes in the mesh through reserved and vendor-specific values for the Congestion Control Mode Identifier. The Congestion Control Mode Identifier is included in the Mesh Configuration Element, which itself is part of the Mesh Beacon. A Null Protocol value for the Congestion Control Mode Identifier indicates support of no congestion control scheme (see Fig. 11.8).

The default congestion control protocol simply defines the signaling and leaves monitoring and detection of congestion, as well as the trigger for the signaling, out of the scope. Moreover, it does not specify to which nodes the signaling message is transmitted. Also, the action that the recipient(s) shall take in response to receipt of a congestion control signaling, i.e., the rate control mechanism, is left unspecified.

The congestion signaling defined in the default protocol consists of a Congestion Notification message. The Congestion Notification message is transmitted by a node that detects congestion and informs the recipient(s) of the estimated time that the congestion is going to last for each of the access categories. Figure 11.9 illustrates the Congestion Control Notification frame and the Congestion Notification element. The Congestion Control Notification frame allows for inclusion of different Congestion Control Elements. The Congestion Control Element used

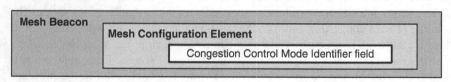

Fig. 11.8 Illustration of the relation of congestion control mode identifier field, mesh configuration element, and mesh beacon

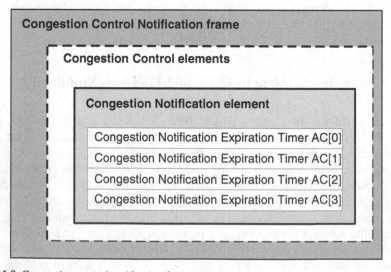

Fig. 11.9 Congestion control notification frame structure

with the default protocol is the Congestion Notification Element. Other Congestion Control elements may be used when the Congestion Control Mode Identifier value indicates that a protocol other than the default protocol is being used in the mesh network.

The Congestion Notification Element includes four fields, each containing a timer value. Each timer represents the estimated duration of congestion for a specific Access Category. The default protocol does not specify how these values are set. However, it is expected that a timer value of zero indicates no congestion has been detected in the queue of the specified Access Category. Further discussion on how the signaling protocol might be used by a congestion control protocol is presented in Sect. 11.8.

Alternative congestion control schemes to be defined will use a new Congestion Control Mode Identifier value, chosen from either the values reserved for future use or vendor specific values. A new congestion control protocol may also define new congestion control element(s), which would be included in the Congestion Control Notification frame body. A new Congestion Control Element needs to be defined to include additional information needed by the new congestion control protocol.

11.8 Thoughts for Practitioners

In Sect. 11.7, we presented the framework developed in the IEEE 802.11s draft standard for congestion control in mesh networks. Here we further describe how the default protocol, as well as the framework, can be used with more advanced and alternative congestion control schemes. Additionally, we raise some points worth considering by the developers and implementers of congestion control protocols.

The default protocol specified in IEEE 802.11s draft standard defines the signaling for notification of congestion as well as an estimation of duration of the congestion per Access Category. The estimated duration of the congestion can simply be viewed as a measure of how grave the congestion is. It can be computed by monitoring the length of the queues for each of the Access Categories. In a simplified form it can be used in a quantized form as notification of congestion.

Although, the specified signaling allows for communicating existence of congestion and its estimated duration among nodes, it is insufficient for efficiently dealing with congestion in the network, and can also aggravate the congestion throughout the network and have serious negative impact on performance of the system if not used appropriately.

A critical requirement for a congestion control scheme is definition of a mandated unique behavior in response to the congestion control signaling. If in a network only some recipients of congestion notification take action towards alleviating the congestion and the rest do not reduce their traffic input to the network, the congestion control scheme will not be able to remove the congestion. Additionally, it will also introduce serious fairness and possible starvation into the network, because as long as the congestion is not removed, the conforming nodes receive congestion notification and continue to reduce their transmission rate. The nonconforming greedy nodes can continue to increase their throughput by utilizing the resources released by rate control in the conforming nodes.

A nonstandardized triggering mechanism for the congestion signaling can also provide an opportunity for malicious nodes. A malicious node can request other nodes in the network to reduce their traffic input to the network so that its throughput is increased by transmitting fake congestion notification messages.

Another point that requires careful consideration is the stability of the network performance in the presence of a congestion control function. Finding stability conditions for a congestion control scheme through theoretical analysis that would also hold true in real world deployments is a very difficult task. The system dynamics and the interactions between the system load and network throughput while the congestion control protocol is active should be incorporated in any stability study. Most importantly, the propagation delay for the congestion signaling, as well as the duration it takes for its impact to be distributed through the network and reach the source of the flows, has the most significant impact on stability. If the system does not allow for the impact of each change in congestion and system status to complete before triggering further action, the system can face instability resulting in seriously degraded performance.

Finally, regardless of the congestion status in the network, there are always high priority packets, including the control layer messages and emergency signaling that require prompt handling. A congestion control protocol should allow for instantaneous and unlimited-rate transmission of such high priority signaling messages.

11.9 Directions for Future Research

Although, as described in Sect. 11.2, there have been many theoretical studies on performance of congestion control in wireless mesh networks, a bench-marking approach that quantifies the achievable performance, subject to constraints of a system model that successfully replicates the limitations of a real world network, is missing in the literature.

It is known that in the presence of delay, hop-by-hop congestion control schemes may result in traverse of congestion points through the network. Defining conditions for global stability of a hop-by-hop congestion control algorithm in the presence of delay is another open research topic.

In design of congestion control protocols, the tradeoff between the complexity and the performance has not yet been studied. The performance of congestion control algorithms is highly dependent on interactions among the nodes and transfer of information via signaling. How congestion impacts the ability to perform the signaling and how the signaling interacts with the congestion control protocol are questions that have not been addressed by the existing research work. Additionally, the tradeoff between the amount of information transferred and the performance of different congestion control schemes is yet unknown.

Finally, on a more practical note, development of algorithms for congestion control in mesh networks that use the existing framework in IEEE 802.11s draft standard and satisfy different usage scenarios identified by the Task Group is left for future research. The protocol design includes identifying the triggering condition for invoking congestion signaling, as well as the local rate control mechanism at the nodes.

11.10 Conclusions

In this chapter we studied the problem of congestion control in wireless mesh networks. Congestion control is considered a key functionality for communication networks. The congestion control schemes that are used in traditional networks have proven insufficient for wireless networks, where link layer failures and random access MAC mechanisms impact the congestion control functions.

We defined congestion, investigated its impact on system throughput, and established that imposing limitation on the amount of injected traffic to the network by flows that lose packets because of congestion will save otherwise wasted network resources. These resources can be used by other flows in the network to increase overall system performance. How the resources are shared among the flows and the impact of congestion control on system stability were then discussed. Providing global stability in the presence of delay was identified as a challenge for hop-by-hop congestion control protocols in a wireless mesh network.

In a discussion of the congestion control protocols, we identified the three functional building blocks of such protocols, i.e., congestion detection, signaling, and resolution. We also presented the framework for congestion control in IEEE 802.11s draft standard and provided some thoughts for practitioners in this field.

11.11 Terminologies

1. *IEEE 802.11 TGs*. The Task Group in IEEE standardization body responsible for developing an amendment to IEEE 802.11 MAC standard to enable IEEE 802.11-based wireless mesh networking.
2. *Congestion*. The condition of a network where, because of lack of resources, an increase in the offered load does not increase and may result in reduction of the network throughput. Similarly, the state of queue buildup/overflow in a node.
3. *Congestion control*. A mechanism implemented in the network to remove and/or limit the congestion in the network by manipulating transmission and/or forwarding rates of the nodes.
4. *Congestion detection*. Function of monitoring the network use to detect and estimate the state of congestion in the network.

5. *Congestion signaling.* Function of distributing information about the congestion state in the network. This information is used by the nodes to react individually or collectively to resolve the congestion in the network.
6. *Congestion resolution.* Function of limiting the transmission or forwarding rate of the packets at a node to remove/reduce congestion experienced by other nodes in the network.
7. *Target rate.* The maximum allowed transmission rate by a node that does not cause congestion in the network.
8. *Rate control.* Manipulation of transmission rates at nodes.
9. *Resource allocation.* Specification of share of each user/node from network resources to maximize system performance while satisfying some fairness objectives.
10. *MAC transmission rate.* The rate at which packets are released for transmission by the Medium Access Control function in a wireless radio.

11.12 Questions

1. What are the congestion detection, signaling, and resolution functions in IEEE 802.11 MAC?
2. What are the congestion detection, signaling, and resolution functions in TCP?
3. How are IEEE 802.11e MAC parameters used to provide relative priority among different access categories? What parameters would give absolute priority to AC[0]?
4. Assuming that all flows in the IEEE 802.11 network of Fig. 11.10 are always backlogged and the MAC scheduler at node D is round robin, find the target rate of flows f_a and f_b that maximizes network throughput. The capacity of all links is C and the fairness objective requires equal throughput for flows f_a, f_b, and f_c. How does change of offered load of flows f_a and f_b to their target rate impact throughput of flow f_e?
5. Prove that a passive congestion control scheme with instantaneous feedback is self-stabilizing.

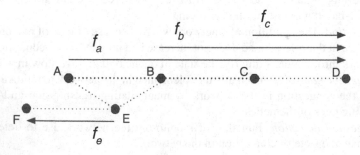

Fig. 11.10 Network scenario of Question 4

Fig. 11.11 Network scenario
of Question 7

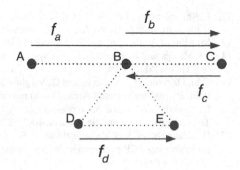

6. What are some of the ways that a malicious node can exploit a congestion control scheme?
7. Compute the utilization of IEEE 802.11-based network of Fig. 11.11 with and without a passive congestion control scheme, assuming all nodes are offered infinite load, the capacity of all links is C, and the flows are served by a round robin scheduler in node B.
8. What are the disadvantages of a centralized congestion control scheme vs. a distributed one? What information should be available to the resource allocation function?
9. Use an example congestion control protocol to show how the IEEE 802.11s congestion control framework can be extended to implement it.
10. What is the stability region for your congestion control scheme defined in Question 9?

References

1. Tropos Networks *http://www.tropos.com*
2. BelAir Networks *http://www.belairnetworks.com*
3. Technology for All *http://tfa.rice.edu*
4. MIT Roofnet *http://pdos.csail.mit.edu/roofnet/doku.php*
5. B. O'Hara and A. Petrick, The IEEE 802.11 Handbook: A Designer's Companion. IEEE Press, New York, NY, (2005).
6. F.P. Kelly, A.K. Maulloo, and D.K.H. Tan, Rate control for communication networks: shadow prices, proportional fairness, and stability. Journal of the Operational Research Society, 49(3), 237–252, (1998).
7. S. Shenker, A theoretical analysis of feedback flow control. Proceedings of SIGCOMM, 156–165, (1990).
8. S. Keshav, A control-theoretic approach to flow control. Proceedings of SIGCOMM, (1991).
9. S. Floyd and V. Jacobson, Random early detection gateways for congestion avoidance. Transactions on Networking, 1(4), 397–413, (1993).
10. J. Hoe, Improving the start-up behavior of a congestion control scheme for TCP. Proceedings of SIGCOMM, (1996).
11. M. Mathis and J. Mahdavi, Forward acknowledgement: Refining TCP Congestion Control. Proceedings of SIGCOMM, (1996).

12. L. Brakmo and L. Peterson, TCP Vegas: End-to-end congestion avoidance on a global internet. IEEE Journal on Selected Areas in Communication, 13(8), 1465–1480, (1995).
13. M. Allman, V. Paxson, and W. Stevens, TCP Congestion Control, RFC 2581, (1999).
14. P.P. Mishra and H. Kanakia, A hop-by-hop rate based congestion control scheme. Proceedings of SIGCOMM, (1992).
15. C.M. Ozveren, R.J. Simcoe, and G. Varghese, Reliable and efficient hop-by-hop flow control. IEEE Journal on Selected Areas in Communications, 13(4), 642–650, (1995).
16. L. Tassiulas, Adaptive back-pressure congestion control based on local information. IEEE Transactions on Automatic Control, 40(2), 236–250, (1995).
17. H. Balakrishnan, V.N. Padmanabhan, S. Seshan, and R.H. Hatz, A comparison of mechanisms for improving TCP performance over wireless links. IEEE Transactions on Networking, 5(6), 756–769, (1997).
18. G. Holland, and N. Vaidya, Analysis of TCP performance over mobile ad hoc networks. Mobile Computing and Networking, 219–230 (1999).
19. L. Wang and C. Lee, A TCP-physical layer congestion control mechanism for the multi-rate WCDMA system using explicit rate change notification. Proceedings of the 19th International Conference on Advanced Information Networking and Applications, (2005).
20. M. Chiang, To layer or not to layer: balancing transport and physical layers in wireless multi-hop networks. Proceedings IEEE INOFOCOM, (2004).
21. X. Lin and N.B. Shroff, Joint rate control and scheduling in multi-hop wireless networks. Proceedings of the IEEE Conference on Decision and Control, (2004).
22. L. Chen, S.H. Low, and J.C. Doyle, Joint congestion control and media access design for wireless ad hoc networks. Proceedings IEEE INOFOCOM, (2005).
23. X. Lin and N.B. Shroff, The impact of imperfect scheduling on cross-layer congestion control in wireless networks. IEEE/ACM Transactions on Networking, 14(2), 302–315, (2006).
24. Y. Yi and S. Shakkottai, Hop-by-hop congestion control over a wireless multi-hop network. Proceedings of IEEE INFOCOM, (2004).
25. G. Zhang, Y. Wu, and Y. Liu, Stability and sensitivity for congestion control in wireless mesh networks with time varying link capacities. Ad Hoc Networks, 5(6), 769–785, (2007).
26. G. Bianchi, Performance analysis of the *IEEE* 802.11 distributed coordination function. IEEE Journal on Selected Areas in Communications, 18(3), 535–547, (2000).
27. A. Tsertou and D.I. Laurenson, Insights into the hidden node problem. Proceedings of the 2006 International Conference on Wireless Communication and Mobile Computing, 762–772, (2006).
28. M. Gerla, R. Bagrodia, L. Zhang, K. Tang, and L. Wang, TCP over wireless multi-hop protocols: Simulation and experiments. Proceedings of IEEE ICC, (1999).
29. V. Gambiroza, B. Sadeghi, and E. Knightly, End-to-end performance and fairness in multi-hop wireless backhaul networks. Proceedings of ACM MobiCom, (2004).
30. B. Sadeghi, A. Yamada, A. Fujiwara, and L. Yang, A simple and efficient hop-by-hop congestion control protocol for wireless mesh networks. Proceedings of the Second annual International Workshop on Wireless Internet, (2006).
31. P. Ferguson and G. Huston, Quality of Service: Delivering QoS on the Internet and in Corporate Networks. Wiley, New York, NY, (1998).
32. A. Yamada, A. Fujiwara, L. Yang, and B. Sadeghi, EDCA based congestion control for WLAN mesh networks. Proceedings of IEEE VTC, (2006).
33. J.-D. Kim and C.-K. Kim, Performance analysis and evaluation of *IEEE* 802.1 1e EDCF. Wireless Communications and Mobile Computing, 4(1), 55–74, (2004).
34. J. Mo and J. Walrand, Fair end-to-end window-based congestion control. IEEE/ACM Transactions on Networking, 8(5), 556–567 (2000).
35. IEEE 802.11-04/54r2, PAR for IEEE 802.1 1s ESS Mesh.
36. M. Gerla, K. Tang, and R. Bagrodia, TCP performance in wireless multi-hop networks. Proceedings of *IEEE* WMCSA, (1999).

Chapter 12
Wireless Mesh Networks-Based Multinetwork Convergence and Security Access

Jianfeng Ma and Chunjie Cao

Abstract Every kind of wireless network has its own advantages and disadvantages, and the provided services, traffic and coverage of various wireless networks differ a lot. To fully use the advantages of various wireless networks and provide better Quality of Services (QoS), the wireless network convergence based on IP technology aroused great interest in recent years. At the same time, different wireless access technologies have their own standards within mesh network architecture. Accordingly, various mesh network technologies are prosperous. However, the requirement to access network anywhere anytime needs the interconnection of various networks. Therefore, the convergence of wireless mesh networks (WMNs) with other wireless networks becomes an extremely important research subject. In this chapter, WMNs-based multinetwork convergence technology is introduced. With the new technology, wireless terminals can access heterogeneous wireless networks and Internet through the backbone WMN using a multimode manner, which help the wireless terminals realize seamlessly roaming among different wireless networks.

12.1 Introduction

12.1.1 Multinetwork Convergence

Extending IP connection to "last one mile" is an open problem being studied, however, no satisfying solution has been proposed so far. Although there are many possible solutions, such as all connected end to end optical networks, the deployment of these solutions needs large amounts of cables. The difficulties to deploy in some environments (urban area, the wild) also stunt the wide application of these access networks.

J. Ma (✉)
Key Laboratory of Computer Networks and Information Security of the Ministry of Education, Xidian University, Xi'an 710071, People's Republic of China
e-mail: jfma@mail.xidian.edu.cn

S. Misra et al. (eds.), *Guide to Wireless Mesh Networks*, Computer Communications and Networks, DOI 10.1007/978-1-84800-909-7_12,

Wireless mesh networks (WMNs) are multihop ad hoc networks composed of mobile or static nodes connected through wireless links [1]. The infrastructure of WMN is self-organized, self-optimized and fault-tolerant. It facilitates the extension of IP connection to areas that any other single access technology can not reach. Compared with optical networks, WMN has the advantages of small investment and easy configuration. At present, many companies, including Nokia [2], Microsoft [3], Motorola [4] and Intel [5], are actively promoting WMNs as a perfect IP solution. It is shown that the WMN has a large potential in practice [6–8] and it can converge the existing and emerging wireless networks, including the cellular networks, ad hoc networks and sensor networks. The application of these research results contributes a lot to the development, realization of WMN.

With the development of computer and communication technologies, various wireless networks, wireless wide area networks (such as 3G, 802.22), wireless metropolitan area networks (such as IEEE 802.16), wireless local area networks (such as IEEE 802.11), satellite communication networks, Bluetooth networks and so on, are substituting traditional wired networks as last hop to access Internet. The convergence of all these wireless networks and provision of wide coverage, high bandwidth, high mobility and low cost Internet access are the trends of the next generation communication systems (such as B3G, 4G).

WiFi, WiMAX, WBMA and 3G play an important role in the area of high-speed wireless data communication, but they are designed to solve different application problems. Different requirements of these technologies in applications will lead to their own living space in the increasingly refined market. Therefore, their relationships are not so much competition as symbiosis. 3G's convergence with the other wireless networks is increasingly thought of as an irreversible trend. Using mobile phones, users can access network services through WLAN when at company, enjoy 3G network services out of door, call the others through Bluetooth at home, and access ground digital broadcasting services. Seamless network access leads to continuous phone calls, interruption during the call will never happen while across network boundaries.

At present, the standards of WiFi, WiMAX, WBMA and 3G progress rapidly. With the increasingly perfect standards, increasingly advanced technologies and increasingly practical products, their unchanged requirements are higher bandwidth, stronger security and better compatibility. According to the trend of 3G's dual-mode or multimode development, there are two forms: one is dual-mode or multimode of various modulation methods, analogous to dual-mode of GSM and CDMA in the 2G era; the other is the convergence between 3G and the other wireless networks, such as WLAN, WiMAX, which is mentioned above. It can be said that the former is to provide smooth transition, for instance, the frequently mentioned dual-mode of GSM and CDMA; the latter is the tendency of future industrial development, for example, the convergence between 3G and the other wireless networks, such as WLAN, WiMAX. First, many telecommunication service providers have provided wireless coverage to facilitate users' access to Internet. Second, some terminals connected to wireless networks, such as PDA, mobile communication devices (mobile phones), can access network through wireless technologies, too.

In the framework of heterogeneous wireless networks convergence, the seamless convergence between 3G and WLAN is especially concerned for their complementary characteristics. Many research institutes, including 3GPP (Third Generation Partnership Project), are dedicated to the study the seamless convergence of 3G and WLAN. Recently, loose convergence model has been widely advocated as the most prefer model of B3G and 4G [9]. But, there are still a series of problems to be solved to converge the heterogeneous wireless networks [10].

The tendency of future wireless networks will focus on IP and integration of various wireless access technologies. Compared to homogeneous networks, a unified structure and easy management, heterogeneous networks possess much more complex characteristics. Therefore, authentication, secure data transition and the corresponding accounting problem come out when users roam among different networks. A unified heterogeneous network architecture is required to realize truly heterogeneous convergence, in which secure access, authentication and accounting scheme independent of various low layer wireless technologies are provided, to guarantee secure and reliable communication from any access points [11].

12.1.2 Design Principles of Multimode Security Access

Adequate universality and flexibility are required in practical applications for multimode security access in the environment of heterogeneous network. To achieve this goal, we are expected to adhere to the following principles in designing multimode security access and authentication schemes.

- *Universality*. Integrated multimode security access system should support various wireless standards, such as 802.11b/g, 802.15, 802.16, 802.20, 3G and so on.
- *Security*. Integrated security access system for heterogeneous networks enables the wireless terminals to access various wireless networks securely and adaptively. Central security solution has to be provided for wireless networks. Besides, we have to coordinate different security policy among heterogeneous networks, and guarantee consistency of multimode security access policy of the wireless terminals to reduce the risks resulted from inconsistency of the security policies.
- *Reliability*. The status of integrated security access system for heterogeneous networks is very important for normal operation of the entire wireless networks. Therefore, we have to adopt the hardware and software systems of high reliability.
- *Efficiency*. Because of the limitation of computation and communication abilities of wireless devices, fully consideration has to be given to efficiency of the access system. The fast handover and roaming efficiency can be achieved by recurring to existing technologies (such as 802.11r). In the premise of guaranteeing basic security requirements, we should improve efficiency as much as possible.
- *Scalability*. We shall give full consideration to scalability in two aspects in the process of design. One is that the function of the system can be extended and

improved as different requirements, so the function of update and interface of system extension must be provided; the other is the concept of modularizing design: the additional functional modules could be installed when necessary.

12.2 Background

Future mobile communication will be composed of various and heterogeneous wireless access networks. Heterogeneous access is one of the most important problems that must be solved before heterogeneous network convergence. It involves many different aspects, including the discovery and selection of the optimal access network, execution of the vertical handover, update of terminal's location, authentication, accounting, etc. One of the greatest challenges confronting heterogeneous access research is how to design the seamless vertical handover technology [12]. Traditional internal handover decision is performed based on the strength of received signals and then a new connection will be setup. Vertical handover involves different types of wireless access technologies, some new considerations will be introduced in its design process, such as the configuration of access technologies on different access points, QoS support, overall loads, user preference, security and cost.

The design of effective access architecture and vertical handover scheme to realize the convergence of heterogeneous networks become the focus, a number of related works have been discussed in this area. In the study of e-Japan plan MIRAI [13, 14], a new type of heterogeneous access scheme, Common Access, is proposed. It aims to develop technologies for seamless integration of heterogeneous wireless networks. Another solution is based on proxy and dynamic packet buffering queue [15], in which data streams are sent to both the original and the new access network during the handover. To cooperate above link layer and reduce packet loss [16], the concept of generic link layer (GLL) is introduced. On the basis of GLL and WMNs, a multinetwork convergence architecture and multimode security access system [17] is proposed.

12.2.1 Common Access

The Communications Research Laboratory (CRL) has been conducting the MIRAI research project [13, 14], a national project under the e-Japan Plan. MIRAI is an acronym of Multimedia Integrated network by Radio Access Innovation. Its goal is to develop technologies for the seamless integration of heterogeneous wireless networks. The common access, a new type of heterogeneous access scheme is one of their technologies. It is based on three major entities: common core network (CCN), basic access network (BAN) and multiservice user terminal (MUT).

CCN provides a common platform through which all MUTs can communicate with correspondent nodes in the Internet. Most access points of radio access

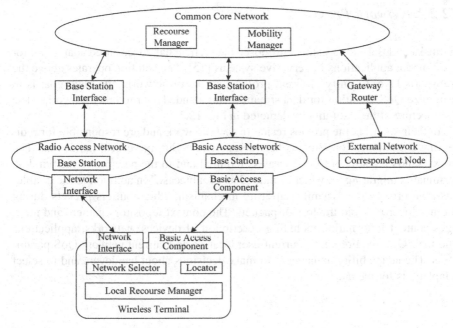

Fig. 12.1 The architecture of MIRAI

networks (RANs) are connected to CCN, which provides QoS-guaranteed routing and seamless handover among RANs. BAN provides a common control/signaling channel to enable all MUTs to access the common platform. The MUT is equipped with a multiradio system and a BAC to communicate with the BAN. Besides, some radio subsystems based on software-defined radio (SDR) technologies are also equipped to access the CCN.

The architecture of MIRAI is depicted in Fig. 12.1. It is composed of four major building blocks: the wireless terminal, the RAN, the CCN, and the external network.

In the architecture, the gateway router (GR) is responsible for the connection between the external network and the CCN. The resource manager (RM) and the mobility manager (MM) of the CCN mainly deal with traffic distribution and mobility-related problems. The base station (BS) deals with wireless access problems in the normal link layer. It communicates with the CCN via base station interface (BSI), which provides a uniform access mechanism for the base stations to access the CCN. The network interface (NI) is used by BS to access the network. All wireless terminals have a BAC to communicate with the BAN and an NI, which is based on SDR technologies to access RANs. The network selector (NS) communicates with the RM to tune the radio for the RANs to use and the network selection control protocol is used to enable the proper selection of an access network. The locator (LOC) provides the RM with information on the location of wireless terminals. The local resource manager (LRM) deals with the local resources of the terminal and interacts with the RM at the CCN.

12.2.2 Network Proxy

Indulska and Balasubramaniam proposed the vertical handover-based adaptation for multimedia applications in pervasive systems [15]. The solution operates above the transports layer to unify different protocol stacks of networks. Its main goal is to minimize QoS violation for data streams being handed over to other networks. The architecture of their solution is depicted in Fig. 12.2.

In their solution, the proxies reside in each network and are responsible for redirecting communication streams between networks during vertical handovers. The proxy sends the stream to the wireless terminal and to the new proxy the wireless terminal is migrating to, which is named as "doublecast." In addition to the doublecast, the proxy uses dynamic buffering mechanism, which buffers packets during the handover to avoid the loss of packets. The context repository gathers and manages context information including description of devices, networks, applications and their QoS requirements, current user location and current network QoS parameters. The adaptability manager is to make decisions about handovers and to select adaptations for the data stream.

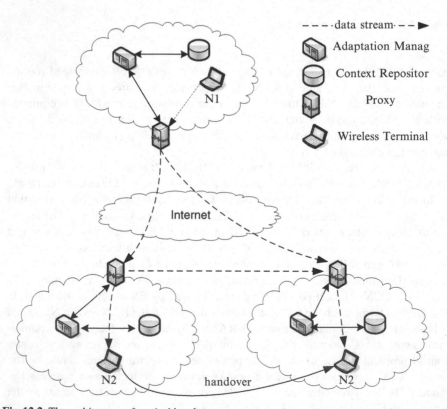

Fig. 12.2 The architecture of vertical handover

12.2.3 Generic Link Layer

GLL [16] was proposed to provide that old configurations and states of wireless link layer can be transferred to the new one if the wireless link layer is designed in compatible and generic mode, thus seamless data transmission can be realized during the handover. GLL enables a cooperation of different access networks at the link layer to overcome the packets loss problem during a handover. In Fig. 12.3, a simplified protocol stack is depicted, in which a mobile terminal communicates with a correspondent node.

GLLs bridges radio physical layer and higher network layer. To achieve this goal, they should perform several very important functions, such as queuing of incoming data, scheduling of data for transmission and header compression of higher layer protocol fields. In addition, to allow seamless and lossless cooperation between different access networks, the GLL should support three additional functions in order: cooperation with mobility management functions and protocols, context transfer at handover and lossless reconfiguration.

The GLL has three interfaces (depicted in Fig. 12.4): the interface to higher protocol layer, the interface to physical layer and the interface to configuration/control manager. With these interfaces, the GLL can complete normal data transmission and control functions.

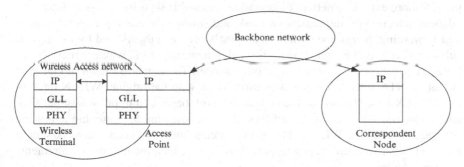

Fig. 12.3 Generic link layer reference model

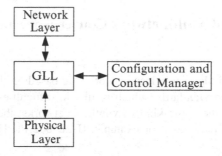

Fig. 12.4 Generic link layer functions and interfaces

To allow seamless data transmission, the GLL needs to enable lossless reconfiguration at intersystem handover. It requires the following steps [16]:

- The link layer triggers the execution of the handover;
- At handover, the existing GLL transmission context is transferred to the new
  point of execution;
- The GLL entities are reconfigured according to the new radio access technology.
  The GLL receiver maintains the old transmission context;
- The outstanding part of higher layer data from the old transmission context is
  transmitted from the reconfigured GLL transmitter via the new radio connection.
  The receiver reconstructs the data from the old receiver context;
- After all outstanding data of the old GLL transmission context have been
  delivered, the transmission continues in normal operation via the new radio
  connection.

12.2.4 WMNs-Based Multinetwork Convergence

WMNs-based multimode security access (WMN-MSA) system [17] is an extensible access and authentication system, which adopts the GLL as a building block and
can deal with different authentication schemes of different access networks. It can
provide more extensive network integration, extensible security access architecture,
adaptive selection of authentication mode and strong protection of internal routing
and forwarding. Multimode wireless terminals can be authenticated by its original
subnet and access the subnet again through a new subnet via wireless mesh backbone network when it migrates to another subnet. The scenario shown in Fig. 12.5
is that the STA authenticates and communicates with the original WLAN through
the WiMAX base station (or the cellular network base station, the sensor network
sink node) and wireless mesh backbone network when the STA moving out of the
coverage of its home WLAN. Therefore, wireless mesh backbone network should
be able to deal with different access technologies, which reduces the requirements
of access devices.

12.3 WMNs-Based Multinetwork Convergence and Security
   Access

At present, different kinds of wireless networks are based on the different authentication and encryption standards, which results in difficulties in the users' network access. Even in the same kind of wireless networks, there exist more than
one authentication technologies, for example, IEEE 802.11i [18] and WAPI [19]

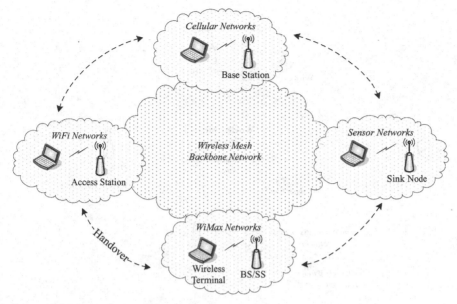

Fig. 12.5 The architecture of WMN-based multinetwork convergence

in WLAN. So the mismatch of authentication technologies causes that the network denies the user's access. The WMN-MSA technology aims to solve this problem.

12.3.1 Introduction

WMN-MSA can realize more extensive network integration, scalable security access architecture, adaptive selection of authentication mode, strong protection of internal routing and forwarding. In WMN-MSA, the security access of all kinds of wireless networks such as WLANs, WiMax networks, cellular networks, sensor networks and satellite networks, can be successfully completed as long as the mesh gateway is designed according to the architecture of this system and the wireless terminal is installed the corresponding security access module. In addition, the user can design authentication modules on his own and install the module on the system. It is an interesting characteristic to the stronger security requirement applications. Moreover, WMN-MSA provides protection mechanisms for reliably routing and retransmitting in the mesh backbone network, which greatly strengthens the security of the entire network.

In the architecture of the WMN-MSA, the GLL technology and the plug-in technology are adopted. The former guarantees that the wireless mesh backbone network can interconnect with existing wireless access networks in link layer. The latter guarantees the adaptability and the scalability of the authentication module, which reduces the overheads of system maintenance.

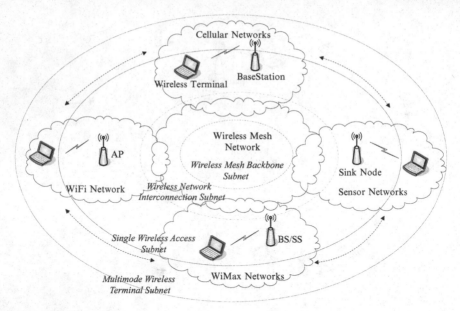

Fig. 12.6 The architecture of WMN-MSA

12.3.2 The Architecture

The architecture of WMN-MSA, depicted in Fig. 12.6 can be divided into wireless mesh backbone subnet, wireless network interconnection subnet, single wireless access subnet and multimode wireless terminal subnet.

Wireless mesh backbone subnet provides an infrastructure among different wireless networks to realize the convergence of different wireless networks. Wireless network interconnection subnet is the border of the wireless mesh backbone network and other wireless access networks, which mainly includes mesh network portals (MPPs) and some base stations of single wireless access network. The single wireless access subnet is the network that provides access services for wireless terminals with the existing or emerging access technologies. It includes the base stations of all the wireless access networks. The multimode access wireless terminal subnet is mainly made up of wireless terminals that are equipped with an integrated authentication platform and can roam among different wireless networks.

12.3.2.1 Wireless Mesh Backbone Subnet

The main entity of the wireless mesh backbone subnet is wireless mesh point (MP), which adopts the MAC and PHY protocols of WMNs to communicate with each other. The protocol stack of MP is depicted in Fig. 12.7. The functions of an MP are achieved by following four modules: security management module, data forwarding module, neighbor management module and routing management module.

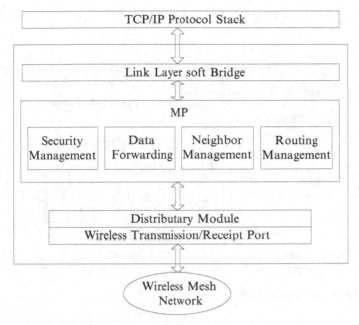

Fig. 12.7 The protocol stack of MP

The security management module provides a variety of authentication and encryption techniques, such as IEEE 802.11i and WAPI, to support the establishment of secure channel between MPs.

On receiving a data frame, the data forwarding module of an MP sends it to the higher layer if the destination of the data frame is this local machine; else delivers it to the lower layer, which forwards the data frame based on the result of routing.

The neighbor management module controls the join and leave operations of neighbor MPs, and maintains all kinds of state information of the accessed MP.

The routing management module executes specific functions of routing protocols, such as the route establishment, selection and maintenance. In addition, it supports the query of routing information from the data forwarding module. This module is independent of specific routing protocols.

The wireless transmission/receipt port is concerned with sending and receiving of data frames. The distributary module is concerned with the distribution of data frames from the wireless transmission/receipt port to different modules of higher layer.

12.3.2.2 Wireless Network Interconnection Subnet

Wireless network interconnection subnet is the border between wireless mesh backbone network and different wireless access networks. It mainly includes the mesh

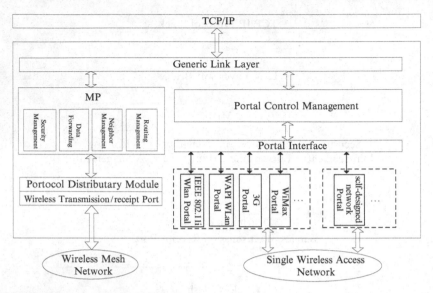

Fig. 12.8 The protocol stack of MPP

point with portal (MPP) and some base stations of wireless access networks. MPP is an MP node with the functions of a portal and can access different base stations. The GLL and plug-in technologies are adopted when designing the MPP. The protocol stack of MPP is depicted in Fig. 12.8.

The GLL can bridge most kinds of wireless or wired networks, thus the interconnection of heterogeneous networks is realized. Future networks can also be supported by GLL.

The portal control management module is concerned with installation, activation, uninstallation and configuration of the existing portal modules. The portal control management module is very important for achieving multimode access. On receiving the instructions of the data link layer, the right portal module is chosen and then configured, started or uninstalled.

Portals have the unified interface. Different portal modules can be automatically installed and uninstalled through the plug-in technology, which makes the system flexible and scalable. When a new kind of network occurs, a new portal module should be coded according to the interface criterion and added to portal database. The interface criterion not only supports the interfaces of standard access method, such as WLANs, Cellular networks, WiMax networks, sensor networks and satellite networks, but also supports interfaces of self-designed and future access mode.

Another important communication entity in wireless network interconnection subnet is the access point or base station of wireless access networks, such as access point in WLAN. MPP can connect with the access network, which makes network convergence possible. To ensure the compatibility of the existing access technologies, the original technologies and network services of such nodes should be reserved.

12.3.2.3 Single Wireless Access Subnet

Single wireless access subnet is the wireless network that uses some existing or emerging wireless network access technology individually. All the wireless access networks should maintain the existing network services to be compatible with the existing terminals.

12.3.2.4 Multimode Wireless Terminal Access Subnet

The multimode mobile terminal access subnet consists of wireless terminals with the function of multimode access authentication, which allows users roaming among different wireless networks. The system framework of the wireless terminal, depicted in Fig. 12.9, includes the following subsystems: management subsystem, security access subsystem, authentication protocol subsystem, implementation subsystem and external security support subsystem.

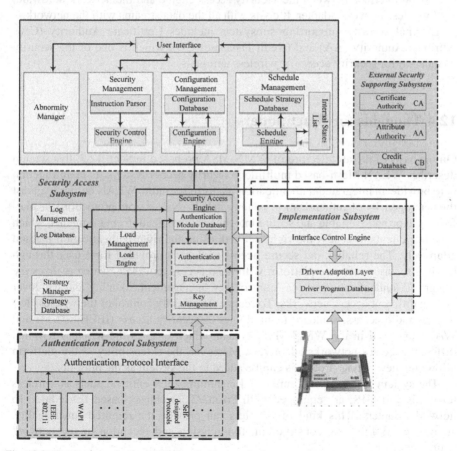

Fig. 12.9 The framework of multimode wireless terminal

Management subsystem is concerned with configuration of system's functions and security. Through user interface, management subsystem can carry out security management, schedule management, configuration management and abnormity management.

Security access subsystem should complete the entire process of authentication with corresponding wireless network. On receiving the instructions of the management subsystem, the security access subsystem interacts with the wireless access network using the correct authentication module to achieve the authentication and security access. All the functions of the security subsystem are realized by the co-operation of the security access engine, load management module, strategy management module, and log management module.

Authentication protocol subsystem is an open and scalable system, which new security protocols can be easily extended into. Both the standard authentication protocols and self-designed authentication protocols can be supported by the authentication protocol subsystem.

Implementation subsystem is the lowest layer of the system framework, and it is also the interface between the security access engine and the network hardware, e.g., wireless network adapter. It deals with all the data streams with the network.

External security supporting subsystem includes Certificate Authority (CA), Attribute Authority (AA) and Credit Database (CD), which is one of the security foundation for security access of wireless terminals.

12.4 Thoughts for Practitioners

On the basis of the architecture of WMN-MSA proposed in Sect. 12.3, this section describes WLAN Mesh-based multimode security access system. Its main function is to provide an integrated authentication platform for multi-WLAN access. Against the heterogeneity of authentication systems in WLANs' security protection, the platform is to provide reliable and secure integrated authentication environments, which facilitate clients to access the right network using the right authentication method adaptively. The reliable and secure access method is provided for a user through binding security software platform with radio access terminals, which realizes the integrated authentication, access control, key management, roaming and handover among different access networks. To realize the security control of the wireless terminal's radio access, standard technologies of security access (IEEE 802.11i and WAPI) are used in the WMN-MSA system, which enables the wireless terminal different accesses networks adaptively. Moreover, the extensibility of the platform allows the new access protocols can be added to the system as the plug-ins.

The system is based on Linux, and communication entities, such as wireless terminals and MPs, are equipped with Prism2/2.5/3 chipset-based IEEE 802.11b network adapters. This kind of adapters can provide the required functionality, such as controlling the construction, transmission and reception of the network frames, etc.

The system's basic functions are implemented in the driver, which provides a set of system interfaces being called by user programs. These functions are time constrained, such as the transmission, reception and frames forwarding of frames, timed sending of Beacon frame, etc. On the other hand, the functions of neighbor management, authentication and encryption, and route maintenance are implemented in user programs.

12.4.1 Interconnecting Devices

12.4.1.1 Protocol Stack

According to various communication functions, interconnecting devices appeared in WLAN mesh can be divided into following categories: MP, MAP (mesh AP), and MPP. All the three kinds of nodes have to be redesigned so that they can support the interconnection with other networks. Because most functions of WLAN Mesh are implemented on MAP and MPP, we will focus on the design of the MAP with Portal.

The framework of the MAP with Portal, depicted in Fig. 12.10, consists of following four modules:

– *AP module*. This module involves all functions of an AP, which is specified by IEEE 802.11, including security management, STA management and packages forwarding.
– *MP module*. The functions of this module include security management, neighbor management, package forwarding and routing management.
– *Portal module*. This module is used by an MPP to interconnect with other wireless networks.

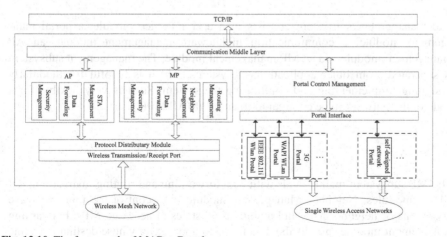

Fig. 12.10 The framework of MAP + Portal

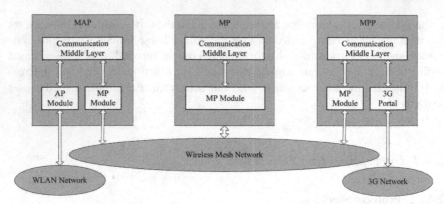

Fig. 12.11 The system interconnection model

– *Communication middle-layer module.* The function of this module is to bridge AP, MP and Portal in link layer.

The following Fig. 12.11 depicts the interconnection model in the system. In the MPP, different types of portals take charge of the interconnections with other wireless networks.

12.4.1.2 Functions of Subsystems

The system is software running on each MP and then adaptively installs corresponding modules on demands. The framework of the MAP + Portal system is depicted in Fig. 12.12.

Management Subsystem

The management subsystem consists of a user interface module, a portal management module, a communication middleware management module, an AP management module and an MP management module. The management subsystem is separated from other application subsystems, which is helpful for modularly designing, implementing and configuring the system. So it is very flexible for a node of the network to be configured to an MP/MAP/MPP.

MP Subsystem

The MP subsystem mainly deals with data forwarding and routing maintaining. Of all the modules, Neighbor Management module is to control the join and leave operations of neighbor MPs and maintains all states of accessed MPs. Forwarding management module submits the data frame to upper layer whose destination is the local host, or re-encapsulates the data frame according to the routing information

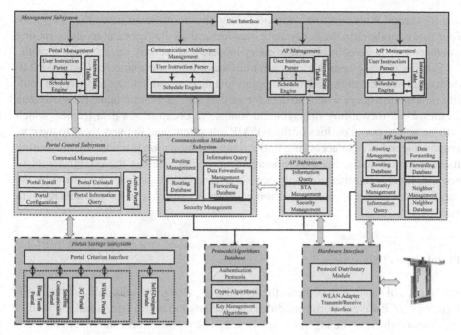

Fig. 12.12 The framework of MAP + Portal system

from routing management module and delivers it to the lower layer to forward. Routing management module maintains routing information and responses to the query of forwarding management module. Security management module provides several kinds of access methods, which enable the MP be accessed under different modes. Also, this module provides services of encryption/decryption for frames. Information query module provides the running states of the MP for the management subsystem.

AP Subsystem

The AP subsystem mainly provides the access services for wireless terminals (STA) in WLAN. It includes an STA management module, a security management module and an information query module. The STA management module handles the join and leave operations of wireless STAs and maintains the states of accessed STAs. Security management module and information query module provide similar functions as the same modules in MP subsystem.

Communication Middleware Subsystem

The communication middleware subsystem is used to bridge different kinds of wireless or wired networks in the link layer and it is scalable for future networks. Frames are forwarded with uniform message format among different networks. There are four major modules in the subsystem: data forwarding management module, routing management module, information query module and security management module.

Routing management module maintains the routing information of Portal/AP/MP entities connected to the communication middleware, which can support the routing information query when the data forwarding management module forwards data frames. Data forwarding management module re-encapsulates data frames, which will be sent to portal entities or the AP/MP subsystem, and delivers them to the corresponding subsystem according to the routing information provided by the routing management module. Security management module enables the establishment of security links with other nodes if necessary. Information query module provides the running states of the communication middleware subsystem for the management subsystem.

Portal Control Subsystem

Portal control subsystem controls the installation, activation, uninstallation and configuration of different portal modules. It executes the above operations on receiving commands from management subsystem. The portal control subsystem includes following modules:

- *Command management module.* It executes portal control commands of the management subsystem and sends running states of the subsystem to the management subsystem.
- *Portal management module.* On receiving the command of command management module, it accomplishes the corresponding operations, such as installation, uninstallation, configuration and query.
- *Active portal database.* It is a set of portal modules being in use, and these portal modules are controlled by the portal management module.

Portal Storage Subsystem

All the portal modules are stored in the portal database by the portal storage subsystem, which is controlled by the portal control subsystem. When a new access network occurs, a corresponding portal module should be coded according to the interface criterion and stored in the portal database. Then the portal control subsystem can load the new portal module from the portal database if necessary. The portal storage subsystem includes the portal criterion interface and the portal modules database.

1. *Portal criterion interface.* It consists of a portal installing criterion interface, a portal uninstalling criterion interface, a portal configuring criterion interface and a portal information querying criterion interface. These criterion interfaces pro-

vide integrated specifications for multimode access system, and then a new portal module can be easily extended into the system.

2. *Portal modules database.* The database supports two kinds of portal modules, one is the standard portal modules, such as 3G, BlueTooth, and so on; the other is self-designed portal modules. The latter is aimed at special applications and can meet some special requirements that the standard portal module cannot achieve.

Algorithms and Protocols Library

The library is an independent module, which can support the communication middleware subsystem, AP management subsystem and MP management subsystem with all kinds of authentication algorithms, encryption algorithms, key management protocols, etc.

Hardware Interfaces

This subsystem contains the distributary/aggregation equipment and the transmission/receipt interface of network adapters. The former mainly has two functions: one is the distribution and encapsulation of data frames received from lower layer, and then sends these frames to corresponding subsystems to deal with; the other is the encapsulation of frames received from some subsystems with uniform format and then delivers these frames to transmission/receipt interfaces of network adapters. The latter also has two functions: first, it sends out frames received from the distributary/aggregation equipment through hardware interface or vice versa. Second, it provides a uniform interface for sending and receiving frames, which can screen the differences of hardware.

12.4.2 Wireless Terminals

This section introduces the multimode security access system for WLAN terminals. It is based on IEEE802.11b network adapter and can provide the integration of different security authentication protocols. This system can be used in various WLAN access networks and ensure that wireless terminals adaptively access wireless networks with corresponding authentication schemes. Moreover, the system is scalable and can be reconfigured with a new security access scheme through the plug-in technology. The framework of the multimode security accessing system of the wireless terminals is depicted in Fig. 12.13.

The system consists of following functional modules: main procedure module, authentication module, authentication module scheduler, driver module scheduler, system control interface and driver adaption layer.

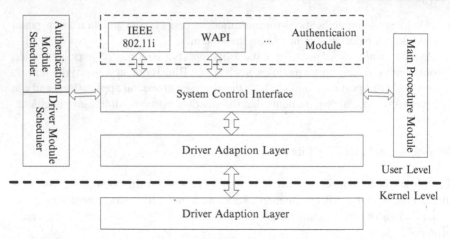

Fig. 12.13 The framework of the multimode security accessing system

Main procedure module. This module realizes the functions of the user interface, security management, configure management, log management and abnormity management in the framework of multimode wireless terminals. Concretely, it should provide the detection and load of authentication module plug-ins, analysis of configuration log, assignment and release of the resources, and load of different authentication module plug-ins.

Schedule module. The scheduling module completes functions of the schedule management module in the framework of multimode wireless terminals. Concretely, the authentication module scheduler achieves the adaptive schedule of authentication module plug-ins according to the contents of user's configuration files. When the access network is changed, the right authentication module is loaded if it has been configured in the configuration file. Otherwise, the authentication module scheduler analyzes the access network and selects the right authentication module.

System control interface. This module realizes the functions of authentication protocol interface in the framework of multimode wireless terminals. The interaction between main procedure and authentication plug-ins depends on the system control interface, which is a logic interface and partly realized both in main procedure and plug-ins. It can be divided into register interface, initialization interface, communication interface and functionality interface. The register interface is concerned with the registration and logout of the authentication modules. The initialization interface is concerned with the load of authentication modules and assignment of the resources. The communication interface is concerned with the communication among different modules. The functionality interface is to call the different system functions. The system control interface can not only load the existent functionality modules to the system but also import a new functionality module, which provides the scalability for the system.

The system control interface is very important and all the data streams are delivered through it. So the design of the interface should be very careful to guarantee the scalability and efficiency of the system.

Authentication module. The authentication module belongs to the authentication protocol subsystem in the framework of multimode wireless terminals. It deals with standard authentication protocols and self-designed authentication protocols. Its scalability is realized through the dynamic link lib (DLL) and each authentication protocol is designed to be a DLL and coded to be a plug-in. The interaction between the authentication module and other modules is based on the system control interface. When a new authentication protocol occurs, it should be coded according to the interface specification so that it can be compatible by the system. All the authentication plug-ins are scheduled by the authentication module scheduler.

Driver adaption layer. This module realizes functions of driver adaption layer in the framework of multimode wireless terminals. For the messages received from different drivers, this module encapsulates these messages for the authentication protocol according to the protocol message format. For the messages of the authentication protocol will be sent, this module encapsulates the messages for the driver. Therefore, this module enables the system to be compatible with other platforms.

The framework of driver adaption layer is depicted in Fig. 12.14. For each authentication protocol plug-in, there is a corresponding operation set that formats all drivers supporting the authentication protocol. For example, 802.11i_driver_ops is the operation set of 802.11i and driver hostap supports both 802.11i and WAPI. Then the driver hostap should be formatted by 802.11i_driver_ops and WAPI_driver_ops, respectively.

The scalability of the driver adaption layer lies in the following two aspects. First, when a new authentication protocol plug-in module is added, the system inserts its information into the corresponding driver module. If the system cannot find the

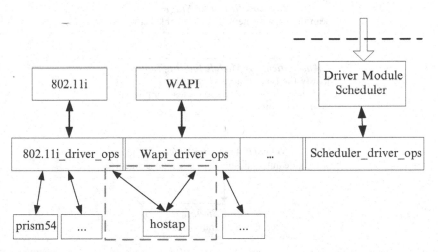

Fig. 12.14 The framework of driver adaption layer

corresponding driver, it should be assigned by the user. Second, when a new driver is added, the system registers an operation in the authentication protocol operation set, which is supported by the driver. Then the driver can be loaded after restarting the system.

The work flow of the multimode security accessing system for WLAN terminals has the following five stages that are depicted in Fig. 12.15.

Initializing stage: Once the operating system is started, the network adapter's driver is called and the work mode of the adapter is set to hostap mode. Then the adapter is started and main procedure initializes all modules.

Network scanning stage: After the initializing stage, the system will scan the assigned or available access devices to attain the required access information, e.g., the authentication protocol.

Authentication module loading stage: The system analyzes the information provided by the network scanning stage and identifies the type of the access network. According to the type of the network, the schedule module selects and loads the corresponding authentication protocol plug-in. At the same time, the schedule module should record state information of the authentication module.

Network accessing stage: In this stage, the system should complete the authentication and key management via the corresponding authentication module. Once the authentication is successful, the wireless terminal can access the network.

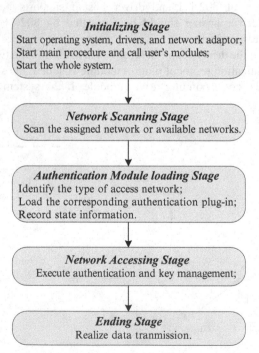

Fig. 12.15 The work flow of wireless terminals

Ending stage: the system should unload the corresponding modules and release occupied resources. Then a secure channel between the wireless terminal and access device is established and the normal communication is realized.

12.5 Directions for Future Research

In the future, there will be a multitude of wireless access networks based on different standards and technologies. Then the seamless communication across these heterogeneous networks would be the basic requirement. Although WMNs-based network convergence is a very flexible and scalable solution, there are some directions remaining to be further studied.

Multichannel communication. Currently, the WMN-MSA system is based on single channel communication mode. To extend the capacity of the network and improve the efficiency of the communication, the multichannel communication mode should be adopted.

Evaluation in practice. The WMN-MSA system is a prototype system, so the evaluation of this system should be carried through in practice.

Enlarging the coverage. The wireless mesh backbone in the WMN-MSA system is WLAN mesh, so the coverage of the system needs to be enlarged through other wireless mesh technologies.

Besides, a number of other research issues have unsatisfactory solutions currently, such as the fast seamless vertical handover, QoS support, secure routing algorithms and access authentication protocols, the strategies and algorithms of dynamic bandwidth allocation in heterogeneous networks, trusted and survivable architecture, etc.

12.6 Conclusions

As a necessary technology in the next generation wireless network, WMN technologies will draw much more attention. The convergence of different wireless networks based on WMN is an effective approach. In this chapter, WMNs-based multinetwork convergence technology is introduced. With the new technology, wireless terminals can access heterogeneous wireless networks and Internet through the backbone WMN using a multimode manner, which helps the wireless terminals realize seamlessly roaming among different wireless networks. Based on the technology, a WLAN mesh-based multimode security access system is implemented. The system is a suite of softwares installed in wireless terminals, mesh access points (MAPs) and MPPs. It is scalable for the new or self-designed security technologies without any change to hardware. So it is very suitable for the circumstances where high flexibility and robusticity are required.

12.7 Terminologies

1. *Network convergence.* Converged networking provides an integrated network, which is provision of wide coverage, high bandwidth, high mobility and low cost Internet access for network terminals.
2. *Multimode access.* A wireless terminal can access all the converged heterogeneous networks via the access technologies corresponding to the access networks. These access technologies can be adaptively selected by the wireless terminal and access networks.
3. *Vertical handover.* Vertical handovers refer to the automatic conversion from one technology to another to maintain communication. This is different from a "horizontal handoff" between different wireless access points that use the same technology in that a vertical handoff involves changing the data link layer technology used to access the network.
4. *Seamless network access.* Seamless network access leads to continuous phone calls, interruption during the call will never happen while across network boundaries.
5. *Mesh point with portal (MPP).* An MPP is a mesh point that sits between a WMN and an external network, such as a cellular network, with some defined border functionality.
6. *MIRAI.* MIRAI is a project of e-Japan Plan and it aims at developing new technologies to enable seamless integration of various wireless access systems for practical.
7. *Doublecast.* Doublecast refers to a data stream transmission mechanism that the proxy sends the stream to the wireless terminal and to the new proxy the wireless terminal is migrating to.
8. *Generic link layer (GLL).* GLL provides that old configurations and states of wireless link layer can be transferred to the new one if the wireless link layer is designed in compatible and generic mode, thus seamless data transmission can be realized during the handover.
9. *WMN-MSA.* WMN-MSA is a security access system for WMNs-based multi-network convergence. In WMN-MSA, the security access of all kinds of wireless networks can be successfully completed as long as the mesh gateway is designed according to the architecture of this system and the wireless terminal is installed the corresponding security access module.
10. *Plug-in.* Plug-in is a computer program that interacts with a host application to provide a certain, usually very specific, function "on demand."

12.8 Questions

1. Explain the reasons of the convergence of different wireless networks.
2. What are the design principles of multimode security access system?
3. What entities are involved in MIRAI?

4. What is doublecast?
5. What are the functions of the dynamic buffering mechanism used in Indulska and Balasubramaniam's vertical handover solution?
6. List the steps of handover in GLL.
7. Briefly describe the WMN-MSA system.
8. What subnets consist of the architecture of WMN-MSA?
9. To realize the convergence of different wireless networks, the WMN-MSA system should be installed in which kind of nodes?
10. Which kind of functions of WMN-MSA is implemented in the driver and user program, respectively?

References

1. R. Bruno, M. Conti, E. Gregori, Mesh networks: commodity multihop ad hoc networks. IEEE Commun Mag **43**(3): 123–131 (2005).
2. Nokia, http://www.iec.org/events/2002/natlwireless_nov/featured/tf2_beyer.pdf. Accessed 6 May 2002.
3. Microsoft, http://www.research.microsoft.com. Accessed 16 September 2005.
4. Motorola, http://motorola.canopywireless.com/. Accessed 23 June 2007.
5. Intel, http://www.intel.com. Accessed 8 July 2007.
6. K. Rayner, Mesh wireless networking. Commun Eng **1**(5): 44–47 (2003).
7. B. Schrick and M. J. Riezenman, Wireless broadband in a box. Spectrum **39**(6): 38–43 (2002).
8. P. Whitehead, Mesh network: a new architecture for broadband wireless access systems. *RAWCON 2000* (2000).
9. M. Buddhikot et al., Integration of 802.11 and third-generation wireless data networks. *INFOCOM 2003* (2003).
10. H. Suk Yu and Y. Kai Hau, Challenges in the migration to 4G mobile systems. IEEE Commun Mag **41**(12): 54–59 (2003).
11. Detailed Technical Specification of Security for Heterogeneous Access(Version 1.0). Available via http://www.ist-shaman.org/, 2002.
12. J. McNair and Z. Fang, Vertical handoffs in fourth-generation multinetwork environments. IEEE Wireless Commun Mag **11**(3): 8–15 (2004).
13. W. Gang, M. Mizuno, and P. Havinga, MIRAI architecture for heterogeneous network. IEEE Commun Mag **40**(2): 126–134 (2002).
14. M. Inoue et al., MIRAI: a solution to seamless access in heterogeneous wireless networks. *ICC 2003* (2003).
15. I. Jadwiga and B. Sasitharan, Vertical handover based adaptation for multimedia applications in pervasive systems. *IDMS/PROMS2002* (2002).
16. J. Sachs, A generic link layer for future generation wireless networking in Communications. *ICC 2003* (2003).
17. J. Ma and C. Cao, Multimode access platform of wireless networks. Technical Report (2006).
18. IEEE Standard for Information technology – Telecommunications and information exchange between systems – Local and metropolitan area networks – Specific requirements Part 11: Wireless LAN Medium Access Control (MAC) and Physical Layer (PHY) specifications Amendment 6: Medium Access Control (MAC) Security Enhancements. IEEE Std 802.11i-2004.
19. National Standard of the People's Republic of China. GB15629.11-2003. "Information technology-Telecommunications and information exchange between systems – Local and metropolitan area networks – Specific requirements – Part 11: Wireless LAN Medium Access Control (MAC) and Physical Layer (PHY) Specifications", 2003.

Chapter 13
Scalability in Wireless Mesh Networks

S. Srivathsan, N. Balakrishnan, and S.S. Iyengar

Abstract Wireless mesh networks (WMNs) is emerging as a prominent technology for next-generation broadband wireless networking. They have distinct advantages over current wireless networks and because of the promise they hold, research and development in this area is progressing rapidly. However, studies in the recent past have noted that WMNs have shortcomings and the technical issues are being addressed both in the industry and academia for successful deployment for a variety of applications.

This chapter provides a detailed survey about scalability – one of the major deciding factors for any new networking technology to be accepted, deployed and to evolve continuously. It is now well known that available ad hoc wireless networking protocols are not scalable for WMNs. This chapter discusses the various factors that impact scalability of WMNs and presents a detailed study of various current and ongoing research efforts in the different aspects of networking such as network architecture, physical layer, MAC layer and network layer that aim to improve scalability in large-scale WMNs. Various improvements to existing wireless ad hoc protocols, wireless sensor network protocols and revisiting of protocol designs from the perspective of WMNs and scalability are discussed with an emphasis on open research issues among the protocols. Other aspects in network design and deployment that aim to improve scalability are also discussed.

13.1 Introduction

A new form of distributed wireless networks called wireless mesh networks (WMN) looks to be a promising candidate for next generation wireless broadband networking. It is believed to play an important role in future wireless network

S. Srivathsan (✉)
Louisiana State University, 164 Coates Hall, Tower Dr, Baton Rouge, LA 70803, USA
ssrini1@lsu.edu

S. Misra et al. (eds.), *Guide to Wireless Mesh Networks*, Computer Communications and Networks, DOI 10.1007/978-1-84800-909-7_13,

applications ranging from civilian wireless Internet applications to military and emergency response applications besides indoor applications. WMNs represent a paradigm shift away from the liabilities of traditional wireless formats such as rigid structures, meticulous planning requirements of the wired backbone and toward a real-time plug-and-play deployment model that is up to the challenge of today's rapidly changing connectivity environment. Because of technological advances and breakthrough approaches, WMNs are poised to make the leap from localized hotspots to fully wireless hot-zones with building-wide or campus-wide coverage and even hot regions that span an entire metropolitan area.

WMNs are characterized by inherent tolerance against network failures, rapid deployment and flexible coverage areas. Its dynamic self-organization and self-healing capabilities help in lowering the deployment costs and enabling ease of maintenance. The diverse capabilities and advantages of WMNs have inspired many industries, sparked interest in many researchers and have led to rapid commercialization in a slew of indoor and outdoor applications such as community networking, building automation, broadband home networking, high-speed metropolitan area networks and enterprise networking. They are also appearing in several vertical markets such as mining, manufacturing, transportation, and other enterprise settings.

Though WMNs have many advantages and have become popular in recent years for both urban and rural applications, network providers are realizing the issues associated with it. One of most apparent problem is the scalability of such networks (often seen in any multihop network) both in terms of increased geographical area coverage and number of users. Under those conditions, the network has to grow efficiently and cost-effectively. But, as more nodes are deployed and as more users participate, the usual advantages of multihop wireless networks begins to fail and would degrade the performance. The denser the users become, the more will be the cochannel interference that would deteriorate the performance and scalability [20, 24].

This chapter presents a survey of recent advances in algorithms and protocol design to maintain acceptable levels of scalability for large-scale WMNs and also gives some insight to some critical factors influencing protocol design. The rest of the chapter is organized as follows. Section 13.2 discusses background in the area of WMNs and scalability in WMN. Section 13.3 provides a brief introduction to similar technologies. Section 13.4 describes in detail about scalability in WMNs. Section 13.5 mentions about some thoughts for practitioners. Possible future directions in research and open issues are discussed in Sect. 13.6 and concluding remarks are given in Sect. 13.7.

13.2 Background

WMNs were originally developed to give soldiers reliable broadband communications anywhere in the battlefield. It was developed with a set of key requirements in mind – broadband capacity along with quick deployment in infrastructure-less areas, quick tear-down, automatic configuration and reconfiguration capabilities without

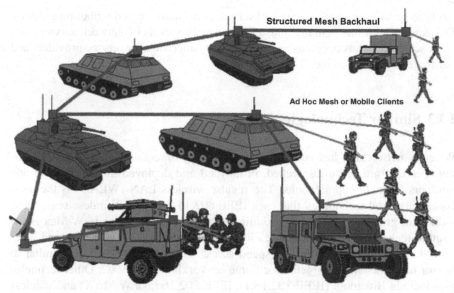

Fig. 13.1 An example of military communication using mesh networks. Image source [25]

the need to predeploy large towers or antennas. Every soldier's radio powered the network creating an ad hoc interconnected wireless system that increased tactical situational awareness by adding support for real-time data and video connectivity to individual soldiers and battlefield commanders [41] (Fig. 13.1).

From military applications, WMNs have gradually trickled into the civilian space. Initially, it started off for mission-critical law enforcement agencies, fire departments, emergency medical service providers and other public safety organizations. In the past, the first responders were limited to voice communications with a centralized command and control. More recently, the available options have expanded to text-messaging, satellite video and other forms of communications. The advent of licensing the 4.9 GHz spectrum for the exclusive use of public safety agencies and the technological/economical advantage WMNs promise will bring unprecedented communication capabilities. From public safety applications, WMNs are now moving into community wireless broadband services to support data, voice and video [44].

By eliminating the expense of wiring and the impediments involved in that process, recent adaptations of existing wireless and switching/routing technologies have now made WMNs practical and affordable in more situations than ever before. To achieve this paradigm shift, the wireless mesh takes full advantage of several proven technologies. The ubiquity of wireless networks due to the cheap and reliable Wi-Fi and Bluetooth products have enabled new markets have been developed. Current wireless networks form isolated communication islands without any interconnection among them.

Multimedia data services have grown at a remarkable rate in recent years and there is considerable demand for such services globally. The increase in demand is likely to grow even faster because of advances in multimedia distribution services. To keep up with this demand and offer acceptable levels of network service, network scalability thus becomes an important consideration for service providers and hardware/software vendors.

13.3 Similar Technologies

Wireless networking has been around for a long time and there are many technologies that are being researched, prototyped and deployed in both indoors and outdoors for many applications. The regular wireless LAN (WLAN) is the most commonly used technology that uses IEEE 802.11 (a/b/g/n). Wireless sensor networks are gradually coming to prominence after years of research. WiMax is yet another technology where a whole city can skip building infrastructure and jump onto wireless mobility with high-speed access. Zigbee and 802.15.4 is similar to sensor networks and is targeted for home networking and HVAC. Other technologies include Bluetooth (IEEE 802.15.1), IEEE 802.16 (aka WiMAX) and wireless personal area networks (WPANS) are some of the current networks that are operational worldwide.

WMNs consist of a mish-mash of network components such as wireless routers, gateways, printers, servers, mobile clients (bus, subway, airplanes, etc.), and stationary clients. The users could use 802.11x, Bluetooth or any other proprietary technology and would not be a bottleneck if properly designed. The router to router links are wireless and usually the cause for bottleneck. In most applications, the user–user data flows constitute only a small fraction of the data flow and the majority is because of the user–internet data flow.

Typically, in WMNs, the nodes are fixed and some are mobile. It relies on some infrastructure and most traffic that runs through it is user-to-gateway. On the contrary, in ad hoc networks, the nodes could be mobile and most of the traffic is user-to-user. The bandwidth of WMNs is very much greater than what wireless sensor networks can provide and the nodes in WMNs generally do not have severe energy limitations like those of sensor networks.

13.4 Scalability

Adoption of distributed wireless network services is becoming prevalent and this drives the efforts to make such networks grow seamlessly with the addition of new nodes and new users without any noticeable degradation in its throughput.

It is shown that traffic pattern and topology plays a role in whether a network's per node capacity would scale to large networks [32–34, 38]. The authors of this paper also show that for a total capacity to scale with network size, the average distance between source and destination nodes much remain small as the network grows. The analysis by Gupta and Kumar [22] provides an estimate of per node capacity in wireless ad hoc network. In that paper, two network models were considered. The first called *arbitrary network* model, which has N immobile nodes with no restrictions on their locations, have different transmission powers and all nodes have omnidirectional antennas that can either transmit or receive (not both) at a given time. The authors showed that that the throughput per source–destination pair is $O(1/\sqrt{n})$ as $n \to \infty$ where n is the number of nodes in the network. The second model considered is called the *random network* model, which has additional constraints such as random locations, traffic patterns and all nodes use a fixed transmission power. The total amount of data that can be sent simultaneously for one hop increases linearly with the total area of the network provided the sender–receiver pairs are sufficiently separated to avoid interference. This result can roughly be equated to mesh networks in the sense that when the node density is constant and if there are multiple hops, the end to end capacity is roughly $\Theta(1/\sqrt{n\log n})$ for arbitrary networks and the end to end throughput at each node is $\Theta(1/\sqrt{n})$ for random networks. This study points to the notion that ad hoc networks are fundamentally nonscalable, which may not reflect reality. This study was based on the assumption that the nodes communicate at random and that each node is equally likely to communicate.

Another research study by Gupta, Gray and Kumar [23] presents a new experimental result, which states that the per node throughput decays like $O(1/n^{1.68})$. Onur Arpacioglu and Zygmunt J Haas [2] provide theoretical results that demonstrate the dependencies among the maximum achievable per node end-to-end throughput, the number of nodes in the network and other parameters of the wireless network. They also determine the implications of these dependencies on the scalability of the wireless network. Their analytical result shows that the end-to-end per node throughput is $\Theta(1/n)$. The authors prove that this throughput scaling could be achieved even without simultaneous transmissions. Their work is primarily for peer-to-peer wireless networks such as WMNs and ad hoc networks, but the results also holds for infrastructure based wireless networks. Their work considers temporal variation in transmission powers and simultaneous per-node multiple transmissions and receptions.

The important fact to note is that of scaling relationships: As number of nodes increases, the load increases and load increases as the communication distance increases and the total one-hop capacity increases as the coverage of the network increases [36, 37]. The authors in [22] have also stated that traffic pattern does impact on scalability. The study in [38] provides a clear idea of how traffic patterns and locality of the traffic could impact scalability. If the size of the network increases and if the traffic pattern remains local, that is, if users tend to communicate within a limited area always (say, a building, or same university), scalability is not affected. But, in general, it is known that scalability degrades when networks and users scale up. The question is "by how much" and how this degradation can be kept to a minimal value. In [28], the authors study the existence of scalable protocols that achieve the capacity limit of $O(1/\sqrt{n})$ for each source and destination pair in a large wireless network where the buffer size does not grow as the size of the network grows. They proceed to show that there is no end-end to protocol capable of this maintaining this limiting throughput with constant buffer space. The authors establish that there exists a protocol that can achieve a slightly smaller throughput of $O(1/\sqrt{n\log n})$ with the nodes having constant buffer space.

13.4.1 Factors Affecting Scalability

There are many reasons for poor network scalability in WMNs. Some of them are listed below:

- Co-channel interference
- Routing protocol overhead
- Half-duplex nature of radio antennas
- Difficulties in handling multiple frequency radio systems
- Deployment architecture
- Medium access control
- Topology (denseness of nodes, degree of nodes)
- Communication pattern (locality and number of hops)

In the following sections and subsections, we shall see how the above factors play a role in affecting the scalability of WMNs. Not all the factors are mutually exclusive and most of them interrelated. Hence, we shall be observing the effects of more than one factor when we are discussing about the impact of another.

13.4.1.1 Network Architecture

Of the many criteria for a successful deployment of a WMN, the need for a careful analysis and consideration of appropriate deployment architecture is essential. WMNs can be deployed in flat, hierarchical or hybrid fashion. Typically, for any emergency deployment, a random flat architecture is used. These architectural design issues [54] are described in the following subsections.

Modern mesh network requirements have evolved from their military origins as mesh networking moved from the battlefield to the service provider, enterprise, and residential networking environments. Today, Internet connectivity is needed more than local peer-to-peer connectivity. Data sources are primarily resident on the Internet, not on a peer. Also, to cover large areas in a cost-effective way, nodes may be placed further away from their eventual connection to the wired network, which implies more relay or "hops" within the mesh until the Ethernet cable (wired network) is reached [13]. Below are the typical wireless network architectures that are used in a variety of applications:

Flat Wireless Mesh

Typically, in a wired or wireless network, clients are data providers and routers are used as data forwarders. But, in a flat WMN, all the nodes are seen as peers.

That is, clients (users), routers, gateways, etc. are all on the same level and there is no clustering or grouping of nodes in any logical or physical sense. In this scenario, the peers communicate with their neighbors not only for data transfer but also for transfer of control data that is, for routing, configuration and provisioning and other application related parameter handshaking. This kind of architecture closely resembles a traditional ad hoc wireless network. Though this kind of architecture is simple, it is well known from Gupta and Kumar [22] that such a topology would not scale well and has the potential to put very high resource constraints. Moreover, addressing schemes and service discovery would prove to be a major bottleneck against scalability.

Hierarchical Wireless Mesh

In this kind of architecture, as the name suggests, the nodes are grouped (logically, geographically, etc.) and form multiple tiers. Each level would have different kinds of nodes with different functionality (but, also functions as peers) and the lowest of them all would constitute the users (clients). One of the upper level nodes would be a router that forms a part of the WMN backbone (or backhaul) network [4, 48]. These backbone nodes do not originate or terminate data traffic like the client nodes, but, enable the data traffic to reach the next hop by observing the required QoS. They also ensure that the backbone network is functional under partial network failure. This is done by rerouting, renegotiating links and discovering new alternate routes in real-time. This is the key to network resilience [6]. When one or more of the backhaul nodes fail, the other backhaul nodes would either recompute alternate routes in real-time (by using reactive routing protocols) or would already have alternate routes in their routing and forwarding tables precomputed (by those using proactive routing protocols).

Hybrid Wireless Mesh Network

Hybrid WMNs, which aims to achieve better capacity than ad hoc networks without infrastructure support, uses other wireless networks for communication. This kind of architecture has (usually) three tiers, which consist of mobile nodes, wireless relay nodes and wired access points. It becomes essential, when one considers scalability, to note that most applications involve traffic flows to and from the Internet in addition to peer-to-peer communication among radio nodes. Furthermore, the addition of infrastructure nodes to ad hoc networks can reduce the mean number of hops from source to destination, thereby improving performance when compared to flat architecture (Fig. 13.2).

Suli Zhao et al. in [56] propose a novel self-organizing hierarchical architecture for improving performance and scalability and it is compared with two well-known classes of ad hoc routing protocols – dynamic source routing (DSR) and ad hoc on-demand distance vector (AODV) and evaluated with flat ad hoc network. They

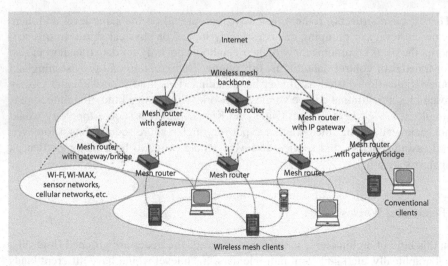

Fig. 13.2 A typical hybrid wireless mesh network. Image source [1]

show that their new three-tier hierarchical architecture shows significant capacity increases and better scalability when the number of forwarding nodes and access points are increased in the right proportions. They also observe that scaling would depend on the spatial distributions and the relative densities of client nodes and other nodes (gateways, backbone nodes, etc.). In [55], the same authors study the scaling properties of a three-tier hierarchical hybrid wireless network that contains a mix of mobile nodes, access points, forwarding nodes.

13.4.1.2 MAC Design

The MAC layer for any protocol for WMNs perform the usual function of finding the channel and using it at the right time. Typically, MAC protocols are designed for homogenous networks where the radios of all the nodes use a single frequency to communicate and use omnidirectional antennas. But, in the case of WMNs, the situation is quite different. The network in heterogeneous, that is, the nodes are different and the radios and channels used would be different. In such cases, the MAC protocols must be able to switch among the available channels to quickly find the unused channel and transmit. Such behavior will improve performance of the network by avoiding collisions and interference during multihop routing. Multiple channels would enable a node to transmit simultaneously or enable neighboring nodes to simultaneously communicate without interference and collisions (Fig. 13.3).

But, designing a good MAC for WMNs is challenging due to many reasons. In case of multichannel MAC protocols, distributed channel selection is a big issue. Besides that, the usual issues with wireless MAC comes into picture – Collisions,

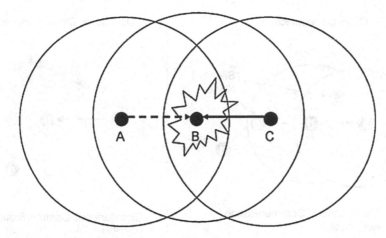

Fig. 13.3 Hidden terminal problem

interference, Hidden Terminal Problem, Exposed Terminal Problem, Deafness problem (if the nodes use directional or smart antenna). In Fig. 13.1 above, nodes C and B are communicating with each other initially. Node A is out of the communication range of C and hence does not hear its signal in the medium and tries to communicate with B. When B gets this signal, there will be collision at B as it receives signals from both A and C simultaneously. Most MAC protocols have not fully solved the scalability issue and only partial problems are solved while raising other new problems. New distributed and collaborative algorithms are being researched to maintain a high level of network throughput as the networks scales up [1]. Huang and Lai [26] have shown that scalability is affected by the physical layer also.

Because a MAC protocol's main task is to provide fair and efficient resource sharing, care should be taken during the design of MAC protocol for WMNs to ensure scalability. Because conventional MAC protocols may suffer from low throughput because of interference and collisions during multihop data relay, multichannel MAC protocols are being considered in recent years. Some of the recent efforts in designing efficient MACs in terms of performance and scalability are the following:

Use of Directional Antennas

Typically, nodes in a wireless network would have omnidirectional antennas that have well known disadvantages. Nearby nodes (in the vicinity of the one that is transmitting) have to remain silent (backoff) until they sense the medium to be clear. This is the basis of CSMA and this can be seen in Fig. 13.2 Omni Communication case. Because this idea would prove detrimental to the any dense deployment of WMN nodes, directional antennas are being investigated currently (Fig. 13.4).

Directional antennas provide a nice way to improve spatial reuse and thereby have the potential to increase the performance and scalability of WMNs [46]. With

Omni Communication | Directional Communication

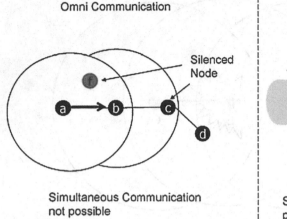 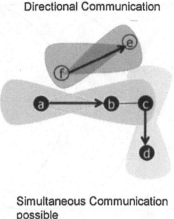

Simultaneous Communication not possible | Simultaneous Communication possible

Fig. 13.4 Comparison between Omni directional communication and communication using directional antenna. Image source [9]

the use of directional antennas, the antennas of the sender and receiver nodes are directed towards each other and the gain of the antenna is tuned towards the receiver and is not uniform in all directions. Because the gain is very little in the remaining directions, the nodes in the vicinity can effectively transmit or receive signals to/from other nodes without interference or having to wait to capture the medium for data transmission. This effectively translates into better latency, improved scalability and performance of the network [53]. As the network grows and the number of users grows, the degradation in the network performance is not drastic like it would be with traditional MACs. In Fig. 13.2 (direction communication), we see that for the same topology, the previously silenced nodes are able to communicate to different nodes because of directional gain and the absence of interference.

Though directional antennas bring lots of promise in performance and scalability, it also has its own set of unique issues. The hidden terminal problem is manifested in a different way called "deafness" and due to the high gain, there would be greater interference along the direction of transmission. Moreover, directional transmission, which has a high gain in the direction of transmission, introduces new hidden nodes. This introduction of new hidden nodes would result in increased number of collisions and has the potential to reduce throughput. Therefore, a good MAC protocol design to exploit the advantages of directional antennas must strike a balance between spatial reuse and increased number of hidden nodes (Fig. 13.5).

Multichannel MAC Protocols

Usually, nodes in a network tend to use the same channel during multihop communication. This behavior would degrade the performance as the number of hops

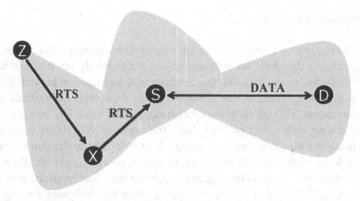

Fig. 13.5 Deafness problem in directional antennas – Node Z would not get a CTS message from node X. Image source [9]

increases. Multiple nonoverlapping channel usage is specified in the current 802.11 standards and nodes can use different (preferably orthogonal) channels to facilitate simultaneous transmissions. Here, the number of available channels is limited and the nodes must select the appropriate channel dynamically to enable spatial reuse.

There have been some recent developments from the standards community to provide a standardized MAC for WMNs. Because the traditional wireless MAC (802.11) and related enhancements were geared towards single-hop networks, the requirement to provide support for multihop networks has become all the more important to improve the scalability and performance. Developing such multichannel MACs, which utilize the available spectrum to achieve better efficiency, is a more challenging task than designing a single-channel MAC.

The newly formed IEEE 802.11 task group drafted an initial version of 802.11s that supports the original 802.11 DCF protocol and the 802.11e EDCA protocol with some additional MAC features like intramesh congestion control scheme to alleviate congestion at the mesh nodes, an optional multichannel MAC protocol (otherwise called CCF, common channel framework) and an optional mesh deterministic access (MDA) scheme, which aims for better quality of service. This initial draft is being revised for further improvements and added features. In [27], the authors propose a multichannel ring-based WMN where they design an appropriate MAC with distance based rate adaptation with multihop connections. To maximize the capacity and service coverage, application of mixed-integer nonlinear programming optimization is used. In [29], a novel scheme of integrating the use of directional antennas and time division duplex (TDD) protocol is used to direct and coordinate simultaneous transmissions in a WMN. This scheme aims to maximize spatial and spectral reuse to offer better scalability. This requires a common timing source such as the one provided by GPS. TDD MAC protocol will then coordinate precisely directional transmit/receive signals across the mesh topology.

13.4.1.3 Routing Protocols

Wireless Ad Hoc networks and WMNs are similar in many ways, but, the architectures and protocols designed for wireless ad hoc networks do not perform well for WMNs. This is due to the fact that the fundamental architecture behind these two networks is different, which is due the application, deployment objectives and the resource constraints. Though both these networks use the same key concept of communicating among nodes over multiple hops, the nodes in WMNs are usually static and those in ad hoc networks are assumed to be mobile. The nodes can be mounted on rooftops, light-posts, etc. where there is sufficient power supply and these nodes usually have multiple wireless interfaces to improve the capacity of the network.

Scalability in WMNs is very much dependent on the design of routing protocols that relay traffic from the source to the destination in the most reliable manner satisfying a set of quality of service parameters under changing environmental conditions and network topology. Such routing protocols are the key to providing the advantages WMNs promise to have – network discovery, self-configuration and self-healing capabilities. Because most of the nodes in WMNs are stationary, the focus of routing algorithms for WMNs is on improving the throughput rather than dealing with mobility and highly dynamic topology. Such tasks must be accomplished by the routing protocols with minimal overhead to provide good scalability [49]. The scalability of various routing protocols and their comparison is given in [50].

There are different types of routing protocols, which are generally categorized into reactive, proactive and hybrid routing protocols. The reactive types like AODV would compute the next hop node only when necessary whereas the proactive ones would already have a next hop neighbor, that is, all nodes would know the next hop all the time. The reactive protocols would introduce a small delay (even though the cost of overhead is reduced) for the first packet to get relayed whereas the proactive ones would not have such a delay. In terms of scalability, it can be seen that the reactive protocols would deteriorate gradually as the number of hops increases. And, for the proactive ones like OLSR, QOLSR, TBRPF (topology broadcast based on reverse-path forwarding routing protocol), HSLS (hazy sighted link state routing protocol), MMRP (mobile mesh), etc., the cost of overhead (in the control channel) for maintaining the list of routes at each node would be detrimental as the network scales.

Another class of protocol called hybrid protocols aims to infuse the advantages of reactive and proactive protocols to achieve better performance and scalability. This aims at providing a mechanism to adopt proactive routing for nearby nodes and/or for those routes that are used very often. The reactive mechanism is used for nodes that further away and for those nodes that are seldom used for data relay.

ad hoc on-demand distance vector routing protocol (AODV) is well-known protocol that has been studied extensively in the recent past. This reactive routing protocol is popular for MANETs and has been standardized in the IETF as experimental RFC 3561. Adaptations of this protocol have been used for WLAN mesh networking. This protocol employs a straight-forward request-reply messaging for route discovery and uses periodic hello messages to maintain neighbor connectivity. This reactive routing protocol reacts quickly to topological changes and updates only those nodes that are affected, but, it produces a large amount of overhead because of the flooding of control messages triggered by link failures. Hence, AODV does not scale well in mobile networks or networks with heavy load, but, seems to be a good candidate (with some improvements) for WMNs.

Cross Layer Routing

To reduce the interference (and thereby improve performance) in WMNs, the transmission power can be regulated according to the immediate needs. If the transmission power is high, reliability increases, but at the cost of greater interference and lesser throughput, which directly translates into poor scalability.

If the routing layer has enough information about the PHY layer, the protocol can utilize that knowledge to find better routes in terms of bandwidth and other service parameters.

Several research efforts to improve throughput and reduce routing overhead in large-scale mesh networks have been made. Weirong Jiang et al. [31] study the effect of long paths and routing overhead in WMNs and provide a routing scheme (based on Link State Routing) that uses a carefully computed routing metric (in multichannel routes) and control messaging that is designed to minimize the routing overhead.

It is well known that ad hoc routing protocols in general do not scale well in heterogeneous networks. Typically, the protocols used in such networks do not differentiate among the transmission capabilities or the channel access scheme of the nodes when they do routing computations despite the fact that many of the nodes would have multiple interfaces. In [21], the authors observe that the typical OLSR send out control messages to all the available interfaces and this causes very high control overhead. The authors propose optimizations to OLSR to limit this overhead, which improves protocol scalability in large scale heterogeneous wireless networks.

Routing metrics also play a crucial role in maintaining the scalability in WMNs and several metrics have been proposed [17, 52] and [12]. Draves, Padhye and Zill in [17] have come up with a new routing metric called weighted cumulative expected transmission time (WCETT) for multiradio multihop mesh networks. Their scheme achieves good tradeoff between delay and throughput. This scheme (MR-LSR) also is built on top of LSR (link state routing) protocol and assumes short paths where all the hops with the same channels interfere with each other. It is critical to efficiently choose the right path with the right set of channels along the path to achieve acceptable levels of throughput and scalability. The authors in [31] propose a new LSR protocol called timer-hit-use OLSR (THU-OLSR), which aims at reducing control overhead. Controlling the routing overhead directly impacts the scalability favorably. In [30], the authors study large-scale WMNs and propose an interference-aware routing metric called exclusive expected transmission time (EETT) that selects multichannel routes with least interference to maximize the end-to-end throughput. It is to be noted that for large-scale WMNs, the end-to-end paths are very long and the channel distribution has a significant impact on the route performance [31].

Fisheye state routing (FSR) [45] uses different control message broadcasting intervals for different routing table entries. This scheme aims to reduce the control message size and the frequency of transmissions and as a result, this scheme would scale well to large networks with mostly static nodes. This proactive routing protocol, which is based on link state routing, maintains a full topology map at each node (which has certain disadvantages discussed previously). FSR maintains up-to-date information about nearby nodes and the accuracy of information decreases as the distance from the nodes increases.

Optimized link state routing (OLSR) [10] is an improvement over the traditional link state routing protocol to address the issues in wireless ad hoc networks. This protocol exploits the multipoint relays (MPRs), reduces the size of the control messages by declaring only those neighboring nodes that are in its MPR selectors. The concept of MPR is to reduce flooding of broadcast messages by shrinking the number of nodes that have to retransmit those control messages. This reduces the number of retransmissions and has significant scalability improvement over regular flooding procedure. OLSR does not scale well because the routing information is sent to all the nodes in the network irrespective of whether all of them needs it. In [8], the authors study two predominant routing protocols considered in IEEE 802.11s (AODV and OLSR) and integrate the advantages of FSR and OLSR to get better scalability in terms of traffic loads than AODV. This extension of OLSR called optimized Fisheye link state routing (OFLSR) reduces routing overhead and also improves end-to-end latency.

The frequency with which control messages are sent out is critical to scalability of the network. If the timings of control messages are adaptive, then, it would prove to be very beneficial for large-scale networks. FSR and OLSR works on constant control message intervals irrespective of the path chosen, traffic patterns and other known parameters, which can be exploited. In the case of large-scale WMNs with relatively few mobile nodes, frequent generation of control messages would create a lot of overhead, suppress the transmission and reception of useful data thereby reducing throughput and impacting scalability adversely. This is exactly what THU-OLSR does. It dynamically adjusts the transmission of control messages and the theoretical analysis and simulation results show that it performs better than OLSR in both dynamic/mobile ad hoc network and in static/hybrid WMN.

Dynamic addressing is shown to alleviate the overhead problem as network scales up. Study in [18] shows that dynamic addressing can enable scalable routing where each node has a unique permanent identifier and a transient routing address. The research study examines how dynamic addressing could provide a basis for scalable routing architecture and presents address allocation, routing and address lookup algorithms. The inherent problem with dynamic addressing is the binding of address with topological location. Furthermore, address lookup will also be an issue and this paper addresses these issues.

In [47], the authors propose an *anycast* routing protocol that can scale well with network size. In this study, the authors have used a novel routing strategy using potential fields (which have been of interest in recent years). The idea is to come up with a proper scalar field on the network that assigns a scalar value to every node in the network where the destinations have the maximum scalar value. The routing strategy is to route packets to those nodes that have a higher scalar value that the sender/forwarder. These paths are often called the steepest-gradient paths. Here, in this paper, every node has a temperature value and packets are routed along those nodes that have increasing temperature value. The scheme routes data packets towards an Internet gateway that are modeled as heat sources. This scheme does not require flooding of control messages, thereby improving scalability of this protocol. The node that needs to send data packets would only consider the temperature of its one-hop neighbor. This scheme displays a constant overhead when the network size is increased (and by keeping the average node degree to be constant). OLSR performs slightly poorer. When it comes to assessing the scalability with node density (the average node degree was not kept constant, the coverage area was kept constant), the performance of both OLSR and HEAT remained mostly the same, that is, both their performance slightly degraded as both the protocols (when node degree increases) had to increase the HEAT beacons or Hello messages, which resulted in more interference.

13.4.1.4 Effect of Radios

To improve scalability, robustness and performance by reducing interference, a lot of recent research is in the area of designing nodes with multiple radios and efficiently using those radios for scanning the spectrum and utilize the available spectrum. Though it is known that nodes that utilize multiple radios to communicate on different channels would not experience as much interference as single channel nodes, it should be noted that the routing protocols must have knowledge about these extra parameters in looking for the most appropriate route considering the fact that different channels would have different transmission rates and ranges [16]. This would affect the way in which shortest path algorithm (or any routing algorithm) would be implemented as more and more constraints and factors come into picture.

Dual-radio mesh nodes are being studied which have two radios operating on different frequencies. One radio is used for client access and the other to support the wireless backhaul. Because they operate in different frequency, they can function simultaneously without interference issues. However, in this scenario, because the wireless backhaul is also a shared medium network, it is subjected to the same problem as a single-radio wireless network that would have contention issues and will limit the capacity, introduce latency and would not be good for voice traffic.

Mutliradio WMNs (MR-WMN) are actively studied currently to mitigate the effects of interference, but they continue to face several challenges such as dynamic management of spectrum resources, efficient management of multiple radio interfaces and adjacent radio interference. The multiradio approach eliminates the wireless backhaul bottleneck by providing dedicated wireless links for both ingress and egress traffic. Radios, which are adjacent to each other in close proximity, cause interference to other interfaces on the same node. Careful mechanical design of the nodes and enough separation of antennas or network interface cards are required.

13.5 Thoughts for Practitioners

Though the mentioned issues related to scalability in WMNs have been addressed or actively pursued in many research papers and in several industries in the recent past, none of them have been tested in real deployment scenario with large number of nodes. For example, the MIT Roofnet project [9, 42] is run with around 40–50 nodes. The OLSR setup in Amsterdam consists of about 75 nodes, Dharmasala [15] with greater than 30 nodes, the Berlin OLSR [5] project has about 180 nodes, and CuWin [11] project is working with around 50 nodes and more. There are many commercial deployments from BelAir Networks [7], Tropos Networks [51], Meraki [39], Mesh Dynamics [40] and other companies. But, we do not have the exact deployment topology and information regarding their success (or issues they have faced).

It is evident that a new breed of WMN is needed to solve this multihop dilemma. The new network needs to employ a modular, radio-agnostic, multiradio, multichannel mesh network system that exhibits superior system backhaul. This will allow for the deployment of high-performance, highly scalable networks that support real-time voice, video and data applications.

The density of users translates into issues in determining the frequency spectrum allocation. Most WMNs today operate in the unlicensed industrial, scientific and medical (ISM) and this fact would not be acceptable for certain mission-critical applications such as disaster recovery, military, emergency response communication, etc. because the ISM band shares the spectrum with other WLANs and a range of other wireless devices, which could potentially interfere and aggravate the scalability problem. Because the skeptics are concerned with the ISM band use for mission-critical communications, the US FCC has made the licensed 4.9 GHz band available for public safety and Homeland Security applications [14]. Tropos networks [51], Proxim Wireless, Firetide [19], MTI wireless and other vendors have provided support in terms of hardware and software to realize this effort by FCC (Fig. 13.6).

Relatively speaking, the WMN concept is still in its infancy though it may be a sound strategic choice for wireless broadband services for various community and tactical deployments. Such scenarios would be the perfect battleground for both the technical challenges WMNs pose as well as the public policy debates over the role of governments that compete with local DSL and cable providers. Currently, several deployments have been made commercially by Tropos Networks, BelAir Networks, Motorola (acquisition of MeshDynamics) [40, 43], FireTide [19], Nortel and other companies. But, these deployments are not large-scale and use proprietary protocols for their communication. WMNs are in a stage today where potential service providers would have to decide (based on their short-term and long-term goals) to

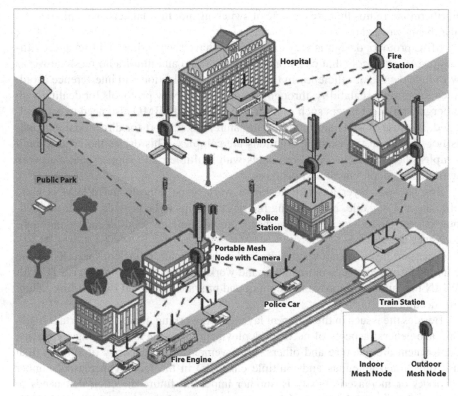

Fig. 13.6 A typical municipality scenario displaying a public safety operation with the aid of data and video communication by the wireless mesh nodes. Image source [19]

either go with currently available solutions in the market or wait it out until the technology matures and standards emerge.

13.6 Directions for Future Research

Though WMNs hold a lot promise for the future community wireless access applications, there are several challenges that needs to be addressed. There are many open issues in all the layers of networking. Ad hoc communication is constantly evolving and it is not practical to provide a comprehensive coverage of all the developments in this subject. We have not covered all the possible factors that impact scalability, but, many of prominent work in this area are presented with an emphasis on open issues and future directions. In the following paragraphs, we briefly describe the challenges that lay ahead in different aspects of WMNs.

In the physical layer, development of new radio technologies that can provide high data rates under very high interference level is very much wished for.

Furthermore, radios that are capable of switching among channels at a rapid rate are also being worked on.

MAC protocol design is very crucial as we have seen earlier in this chapter. Efficient MAC protocols that can work for single-radio and multiradio mesh networks with high degree of efficiency in terms of channel selection and interference avoidance to improve scalability, throughput and latency. New protocols for dealing with emerging physical layers such as ultra-wide band and MIMO also need to be developed. Because WMNs are multihop in nature, the support for directional antennas is not trivial and the existing solutions are complex. This drives the need to built simpler, more efficient MAC protocols with multiple radio support that can work with flexible deployment architectures.

The network layer has quite a bit of work to be done to make WMNs more efficient and scalable. New routing metrics that aims to be provide higher throughput, schemes to reduce routing overhead, provide efficient load-balancing and fault-tolerance are needed.

Transport layer has not been addressed in this chapter for lack of space. There are existing unsolved issues in this layer and work is being done to improvise TCP for WMN that aims to improve throughput and provide quick and explicit link failure notifications.

Besides the issues in the different layers of networking, there are important issues in some general aspects of network deployment. They include coverage, capacity, deployment architecture and others. Improvement in the capacity of WMNs (that use directional antennas and multiple channels) in the face of increased number of nodes or increased density is another important future direction that needs to be considered. The initial work on improving capacity and scalability in WMNs that support multichannels have been studied well in the recent past. Theoretical modeling and analysis of MAC protocols that support multiple radios is very much needed for better understanding of the performance, limits and for providing useful insights for designing better protocols [54].

When the coverage area is increased by adding more mesh nodes in the network, the average path length to the destination increases (if the users do not confine themselves to local communication). As the path length increases, throughput decreases with every hop in the network. This behavior severely limits the scalability of WMNs in terms of network size and diameter. To address this issue, enhancements and refinements to MAC protocols and routing protocols have been studied extensively and have promising simulation results. However, most of them have not been put to real-life deployment tests with a very large number of nodes.

13.7 Conclusions

The advent of broadband wireless networking and Wi-Fi is now poised to be the network of choice within both residence and enterprise market, and this advancement in technology has made it easier and less expensive to provide "triple play" data, voice and video communications throughout an entire metropolitan or rural area. This is evident from the growing number of community wireless mesh projects around the world. The ubiquity and low cost of Wi-Fi, along with the robust market opportunities continuing to drive its technological evolution, make Wi-Fi mesh networking the vehicle of choice for many mesh applications.

As WMNs become more popular and their size and complexity continues to grow, mesh networks that contain multiple hops are increasingly vulnerable to problems such as throughput degradation, cochannel interference and end-to-end network latency. Thorough investigation and analysis has proved that single-radio systems are incapable of supporting WMNs because the capacity decreases almost by 50% at every hop. The selection of single, dual or even some multiradio mesh networks will result in a low performance network that will not scale and which cannot support a wide range of applications.

This chapter presented a survey of scalability for WMNs and the factors that are involved in affecting the scalability in WMNs. There are active research efforts both in academia and in the industry to deploy WMNs that efficiently scale as the network grows both in terms of coverage area and number of users. From the earlier sections and past studies, it is well understood that there are many challenges that have to be addressed to enable the rapidly growing interest in WMNs to achieve its full potential.

For successful deployment and revenue generation, the service providers must ensure an expandable overall system capacity in terms of number of users supported, density of users, geographical coverage area, the data rate, etc. In short, scalability of WMNs is a very important factor in deciding its acceptance and deployment for any particular application. Also, for multimedia services, the deployed mesh must be able to support any new services that might be added in the future. A scalable network provides an economical means of expanding an existing network to accommodate future service upgrades with minimal interruption to service availability.

The economies of scale would also play a part when deploying large-scale WMNs by service providers. Despite the fact that they are easy to install and configure, WMNs are not in a stage where the system would scale to meet growth needs at a reasonable cost. Although a few initial pilot installations would be necessary and be considered as a fundamental step for any large-scale rollout, scalability is the most critical attribute that the service providers must ensure. Rigorous testing with real-life deployment (before they go live) with plenty of test-cases in all aspects of networking including different architectures for different applications, spectrum utilization, multiradio efficiency, routing protocols, load-balancing, throughput and other performance metrics must be at acceptable standards.

When these critical issues are addressed successfully and with the help of better technologies, (from [3]), "Wi-Fi mesh networks would really have the potential to bridge the digital divide, foster economic development, increase public safety organization's efficiency and effectiveness, and connect communities. This emerging technology is making it possible, for the first time, to create truly unwired cities."

13.8 Terminologies

1. *Throughput*. Throughput refers to the amount of data transferred from one node to another in a specific amount of time. Throughput is usually the term that is used to express performance in communication networks.
2. *Latency*. The delay involved in transmission of a unit of data from the source to the destination. Usually, latency degrades as the number of hops to the destination increases.
3. *Scalability*. The ability of a network to maintain its performance (throughput, latency and other QoS parameters) when the network grows both in coverage area (addition of new nodes) and number of users.
4. *MAC*. Medium access layer. It is the second layer in the traditional network layer. MAC protocols ensure that it acquires the wireless channel to successfully send data packets across the medium.
5. *Cochannel interference*. Homogenous nodes operating in the same frequency band and that are sufficiently close by tend to suffer from cochannel interference. It is the interference caused by another signal operating in the same channel.
6. *Hidden terminal problem*. The collision at a receiver node caused because of a transmission from a third node that is out of the range of another (invisible) sender node.
7. *Exposed terminal problem*. This problem occurs when a node is prevented (because of its vicinity to a neighboring transmitting node) from communicating with another node.
8. *Deafness*. When nodes equipped with directional antennas, deafness occurs when a node A initiates communication to a node B and B doesn't respond to A because B is beamformed towards some other node C.
9. *Overhead*. It is control messages that are exchanged among the nodes in a network for various purposes like topology change, maintaining connectivity, exchange of routing table information and more.
10. *Backhaul network*. In multiradio mesh networks, there is a set of nodes that interconnects all the nodes in the respective cluster and handles traffic among backhaul nodes. Such a network (usually operates in a different frequency) that carries data from the clients is called backhaul network.
11. *SINR*. Signal to interference plus noise ratio. Typically, in wireless networks, SINR is used to decide the power level required for transmission. The interference here is the sum of the interferences from all the nodes and the Noise is the ambient noise in the system. Note that there are many SINR models available in the literature.
12. *MPR*. Multipoint relays. MPR is a set of selected nodes that forward broadcast messages during the flooding process. OLSR specification (RFC3626) introduced this concept.

13.9 Questions

1. Why is WMN seen as a promising technology for future wireless broadband access?
2. Why is scalability seen as a very important factor in deciding WMN's successful deployment?
3. What are the various factors that affect scalability in a typical wireless network?
4. How is the capacity of a network affected at each hop? What is the per node throughput in a multihop network? What are the various mechanisms that aim to improve this throughput?
5. What are the advantages of multiradio multichannel WMNs? How is it better than single-radio systems? What are the challenges involved in using multiradio multichannel WMNs?
6. How does appropriate choice of routing metrics help in efficient and scalable routing? Give some examples.
7. What are the advantages of directional antennas over onmi-directional antennas? What are the challenges in using directional antennas?
8. How is latency and scalability affected by using a poor MAC protocol? What are the reasons that traditional ad hoc wireless MAC protocols are not suitable for WMNs?
9. How does routing overhead (control message) affect a network's performance? What are the current efforts in reducing the number of control messages in the face of growing number of users and number of nodes?
10. How does the deployment architecture affect the performance of WMNs?
11. Name some important design decisions when considering a complete solution for WMNs. Choose any common application scenario for your solution.

References

1. IF Akyildiz and X Wang (2005) A survey on wireless mesh networks. IEEE Radio Communications.
2. O Arpacioglu and ZJ Haas (2004) On the scalability and capacity of wireless networks with omnidirectional antennas. Proceedings of the IPSN'04, Berkeley, California, USA.
3. M Audeh (2004) Metropolitan-Scale Wi-Fi Mesh Networks, Computer, vol. 37, no. 12, pp. 119–121.
4. EM Belding-Royer (2003) Multi-level hierarchies for scalable ad hoc routing. ACM/Kluwer Wireless Networks (WINET), vol. 9, no. 5, pp. 461–478.
5. Berlin OLSR Wireless Mesh Network. http://start.freifunk.net/Accessed 16th Nov 2007.
6. J Bicket, D Aguayo, S Biswas and R Morris (2005) Architecture and evaluation of an unplanned 802.11b mesh network. Proceedings of MobiCom'05, Cologne, Germany.
7. Capacity of Wireless Mesh Network – Understanding single radio, dual radio and multi-radio wireless mesh networks. White paper by BelAir networks. http://www.belairnetworks.com/resources/pdfs/Mesh_Capacity_BDMC00040-C02.pdf Accessed 7th Dec 2007.
8. J Chen, Y-Z Lee, D Maniezzo and M Gerla (2006) Performance comparison of AODV and OFLSR in wireless mesh networks. IFIP Fifth Annual Mediterranean Ad Hoc Networking Workshop (Med-Hoc-Net '06).

9. RR Choudhury (2005) Utilizing directional antennas for wireless multi-hop networks. Presentation at UCSD. http://www.crhc.uiuc.edu/~croy/presentation.html. Accessed 23rd Dec 2007.

10. T Clausen and P Jacquet (2003) Optimized Link State Routing Protocol (OLSR). RFC 3626.

11. Community Wireless Solutions. http://www.cuwireless.net/Accessed 14th Nov 2007.

12. DSJD Couto, D Aguayo, J Bicket and R Morris (2003) A high-throughput path metric for multi-hop wireless routing. Proceedings of MobiCom'03.

13. F daCosta (2004) Performance analysis of three competing mesh architectures. http://www.meshdynamics.com/performance-analysis.html Accessed 23rd Dec 2007.

14. FCC Designates 4.9 GHz Band for Public Safety and Homeland Security. http://www.fcc.gov/Bureaus/Wireless/News_Releases/2002/nrwl0202.html. Accessed 12th Dec 2007.

15. Dharmasala Community Wireless Mesh Network. http://drupal.airjaldi.com/node/56 Accessed 25th Nov 2007.

16. O Dousee, F Baccelli and P Thiran (2003) Impact of interferences on connectivity in ad hoc Networks. Proceedings of IEEE INFOCOM, San Francisco, CA.

17. R Draves, J Padhye and B Zill (2004) Routing in multi-radio, multi-hop wireless mesh networks. Proceedings of MobiCom'04.

18. J Eriksson, M Faloutsos and SV Krishnamurthy (2007) DART: dynamic address RouTing for scalable ad hoc and mesh networks. IEEE/ACM Transactions on Networking, vol. 15, no. 1.

19. Firetide Networks. http://www.firetide.com/Accessed 10th Nov 2007.

20. B Fong, N Ansari, ACM Fong, GY Hong and PB Rapajic (2004) On the scalability of fixed braodband wireless access network deployment. Proceedings of IEEE Radio Communications vol. 42, no. 9, pp. S12–S18.

21. Y Ge, L Lamont and L Villasenor (2005) Hierarchical OLSR – A scalable proactive routing protocol for heterogeneous ad hoc networks. Proceedings of the IEEE International Conference on Wireless and Mobile Computing, Networking and Communications.

22. P Gupta and PR Kumar (2000) The capacity of wireless networks. IEEE Transactions on Information Theory, vol. 46, no. 2, pp. 388–404.

23. P Gupta, R Gray and P Kumar (2001) An experimental scaling law for ad hoc networks. University of Illinois at Urbana-Champaign.

24. G Held (2005) Wireless mesh networks. Auerbach Publications, Taylor & Francis Group, Boca Raton, FL.

25. B Henderson and F DaCosta (2007). Third-generation structured mesh for military environments. Invited presentation made to the Army Sciences Board. http://www.meshdynamics.com/documents/MDThirdGenerationMesh.pdf Accessed 26th Nov 2007.

26. LF Huang and T-H Lai (2002) On the scalability of IEEE 802.11 ad hoc networks. Proceedings of the Third ACM International Symposium on Mobile Ad Hoc Networking and Computing (MobiHoc'02), Lausanne, Switzerland.

27. J-H Huang, L-C Wang and C-J Chang (2005) Coverage and capacity of a wireless mesh network. International Conference on Wireless Networks, Communications and Mobile Computing, pp. 458–463.

28. PR Jelenković, P Momčilović and MS Squillante (2007) Scalability of wireless networks. Proceedings of IEEE/ACM Transactions on Networking, vol. 15, no. 2, pp. 295–308.

29. B Jenkins (2006) Synchronous mesh offers scalability. Network World article (05/22/06) http://www.networkworld.com/news/tech/2006/052206-synchronous-mesh.html Accessed 12th Dec 2007.

30. W Jiang, S Liu, Y Zhu and Z Zhang (2007) Optmizied routing metrics for large-scale multi-radio mesh networks. Proceedings of the IEEE International Conference on Wireless Communications, Networking and Mobile Computing (WiCom'07).

31. W Jiang, Z Zhang and X Zhong (2007) High throughput routing in large-scale multi-radio wireless mesh networks. Proceedings of the ECNC 2007.

32. A Jovicčić, P Vishwanath and S Kulkarni (2004) Upper bounds to transport capacity of wireless networks. Proceedings of IEEE Transactions of Information Theory, vol. 50, no. 11, pp. 2555–2565.

33. J Jun and M Sichitiu (2003) The nominal capacity of wireless mesh networks. IEEE Wireless Communications Magazine, vol. 10, no. 5, pp. 8–14.

34. UC Kozat and L Tassiulas (2003) Throughput Capacity of random ad hoc networks with infrastructure support. Proceedings of the 9th Annual International Conference on Mobile Computing and Networking (MobiCom'03), pp. 55–65.
35. S Kulkarni and P Vishwanath (2004) A deterministic approach to throughput scaling in wireless networks. Proceedings of IEEE Transactions of Information Theory, vol. 50, no. 5, pp. 748–767.
36. RK Lam, D-M Chiu and JCS Lui (2007) On the access pricing and network scaling issues of wireless mesh networks. IEEE Transactions on Computers, vol. 56, no. 11, pp. 1456–1469.
37. O Leveque and E Telatar (2005) Information theoretic upper bounds on the capacity of ad hoc networks. Proceedings of IEEE Transactions of Information Theory, vol. 51, no. 3, pp. 858–865.
38. J Li, C Blake, DSJ De Couto, et al. (2001) Capacity of ad hoc wireless networks. Proceedings of 7th ACM International Conference on Mobile Computing and Networking, July 2001.
39. Meraki Networks. http://meraki.com/
40. Mesh Dynamics website, http://www.meshdynamics.com.
41. Mesh Networks – Motorola technology position paper. http://www.motorola.com/mot/doc/6/6007_MotDoc.pdf Accessed 12th Nov 2007.
42. MIT Roofnet Homepage. http://pdos.csail.mit.edu/roofnet/doku.php?id=roofnet Accessed 20th Dec 2007.
43. Motorola (acquisition of Mesh Dynamics). http://www.motorola.com/mesh/ Accessed 18th Nov 2007.
44. D Niculescu, S Ganguli, K Kim and R Izmailov (2006) Performance of VoIP in a 802.11 wireless mesh network. Proceedings of IEEE INFOCOM'06.
45. G Pei, M Gerla and TW Chen (2000) Fisheye state routing in mobile ad hoc networks. Proceedings of IEEE ICDCS Workshop.
46. C Peraki and SD Servetto (2003) On the maximum stable throughput problem in random networks with directional antennas. Proceedings of ACM MobiHoc, Annapolis, MD.
47. B Rainer, H Simon, L Vincent and M Martin (2007) HEAT: Scalable routing in wireless mesh networks using temperature fields. Proceedings of IEEE WoWMoM'07, pp. 1–9.
48. TS Rappaport (1996) Wireless Communications, Principles and Practice. Prentice Hal, NJl.
49. CA Santiváñez, R Ramanathan and I Stavrakakis (2001) Making link-state routing scale for ad hoc networks. Proceedings of the 2001 ACM International Symposium on Mobile Ad Hoc Networking & Computing (MobiHoc'01), Long Beach, CA.
50. CA Santiváñez, B McDonald, I Stavrakakis and R Ramanathan (2002) On the scalability of ad hoc routing protocols. Proceedings of IEEE INFOCOM'02, New York.
51. Tropos Networks. http://www.troposnetworks.com Accessed 16th Nov 2007.
52. Y Yang, JWang and R Kravets (2005) Designing routing metrics for mesh networks. Proceedings of WiMesh'05.
53. S Yi, Y Pei and S Kalyanaraman (2003) On the capacity improvement of ad hoc wireless networks using directional antennas. Proceedings of ACM MobiHoc, Annapolis, MD.
54. Y Zhang, J Luo, H Hu (2007) Wireless Mesh Networking – Architectures, Protocols and Standards. Auerbach Publications, Taylor & Francis Group, Boca Raton, FL.
55. S Zhao and D Raychaudhuri (2006) On the scalability of hierarchical hybrid wireless networks. Proceedings of the Conference on Information Sciences and Systems (CISS'06), Princeton, NJ.
56. S Zhao, I Seskar and D Raychaudhuri (2004) Performance and scalability of self-organizing hierarchical ad hoc wireless networks. Proceedings of the IEEE Wireless Communications and Networking Conference (WCNC 2004).

Chapter 14
Mobility Management in Wireless Mesh Networks

Vinod Mirchandani and Ante Prodan

Abstract A viable support of an on-going or a new session for a subscriber on the move requires an effective scheme for *Mobility Management*. To this end, an array of protocols such as MIPv4, MIPv6, HMIPv6, FMIPv6 have been proposed for the wired Internet. Unfortunately, the wireless connectivity in the wireless mesh networks (WMNs) gives rise to several issues that limits the direct applicability of these mobility management protocols for the wired network. We have contributed to this chapter by identifying and explaining these issues and then giving a critical review of some of the key research proposals made in this area. The literature review also shows that the proposals offer a limited support for mobility management in multiradio wireless mesh networks (MR-WMN). Thus, we have further contributed, by proposing a scheme to carry out a seamless mobility management in WMN as well as MR-WMN. We have taken into account the lessons learnt from the proposals made in the literature. This chapter has been written in a simple way such that students as well as professionals including those who are new to this area should be able to significantly benefit from reading it.

14.1 Chapter Overview

Mobility management involves managing two forms of mobility. (1) Terminal mobility – where the mobile terminal (MT) moves within and across network domains while continuing to receive access to telecommunication services without any data packet loss and with a minimum handover delay and (2) Personal mobility – where the subscriber obtains services in a transparent manner within any network

V. Mirchandani (✉)
Faculty of Information Technology, University of Technology, Sydney, PO Box 123, Broadway, NSW 2007, Australia
e-mail: vinodm@it.uts.edu.au

S. Misra et al. (eds.), *Guide to Wireless Mesh Networks*, Computer Communications and Networks, DOI 10.1007/978-1-84800-909-7_14,

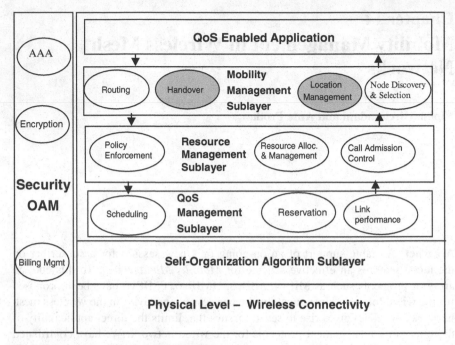

Fig. 14.1 Layered architecture for MR-WMN

and on any terminal, on the basis of subscriber identification and networks ability to provide the concerned services. Terminal mobility management in the multiradio wireless mesh networks (MR-WMN) is the area of focus in this chapter.

Figure 14.1 illustrates a simple view of MR-WMN layered architecture that is composed of key entities. The shaded ellipses indicate the areas, which have been covered in this chapter.

Internet engineering task force (IETF) has proposed protocols for mobility management such as: mobile IPv4 (MIPv4), mobile IPv6 (MIPv6), fast mobile IP v6 (FMIPv6) and fast hierarchical mobile IPv6 (FHMIPv6). However, these protocols are more suitable for networks with a wired infrastructure, i.e. the connectivity between the nodes of the network is wired and wireless connectivity is provided to the mobile subscriber only at the edges. In contrast, the infrastructure in a wireless mesh network (WMN) has wireless connectivity among the mesh nodes of the network as well as between the access points and the subscriber. The wireless nature of the connectivity in MR-WMN gives rise to various issues, discussed in this chapter, which limits the direct use of the above stated wired infrastructure networks based mobility management protocols. Although, there is no definite solution for the mobility management problem in WMNs environment, well-established solutions for the wired networks may be used as guidelines. Further, the work published in literature to date for seamless mobility management in WMN is not explicitly based on mesh nodes that offer a *multiradio connectivity*.

14.2 Introduction

The objective of mobility management is to offer a *seamless support* of real-time as well as nonreal-time services for a subscriber who is on the move. Seamless support refers to obtaining a low handover latency and packet loss. Examples of real time service that mobility management should support are interactive voice/video and streaming audio/video whereas nonreal-time service include email, file transfer, and web-browsing. It will be tightly coupled with quality of service (QoS) so as to satisfactorily support real-time services under dynamic network conditions.

The support of mobility management process entails the use of suitable solutions for (1) *handover management*, (2) *location management*, and (3) *route optimization* [1]. We first broadly define these terms:

(1) The term *handover* (RFC3753 [2]) refers to the process by which the mobile terminal changes its point of attachment to the network. *Handoff* and *handover* terms are used interchangeably to refer to the same process. Handover management thus deals with maintenance of an ongoing communication session with a roaming subscriber on the move.
(2) Location management refers to the process of finding out the connectivity location of the roaming subscriber's mobile terminal within the geographical region; security and authentication information and QoS capabilities.
(3) Simply put, route optimization is the process by which a route is created efficiently between the calling person's mobile terminal and the called person's mobile terminal.

Handover management can be further divided into layer 3 (IP layer) handover and L2 (link layer) handover.

- L2 handover – It occurs when the mobile terminal (MT) moves out of the satisfactory transmission/reception range of an access point (AP)/base station, which triggers an implementation specific mechanism to reassociate with a new AP/base station. The mechanism for instance could be based for example on signal to noise plus interference ratio (SINR) or received signal strength (RSS) or a lack of substantial number of ACKs. L2 handover is illustrated in Fig. 14.2 where the MT moves away from the satisfactory range of base station (BS) 'A' when it associates with BS 'B' and de-associates with BS 'A'.

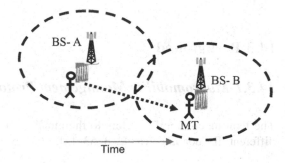

Fig. 14.2 Handover concept –
An overview

- L3 handover – When the L2 handover occurs the point of attachment of the client will change. This means that the path of the packets from the correspondent node (CN) needs to be switched to the new AP/base station. The peer node with which the mobile node is communicating is termed as the correspondent node (CN) that may be mobile or stationary. As the process of L3 handover is concerned with a routing update, which involves layer 3 of the OSI protocol stack, therefore it is termed as a L3 handover.

Although the aim of L3 handover is to be technology independent, developing L3 handover independently, i.e. without considering L2 will result in severe performance degradation and considerable increase of handover latency [3]. To reduce the handover latency, a well-defined co-ordination between L2 and L3 is required. Ideally, L3 handover should occur concurrently with L2 handover, resulting in handover latency being equal to either L2 or L3 handover time.

There are several issues that need to be addressed by mobility management some of which are:

- How to cater for both the fast and slow moving subscribers?
- How to address intradomain and interdomain handovers?
- How to address handovers of roaming subscribers between networks of the same type and those that are heterogeneous?
- How to reduce the signalling load when the mobile terminal experiences frequent handover across smaller cells?
- How to carry out an effective network discovery and selection with the constraint of limited battery power in the mobile terminal?

The background information on the mobility management protocols necessary to understand this chapter is first provided in Sect. 14.3. Then in Sect. 14.4, we explain the issues because of which the mobility management protocols discussed in Sect. 14.3 for the wired Internet can not be used directly in the wireless mesh domain. Following which, we concisely review some of the key mobility management proposals in the literature for WMNs and then discuss our proposed mobility management scheme. The key differentiators between our mobility management scheme and the reviewed proposals are also stated in the section. The thoughts for practitioners and directions for future research are provided in Sects. 14.5 and 14.6, respectively. Conclusions that can be made from this chapter are given at the end in Sect. 14.7.

14.3 Background

14.3.1 Macromobility Management Protocols

The term macromobility refers to the mobility of a mobile terminal (MT) among different IP domains (RFC 3753 [2]), for example mobility across different sub-networks – such as between the home network and the visited network. Popular macromobility protocols are MIPv4, MIPv6, FMIPv6, and HMIPv6.

14.3.1.1 MIPv4

Mobile IP [4,5] is a mobility management solution proposed to resolve the macro-mobility problem in the wired Internet, where the wireless connectivity is provided to the subscribers at the network edges. With mobile IP, roaming subscribers enjoy Internet connectivity in a transparent manner without any manual configuration. The main motivation behind the creation of MIPv4 was to [6]:

- Enable a user to change their point of attachment on the Internet, i.e. in effect change their IP address.
- Maintain the existing TCP connection for the ongoing session.

However, the above two points oppose each other because the TCP connection can only be maintained if the IP addresses of the connection end-points remain the same. TCP connection information involves a pair of IP addresses of the two end-points involved in the session and the port numbers. Whereas, a change in the point of attachment on the Internet necessitates a change in the IP address.

Mobile IP addresses the above explained conflict by allocating two IP addresses to the MT – a home address and the other is a temporary care of address that represents the current location of the MT. An association is created between these two addresses, which is called as *binding*.

The mobile terminal is identified by its home address (HoA), irrespective of where it may be attached to the Internet. The HoA is an IP address and is assigned to the MT permanently based on the network where the subscriber is a resident. The network of which the MT (subscriber) is a resident is called as the *home network*. Any network other than the home network to which the MT may be connected while roaming is called as the *visited network*.

The operation of MIPv4 is shown in Fig. 14.3 as well as quickly explained by listing the key steps involved as explained in RFC 3344 [7]:

- The presence of the foreign and home agent (HA), which are essentially routers, is made known to the MT by means of agent advertisement messages. Alternately, the MT may solicit an agent advertisement message through an agent solicitation message. By using the agent advertisement messages the MT can determine if it is in a home or a visited network.
- If the MT finds that it has moved into a visited network it will obtain a Care-of-address. This can be obtained from the foreign agents advertisements or can be provided via a DHCP mechanism.
- While in a visited network a roaming MT is required to register with its home agent the current care-of-address (CoA) to access the Internet. Optionally, the registration may occur via the foreign agent.
- Home agent redirects the packets received for the MT while it is away from the home network, to the current CoA of the MT. The home agent intercepts

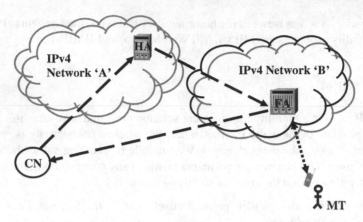

Fig. 14.3 An overview of MIPv4

the packets for the MT and then tunnels them to the care-of-address of the MT. Tunnelling is essentially the process of encapsulating one IP packets as a payload of another IP packet, i.e. IP-within-IP [6]. The tunnelled packets end point could either be the foreign agent (FA) or the MT itself, and is finally delivered to the MT.

- In the reverse direction, the MT can send the packets to the CN directly or via the foreign agent using the conventional IP routing mechanisms.

Thus in the MIPv4 the home agent routes the packets received from the CN for the MT outside of its home network. However, packets from the MT to the CN are routed directly. This results in triangle routing as shown in Fig. 14.3, which is a major drawback of MIPv4.

14.3.1.2 MIPv6

MIPv6 addresses the above explained issue of triangular routing by means of route optimization [8] to the CN as shown in Fig. 14.4. (*Note*: Route Optimization was defined in the Introduction).

Packets addressed to the mobile host are delivered using regular IP routing to the CoA thereby offering a transparent, simple, and scalable global mobility scheme. Even though network support for seamless mobility was not considered when Mobile IP was originally developed, it finds applicability in the wireless environment through the endeavours of Mobile IP Working Group.

A drawback of mobile IP is that in a wireless environment a MT frequently changes its point of attachment, i.e. performs handovers to initiate or continue communication sessions with other nodes in the network. Because a local CoA must be obtained and communicated to the HA and the CN after every migration, the significant latency introduced by the mobile IP causes considerable packet loss during the handover period (especially if the home and foreign network are far apart), rendering real-time data transfer useless until the CN is notified of the new CoA.

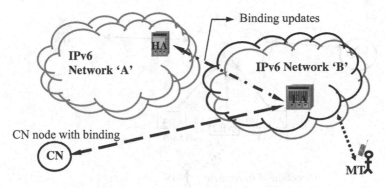

Fig. 14.4 A conceptual view of MIPv6 with *route optimization*

14.3.1.3 MIPv4 and MIPv6 Compared

As MIPv6 was developed after MIPv4 therefore it makes use of the lessons that were learnt from MIPv4 as well as it shares many features with MIPv4. In particular, MIPv6 makes use of the mobility features that have been integrated within IPv6. A short comparison between MIPv6 and MIPv4 is listed below based on RFC3775 [9]:

- MIPv6 does not need the provision of FAs as in the case of MIPv4.
- Route optimisation is key attribute of the MIPv6.
- Route optimization in MIPv6 can operate securely.
- Routing overhead in MIPv6 is reduced relative to the MIPv4. This is because the packets sent to the MT while away from its home network are sent in the IPv6 routing header rather than IP encapsulation.
- MIPv6 is more robust than IPv4. For example, its operation is not based on any specific link layer.

14.3.1.4 Motivation for HMIPv6 and FHMIPv6

For MIPv6 (RFC 3775 [9]) it has been found that if the MT is some distance away from the home network, then it might take up to 100 ms to send the binding update (BU) after handover. BU essentially binds or maps the assigned IP addresses of the MT.

This will result in many packets addressed to the MT being dropped during that period [10]. With smaller cell sizes for high-data rate access, the handover rate will also increase considerably and the nodes with fast mobility will contribute to the signalling overhead, causing an inefficient spectrum use. Moreover an increase in handover decision criteria with divergent user preferences will create bottlenecks within the networks. Hierarchical mobile IPv6 (HMIPv6-RFC 4140) [11] and Fast mobile IPv6 (FMIPv6-RFC 4068) [12] have been developed as extensions of the Mobile IPv6 by the IETF to reduce these conditions by localizing the signalling traffic within the proximity of the MT.

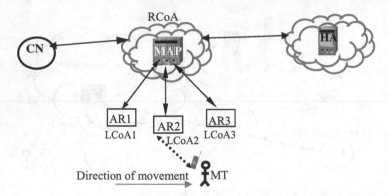

Fig. 14.5 HMIPv6 Overview – adapted from [11]

14.3.1.5 HMIPv6

Hierarchical Mobile IPv6 (HMIPv6 [11]) incorporates a mobility anchor point (MAP), which is a router in the network that a MT is visiting. MAP essentially plays the role of a local home agent for the MT. This protocol generally has a multitier system but it can also have a distributed system (RFC 4140 [11]). HMIPv6 facilitates to reduce the signalling overhead and delay associated with the location updates. This it does by enabling the MT to send a binding update (BU) to the local MAP, rather than to the distant HA and the CN. Because the MT uses a regional care of address (RCoA) [13] as its global care of address (CoA) for the domain and updates the on-link CoA (LCoA) with the MAP after every handover, all local movement within the domain is hidden from the HA and the CN. The RCoA is an address on the MAPs subnet whereas the LCoA is an on-link CoA configured based on the MTs interface. This is more clear from Fig. 14.5 in which the ARs are the access routers.

The BU essentially binds or maps the RCoA and LCoA of the MT in the MAP. The HA and CN use the RCoA of the MT to direct the packets to the MAP, which has the binding information for the MT. The MAP then encapsulates the received packets for the MT and uses the LCoA of the MT as the destination address. The reverse process occurs when packets are sent from the MT to the CN. Thus by localizing the signalling in the form of binding updates to the local MAP handover latency is considerably reduced.

14.3.1.6 FMIPv6

Fast handover for mobile IPv6 (FMIPv6-RFC 4068 [12]) uses bidirectional tunnels between new access router (nAR) and previous AR (pAR) to transfer the traffic to and from the MT while the actions of (1) L3 handover, i.e. the BU to MAP (2) Binding acknowledgement, and (3) route update (L3 handover) are taking place.

Fig. 14.6 Bidirectional tunneling in FMIPv6. Adapted from [12]

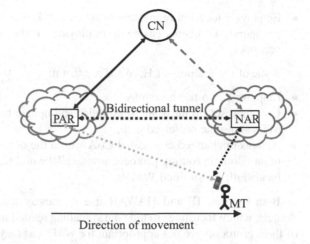

This can be used in conjunction with HMIPv6 process or it could be used independently. FMIPv6 thus allows a fast handover without the usual signalling overhead and latency resulting from a typically far away HA.

Even though these macromobility solutions discussed in literature reduce the handover latency considerably, they fail to address such issues as passive connectivity and paging. This is particularly evident when the subscriber, while registered in the domain, roams greater distances without initiating a communication session. It then becomes imperative to know the approximate location of the idle subscriber and devise a scheme to efficiently search and find (known as paging) these users in a scalable and timely manner when data needs to be forwarded to them. Such passive connectivity reduces the load over the radio interface and the core network and allows preservation of battery power in the MT (Fig. 14.6).

14.3.2 Micromobility Management Protocols

Micromobility refers to the mobility of the MT within an IP domain (RFC 3753 [2]), for example across different access points/base stations within the same subnetwork, which could be the home network or the visited network. Some of the commonly used micromobility protocols are handoff aware wireless access Internet infrastructure (HAWAII) and Cellular IP.

Cellular IP [14, 15] and (HAWAII) [16, 17] are based on the IP design principles and have been proposed as possible micromobility solutions optimised to provide access to a Mobile IP enabled Internet, addressing both passive connectivity and paging. Some of the features of Cellular IP are that it:

- Employs a hop-by-hop routing mechanism.
- Has soft-state routing cache entries for recently active MTs.
- Can operate at layer two or layer three.

- Employs a location management scheme of Ethernet switches and has minimal configuration, thereby easing the deployment and management of wireless access networks.

Some of the features of HAWAII are that it:

- Employs a two-tier hierarchy.
- Uses path set-up message to establish and update host-based routing entries for MT's along the preferred path.
- Assigns unchanged collocated CoA within the domain and assumes some form of intradomain routing protocol among all the nodes. This makes it inefficient for bandwidth constrained WMNs.

Both Cellular IP and HAWAII use a gateway foreign agent (GFA) for each domain, which facilitates to hide the signalling related mobility messages [18]. Both of these protocols are not appropriate for WMNs as they involve the mobile hosts in the mesh backbone routing and use host-specific routing protocols. Consequently, it makes the deployment of the mobility management difficult in the WMN.

14.4 Wireless Mesh Networks Based Mobility Management

In Sect. 14.3, we had given a concise review of mobility management schemes that are primarily applicable to the wired Internet with wireless connectivity at the edges. We have conducted an extensive research for mobility management schemes in WMNs and given a review in this section of some of the key works that we have identified. Figure 14.7 shows a 802.11 mesh infrastructure to facilitate broadband

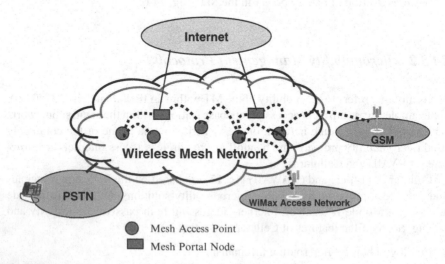

Fig. 14.7 A simple overview of a multiradio wireless mesh network

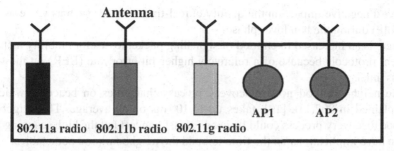

Fig. 14.8 A mesh access point node

wireless connectivity to the heterogeneous access networks such as GSM, WiMax, Cdma. The wireless connections in Fig. 14.7 are shown by means of the dashed lines and the solid lines indicate wired connectivity. Nodes in a WMN are generally static but the clients may be mobile or static. The root node in the mesh networking terminology is known as the *mesh portal* node and is shown in Fig. 14.7.

Each of the mesh nodes, i.e. mesh routers, which also have an access point functionality, is termed as the *mesh access point* and its internal structure is shown in Fig. 14.8. The mesh access point node essentially multihops the traffic to and fro between the access networks and the wired Internet.

To a reader who is not so familiar with the area of mobility management in WMN it may appear at first that the mobility management protocols for the wired networks could as well be easily applied in the wireless domain of WMN. As such, we first state and explain the reasons as to why this is not so to comprehend the issues that exist in this regard for WMN.

14.4.1 Issues

The principal reasons that limit the use of wired mobility management protocols, which were described in Sect. 14.3, for a WMN are:

- HMIPv6 implementation needs the construction of a hierarchical tree, which is relatively more difficult in the unplanned graph topology of the WMNs as compared to a network with fixed stable links. This makes the decision to place the MAPs relatively more challenging during the network layout at deployment of the WMNs.
- In a WMN the path between two nodes, which maybe geographically close to each other, may involve several hops. This could be because there is no direct wireless connection existing among them.
- The time to transfer the BUs is more or less fixed in a wired topology. Whereas in the WMN the dynamic nature of the wireless connectivity can easily cause the time to transfer the BUs to the MAP to vary because of route changes. This will

have a negative impact on the quality of real-time services such as voice over IP (VoIP) during the handover phase.

- There is an increased likelihood of signalling packet losses for mobility management protocols because of a relatively higher bit error rate (BER) on the wireless links.
- The neighbourhood node discovery process that relies on beacons, which is explained in 802.11k [19], takes up to 100 ms on an average. The neighbourhood discovery process could be used for fast handoff but the high scanning time can have implications on the time to build the neighbourhood node tables.

On the positive side, in a WMN the wireless nature of the links enables the nodes interfaces operating on the same channel to listen for packets of another neighbouring node. As such, the packets of the neighbouring node can be cached, which may then be used to offer a handover with a lower packet loss to a MT if it moves toward the node caching the packets. This can alleviate the handover performance in WMN. Such an approach has been suggested by [20], which we review further in Sect. 14.4.2.

14.4.2 Related Works: Literature Review

Reference [21] has shown that in general HMIPv6 has a better performance than MIPv6 in the wireless domain. The impact of the performance on handover delays in the wireless domain is influenced by the placement of MAPs, that are used in a HMIPv6 protocol [21].

To draw a similarity of this problem in the wired domain we consider the case of placement of domain name servers (DNS) in the Internet. DNS are one of the most critical part for the smooth running of the Internet. DNS helps to resolve the domain name such as www.uts.it.edu.au to an IP address. The DNS is an inverted tree hierarchical structure – this means that the root servers are at the top most level. Currently there are 13 root servers spread around the world. Below these 13 root servers are 11 generic top level domains (gTLDs) that help in directing the query for domain names to the appropriate domains under them.

Several mirrors of the root servers have been deployed around the world so as to enhance the smooth running of the Internet even when denial of service (DoS) attacks are directed towards the other root servers. With the aim of reducing the impact of DoS attacks the process of providing redundant root servers is set to continue. A lot of studies are being conducted to determine the possible placement of DNS. This is done by sending probe packets specifically addressed to different root servers and determining the probe query response time. This would be done for different times of the day and periods. Also, stochastic models of the DNS system are used to determine the optimal locations of future mirror root servers.

Reference [21] presents a mathematical solution for the placement of MAPs in a WMNs. The solution in itself seems intractable to translate into real world WMNs because it relies on the a priori 'mobility information pattern' of the mobile terminals. However, the main feature that can be extracted and used from the solution is that essentially the mesh nodes that have a high degree of *closeness centrality* are better suited for MAP placement. The term *closeness centrality* means smallest average half-round trip delay time to the neighbouring nodes. However, [21] has not conducted any traffic studies of the visiting MTs through the MAP domain. This will have a bearing on the number of visiting mobile terminals that could be satisfactorily handled by the MAPs.

Other possible alternatives suggested in [21] is to collocate the MAPs in the root gateways, i.e. the networking elements that link the WMN to the Internet. However, the number of gateways to be used and their placement is left as an open issue. In general the cost of MAPs is lower than that of the gateways so the number of gateways to be used should be as few as possible. A possible scheme in this regard has been worked on in [22, 23]. The other option suggested in [21] is for nodes in WMN to randomly be selected as MAPs.

So in conclusion the placement of MAPs in the WMN has an important bearing on the handover delay and in summary the three possibilities of their placement are:

- Use the closeness centrality approach of [21].
- Random selection of WMN nodes to operate as MAPs.
- Collocate the MAPs with the gateways.

Reference [24] claims to be the first work that has been conducted to offer seamless services in the WMN. It has proposed a fast handoff for WMN in which the MTs are transparent to the backbone infrastructure of the mesh nodes, i.e. they are unaware if its wired or wireless. The transparency feature is in terms of mobility management protocol – the MTs do not have to incorporate any mobility management protocol in their stack. As such they can support mobility in any heterogeneous network.

Although, the transparency feature is useful but we believe that on the other hand it will limit the MTs mobility operation in the planned 4G networks, which allow a service to be provided to a subscriber *anywhere and anytime*. The 4G networks will be a mix of wired and wireless networks including WMNs. Many of the wired networks will use mobility management protocols such as HMIPV6 or FMIPv6. From this aspect if a transparent MT of [24] can have seamless mobility in WMN once it moves over to a neighbouring network of 4G architecture it will not be able to make use of the much advocated mobility management protocols of the wired networks. The wired mobility management protocols require the MT to have these protocols incorporated within it.

In S-mesh [24] each client has two multicast groups associated with it – client control group (CCG) and the client data group (CDG). The nodes in the vicinity of the MT form a CCG based on the signal strength received from the MT. In effect, if two nodes determine that they have the same signal strength from the MT then they can both be part of the CCG. The CCG is essentially a multicast tree so that all the members of the CCG can keep each other informed of the new nodes joining or leaving the CCG. The nodes that form the CCG then become members of CDG if they believe they have the best connectivity to the client. It may so happen that more than one node could believe it has the best connectivity in which case the duplicate packets will arrive at the client because of the multicasting of data packets to and from the nodes in the CDG. Although this approach will provide for availability of the nodes for the MT but in this instance the duplication may reduce its efficiency.

Although, the concept of multicasting improves the handover performance in terms of handover delay and packet loss but it does this at the cost of increase in bandwidth use. Furthermore, S-Mesh assumes that all the nodes operate at the same channel whereas in MR-WMN this is not the case. It should also be mentioned here that the concept of using multicasting to increase the performance of handover has been reported in IST's Daedalus project [25] in 1996. The handoff delay obtained in their implementation was in the range of 8–15 ms with zero packet loss on a 2 Mbps link.

Like S-Mesh, I-Mesh [26] also has the primary goal of a mobility management scheme with 'client side transparency'. The drawback of such a feature has been explained above while discussing the S-mesh approach. Another aspect of I-mesh is that it demonstrates through experimental results that the performance of handover latency while using a flat-routing scheme is much better than a traditional layer-3 handover technique such as transparent mobile IP. The layer-3 latency for routing is faster by a factor of about 3–5 times.

In I-mesh the MT uses probe requests to assess the strength of the channels from different nodes in its vicinity. Based on the SINR value of the probe responses received from the neighbouring nodes on each channel the MT then selects the node interface that offers the best SINR value. The reason that I-Mesh uses probes to *associate* with the mesh node instead of beacon signals from the nodes is that beacon intervals can often be as high as 100 ms. Furthermore, there may not be any nodes to associate with on the current channel of the MT. This does not mean that probing alleviates the handover delay as studies conducted by [27] have shown that one of the major factors in the handoff delay is the time spent in probing and waiting for the probe responses. In particular [28] have suggested optimising the probe feature by the use of probing on a small set of channels based on prior knowledge.

In [18] authors have proposed a network-based mobility management scheme, which they have termed as *Ant* for WMN. Like I-mesh and S-mesh, *Ant* also offers a client side transparency, i.e. no software upgrades are required in the mobile hosts. Ant aims to decrease the handover latency and packet loss during handoff in the architecture. It reduces handover latency by a scheme very similar to that of fast handoff (RFC 4068) by using bidirectional tunnels that are formed between the previous mesh node and the new mesh node following the handover. The way in which the new mesh node determines the previous mesh node's IP address is by means of a location server or through the neighbourhood mesh node list that each mesh node creates. The location server maintains a binding among the MT interfaces MAC address, IP address of the MT interface and the IP address of the mesh node to which the MT is linked. In [18] packet loss is decreased by the previous mesh node, which starts to buffer the packets upon detecting the MTs MAC layer de-association event. The packets buffered by the previous mesh node are the ones that are sent from the CN and destined to MTs IP.

The work proposed in [18] has also been implemented in a small testbed of three mesh nodes with two 802.11b cards each [14], in which it has been shown the layer-2 handover latency to be around 29.1 ms and layer 3 handover latency of 3.4 ms. The total handover latency realized is thus 32.5 ms, which is good enough for real time traffic such as VoIP.

Reference [20] also proposes network-layer based mobility management protocol. Two types of *data caching mechanisms* have been proposed in [20] to decrease the handover packet loss to offer a seamless handoff support in WMN. The caching mechanisms are useful as in the process of route changes during a handover some of the packets may get lost. Lost packets may affect the performance of real-time applications such as VoIP and video or it can decrease the TCP throughput.

The caching mechanisms proposed are termed as *En-route and Promiscuous*. En-route caching occurs in the nodes that are in the current flow route. The en-route node checks the destination address in the data packets and if the destination node is a neighbour of the en-route node then it will cache the data packets for that destination. As a result if the MT were to associate with the en-route node then the handover will result in a low packet loss. Promiscuous data caching occurs in all the neighbouring nodes that can overhear the transmission between the MT and the currently linked node. If the MT were to move towards one of the promiscuous neighbouring nodes and associate with it then the handover process will be seamless.

Reference [20] have conducted experiments over a small testbed of 14 nodes with the backbone connectivity provided by 802.11a links. 802.11b is used for the connectivity between the MT and the mesh nodes. The conclusion from the results of these experiments is that overall the promiscuous caching gives the best results for packet loss rate and average packet delay.

Although, we acknowledge that the above caching mechanisms will be useful in a WMN but a reasonable buffer size needs to be estimated. As too big a buffer will not be useful for real-time applications. Furthermore, in a MR-WMN most of the connectivity around the neighbouring nodes will use different channels to decrease the mutual interference. As such, the neighbouring nodes will not be able to overhear the packets transmission between the MT and the current node. This means that the cache hits for promiscuous mode caching will be very low. Therefore in a MR-WMN we believe that en-route caching will be more suitable than a promiscuous caching.

The increase in deployment of 802.11 based networks coupled with client devices, such as laptops, palmtops and mobiles phones that can operate over the WLANs, has created a need to support real-time services for the mobile hosts on the move. IEEE 802.11 TGr (r-roaming) was created to address the *roaming issues* that arise for a mobile client that use real-time applications, which make use of 802.11i (security) and 802.11e (QoS) enhancements. The issues due to 802.11r and 802.11e arise by way of increase in overhead because of multiple management frame exchanges. This increase in overhead results in the delays of basic service set (BSS) transition during roaming, which can be of up to hundreds of milliseconds or even up to a sec [29].

In 802.11r the latency during the transition process is decreased by ensuring that most of the authentication processes are carried out by the MT before it begins roaming. Furthermore, the 802.11e based TSPEC negotiations are completed during re-association phase instead of just before data transfer – 802.11r natively supports 802.11e [29]. This enables a MT to roam from one AP to another and support high quality voice calls.

Handoffs can be of two types – Horizontal and Vertical. As per RFC3753 [2] Horizontal handover occurs when the MT moves among the Access points of the same technology type. Whereas vertical handover occurs when the MT moves among APs of different technology types such as between UMTS and WLAN. However, in some cases if a handover is vertical or a horizontal maybe a bit vague. For example as per RFC 3753 [2] the handover between an 802.11a AP and 802.11b AP is considered as vertical even though the access protocol, i.e. CSMA/CA is used in both the cases. Handovers among heterogeneous networks can be challenging because they may have different QoS, security and power management requirements. The emerging IEEE 802.21 standard tries to address the challenges of vertical handover by means of media independent handover framework (MIHF) [30]. In particular, the 802.21 facilitates vertical handover through the process of network discovery and selection. This enables a mobile to connect to the most suitable network based on operator policies and/or subscribers service profile.

Reference [30] has described the 802.21 as well as carried out experiments that implement certain aspects of 802.21 framework. The results obtained demonstrate the usefulness of 802.21 for a vertical handover in which the quality of interactive VoIP continues to be acceptable.

14.4.3 Proposed Mobility Management Scheme

The objective of our proposed mobility management scheme is to offer a seamless handoff to the mobile client in a MR-WMN, which was shown in Fig. 14.7. The mobility management aspects that we have dealt with herein are related to the mechanism for handover and location management. Our work currently is not concerned with the schemes to maintain the QoS and carry out an effective routing during the handoff process. So these are not discussed in the chapter.

Some of the attributes of the MR-WMN architecture for which we propose the mobility management scheme are:

- Wireless mesh modes used are independent of any radio technology.
- Mechanisms proposed in the architecture are distributed.
- Power efficient algorithms are used in the mesh nodes.
- Wireless mesh nodes are considered to be stationary and can be heterogeneous. By heterogeneous we mean that they could be of different wireless technologies such as 802.11, WiMax.
- VoIP will be the main service offered as it is low cost hence affordable.

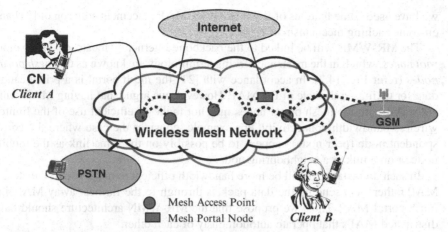

Fig. 14.9 Multiradio wireless mesh network architecture with clients

- WMN should be maintenance free as much as possible – essentially plug and play.

The WMN is expected to serve stationary as well as mobile subscribers under dynamic network conditions and offer any type of service. Figure 14.9 shows the diagram of WMN in which an end-to-end connectivity is provided between two clients across the WMN.

In the MR-WMN architecture of Fig. 14.9 there are three possibilities of handover:

- Handover of a client such as Client B from the interface of one MR-WMN node to a neighbouring node interface in the MR-WMN.
- Handover of the mobile client from one of the access networks to the node interface in the MR-WMN. The access networks use the WMN as a backbone to interconnect with the Internet.
- Handover of a mobile client within one of the access networks linked to the MR-WMN.

In accordance with the aim of this chapter, we consider only the handover of the mobile clients within the MR-WMN. Further, in Fig. 14.9 both the peer communicating clients A and B could be located such that:

- Client A and Client B are both within the MR-WMN.
- Client A could be in MR-WMN and Client B could be linked directly to the Internet outside of the MR-WMN.
- Client A could be in MR-WMN and Client B could be linked to a node within one of the access networks of the MR-WMN.

Our mobility management proposal incorporates some of the ideas made by other proposals and makes it suitable for use in a MR-WMN architecture. In particular, we have used some features of IETF HMIPv6, MAP placement solution of [21] and en-route caching mechanism of [20].

The MR-WMN will be linked to the backbone Internet by means of one or more *root nodes*, which in the mesh networking terminology are known as the *mesh portal nodes* (refer Fig. 14.7). In accordance with [21] the mesh portal is a prime candidate for taking on the role of the MAP. However, we argue that having MAPs only colocated with the mesh portal nodes may not make an efficient use of the limited wireless bandwidth connectivity. For example, consider the case where the correspondent node (peer node) happens to be possibly on the same link as the mobile node or on a link in a neighbouring node.

In such an instance it will be more bandwidth efficient to make use of the closest MAP rather than sending the data packets through to the furthest away MAP, i.e. mesh portal MAP. Thus, we propose that the MR-WMN architecture should have distributed MAPs that operate autonomously of each other.

In our paper [31], we had detailed the initialisation process of channel assignment in the MR-WMN. It begins by building a spanning tree from a root interface (mesh portal) that spans an area of the mesh network. The spanning tree nodes we had termed as the seed nodes, which build a cluster of nodes around itself. The seed nodes are candidates for MAPs in the MR-WMN as shown in Fig. 14.10.

RFC 4140 [11] allows the overlapping of MAP domains so this means a node in MR-WMN could potentially be registered with two neighbouring MAPs. The literature reviewed by us does not indicate any similar approach for a distributed MAP environment within a MR-WMN.

Alternately, if we used the random initialisation process as explained in [32] for channel assignment the MAPs could be collocated with the nodes based on a suitable election mechanism. For example more MAPs could be located near the hotspot

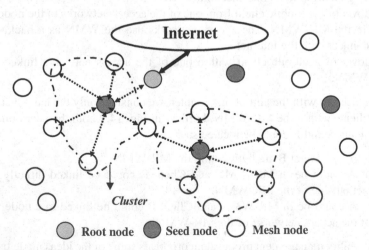

Fig. 14.10 Proposed cluster formation in the MR-WMN – adapted from [31]

areas or at any other areas based on the discretion of network operator. A suitable election mechanism could be developed to elect the MAP(s) within each domain.

The other possibility of collocating the MAPs with the nodes would be based on the closeness criteria of the solution proposed in [21]. The seed nodes of the spanning tree would satisfy this criteria.

An advantage of having the MAPs sprinkled around the MR-WMN is to carry out a 'distance based' selection of a MAP by the mobile client. For example, if a mobile client is travelling fast then it will be more suitable for it to register with a farthest away MAP such as the mesh portal MAP so that it does not need to frequently initiate new registrations with new MAPs in the administrative domain. This will improve on the handover performance as the mobile client does not need to inform the CN of a change in its RCoA address. The process of using MAPs for mobility management was explained in Sect. 14.3.1.5 on HMIPv6.

Within the MR-WMN the mesh-portal MAP would be used for channelling the communication between a CN on the Internet and the mobile client in the MR-WMN or for communication between two clients that may be in different administrative domains.

14.4.3.1 Location Management

We also propose a distributed database scheme to facilitate the location management during the call set up phase. Such a location management system will also assist in the setting up of the VoIP calls. For example, when a mobile client within the MR-WMN needs to establish voice connectivity or establish a packet flow with another client it first of all needs to know the node to which the CN (called person) is located. We can either have a two level or a three level hierarchical distributed database system. The first level databases would be located at each of the APs in the mesh nodes. The second level would either be colocated along with the distributed MAPs (explained earlier) or at the seed nodes. The third level could be colocated with the

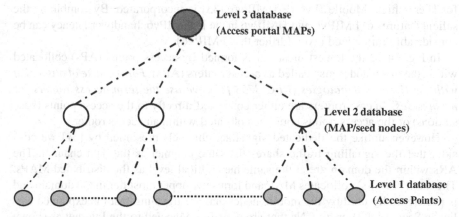

Fig. 14.11 Distributed data base system for location management

access portal MAPs. Figure 14.11 illustrates the topology of the distributed data base system.

The location database entry creates a mapping of the form {CL-IP, CL-Node interface-IP, CL-LCoA, CL-RCoA}. The following two entries will be used if the mobile client is in its home domain:

CL-IP: Will help to identify the network interface of the client.
CL-Node interface –IP: Will help to identify the node interface in the MR-WMN that is linked to the client.

The remaining two entries will be present only when the client is visiting a domain in the MR-WMN:

CL-LCoA: On-link care-of-address (HMIPv6 Terminology – RFC4140 [11])
CL-RCoA: Regional care of address – Address of the MAPs subnet.

The CN (called mobile's) location will be determined through database interrogation at successively higher levels. In a distributed database the search is more efficient than a centralized server because the search can be conducted in a ripple like fashion. That is the area close to the CN is searched first for the MT and then the search progressively extends to cover larger distances. Furthermore, a centralized server can become a bottleneck and a single point of failure.

The functionality of the distributed database of Fig. 14.11 for location management is explained when we walkthrough the operation of establishing a voice call between the mobile client and the CN in Sect. 14.4.3.3. We also use en-route caching mechanism proposed in [20], which was explained earlier in Sect. 14.4.2. The reason for not using the promiscuous mode of caching was also explained in Sect. 14.4.2.

14.4.3.2 An Architecture for Proposed Mobility Management Scheme

An underlying hierarchically distributed structure, as shown in Fig. 14.12 is adopted to facilitate seamless mobility in MR-WMN architecture in which a Fast Handover for Hierarchical Mobile IPv6 (F-HMIPv6) [33] is incorporated. By combining the salient features of FMIPv6 and HMIPv6 in the F-HMIPv6, handover latency can be considerably minimized (even further than FMIPv6).

In Fig. 14.12 the lowest most tier is formed by access points (APs) collocated within the mesh nodes also called as access routers (ARs). *For the sake of alignment with the IETF terminologies (RFC 3753 [2]), we use the term access routers for mesh nodes.* Access routers are either connected directly to the access points (base stations) or the access points could be colocated within an access router.

However, unlike the dedicated signalling channels presumed by [34] we consider that the signalling traffic shares the same channel as the data channel. The ARs within the domain are at the same hierarchical level as the distributed MAPs. The mesh portal(s) collocates MAP and forms the upper most tier in the domain and is preferentially used by the mobile nodes that are traveling fast (as explained earlier in Sect. 14.4.3) or by CNs that are directly connected to the Internet as shown

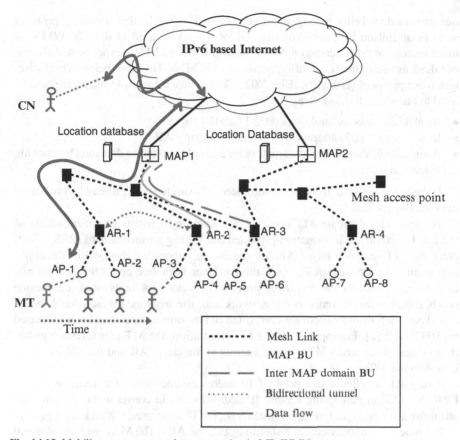

Fig. 14.12 Mobility management scheme operation in MR-WMN

in Fig. 14.12. Whereas, the lower MAPs in the MR-WMN are used by relatively slower mobile clients. Compared to other MAPs in the MR-WMN it is almost certain that the mesh-portal MAPs have a high degree of connectivity to the underlying mesh nodes (ARs). Our random-initialisation process described in [32] ensures this because the mesh portal nodes need to carry bulk of the traffic between the Internet and the MR-WMN. All overlapping MAP domains may access the Internet through the MAPs colocated with the mesh portal nodes. An administrative body may administer a single MAP domain or several MAP domains.

As it may have been apparent from the overview of HMIPv6 in Sect. 14.3.1.5 a roaming MT is identified by three IP addresses: (1) MT's home address (2) MAP's IP address or Regional CoA (RCoA) and (3) Local CoA (LCoA) assigned by DHCPv6 enabled AR. The autoconfiguration feature of AP-AR pair would make it very easy to create connectivity with a mobile client. The ARs can communicate among themselves through 802.11f based protocols to formulate a list of resources offered by the ARs in different parts of the MR-WMN. (Note: We make the assumption here that the ARs (mesh nodes) are using 802.11 based radios.) We consider in our proposed mobility management scheme of Fig. 14.12 that a routing protocol such as optimised link state routing (OLSR) could be used in the MR-WMN. A main reason for considering OLSR is because IEEE 802.11s has advocated the use of OLSR as one of the two routing protocols in WMNs. The other being hybrid wireless mesh protocol (HWMP). IEEE 802.11k [19] based reporting methods could be used to broadcast the following list:

- Available access technologies (802.11a/802.11b, etc.)
- IP and layer 2 (L2) addresses of the neighbouring APs
- A limited number of QoS parameters for example supported data rate, bandwidth, video coding rate

If the list is too long to be accommodated in a single frame, it may be broken up into smaller packets.

As soon as a dormant MT enters a new domain, it listens to the broadcast of the list. Based on the information provided and L2 trigger mechanism (RSS, SNIR, etc.), the MT selects a target AP/AR capable of supporting the QoS of the application. In the proposed architecture the handover is mobile controlled, i.e. mobile initiates and controls the handover process. It is not mobile-assisted, i.e. mobile sends signal measurements to the network and, the network initiates handover, if any. These definitions of mobile-controlled and mobile-assisted handover are based on RFC 3753 [2]. Equipped with the L2 information, the MT generates a registration request that carries MT's home address to the target AR and the AR assigns a LCoA to the MT.

Every AR is allocated a pool of IP addresses and offers the functionality of DHCPv6 [35] in generating unique IP addresses. As in conventional systems, the administrative body assigns the domain a pool of IP addresses, which is then equally or by some prior agreement shared amongst all the ARs. The MAP will be informed by the AR about the new MT within the domain based on which the MAP carries out a BU to the HA on behalf of the MT. The MAP will also authenticate the MT's identity with the HA through a secure mechanism and store the user profile, authentication, security and charging information in two databases – Home subscriber server (HSS) and authentication, authorization and accounting (AAA), located at the MAP. (Note: These two databases are not shown in Fig. 14.12 for the sake of clarity.) Once registration of the MT is accepted by the MAP, the MT will be configured with the RCoA and LCoA by means of the registration acceptance message.

A dormant MT, after initial domain registration, can travel greater distances within the domain without any control signal exchange with other network nodes. The MT configured with LCoA will listen to periodic broadcasts of AR identification after the registration process, even in the sleep mode. Because the LCoA is valid only within the coverage area of an AR, if the MT roams into the coverage area of another AR, then it must be configured with a new LCoA.

14.4.3.3 Mobility Management Operation in the Architecture

We now walkthrough the processes that are shown to occur in the MR-WMN topology of Fig. 14.12 by first considering the process of call set-up. To initiate a voice call for example to a distant CN, the MT makes a call initiation request to the local AR, which then successively interrogates the different levels of databases, as explained earlier in Sect. 14.4.3.1 to retrieve the IP address of the node to which the CN is linked. The interrogation terminates as soon as location information for the CN is obtained. (Note: Session initiation protocol (SIP), which is used for VoIP could also make use of location determination technique described in this chapter.) OLSR routing protocol will then be used to route the packets between the MT and CN assuming first that the MT has registered with the mesh portal MAP of Fig. 14.12. It must be stated here that the MT will be informed about the existence of MAPs their distances and other options by means of router advertisements (Radv). Further details in this regard can be referred to in RFC 4140 [11].

Call reception at the MT involves the CN sending data packets to the MAP by using the RCoA of the MT assuming that route optimisation is used. The MAP then de-tunnels the packets and then encapsulates them with LCoA address of the MT. It then forwards the packets to the AP to which the MT is currently connected but after successfully paging the MT.

There are three types of handovers possible in our mobility architecture: intra-ARs, i.e. assuming that the mesh node (AR) has two APs located in it, inter-AR and inter-MAP. Inter-AR and inter-MAP handovers are the most challenging of the three. As intra-AR handover is trivial so it is not discussed herein.

During an interaccess router handover that is handover among the APs (AP1 to AP3 in Fig. 14.12) of two different access routers (AR-1 and AR-2 in Fig. 14.12) first there will be a context transfer between AR1 and AR2 to enhance the connection. After this the MAP may bi-cast data packets to both of the APs of these AR's, thereby hiding the change of LCoA and local handover process from the distant HA and the CN, which decreases the signalling traffic. The bi-casting will occur only if the route set up can be done quickly. In case of heavy load on the network another possible option is that the previous AR (AR-1) in accordance with FMIPv6 forms a bidirectional tunnel with the new AR (AR-2) until a layer 3 handover occurs that is the route update takes place so that the packets are routed directly to the new AP (AP3) from the MAP. The path of the BU to the MAP by the MT during the hand over to AP3 of AR-2 is shown by solid line in Fig. 14.12.

During an interdomain handover, i.e. inter-MAP handover as per RFC 4140 the MT may send a BU (shown by a line in Fig. 14.12) to its previous MAP with its new LCoA. This will enable the previous MAP to continue to serve the mobile terminal (MT) in the new MAPs domain. It should be noted that this is possible only when both the MAPs are under the same administrative domain. Note: Generally, a new BU to the HA and the CN is required only when there is an interdomain handover, or a dormant MT roams into a new domain. However, in this instance by isolating local signalling and not sending BU to distant HA and CN the signalling overhead can be reduced, which results in a relatively lower handover latency and packet loss.

This eventually results in a smooth inter-MAP handover where the mobile node continues to receive packets.

If both the old MAP and the new MAP are not in the same administrative domain then the MT will need to register with the new MAP and will need to carry out a BU to its HA and CN as well. Until that time the old MAP and the new MAP could set up a bi-directional tunnel between them to transfer the packets to and from the CN and the MT. (Note: RFC4140 has not proposed any solution for this case.)

14.4.3.4 Key Differentiators-Proposed Mobility Management Scheme

The *key differentiators* between the proposed mobility management scheme and those of related literature discussed in Sect. 14.4.2 are summarized below:

- We have proposed a layered and modular MR-WMN architecture with clear definition of functional elements and their interfaces.
- We believe that our mobility management scheme is efficient as it incorporates

  - Hierarchically distributed structure to reduce the signalling load.
  - Distance based MAP selection to cater for both slow and fast moving subscribers.
  - Use of 802.11k reporting techniques to determine the attributes of the neighbouring nodes.
  - Hierarchically distributed data base mechanism for location management.

14.5 Thoughts for Practitioners

The work presented in this chapter has led us to gather some stimulating thoughts that we recapitulate and list below:

- The handover performance in WMN can be improved by leveraging the wireless nature of the links that enables the node interfaces operating on the same channel to listen for packets of another neighbouring node. As such, the packets of the neighbouring node can be cached, which may then be used to offer a handover with a lower packet loss to a MT if it moves toward the node caching the packets.
- Placement of MAPs in the WMN has an important bearing on the handover delay.
- MR-WMN should be able to support both fast as well as slow moving subscribers efficiently. In this regard, the presented distance based MAP selection is a possible approach.
- The MIHF [30] of IEEE 802.21 standard facilitates vertical handover by means of network discovery and selection could possibly be also made use of during handovers in MR-WMN. It would enable a mobile to connect to the most suitable network based on operator policies and/or subscribers service profile.

- In a MR-WMN the wireless connectivity of the nodes will define its topology and as such a strong correlationship between the mobility management performance and the topology of the network may exist.

14.6 Directions for Future Research

In this chapter, we have given an insight into as open-area of research for mobility management in MR-WMN. To this end, we have also proposed a mobility management scheme for MR-WMN. We believe that the next steps in this research would be to:

- Evaluate the performance of the proposed mobility management scheme when two clients are interacting across the MR-WMN while either one of them or both of them are on the move. The scalability aspects of the scheme would also need to be evaluated.
- Development of a protocol for QoS aware mobility. This protocol would aim to provide QoS during handover process and thus facilitate towards an end-to-end QoS availability in the MR-WMN.
- Evaluation of the implications of real-time services transfer on the asynchronous services because if priority is unfairly given to real-time services then the asynchronous services transfer may suffer in the MR-WMN.

14.7 Conclusions

This chapter has shown that the mobility management protocols that have been proposed for the wired Internet such as MIPv4, MIPv6, HMIPv6 can not be directly used in the wireless realm of WMNs. This is primarily because the wireless links in WMN are dynamic unlike the wired links, which detrimentally affects the transfer of signalling messages that are so crucial to the proper operation of the wired mobility management protocols. Furthermore, the unplanned graph topology of the WMN can not be easily used to support a creation of tree that is required for protocols such as HMIPv6. To get an insight into the methods that have been used to overcome some these issues we provided an extensive literature review of the solid research that has been done.

However, our review showed that there are still substantial weaknesses that are associated with the proposals in the literature, in particular that they will not be able to efficiently support mobility management in a MR-WMNs. We have thus proposed a scheme for mobility management in WMNs including MR-WMN, which is based on the augmentation of some the approaches made in the literature and uses the wired mobility management schemes as guidelines. The details of the operation of our proposal in a typical MR-WMN scenario were explained. In particular, we have focused on handoff and location management challenges in mobility management. A major contribution that differentiates our proposal from the other proposals is that our proposal caters well for slow as well as fast moving subscribers. Further, to enable any interested reader to follow the chapter it was gradually approached and useful references and explanations have been strategically provided within the text.

Acknowledgments This research is performed as part of an Australian Research Council Linkage Grant among Bell Labs, Alcatel-Lucent and the University of Technology, Sydney, Australia.

14.8 Terminologies

1. *Mobility management.* A *seamless support* of real-time as well as non real-time services for a subscriber who is on the move. Seamless support refers to obtaining a low handover latency and packet loss.
2. *Handover/handoff.* A process by which the mobile terminal changes its point of attachment to the network. *Handoff* and *handover* terms are used interchangeably to refer to the same process.
3. *Location management.* A process of finding out the connectivity location of the roaming subscriber's mobile terminal within the geographical region; security and authentication information and QoS capabilities.
4. *Route optimization.* A process by which a route is created efficiently between the calling person's mobile terminal and the called person's mobile terminal.
5. *L2 handover.* It occurs when the mobile terminal (MT) moves out of the satisfactory transmission/reception range of an access point (AP)/base station, which triggers an implementation specific mechanism to reassociate with a new AP/base station.
6. *L3 handover.* It occurs as result of the L2 handover and requires a routing update, which involves layer 3 of the OSI protocol stack, therefore it is termed as a L3 handover.
7. *Macro-mobility.* Mobility of a mobile terminal (MT) among different IP domains (RFC 3753 [2]), for example mobility across different sub-networks – such as between the home network and the visited network.
8. *Micromobility.* Mobility of the MT within an IP domain, for example across different access points/base stations within the same subnetwork, which could be the home network or the visited network.
9. *Binding.* Mobile IP allocates two IP addresses to the mobile terminal – a home address and the other is a temporary care of address that represents the current location of the MT. An association is created between these two addresses, which is called as *binding*.
10. *Mesh access point.* Each of the mesh nodes, i.e. mesh routers, which also have an access point functionality, is termed as the mesh access point.

11. *Mesh portal nodes*. Multi-radio wireless mesh network (MR-WMN) is linked to the backbone Internet by means of one or more root nodes, which in the mesh networking terminology are known as the mesh portal nodes.
12. *Horizontal handover*. Mobile terminal moves among the Access points/base stations of the same technology type.
13. *Verical handover*. MT moves among access points/base stations of different technology types such as between UMTS and WLAN.
14. *Multi-radio wireless mesh network (MR-WMN)*. In a MR-WMN the mesh nodes essentially multi-hop the traffic to and fro between the access networks and the wired Internet. As the nodes have multi-radio interfaces therefore a multi-radio connectivity is offered in the MR-WMN. MR-WMN offer a higher capacity as compared to a single radio WMN.
15. *Mobile controlled handover*. The mobile initiates and controls the handover process.
16. *Mobile-assisted handover*. The mobile sends signal measurements to the network and, the network initiates handover, if any.

14.9 Acronyms

AAA – Authentication, authorization and accounting
AP – Access point
AR – Access router
BER – Bit error rate
BS – Base stations
BU – Binding update
CCG – Client control group
CDG – Client data group
CN – Correspondent node
CoA – Care-of-address
DoS – Denial of service
DHCPv6 – Dynamic host configuration protocol version 6
DNS – Domain name servers
FHIPv6 – Fast hierarchical mobile IPv6
F-HMIPv6 – Fast handover for hierarchical mobile IPv6
FMIPv6 – Fast mobile IP v6
GFA – Gateway foreign agent
gTLDs – Generic top level domains
HA – Home agent
HAWAII – Handoff aware wireless access Internet infrastructure
HoA – Home address
HSS – Home subscriber server
IETF – Internet engineering task force
IP – Internet protocol

LCoA – On-link CoA
MAP – Mobility anchor point
MIHF – Media independent handover framework
MIPv4 – Mobile IPv4
MIPv6 – Mobile IPv6
MR-WMN – Multi-radio wireless mesh networks
MT – Mobile terminal
OLSR – Optimised link state routing
QoS – Quality of service
RCoA – Regional care of address
RFC – Request for comments
RSS – Received signal strength
SINR – Signal to noise plus interference ratio
VoIP – Voice over IP
WMN – Wireless mesh networks

14.10 Questions

1. What is the difference between a vertical and horizontal handover ? State an example for each type of handover.
2a. Distinguish between L2 and L3 handover?
2b. What does the term *seamless handover* mean?
3. Explain the main challenges in the concept of mobile IPv4?
4a. State 3 differences between MIPv4 and MIPv6?
4b. Briefly explain the terms BU, HoA, LCoA and RCoA?
5. Distinguish between the terms *micromobility* and *macromobility*.
6. What are the three principal components of mobility management?
7. How does hierarchical mobility management in HMIPv6 help as compared to MIPv4/v6?
8a. State two issues in wireless mesh networks (WMN) because of which wired mobility management protocols can not be directly used?
8b. Select a routing protocol from the list below that 802.11s has advocated for WMN?

   a. OSPF
   b. RIP
   c. OLSR
   d. BGP
   e. None of the above

9a. What is the purpose of router advertisements (Radv)?
9b. To support terminal mobility within a WMN what additional functionality the mesh nodes should support?

10a. How does *promiscuous mode caching* in a WMN facilitate better mobility management as compared to MR-WMN?

10b. How do you think that having a distributed database helps to locate a CN or a MT as compared to having a centralised location management server?

References

1. S.Y. Hui et al., Challenges in the migration to 4G mobile systems, IEEE Communications Magazine, 41(12), 54–59 (2003).
2. J. Manner et al., RFC 3753- Mobile related terminology http://www.faqs.org/rfcs/rfc3753.html Accessed Nov 2007 (2004).
3. G. Hoo et al., QoS as a middleware: Bandwidth brokering system design (1999).
4. C. Perkins, "IP Mobility Support," IETF RFC 2002, October 1996.
5. D.B. Johnson et al., (2003), "Mobility Support in IPv6," IETF Draft, draft-ietf- mobileip-ipv6-24.txt.
6. D. Ghosh, I.P. Mobile, http://www.acm.org/crossroads/xrds7-2/mobileip.html Accessed Nov 2007.
7. C. Perkins, "RFC3344 – IP Mobility Support for IPv4", 2002 http://www.faqs.org/rfcs/rfc3344.html Accessed Nov 2007.
8. D.B. Johnson et al., "Route Optimization in Mobile IP," IETF Draft, draft- ietf- mobileip-optim-07.txt (1998).
9. D. Johnson et al., "RFC3775- Mobility Support in IPv6", http://www.faqs.org/rfcs/rfc3775.html Accessed Nov 2007 (2004).
10. H. Matsuoka et al., A robust method for soft IP handover, IEEE Internet Computing, 7(2), 18–24 (2003).
11. H. Soliman, et al., RFC 4140 – Hierarchical Mobile IPv6 Mobility Management (HMIPv6), http://www.faqs.org/rfcs/rfc4140.html Accessed Nov 2007 (2005).
12. R. Koodli RFC 4068 – Fast Handovers for Mobile IPv6 http://www.faqs.org/rfcs/rfc4068.html. Accessed Nov 2007 (2005).
13. T. Jari et al., Mobile IPv6 Regional Registration, IETF Draft, draft-malinen- mobileip-regreg6-01.txt (2001).
14. A.T. Campbell et al., An overview of cellular IP, IEEE Wireless Communications and Networking Conference (WCNC). 2, 606–610 (1999).
15. A.T. Campbell et al., Design, implementation and evaluation of Cellular IP, IEEE Personal Communications, 7(4), 42–49 (2000).
16. R. Ramjee et al., HAWAII: a domain-based approach for supporting mobility in wide-area wireless networks, IEEE/ACM Transactions on Networking, 10(3), 396–410 (2002).
17. R. Ramjee et al., IP-based access network infrastructure for next-generation wireless data networks, IEEE Personal Communications, 7(4), 34–41 (2000).
18. H. Wang et al., A network based local mobility management scheme for wireless mesh networks, IEEE WCNC, 3795–3800.
19. IEEE 802.11k draft 7.
20. H.Y. Wei et al., Seamless handoff support in wireless mesh network, 1st Workshop on Operator assisted community networks, 1–8 (2006).
21. W. Lei et al., On the problem of placing mobility anchor points in wireless mesh networks, 1st Workshop on Operator assisted community networks, 1–8 (2006).
22. L. Qi, Optimising the placement of integration points in multi-hop wireless networks IEEE ICNP (2004).
23. Y. Bejerano, Efficient integration of multi-hop wireless and wired networks with QoS constraints ACM Mobicom (2002).
24. Y. Amir, Fast handoff for seamless wireless mesh networks, MobiSys (2006).

25. S. Seshan et al., Handoffs in cellular wireless networks, The Daedalus implementation and experience (1996).
26. V. Navda, Design and evaluation of iMesh: An infrastructure-mode wireless mesh network, IEEE WoWMoM, 164–170 (2005).
27. A. Mishra, An empirical analysis of the IEEE 802.11 MAC, layer handoff process, SIGCOMM 33, #2, 93–102 (2003).
28. S. Shin et al., Reducing MAC layer handoff efficiency in IEEE 802.11 wireless Lans, ACM MOBIWAC, 19–26 (2004).
29. S. Bangolae, Performance Study of Fast BSS Transition using IEEE 802.11r. IWCMC, Vancouver, (2006).
30. A. Dutta et al., Seamless handover across heterogeneous networks – An IEEE 802.21 centric approach. Proc. WPMC (2005).
31. V. Mirchandani and A. Prodan et al., A method and study of topology control based self-organization in mesh networks, IEEE AccessNets, Ottawa (2007).
32. V. Mirchandani, A. Prodan et al.. Impact of topology control on the performance of a self-organization scheme for wireless mesh networks, Proc. ACM WICON, Austin, USA (2007).
33. H.Y. Jung et al., Fast handover for hierarchical MIPv6 (F-HMIPv6) IETF Draft, draft-jung-mobileip-fastho-hmipv6-03.txt (2004).
34. B.J. Ko, D. Rubenstein, A distributed, self-stabilizing protocol for placement of replicated resources in emerging networks, 11th IEEE International Conference on Network Protocols (ICNP), Atlanta, GA, USA (2003).
35. J. Bound et al., Dynamic host configuration protocol for IPv6 (DHCPv6), IETF Draft, draft-ietf-dhc-dhcpv6-20.txt, October (2001).

Chapter 15
Low Latency in Wireless Mesh Networks

Robert McTasney, Dirk Grunwald, and Douglas Sicker

Abstract Multimedia requirements of the 1990s drove wired and optical network architects to examine how to combine the advantages of packet switching with the long proven methods of circuit-switching to implement traffic engineering to reduce variance in end-to-end delay. Methods, such as asynchronous transfer mode (ATM) and multiprotocol label switching (MPLS), have been used to create virtual circuits. Because both are mature and proven technologies for wired and optical network architectures, much research has been done to apply these methods to wireless mesh networks (WMNs). But as these are applied, optimal performance improvement eludes WMN designers because of the inherent shortcomings of contention-based WMNs and the differences between the wired/optical and wireless environments in the provision of noninterfering unidirectional internodal links. This chapter will present issues regarding the development of such low-latency WMNs to include multiple orthogonal channels, virtual cut-through and wormhole switching, physical layer circuit switch design, and reservation protocols.

15.1 Introduction

Why do 802.11-based Wireless Mesh Networks (WMNs) provide poor performance? To better understand the source of this poor performance, let's begin by considering the architecture of these networks. WMNs consist of wireless nodes that many times take on the combined functions of a single network (wireless broadcast) interface, router and in many cases also host. Also consider that the network interface accesses a shared radio frequency (RF) broadcast medium commonly using an omnidirectional antenna. The use of such a shared RF environment provides many

R. McTasney (✉)
Department of Computer Science, University of Colorado at Boulder, UCB 430, Boulder, CO 80309, USA
e-mail: robert.mctasney@colorado.edu

S. Misra et al. (eds.), *Guide to Wireless Mesh Networks*, Computer Communications and Networks, DOI 10.1007/978-1-84800-909-7_15,

challenges because of the unpredictable nature of RF channels with regard to reception and interference range. This lack of control of the shared media makes it much more difficult to share/allocate bandwidth compared to wired-optical network architectures.

The WMN application that makes research on these networks appealing is when a wired network interface is included so the wireless node serves as a gateway into a wired backbone network. But given this motivation, why do these WMNs provide such poor performance? Performance is measured with regard to network capacity in the form of channel utilization and throughput. Li, Blake, De Couto et al. [1] present simulation results that provide some insight into why 802.11-based WMNs do not scale well with regard to such performance measures. Their conclusion, based upon simulation results on single-chain, lattice and random topology networks, is that these 802.11 based WMNs do not scale with regard to capacity unless the average distance between source and destination remains small. They show that throughput decreases significantly as the chain length and number of nodes increases for the single-chain and lattice topologies. Two reasons are given for this. First the 802.11 media access control (MAC) layer fails to achieve the optimum schedule (based upon the topology and traffic), because an 802.11 node's ability to send is affected by the amount of competition it experiences for the medium. Also, a small but significant percent of the time spent unable to send a packet is due to 802.11 distributed coordination function (DCF) backoff working badly with ad hoc forwarding due to two-hop interference.

In addition to these findings, Ramanathan in his MOBICOMM 05 paper made the following argument as to why WMN performance lags behind that of wireline networks [2]. "[WMN] operations are hop-centric in that processes are terminated and re-initiated at every hop. This contributes to the end-to-end delay because of processing through the physical, MAC/link, and network layers, re-queuing at the network and MAC layers and recontention for channel access." The thirst for increased bandwidth in these networks will result in WMNs operating at higher frequencies with inherent shorter range. This will result in even more hops from source to destination increasing end-to-end delay through aggregation of the hop-centric delays described above. As more bandwidth is obtained, higher data rates are achieved, which means delay has even more of an impact on performance.

How can performance be improved in these WMNs? A review of how performance has been improved in wired or optical routing networks may provide some answers.

Improving Performance in Wired/Optical Routing Networks. Over the past 10–15 years, network architects have begun to adopt mechanisms that make packet-based networks act like circuit switched networks of the past with improved throughput and reduced delay. Asynchronous transfer mode (ATM) and multiprotocol label switching (MPLS) can provide these features through virtual circuits (VCs) or LSPs where packets can arrive on one wired or optical unidirectional interface and be effectively circuit-switched by departing on another wired or optical unidirectional interface based upon a predetermined switched setting. The switch is set when the end-to-end path is established and resources such as bandwidth and buffer space are reserved at each intermediate node. Throughput is increased and delay is reduced by avoiding the processing up and down the communications stack to determine how incoming packets should be routed to outgoing interfaces.

Why can't technologies such as ATM or MPLS just be implemented in the current wireless environment? In fact, many in the research community have done this with some success in terms of improved performance. But as traffic increases on these networks, performance begins to decline because of a few fundamental differences between the wired/optical environment and the wireless environment. In the current wireless environment, particularly in contention-based protocols such as the 802.11 MAC layer, there is no real ability to implement noninterfering unidirectional internodal links as can be done in the wired or optical environments. The key problem is that there is still competition for the broadcast medium. Therefore WMNs can never get the same performance improvements as wired or optical routing networks when implementing circuit-switching using either (or both) ATM or MPLS. To overcome this, we believe that wireless networks must behave more like their wired counterparts.

Making Wireless Wire-Like. Noninterfering unidirectional internodal links can be provided in a wireless environment to support circuit-switching by implementing orthogonal channels using either one, all, or combinations of time, space, or frequency multiplexing (keeping in mind that space does not necessarily mean directional, it also means omnidirectional distance allowing frequency reuse). Given multiple orthogonal channels, some can be used to exclusively transmit and the others can be used to exclusively receive simultaneously so that these unidirectional links could be established between each node. In the wired environment, each orthogonal channel could be referred to as a separate interface. Because the links are noninterfering, there is no contention for the shared broadcast media and no need for an 802.11 MAC layer. In essence, the multiplexing method(s) used along with scheduling, channel assignment, transmit power, transmit and receive gain now become the MAC. Given the ability to produce orthogonal channels, all one needs is the switching mechanism at each node to implement the circuit switched path. By making the wireless environment wire-like and by implementing a switching mechanism, traffic engineering in a wireless network becomes possible and WMNs become a more viable and appealing technology.

Wireless Traffic Engineering. The purpose of *traffic engineering* (TE) is to improve network performance through more efficient use of network resources and matching resources better with traffic demands. There is some ability to allocate nodal resources in the 802.11 environment with regard to quality of service (QoS) parameters in terms of buffer space and queue servicing (bandwidth) and the 802.11i standard prioritizes traffic admission on a single link. But as stated earlier, WMNs could not support QoS requirements or even further, traffic engineering, because of contention with other nodes also using the shared medium, particularly across multilink paths. With the shared medium problem overcome through the use of orthogonal channels, we would have enough control to really reserve resources to support traffic requirements, as we do with wired or optical networks. This would make traffic engineering mechanisms such as label-switching, proven to work in wired and optical networks, work in wireless mesh networking architectures.

As in wired label-switching mechanisms, such as MPLS, resource reservation protocols, such as resource reservation protocol (RSVP) and Label Distribution Protocol (LDP), are needed to reserve the nodal resources to support a QoS requirement for a traffic flow. In the wireless mesh environment, they would also need to act as the MAC by assigning channels provided through frequency, space and/or time multiplexing for the traffic flow.

15.2 Background

In Raj Jain's vision paper "Internet 3.0: Ten Problems with Current Internet Architecture and Solutions for the Next Generation," he addresses the question "is this the way we would design the Internet if we were to start it now?" [3]. The same could be asked of about the current design of 802.11 based WMNs. Since the late 1800s, circuit-switching (such as with the voice telephone system) provided great performance, but not an efficient use of resources. A little less than a century later, packet switching was fielded and provided better performance for bursty data and efficient use of network resources. But as the demand for data streaming to support multimedia applications began to increase over the past 15 years, it has been increasingly difficult to offer hard quality-of-service guarantees using the reservation protocols layered on top of best-effort packet networks.

In the wired and optical network architectures, this resulted in implementations of bandwidth-reservation mechanisms at the PHY layer in the form of label-switching, using protocols such as MPLS or virtual circuits as provided by ATM. This same trend of "what is old, is new," can be applied to WMNs. In the following subsections, various background materials will be presented to provide familiarity with the issues involved in applying circuit switching solutions to WMNs so that traffic engineering is possible. The next section will define Wireless LAN, Wireless Mesh, Ad Hoc Networks and Mobile Ad Hoc Networks (MANETs). This will be followed by a discussion of what causes latency in WMNs. This latency will be reduced through the combination of three methods; unidirectional noninterfering channels, cut-through and wormhole switching. How these same methods are implemented in existing wired networks through ATM, MPLS, QoS mechanisms and traffic engineering will be addressed next. The final section will address existing mechanisms used to reserve resources and establish paths.

15.2.1 Wireless LAN, Wireless Mesh, Ad Hoc Networks, and MANETs

First we define a wireless LAN, a WMN, an ad hoc network and a MANET. Although there are many forms of wireless network standards, we focus on the

802.11 standards because it is a well know example of a contention-based MAC protocol:

- Wireless LAN: a wireless single-hop, last quarter-mile to one-mile network, consisting of an access point, usually to act as a gateway to wired services and wireless workstations. Commonly referred to in the 802.11 standards as Infrastructure basic set of services (BSS).
- Wireless mesh network (WMN): a peer–peer, self-organizing, wireless multihop network. Its topology can be relatively static or dynamic.
- Ad Hoc Network: a network characterized by temporary, short-lived relationships among nodes. In 802.11 standards these are usually referred to as Independent BSS networks.
- Mobile Ad Hoc Network, MANET: a peer–peer, self-organizing, wireless, possibly mobile multihop network. Its topology can be relatively static or dynamic. The mobile aspect of this term also refers to user mobility throughout the network as in Mobile IP.

The scope of this discussion concentrates on WMNs that are relatively static, in the geographic sense. If this network was supported by current 802.11 hardware, whether it be 802.11 a, b, or g, it would have to be supported by either a reactive routing protocol such as ad hoc on-demand distance vector (AODV) or dynamic source routing (DSR), or a proactive protocol such as optimized link state routing (OLSR). The routing protocols are necessary to allow the network topology to be determined so that traffic can be forwarded hop-by-hop from the source to the destination. Also, 802.11 provides a medium access control (MAC) layer that is a contention-based multiaccess protocol, which allows every node within interference range to compete for the common shared medium (channel).

Taking all of this into consideration, it is evident that nodes in wireless mesh architectures take on the functions of host and router usually with only a single broadcast interface. This is the challenge and reason why 802.11-based (and similar) WMNs do not perform well. The next subsection will expand upon this observation.

15.2.2 Latency in Wireless Mesh Networks

Ramanathan made the following arguments as to why MANETs performance lags behind that of wireline networks [2]:

- "[MANET] operations are hop-centric in that processes are terminated and re-initiated at every hop." This contributes to the end-to-end delay because of processing through the physical, MAC/link, and network layers, re-queuing at the network and MAC layers and recontention for channel access. The thirst for increased bandwidth in these networks will result in MANETs operating at higher frequencies with inherent shorter range. This will result in even more hops from source to destination increasing end-to-end delay through aggregation of the hop-centric delays described above. As more bandwidth is obtained, higher data rates are achieved, which means delay has even more of a negative impact on performance.

- The wireless physical layer is designed for single-hop (to support wireless LAN), not relaying (to support wireless mesh). The physical layer must be redesigned so that relay is a primitive process to avoid the chain of receive-store-process-queue-forward-contend.

Our initial findings through OPNET simulation found the aggregated time spent in the nodal chain of receive-store-process-queue-forward-contend was approximately 11 ms [4]. By using an OPNET relay switch model [5] to implement label-switching, the chain was reduced to receive-forward with a nodal delay of approximately 0.2 ms (a 55-fold improvement). These findings are not far off from Ramanathan's findings during analysis of his similar architecture [6]. He proposes that in the next 3–5 years a 10,000 node network would at most experience 100 ms roundtrip delays. This calculation is made given the diameter of such a network to be 140 hops with an average per hop latency of 0.35 ms. He also states that latency in current MANETs is typically 8 ms per hop. The actual times differ by a factor of 2.5, but an increase in performance still remains (in Ramanathan's analysis about 22-fold).

With these results, there is definite motivation to pursue this performance improvement and the question becomes "how can a fresh physical layer design be developed to support circuit-switching?" The issues involved include multiple channels, circuit-switch design (FFTs and IFFTs), virtual-cut through and wormhole routing and their application to WMNs, low latency wired solutions, QoS, traffic engineering and resource reservation protocols.

15.2.3 Unidirectional Noninterfering Channels

The purpose of a unidirectional noninterfering channel is to transmit traffic from a transmitting node so that it can be received by an intended receiving node without interfering with other nodes. These one-way noninterfering channels can be derived in the wireless environment by using frequency, space and/or time multiplexing (or combinations thereof).

15.2.3.1 Frequency Multiplexing

Traditional frequency division multiplexing consists of dividing up a portion of the radio spectrum and transmitting and receiving on different radio frequencies separated by guard bands to avoid interference. This method of frequency multiplexing does work but is very inefficient because of the amount of the total frequency spectrum being used and the overhead of the necessary guard bands that support no traffic at all.

An alternative method was developed in the 1960s to chop up a single large frequency channel into multiple noninterfering or orthogonal subchannels. This method is referred to as orthogonal frequency division multiplexing (OFDM). In the simplest terms; at the modulator/transmitter, multiple coded signals for each subchannel create a composite waveform through the use of an inverse fast Fourier transform (IFFT) and at the receiver/demodulator, a fast Fourier transform (FFT) is used to extract each of the subchannel signals. IEEE 802.11a and 802.11g use OFDM to provide higher data rates by dividing the data traffic evenly across the 48 subchannels using a single modulation/coding scheme (BPSK, QPSK, 16-QAM or 64-QAM) chosen based upon the channel conditions and then modulating the subchannels into a single OFDM channel. IEEE 802.16 uses OFDM with time-division multiplexing to provide time slots, called OFDM-augmented (OFDMA), to allow for further subchannelization of an OFDM signal.

The future use of OFDM that contributes to this research is that OFDM subchannels could serve as individual low-speed channels instead of aggregating all of the OFDM subchannels to serve as a single high-speed channel. With the addition of future technologies being developed today, such as MEMS-based Hi-Q filters [7] to reduce receive signals from being overpowered by simultaneous transmit power in the same band, it may be possible to send traffic on some subcarriers while receiving traffic on others simultaneously without interference, thus providing unidirectional noninterfering channels.

15.2.3.2 Space Multiplexing

As radio signals travel through space, they degrade. In other words, the signal to noise ratio (SNR) will decrease over distance [8]. This implies that these signals have a certain range beyond which the signal is unusable for passing traffic between a transmitter and receiver. Commonly referred to as path loss, this loss in free-space is dependent upon the frequency of the radio wave. The equation for free space path loss (received power $P_{y,\text{PathLoss}}$ in Watts at receiving node y) is

$$P_{y,\text{PathLoss}} = P_x G_x G_y (\lambda/(4\pi d_{x,y}))^2, \tag{15.1}$$

where λ is the wavelength of the transmit signal (equals speed of light divided by the frequency) in meters, $d_{x,y}$ is the distance in meters between transmitter x and receiver y, P_x is the transmit power of transmitter x in Watts, G_x is the transmit gain and G_y is the receiver (y) gain.

The signal can also be reduced by walls, other obstructions or windows, but also boosted by antennas and amplifiers. TotalLoss in dB can be calculated as

$$\text{TotalLoss} = \text{PathLoss} - \text{ObstacleLoss} - \text{LinkMargin} + \text{RXantennagain}. \tag{15.2}$$

Given all of the above, this understanding of the RF environment presents the space aspect of unidirectional noninterfering channels. Suppose there is a transmitter node, and a receiver node separated by a distance d, that allows for a high enough SNR

Fig. 15.1 Representation of reception distance. Node x is transmitting. The reception distance is d_r

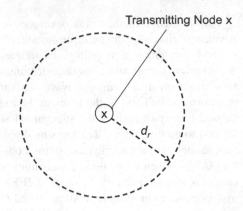

Transmitting Node x

(3) so that unidirectional traffic can be transmitted from sender to receiver (see Fig. 15.1). This distance d_r, is referred to as the reception distance. In this case, both transmitter and receiver are using omnidirectional antennas. If there are no other transmitters in the area, a noninterfering unidirectional link has been established between these two nodes.

$$\text{SNR} = P_y(x)/P_{\text{thermal noise}} \tag{15.3}$$

The variables for the SNR equation (15.3) are $P_y(x)$, the receive power (Watts) of transmitter x at the receiver y and $P_{\text{thermal noise}}$, the background noise (Watts), which is calculated based upon a receiver noise figure, Boltzman constant, temperature and bandwidth. If the free space path loss is being used to estimate the receive power of transmitter x received by the receiver y, $P_y(x) = P_{y,\text{PathLoss}}$.

Suppose another transmitter, z, is transmitting at the same frequency at a distance from the receiver, y, so that the receiver can receive the new transmitter's signal, but far enough away from the sender that the sender could not detect it (if it had a receiver receiving at the same frequency) (see Fig. 15.2). With both transmitters transmitting on the same frequency at the same time, the receiver detects both signals but is unable to discern the data from either transmitter because both transmissions interfere with each other. This is an example of the "hidden node problem."

Up to now, we have basically been presenting issues with regard to a node's reception range and its affects upon a receiver. There is another element that must be taken into account when considering space multiplexing in a wireless environment and that is the interference range.

A node's interference range can be thought of as further than the reception range (too far for a receiver to successfully receive a transmission) but close enough that it could interfere with a receiver receiving a transmission from another node that is within that node's reception range. This is depicted in Fig. 15.3. d_r is the reception range and d_i is the interference range. When taking into account the interference range along with the reception range in the space dimension, a meaningful

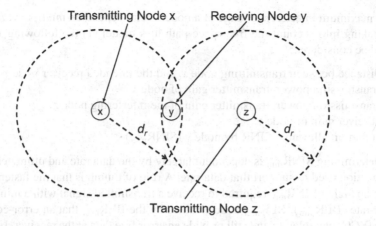

Fig. 15.2 The hidden-node problem

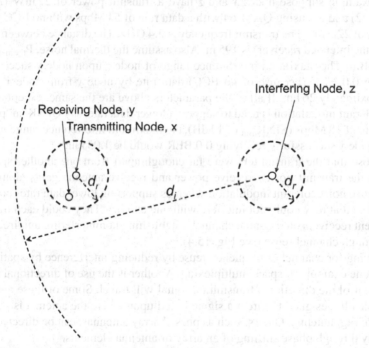

Fig. 15.3 Reception distance versus interference distance

measurement to determine if a transmission can successfully be received is signal-to-interference-noise ratio (SINR). SINR (4) differs from SNR (3) in that it accounts for the interference-noise along with the background noise in calculating signal-to-noise ratio.

$$\text{SINR}_y = P_y(x)/(P_{y,\text{thermal noise}} + P_y(z)) \qquad (15.4)$$

The maximum interference range of a node, y, can depend on multiple variables. If just taking into account the free space path loss model (1) the following variables could be considered:

- Distance between transmitting node x, and the intended receiver node y
- Transmission power, transmitter gain of node x
- Transmission power, transmitter gain of the interfering node z
- Receiver gain of node y
- Minimum allowable SINR by node y ($SINR_{min}$)

Determining $SINR_{min}$ is dependent largely by the data rate and ultimately by the modulation used to support that data rate. A rule of thumb is that the faster the data rate, the higher $SINR_{min}$ required to receive a transmitted signal with a minimal bit-error-rate (BER_{min}). BER_{min} is determined by the BER_{max} that an error-correcting code (ECC) can tolerate and still provide an error free data at the receiver. If no ECC is used at the receiver, then BER_{min} is 0.

For example suppose nodes x and z have a transmit power of 22 mW (P_x and $P_y = 0.022$) and are using QAM-64 with a data rate of 54 Mbps with no ECC with a $SINR_{min}$ of 22.5 dB. The transmit frequency is 2.4 GHz. The distance between nodes x and y (the intended receiver) is 195 m. Also assume the thermal noise, $P_{thermal\ noise}$, is −95 dBm. The maximum interference range of node z upon node y successfully receiving 0.0 BER (because of no ECC) data rate by node y from node x would be approximately 26 km. If all of the parameters above are the same, except a more SINR tolerant modulation is used to support a lower data rate, say QPSK supporting a data rate of 18 Mbps ($SINR_{min}$ of 14 dB), the maximum interference range of node z upon node y successfully receiving 0.0 BER would be 1.05 km.

Suppose that the transmitters were far enough apart from one another spatially and that the transmit power, receive power and receive antenna gains along with interference noise tolerant modulation schemes supporting lower data rates could be adjusted so that they could not interfere with one another. They could each transmit to different receivers on the same channel, establishing noninterfering unidirectional links through channel reuse (see Fig. 15.4).

Allowing for channel or frequency reuse by reducing interference by spatial distance is one example of space multiplexing. Another is the use of directional antennas to control the direction a transmitted signal will travel. Some of these antennas are physically designed to direct a signal based upon where the antenna is pointed, such as a Yagi antenna. Others, such as phased array antennas, can be directed electronically through phase shifting of an array of antenna elements.

Channel reuse is one aspect of space multiplexing that will be used to provide noninterfering unidirectional channels to support this research. Directional antennas can also be used to contribute.

15.2.3.3 Time Multiplexing

Time Division multiplexing involves assigning precoordinated periods of time when various transmitters can transmit traffic over a frequency. A schedule of when each

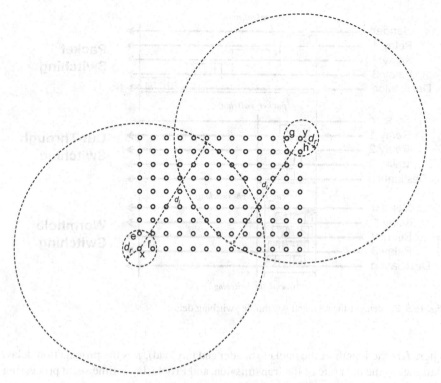

Fig. 15.4 Example of channel spatial reuse. Both nodes x and y are transmitting and have reception range of d_t and interference range of d_i. Both nodes x and y can be transmitting on the same channel to receiving nodes e and f, and g and h, respectively because the other transmitting node's interference ranges are not within range of their intended receivers

transmitter can transmit can be derived so that multiple transmitters can use the same frequency with no interference. If receivers know the schedule they would know when to listen in on the frequency to receive their traffic. This results in another method to provide noninterfering unidirectional channels.

15.2.4 Cut-Through and Wormhole Switching

Current WMNs are hop-centric. Packets are received, stored and then forwarded at every intermediate node from the sender to the destination. This method is usually referred to as packet-switching or store and forward. It causes inherent delay because the packet is received completely at each intermediate node, buffered, processed, and then forwarded to the next intermediate node (see Fig. 15.5). The total delay for an n hop routed path for a packet is

$$[n(p + (L/\text{DataRate}))] + [(n-1) \times \text{ProcDelay}], \tag{15.5}$$

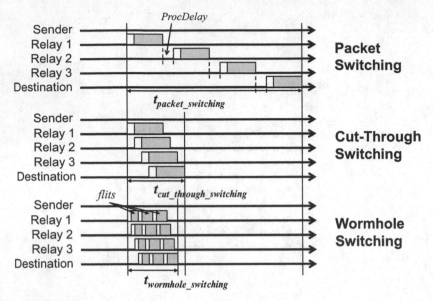

Fig. 15.5 Packet, cut-through and wormhole switching delay

where L is the length of the packet (header and payload), p is the propagation delay, DataRate is the data rate of the transmission, and ProcDelay is time spent processing the packet before forwarding it.

A way to reduce the latency of these mesh networks is to adopt a *modified cut-through switching* or *wormhole networking* approach. Either of these methods could be implemented using the frequency switch proposed above. Cut-through switching is possible by pre-establishing the path from sender to destination by setting each of the switches so that packets are effectively cut-through as they arrive. This eliminates the processing time if the outgoing subchannel decision is based upon the incoming channel. In optical networks, this is the method used because there is no option to buffer photons. If the switching decision is made based upon the header, the processing time is significantly reduced compared to the processing experienced in store and forward packet switching. A calculation of the delay experienced with cut-through packet switching can be seen in Fig. 15.5. The delay experienced with cut-through packet switching is

$$[n(p + (\text{HeaderLength}/\text{DataRate}))] + [(L - \text{HeaderLength})/\text{DataRate}]$$
$$+ [(n-1) \times \text{HeaderProcDelay}], \tag{15.6}$$

where HeaderLength is the length of the packet header and HeaderProcDelay is $\ll$ ProcDelay.

Wormhole switching is another method used to reduce the latency. Packets are further broken up into flits. The flit is a unit of message flow control. Actually, wormhole switching is just cut-through switching using flits. By using this method a packet is pipelined through the network to the point that during the packet's transition through the network it is physically spread across the network, portions of it either propagating among intermediate nodes or residing temporarily in buffers (see Fig. 15.5) [9, 10].

The movement to cut-through or wormhole switching from the current store and forward wireless mesh architecture would reduce end-to-end delay significantly. But there are a few challenges to implementing it in the wireless environment compared to the wired or optical environment.

15.2.5 Low-Latency Wired Solutions: ATM and MPLS

Cut-through switching is well known in the wired or optical world. Since the early 1980s, ATM switching technology was driven by the need to cut down on latency to support multimedia transmission requirements through virtual circuits (VC). ATM switching provided high-speed networking. But with the popularity of Internet and TCP/IP-based applications throughout the 1990s, it was evident that IP routing, though slower, was not going away. The disjoint use of ATM switching to provide the high-speed backbone to support IP routing was the trend. As the 1990s began to come to a close, the networking community was searching for ways to map the IP architecture on the ATM networks. The result was (MPLS) [11].

MPLS is a label-switching protocol that integrates layer 2 switching with layer 3 routing. An MPLS network consists of edge devices known as Label Edge Routers (LERs) and Label Switching Routers (LSRs). A mesh of unidirectional paths, known as label-switched paths (LSPs) are built among the LERs so that packets can enter the network at an ingress LER and be transported to the appropriate egress LER. As a packet enters the network, the ingress router determines which forwarding equivalence class (FEC) the packets belong to. Packets that are to be forwarded to the same egress point in the network along the same path are said to belong to the same FEC. Packets belonging to the same FEC are forwarded with the same MPLS label [12].

Figure 15.6 presents an example of how packets are forwarded using MPLS. Host 1 transmits two packets to its local router (LER A) with one packet addressed to H3 (Host 3) and the other addressed to H2 (Host 2). Each LER and LSR has an already established label switch table that has entries to support unidirectional paths from Host 1 to Host 2 and Host 1 to Host 3. These settings are set by a path establishment or LDP. When the packets arrive at ingress LER A through local interface "in," the header for each is read, a next-label is appended to the header and sent on the appropriate output interface based upon the label switch table entry. The same steps occur at each intermediate LSR until the packet is received at the egress router where the last label is stripped so that the original header remains and the packet is forwarded to its destination host.

The advantage of label-switching is that no time is spent at each router determining the next hop in the route, because only the label is read to do a quick look-up

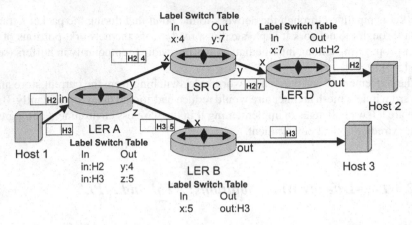

Fig. 15.6 MPLS example

for the outgoing interface and append the next label to the original header. Cut-through switches facilitate this kind of switching method. Also, as long as noninterfering unidirectional links are possible for the internodal links, label-switching can be implemented.

By having the ability to direct traffic through explicit paths and also be able to reserve nodal resources of buffering and bandwidth (in terms of controlling the outgoing service rate through statistical multiplexing), traffic engineering is possible. This will be discussed in more detail in the next subsection.

15.2.6 Quality of Service and Traffic Engineering

With the ability to circuit switch traffic through explicit paths, label-switching provides the missing mechanism to traffic engineer and provide actual QoS to WMNs.

QoS defines the traffic requirements for a traffic flow to support a session. It can be defined in terms of bandwidth, end-to-end delay, jitter (delay variance), packet loss, and priority or class with regard to traffic from other sessions. Providing a level of QoS, requires the translation of traffic requirements into quantifiable nodal and link resources that must be reserved to meet the traffic requirements. These nodal resources are the buffer space used for queuing traffic and the bandwidth, which is a combination of the node's buffer service method and rate along with the link bandwidth to support the service rate. The link bandwidth resource can also be thought of in our paradigm as the channel assignment and the maximum bandwidth it can support.

Traffic Engineering uses QoS mechanisms to achieve the goal of improving network performance (increased throughput, decreased delay) and make better use of network resources by better matching them with the traffic demands. This requires a global view of the network topology and knowledge of all of the traffic requirements upon it to determine which routes should be used to meet these requirements. This goal leads to some interesting challenges. The first is a global view of the network topology, which in WMNs must be constantly updated because of link changes as a result of noise or mobility. The second is a complete inventory of resources that are available and which are reserved. These include channel assignments, nodal buffer space and queue servicing method and rate. Third is gaining knowledge of the traffic requirements. These each can be addressed by a combination existing ad hoc routing protocols, QoS mechanisms, and resource reservation protocols. The next discussion focuses on resource reservation protocols and their interaction with routing and QoS mechanisms to meet these challenges.

15.2.7 Resource Reservation and Path Establishment

Resource reservation protocols are required to implement traffic engineering by reserving the nodal and link resources of the network to meet the QoS requirements of the traffic. These protocols are used to establish an explicit path for traffic to flow from ingress node to egress node. They require some knowledge of the network topology to establish the path. This knowledge usually comes from the routing protocol at the network layer. The resource reservation protocol (RRP) also interacts with multiple layers of the communications stack from the session down to the physical layer (considering the circuit switched label-switching paradigm proposed).

15.2.7.1 Determining the Network Topology Using Ad Hoc Routing Protocols

Resource reservation protocols interact directly with the network layer routing protocols so that the reservation protocol messages can be routed from the ingress node to the egress node. In our paradigm where label-switching is supported by multiple channels (or subchannels), interaction with the routing protocol is necessary to calculate the total network topology in the form of local link states among paired nodes. This is necessary to calculate channel assignment interference graphs (discussed later in this chapter).

There are many ad hoc routing protocols available for use in WMNs. Overall they can be separated into two categories, reactive and proactive. Reactive protocols do not take the initiative for finding a route to a destination until the route is requested. The route query is discovered "on-demand" by flooding its query through the network. This results in reduced control traffic overhead at the cost of increased latency in finding the route. Examples of reactive protocols are AODV [13], DSR [13] and temporally-ordered routing algorithm (TORA) [14]. Proactive protocols are based on the periodic exchange of control messages. Some messages are sent locally to enable a node to know its local neighborhood and some are sent throughout the entire network. This permits the exchange of the whole network topology among all nodes in the network. The advantage of proactive protocols is that they can immediately provide the required routes when needed. The cost is more use of the bandwidth for its control messages. Examples of proactive protocols are destination-sequenced distance vector (DSDV) [13], source tree adaptive routing (STAR) [15], and OLSR [16].

Given that determining the total network topology is a requirement to calculate channel assignment interference graphs, a proactive ad hoc routing protocol is the better choice for our use.

15.2.7.2 The Functions of a Resource Reservation Protocol

The basic functions that a RRP performs are:

- Path Establishment: This involves either label-distribution among Label Switched Routes (LSRs) and/or setting internodal circuit-switches.
- Resource Reservation: The reservation of internodal buffer space and link bandwidth. This is usually a combination of buffer space, queuing service method and rate, and channel assignment.
- Path Tear Down: Once the resources and path are no longer needed to support a traffic flow and its QoS requirements, the path must be torn down (labels released and circuit-switch settings reset) and the resources must be released (buffer space and channel assignment).

15.2.7.3 Two Existing Resource Reservation Protocols: CR-LDP and RSVP

Two resource reservation protocols in existence are constraint based routing – label distribution protocol (CR-LDP) and RSVP [17]. Both are commonly used in wired and optical networks. Below are descriptions of their operation:

CR-LDP. Figure 15.7 presents an example of independent control for downstream on demand mode of LDP. (1) LER A requests a label from label switched

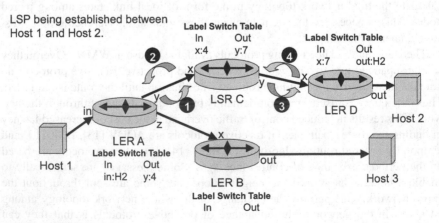

Fig. 15.7 CR-LDP example

router (LSR) C to support a path from Host 1 to Host 2. (2) LSR C selects the label "4," makes an entry in its own label switch table for the label assigned to input inter- face x "x:4," and distributes the label "4" back to LER A. LER A enters the label into its label switch table as the label assigned to output interface y "y:4." (3) LSR C requests a label from LER D to support the path. (4) LER D selects the label "7," makes the entry in its own label switch table for label "7" assigned to input inter- face x "x:7," and distributes the label "7" back to LSR C. LSR C enters the label into its label switch table as the label assigned to output interface y "y:7." LER D also realizes that Host 2 is local, and makes the entry for its local output interface "out:H2."

CR-LDP, in addition to the steps for LDP above, reserves the nodal resources required to support the traffic flow's QoS requirements in LER A, LSR C and LER D as the label requests are made.

RSVP. RSVP is a receiver-based RSVP. Messages that come from the receivers along the path are used to make the reservation. This allows for better support of multicast flows that are diverse or dynamic.

RSVP flows are identified by destination IP, destination port, and in some cases also source IP and source port. When RSVP is used to distribute labels for an LSP, the flows are identified by the labels. The QoS requirement for the flow is identified by the flow specification (flowspec). The flowspec is passed to the routers along the flow's path for examination so each router can identify if the resources are avail- able to determine if the reservation can be made. RSVP with extensions for Traffic Engineering (RSVP-TE), uses labels to identify flows and allows for flows to be established from ingress to egress routers instead of from source host to destina- tion host.

Figure 15.8 presents an example of how RSVP establishes a path to support a traffic flow. (1) An application at host 1 requests an RSVP path be established for a traffic flow with host 2 as the destination with a specific destination port and a specified flowspec. This request results in the sending of a PATH message from host 1 to the next hop in the path to the destination (provided by the routing protocol). The next hop, Router A receives the PATH message, identifies the destination IP

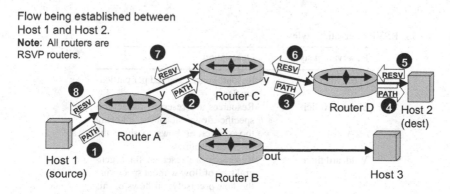

Fig. 15.8 RSVP example

and port and has the routing protocol calculate the next hop. Router A also makes an entry in its PATH state table identifying the flow by its destination IP and destination port (and also possibly the source IP and source port), the IP of the previous hop (Host 1), and the flowspec for the traffic flow. (2) It then copies the PATH message it received and sends it to the next hop (Router C). PATH state table entries are made for the flow with Router A as the previous hop. The same actions take place for step (3). (4) The path message is received by host 2, which determines it is the destination after examining the destination IP and destination port. (5) Host 2 then sends a RESV message to Router D. Router D examines the flow identifiers in the RESV message (destination IP, destination port, and possibly source IP and source port) and refers to its PATH state table to see if it has a matching path state. Once finding the matching path state, it establishes a Reservation State and reserves the resources identified in the flowspec and then copies and sends, step (6), a RESV message to the previous hop listed in the PATH state. The same actions occur at Router C and A, step (7). After Host 1 receives the RESV message from Router A (8), it realizes that all resources have been reserved along the route and it begins sending the flow's traffic.

The above example is an optimistic one and assumes the resources will be available at each node along the route. If a node determines that it can not support a traffic flow, it sends a PATH ERROR or RESV ERROR message to the sender of whom it received the PATH or RESV message and either alternative routes can be attempted to establish the flow or it can travel back to the source host to notify the application that the flow can not be supported.

For networks that are dynamic in nature, RSVP uses a *soft-state* method. Path states and reserve states are only held for a specified period of time. PATH and RESV refresh messages will have to be periodically sent to maintain the path (extending the timeout period). When it is determined that the flow is no longer needed, PATH TEAR and RESV TEAR messages can be sent to delete the path and reservation states at each of the routers and release the reserved resources.

RSVP Styles. Having a separate reservation for every flow at every node can consume resources very quickly in heavily loaded networks. Because of this, there are three different reservation styles provided by RSVP (see Table 15.1).

Table 15.1 RSVP reservation styles

Reservation style	Description
Fixed filter	Resources are reserved per particular flow
Shared explicit	Resources are reserved for several specific flows at once, allowing those different flows to share the reserved resources
Wildcard filter	Resources are reserved for a general type of flow without specifying the flow precisely; all flows of this type share the reserved resources

15.3 Related Work

This section will present practical issues associated with the integration of traffic engineering (TE) and its supporting mechanisms into wireless mesh networking. We will begin with an overview of TE, which is used to improve performance by matching resources with demands. To implement TE in any networking architecture, mechanisms must be integrated into the networking architecture to support it. Related work regarding the resource management of buffer space and bandwidth is presented in the areas of channel assignment, QoS and scheduling. It is followed by a survey of resource reservation protocols used in both wired and wireless networking architectures.

Next, the efforts of applying circuit switching to wireless networks using ATM and label-switching will be presented. This will be followed by work on multichannel wireless mesh next hop routing where multiple radios are used to provide multiple channels to a WMN. After that, will follow recent work by Ram Ramanathan on the use of orthogonal multichannel wireless mesh physical layer circuit-switching to provide ultra low latency WMNs. We will close with a presentation of systems-level work on wireless traffic engineering.

15.3.1 Traffic Engineering Overview

The goal of network traffic engineering is to improve performance (increase throughput, decrease delay) and make more efficient use of network resources by better matching the resources with the traffic demands. The first step is to determine the topology of the network and the state it is in with regard to internodal link bandwidth and nodal buffer space. Next, the traffic demands must be quantified with regard to the traffic load entering and leaving the network. The service constraints must also be quantified in terms of throughput, latency, jitter, ..., etc. The traffic engineering challenge is, given the topology and traffic demands of the network, which routes should be used to meet the service constraints? [18].

Many mechanisms are needed to support traffic engineering. First, resource management of the network in terms of link bandwidth and buffer space to support a certain level of QoS must be possible. Next, the ability to reserve these resources through resource reservation protocols must be available. Additional mechanisms can be used to reduce the amount of delay by eliminating the need for intermediate node routing processing time through methods such as label-switching and circuit-switching.

Defining the service constraints is another critical aspect of traffic engineering. With all of the mechanisms in place to reserve and control the resources of the network, they cannot be reserved correctly if the traffic requirements can not be quantified and translated into the QoS mechanism settings such as buffer size, bandwidth allocation, priority, etc. to support a minimum throughput or maximum allowable delay.

Determining what resources must be reserved to support a defined set of traffic requirements is difficult in a static wired or optical network backbone environment where the topology is relatively unchanging. In a wireless environment, traffic engineering is even more challenging with a constantly changing network topology because of the radio frequency environment and also nodal mobility.

15.3.2 Traffic Engineering Mechanisms

To implement Traffic Engineering in any network (whether it be wired, optical or wireless), the network architecture must be designed with mechanisms to support TE. These mechanisms include resource management, scheduling and reservation protocols.

15.3.2.1 Resource Management

To traffic engineer a network, you need the ability to allocate resources for each nodal link to support the flow. In a wired or single-channel wireless network this usually just consists of allocating bandwidth and buffer space. The common term associated with allocating these resources is QoS. The resource that needs to be allocated and managed in a multichannel wireless network is channel assignment; not only to avoid common channel interference and loops, but also to allocate bandwidth. Once the ability is in place to allocate these QoS and channel assignment resources, it is possible to optimally schedule their settings over time to support traffic requirements.

Channel Assignment. Optimal channel assignment scheduling for broadcast wireless networks (both single and multichannel) has been and continues to be an active research area. Ramanathan and Lloyd [19] discuss broadcast scheduling where links between stations are scheduled to avoid primary and secondary interference. They develop their own channel assignment algorithms for each kind of scheduling based upon experimental investigations that show that radio networks can be adequately modeled by planar or close-to-planar graphs. Based upon this they show that radio networks modeled by trees can be scheduled optimally and planar networks can be scheduled near optimally.

Not only can channel assignments be made to avoid interference, but they can also be made to support expected traffic loads. One example is the Hyacinth centralized channel assignment method [20, 21]. Their argument is that most past research efforts that exploit multiple radio channels require modifications to the MAC protocol and do not work on commodity 802.11 hardware. They propose one of the first multichannel multihop wireless ad hoc architectures that can be built using existing 802.11 hardware where each node has multiple 802.11 NICs operating on different channels. Their channel assignment method uses a greedy algorithm that takes into account some element of load balancing and link bandwidth budgeting where they assign channels based upon which ones currently have the least chance of interference and the most available link bandwidth. Another difference in this method is that they are not implementing unidirectional links. Each 802.11 link is bidirectional, using the 802.11 multiaccess MAC. Even with the lack of noninterfering unidirectional links, Hyacinth is able to increase total network goodput by a factor of up to 8 compared to current 802.11 based single channel architectures. If unidirectional links were supported, total network goodput could be increased even more.

Both of these previous works are centralized algorithms. Work on fully distributed algorithms began with Ma and Lloyd [22]. They developed a fully distributed algorithm for broadcast scheduling (FDAS) that provides interference-free time-slot scheduling with no global network information needed. Information is only needed with regard to its one and two hop neighbors. Another distributed method of channel assignment was applied to OFDMA based wireless ad hoc networks [23]. The authors presented a distributed algorithm for OFDM subcarrier (channel) allocation based upon dividing the network into clusters and superclusters and using a polling and allocation algorithm to allocate the channels. This work did not attempt to build an optimal schedule as much as provide the best wireless backbone connectivity.

Plenty of other research has been done in this field. Overall, the idea is to get the maximum channel use (and reuse) to optimize spectrum efficiency. The problem is getting the pool of channels to assign within a wireless network. Multichannel methods, such as Hyacinth, requires multiple radios, which becomes impractical. Generating orthogonal channels using a single channel radio using modulation techniques such as OFDM or CDMA may be another method.

Qos. QoS is the set of service requirements that must be met by the network while transporting a packet stream from a source to its destination. QoS metrics that guarantee a set of measurable prespecified service attributes to users in terms of end-to-end performance are delay, bandwidth, probability of packet loss and delay variance (jitter) [24]. The two tangible (controllable) resources that can be allocated at a node are bandwidth and buffer space. Flow service requirements can be identified at the packet level by various types of service (ToS) tags that give some kind of priority to the packet (such as from highest to lowest priority; interactive voice, interactive multimedia, streaming multimedia, excellent effort, standard, background, best effort). When a packet arrives at a node, the ToS tag can be accessed from the IP header to determine the amount of bandwidth and buffer space to be allotted for packets with that ToS classification. When the bandwidth and buffer space resources run out, packets are simply dropped.

Adaptive resource scheduling strategies have been developed based upon the classification of packets to manage bandwidth and buffer space. Valeroso et al. [25] presents such an algorithm used within a wired broadband ATM/ISDN network that adaptively reserves resources for three classes of packets depending upon network performance, allowing the system to adapt to changes in the network traffic load.

With the popularity of using label switching architectures to traffic engineer networks, such as MPLS, more methods have been developed to support QoS requirements. MPLS supports QoS requirements through ToS classifications on application or session traffic flows. The two methods developed are *IntServ* (Integrated Services) and *DiffServ* (Differentiated Services). *IntServ* consists of applications specifying the traffic that it will inject into the network, the *TSpec*, and the level of QoS it would like to receive from the network, the *RSpec*. Two different service classes that use the *TSpec* and *RSpec* are identified for *IntServ*, guaranteed service and controlled load. Guaranteed service provides a hard guarantee to an application in terms of bandwidth and delay. Some of the *TSpec* parameters key to setting the hard bound to this guaranteed service are peak rate, maximum packet size, burst size and token bucket rate. The *RSpec* parameter important to the guaranteed service is the service rate or the amount of bandwidth to be allocated to the flow. By identifying these characteristics of the traffic flow, the traffic bandwidth used by an application that generates data at a variable data rate and the maximum delay can be calculated. Controlled load simply tries to ensure that an application receives service comparable to what it would receive if it were running on an unloaded network or adequate capacity for the application (no calculated hard bounds as with guaranteed service). *DiffServ* divides traffic into a small number of classes and allocates resources on a per-class basis. The class of traffic is marked directly in the packet in the differentiate services code point (DSCP). The DSCP is checked at each hop (router) and the queue placement and priority is determined for that node (the Per Hop Behavior, PHB). One could think of *IntServ* as identifying the needs per application session and *DiffServ* as an aggregation of the needs of individual applications into a few common classes [11,12,26].

Scalability is the appeal for *DiffServ*, because *IntServ* QoS requirements are based upon application or session demands (per-flow basis). *DiffServ* is suitable for data-driven or control-driven traffic requirements because QoS requirements are based upon aggregate traffic classes that can be determined by monitoring the traffic at ingress or egress nodes or by service-level agreements, SLAs.

Similar mechanisms to support QoS were being applied to ATM wired and wireless networks. As TCP/IP based applications became more common and gained popularity, it became more apparent that ATM networks would be used to forward IP datagrams. MPLS provided a way to map the IP architecture onto ATM networks and similar traffic engineering network mechanisms could be used to provide QoS [11]. A relevant example of this is the Rapidly Deployable Radio Network (RDRN) QoS Architecture [27]. The RDRN included a high capacity wireless ATM backbone, which supported a cellular like ATM radio network for mobile users. An application specifying its service requirements from the RDRN QoS Architecture would do so with a flow descriptor. The flow specification and filter specification make up the flow descriptor. The flow specification holds two different components: the QoS that is requested by the host (similar to the *RSpec* in *IntServ*) and the traffic that will be generated by the host (similar to the *TSpec* in *IntServ*). The filter specification was simply a label, (source IP, source port, destination IP, destination port, and protocol), used to identify which flow descriptor is bound to which application session.

In 2000, the first Flexible QoS Model for wireless MANETs, FQMM, was developed [28]. The developers of this model took into account the scalability, signaling and router problems of implementing *IntServ* per-flow QoS models with some of the advantages of using *DiffServ* per-class models in the MANET environment. Their decision was to allow highest-priority traffic to use per-flow provisioning of resources while all other priority traffic would use per-class provisioning. Recently, a newer version of this same idea, a hybrid QoS model for mobile ad hoc networks (HQMM) [29], was proposed with a change in the per-flow provisioning from *IntServ* to INSIGNIA [30]. The major improvement here was INSIGNIA's use of a soft-state reservation protocol versus the hard-state protocol which FQMM used. Extending MPLS traffic engineering mechanisms has also been proposed to support QoS for Mobile IP [31, 32].

After reviewing these QoS models for MANETs, one has to realize that these were implemented in an omnidirectional broadcast with multiple access environment. Guaranteeing reserved bandwidth in this environment is extremely difficult, because bandwidth over the single channel cannot be controlled by a node because neighbor and two-hop neighbor nodes also compete for the same bandwidth. However, if a wireless ad hoc or mesh network existed where noninterfering channel assignments could be bound to each flow-link along with the list of required resources, these QoS models could experience considerable performance gains.

15.3.2.2 Scheduling

Scheduling is a method used to divide up the bandwidth by determining when various competing queues of traffic can be serviced/transmitted and for how long.

The RDRN uses time division multiple access (TDMA) scheduling to divide bandwidth between edge nodes and remote nodes. There are no real specifics here except that remote nodes (as many as 64) arbitrate for the available bandwidth of a particular beam formed by an edge node [33].

802.16 provides QoS algorithms for its point-to-multipoint (PMP) mode, the last-mile solution, but no QoS algorithms exist for its mesh mode [34]. Cao et al. [35] propose a QoS mechanism using 802.16's mesh mode coordinated distributed scheduling. During distributed scheduling, request and grant of channel resources (time slot scheduling) are delivered by mesh distributed schedule, MSH-DSCH, message among nodes, while every node sends its available channel resource table to neighbor nodes with MSH-DSCH messages. Requesting resources is basically a 3-way handshake; request, grant, ack. To achieve QoS features in the Mesh mode, the authors have designed a simple slot allocation algorithm for determining a reasonable transmission time by looking up the channel resource table after receiving a request and returning the detail of slot occupation. Their change to the algorithm involves setting two threshold check points to implement proper QoS (if the threshold is met, the network is not congested and treats all traffic equally; if the threshold is not met, pass high-priority traffic, send error messages to low priority traffic).

Significant work has been done in scheduling. But schedules can be developed and optimized only if the traffic requirements are known a priori.

15.3.2.3 Reservation Protocols

With the mechanisms in place to manage resources at the node and link levels, protocols are now needed to reserve these resources to support a traffic flow with certain QoS requirements. These end-to-end protocols are usually transmitted hop-by-hop from the source to the destination (and sometimes in the reverse direction) to establish unidirectional paths with resources reserved at every node in the path to provide the QoS required by a traffic flow. These protocols can be categorized in terms of establishing a hard state where resources at each node are released through a deliberate tear down of the path (initiated by either the source or destination), or a soft-state where resources are released when a time-out value expires. Soft-state reservation protocols require refreshes to keep the nodal resources reserved for the duration of a traffic flow. Because of the nature of these two different categories of reservation protocols, hard-state protocols usually support relatively static backbone link networks and soft-state reservation protocols support dynamic link networks. Another design feature usually included with these protocols is the ability to adapt to changes in the network topology in terms of self-repairing paths when they are broken.

Hard State Versus Soft State. Hard state resource reservation protocols support static backbone link networks. Soft-state resource reservation protocols support dynamic link networks. An established resource reservation protocol (RRP) in the wired and optical world is RSVP [17, 36]. The initial purpose of RSVP was to request specific QoS from the network for demand-driven traffic specified using the *IntServ* method. It is a hard-state protocol that requests resources to support a unidirectional flow by sending PATH messages in the forward direction (establishing a soft-state at each node) and then reserving the resources by sending a RESV message in the return direction along the same path (establishing a hard-state at each node). In addition to reserving resources at each node, RSVP has been extended to support label distribution and label switching protocols such as MPLS and overall traffic engineering. It also has been extended to support *DiffServ* [37].

Lee et al. [30], Mohopatra et al. [24], and Chlamtac et al. [38] provide an example of a (soft-state) flow management protocol designed to support flows in a dynamic mobile ad hoc network. The key component is INSIGNIA, an in-band signaling system that supports fast flow reservation, restoration and adaptation algorithms that are specifically designed to deliver adaptive real-time service in mobile ad hoc networking environments. Dynamic environment network state management is based on soft-state. The authors believe that the soft-state approach is more applicable to ad hoc networks because mobile soft-state relies on the fact that sources send data messages along the existing path. In other words, subsequent reception of a data packet at a router is used to refresh the soft-state reservation. The RRP fielded for the RDRN is based largely upon the INSIGNIA design [39].

Xue et al. [40] argues that a combination of soft-state and hard-state reservation should be used to avoid the waste of extra reservations that are experienced with INSIGNIA. INSIGNIA establishes flows by sending a flow request message with the first and each following packet to establish and maintain the reservation in node along the path. Periodic feedback is supplied by QoS reporting packets that are received out-of-band (along different routes) back to the sender). The QoS reporting packets facilitate application flow-control when QoS resource availability changes. When a bottleneck occurs, reservations change from real-time, requiring a maximum bandwidth, to best effort with a minimum bandwidth. This results in unusable reservations in the nodes preceding the bottleneck that will not be released until the time-out value is reached. Adaptive reservation and pre-allocation protocol (ASAP) requests path establishment and maintenance with SR (soft-reservation) messages in-band with the initial data packet and periodically with the data packets that follow. To reduce extra reservations, HR messages are sent out-of-band from the destination node along the path to the receiver to establish a hard-reservation, notify the sender of any changes in QoS availability along the path and also update/release any extra reservations should they exist in any node along the path.

An example of a hybrid hard/soft-state MAC layer reservation protocol for 802.11-based mesh networks is presented in Carlson et al. [41]. The authors present an end-to-end reservation signaling protocol to support QoS in an 802.11-based mesh network that operates in the existing MAC layer. The distributed end-to-end allocation of times slots for real-time traffic (DARE) protocol supports QoS by reserving time slots in nodes in a completely distributed manner. DARE reserves time slots in all nodes along the route between the source and destination of the real-time flow. Real-time traffic is transmitted only during these time slots so that the traffic can be relayed by each node along the path. Allocated time slots are protected from interference by adjacent nodes that are not part of the path by having those nodes abstain from transmitting during the time slots. The MAC DCF and DARE are used in parallel with data packets coming from nonreal-time applications using CSMA and real-time application packets using DARE. Their study shows that DARE provides constant throughput along with low and stable end-to-end delay for a reserved real-time flow, even for high loads. In contrast, 802.11e enhanced distributed channel access (EDCA), a priority based QoS scheme, only yields higher throughput for low network loads.

Adaptive Path Management. Considering that even a relatively static wireless backbone network is susceptible to noise and outside interference, it is unavoidable that the topology of these networks does change over time. In a dynamic mobile wireless network, even more so. Because of this along with the limited amount of bandwidth the network can support, it is inherent that resource reservation protocols must have the ability to adapt to this changing environment.

Liu and Rachudhuri [42] devised a braided-path method to establish a primary and alternate (backup) label switched path for wireless ad hoc networks called label switched multi-path forwarding. When a break in the primary path occurs, the wireless ad hoc label switching (WALS) protocol forwarder tells the IP layer to send route error messages back to the source of the path based upon the routing protocol. The source then initiates a new flow. The authors also present a method of detecting and suppressing unnecessary duplicated copies of a data packet in multipath forwarding (preventative duplicate filtering). Though this is applied to a layer 2.5 label switched path forwarding method, the same idea could be applied to label switching at the physical layer.

Instead of the reservation protocol establishing primary and backup paths, another alternative is to recover from a link or node failure by linking the upstream and downstream nodes around the point of failure with the intervention of the ingress or egress routers. Flexible MPLS signaling (FMS): provides such a flexible rerouting method in a mobile dynamic network [43]. FMS uses the underlying routing mechanism to establish the alternate path.

Chi-Hsiang, Moutftah and Hassanein [44], propose a graphical reservation scheme for mobile ad hoc networks (GRACE). GRACE uses geographically static cluster heads in regionally defined cells to provide a backbone for reservations to be made. Clusterheads are elected when mobile hosts move in and out of a cell area. If a current clusterhead has to move, another is elected to take its place and databases holding the status of the nodes in its geographical region are transferred. This allows paths to be maintained geographically as intermediate nodes move in and out of a geographical area. An interesting idea that could be applied to a network supported by a large number of wireless nodes.

The resource reservation and adaptive path management protocols mentioned above do work, but they only reserve resources using existing QoS mechanisms available to WMNs. What they do not address is the reservation of channel assignments, which are needed in this kind of networking architecture to provide noninterfering unidirectional links to support traffic engineering. Work on this problem ties the existing research on channel assignment and resource reservation protocols together.

15.3.3 ATM-Based Wireless Networks

The demand for wired digital packet networks to support data, voice and video traffic drove the development of packet network technologies to support circuit-switching. One of the successes of this period is ATM. It combines the best attributes of packet-switching and time division multiplexing, by using short fixed length cells (53 bytes) with abbreviated headers. The headers would be used at the ATM switch to forward the packet to a particular port. ATM is connection-oriented so these switch settings would be based upon a flow establishment (virtual circuits), which includes QoS requirements defined for the flow. The QoS requirements would be supported at each switch by determining when competing cells queued for an outgoing port would be released. A logical next-step to implement label-switching in a wired environment was to use ATM as the switching technology. The same can be said for label-switching in the wireless environment.

The rapidly deployable radio network, RDRN, developed at the University of Kansas in the late 1990s/early 2000s is an example of using ATM switching technology to provide QoS in a wireless environment [27, 33, 39, 45, 46]. The RDRN consisted of edge nodes (EN), which could be used to establish a wireless ATM network, and extend a wired ATM network. The ENs could also provide user service at remote nodes (RN). ENs forming the ATM wireless backbone were interconnected with high-speed (10 Mbps) radios and directional antennas. EN to RN links were established using low-speed (5 Mbps) radios with multidirectional antennas with the EN transmitting at one frequency and each RN transmitting on an different frequency. The network topology was maintained out-of-band by a separate broadcast orderwire network.

The work on RDRN included reservation protocols, adaptive path management and topology change management and though the ATM backbone architecture was bulky, the overall idea met with considerable success. QoS and actual traffic engineering could be provided in this wireless environment by ATM and its inherent QoS scheduling mechanisms, but they were all provided at layer 2 and higher. Delay was still inherent because of the header processing and forwarding table lookup. Pushing the switching technology from the upper layers down to the physical layer was (and still is) where more performance gains could be made.

15.3.4 Broadcast Wireless Mesh Layer 2.5 Label-Switching

With the consumer fielding of low-cost, readily available wireless (WiFi) technology, IEEE 802.11 has become a mature and proven standard. Because it has features to support wireless mesh networking through its Independent BSS configuration, considerable research has been conducted on how to get IEEE 802.11 to support QoS. This has resulted in various recommendations to modify the standard to support broadcast wireless mesh label-switching at layer 2.5.

Acharya, Misra and Bansai [47, 48] present a layer 2.5 label-switching protocol, data-driven cut-through multiple access (DCMA). The authors argue that the IEEE 802.11 medium access scheme was designed implicitly for receiving or transmitting a packet, but not for forwarding operations. To support forwarding operations and label switching, they modify the 802.11 RTS and ACK frames to include the next hop MAC address and label along with implementing label switching tables in the network interface card (NIC). This allows follow-on packets for a traffic flow to be forwarded through the network without delay caused by referencing of the routing information at the higher layers. This method provides a significant improvement in the forwarding of packets in a wireless network, but still the label forwarding information is accessed at layer 2.5 in the NIC card.

Another example of a layer 2.5 wireless mesh label switching protocol using MPLS is Lilith [49, 50]. Though the design addresses the integration of routing and label switched path (LSP) establishment, maintenance and optimization, along with throughput results showing how this method positively impacts the ability to do wireless mesh traffic engineering, it does not address the media access problem in the wireless broadcast environment.

The problems with implementing label switching at layer 2.5 is that inherent delay still exists because of processing up to layer 2.5 at each intermediate hop and also there is still contention for the multiple access medium. If the switching element could be pushed down to the physical layer, the processing delay could be reduced, thus reducing the end-to-end delay. Also, if the MAC could be noncontention based and the network architecture designed so that there is no interference, the internodal links could become unidirectional and wire-like. This would allow for wired solutions that support actual QoS, such as MPLS, to work on WMN architectures.

15.3.5 Multichannel Wireless Mesh Next Hop Routing

Other research to implement QoS and Traffic Engineering in a wireless mesh architecture was to use multiple IEEE 802.11 wireless interfaces per node. This could allow for multiple channel assignments throughout the network to reduce channel interference and contention across the network at the MAC layer.

Hyacinth, a novel multichannel WMN architecture built using IEEE 802.11 technology was proposed by T. Chiueh, A. Raniwala, R. Krishnan and K. Gopalan [20, 21]. Each Hyacinth node has multiple 802.11 compliant NICs, each of which is tuned to a particular radio channel. Channel assignment decisions are based upon reducing the amount of channel interference or contention for the multiaccess broadcast medium and also to balance the traffic load across the network. By using this technique, they could establish a WMN with an increase in total network goodput by a factor of up to 8 compared to the conventional 802.11 single-channel mesh network architecture with each node having just 2 NICs operating on different channels.

Though the performance improvement was significant and the design supported the use of existing commodity hardware, this method is not the optimal way to improve WMNs. First of all, there is still competition for the medium. The Hyacinth channel assignments support multiple bidirectional links between nodes and the 802.11 MAC is required to deal with the contention. Their channel assignment algorithm provides better performance by reducing the interference (compared to that found in a single-channel architecture) and load-balancing. The next inefficiency is caused by the processing time to store and forward packets as they are relayed from one wireless interface to another. If more NICs were added to each node to support noninterfering unidirectional links more performance could be gained because eventually the contention caused by interference could be defeated, but the nodal processing time would still remain. Also adding more NICs becomes cost prohibitive.

15.3.6 Orthogonal Multichannel Wireless Mesh Physical Layer Circuit-Switching (Virtual-Cut through, Wormhole Routing)

The processing delay imposed by layer 2.5 label-switching and multichannel wireless mesh next hop routing still contributes to the overall end-to-end latency in WMNs. Early computer network researchers, such as Kermani and Kleinrock [9], and interconnection network researchers, such as William Dally [51–53], also found this be true in network architectures of the time. Ram Ramanathan's analogy of describing this store-and-forward method as like a "subway design that has passengers get off at every intermediate station enroute to their destination, go outside the station, get in line for a fresh ticket, wait all over again for the next train and board it" presents the inefficiency of this method quite well [2].

In [2], Ramanathan proposes that the conventional IP stack will not be able to adequately support high-bandwidth requirements of MANETs. His approach is to use circuit-switching (virtual cut-through, wormhole routing) to support such requirements. He provides a more detailed explanation of his approach in Ramanathan et al. [6]. He proposes using orthogonal channels at each node to provide circuit-switched paths from end-to-end. To support this he developed a path set up mechanism, path access control, PAC, and a switching mechanism, relay oriented physical layer, ROPL. PAC is an RSVP-like mechanism for reserving resources (floor) multiple hops at a time. ROPL is a cut-through relay that pipelines bits through the receive and transmit chains at the physical layer (a cut-through relay mechanism).

15.3.7 Wireless Traffic Engineering

Though much previous research has been done on developing mechanisms to support traffic engineering in WMNs, little has been published on wireless traffic engineering.

De Greve et al. [54] and Greve et al. [55] present a wireless traffic engineering solution for distributing internet service from various gateways to trains by establishing a wireless backbone network. The wireless nodes have multiple wireless interfaces that can be assigned different radio channels. They propose techniques for improving the throughput within this wireless backbone network by the intelligent distribution of neighbor mesh nodes over the available amount of wireless interface cards and distributed techniques for minimizing link interference by assigning different communication channels to the physical interfaces. Basically, they choose the paths that can support the traffic flow that minimize the number of senders interfering with the path. Though an interesting method, the authors worked within a reduced problem space because train routes and their schedules are predictable, which means the traffic requirements are predetermined.

15.4 Thoughts for Practitioners

Improving performance in WMNs can be summed up as follows:

- The goal is to reduce the internodal processing time when packets are forwarded from one node to another until they reach their destination.

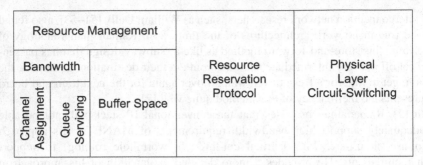

Fig. 15.9 Building blocks for an improved performance wireless mesh network architecture

- One way to reach that goal is to implement methods similar to those used in wired or optical networks, such as label-switching.
- The QoS mechanisms to reserve buffer space and bandwidth are already in place.
- Implementing label-switching in these networks using the existing 802.11 MAC is not a good fit because there is still contention for the common broadcast medium and there is still processing time involved even in layer 2.5 label-switching solutions.
- Implementing label-switching down to the physical layer completely reduces processing time, but it requires MAC involving the use of multiple channels provided by frequency, space and/or time multiplexing (or combinations thereof).
- Resource reservation protocols to establish traffic flows would not only have to reserve nodal resources to meet QoS requirements, they would also act as the MAC by assigning these channels to set up non-interfering unidirectional internodal links throughout the network architecture. Such reservation protocols would cross multiple layers of the communications stack, including those involved with routing and session control. Also, a common network control path would be required in order for these protocols to work (probably the same path used by the routing protocol).

These findings support the combination of methods that serve as building blocks for a low latency WMN architecture (see Fig. 15.9).

15.4.1 Circuit-Switch Design Using FFT/IFFT

As part of Ramanathan's remarks as to why MANETs' performance lags behind that of wireline networks, he proposes that the physical layer must be redesigned so that the relay is a primitive process to avoid the chain of receive-store-process-queue-foreword-contend. By just working within the frequency domain and considering the OFDM modulation technique, the relay function as part of the design of the physical layer can be supported by implementing a frequency-switch. The organization of such a switch is presented in Fig. 15.10.

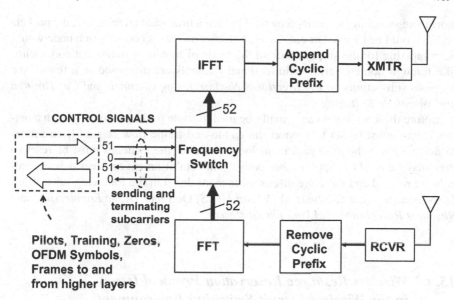

Fig. 15.10 Integration of the frequency switching into an OFDM-based demodulator/modulator transceiver design (relay implemented in the physical layer)

The OFDM demodulator/modulator could integrate such a switch between its FFT (demodulator side) and inverse FFT (modulator side) so that the transceiver could switch incoming traffic from one subchannel to another subchannel, realizing the relay function of the node at the physical layer (see Fig. 15.10). This design also allows for the node to act as a sender or terminator (destination) of traffic by sending and receiving frames to and from the higher layers.

15.4.2 Applying Cut-Through and Wormhole Switching to Wireless Mesh Networks

Cut-Through and wormhole switching have been implemented in wired, optical and interconnection networks for some time now. The traits of these architectures facilitate these methods of switching quite well. Because the wireless architectures are not so wire-like, there are a few challenges that must be overcome to implement these switching methods in wireless networks.

What are the traits of wired, optical and interconnection networks that facilitate cut-through and wormhole switching? First of all there is no contention for the wired or optical unidirectional link between two nodes. Also, there is no interference among links. This is quite different from wireless architectures, especially single-channel wireless broadcast architectures. Every node within interference range competes for the same broadcast medium (or single channel). If cut-through or wormhole switching were implemented in such a broadcast architecture, the packets or flits would be blocked or collide as the traffic increases because each node would be competing for the shared medium. Methods of making wireless networks wire-like through noninterfering unidirectional channels are discussed in a few of the previous subsections on *Unidirectional NonInterfering Channels* and *Cut-Through and Wormhole Switching*.

Before these switches can actually be used to route packets or flits through a network, they must be set to support the end-to-end traffic flow or path from sender to destination. When the path is no longer needed, the switches must be reset so they may be used to support other paths. Path establishment and tear down protocols are needed and these are discussed in more detail in the following subsections *Low-Latency Wired Solutions: ATM and MPLS*, *QoS and Traffic Engineering*, and *Resource Reservation and Path Establishment*.

15.4.3 Wireless Resource Reservation Protocol Issues in the Wireless Circuit-Switching Environment

A number of issues arise when trying to apply such resource reservation protocols in a wireless multiple channel circuit-switching environment. Below are some of the issues that are encountered during development.

Path Dependency Issues. As paths are set up and torn down over time in such a networking paradigm, dependencies occur based upon the current paths established in the network and the new path to be established or next path to be torn down. These dependencies between the paths that must remain and the path added or torn down, require switch settings either be changed or remain the same. This requires the path to be added or deleted to have a knowledge of what existing paths it will intersect that also has its destination (or egress node) in common.

Managing Channel Assignments in a Multichannel Omnidirectional Broadcast Environment. Assigning channels in an omnidirectional broadcast environment using multiple channels requires a considerable amount of management to avoid interference and possible loops. To illustrate some of the inherent management issues, I will begin presenting a notional wireless cut-through switch design and present the mechanics of channel assignment as part of a multiple path establishment scenario.

Figure 15.11 provides a functional diagram of a possible cut-through switch design integrated into the existing 802.11 wireless communications stack. This design has two transceiver interfaces. IFO is a standard 802.11 multiaccess MAC channel used for out-of-band signaling to support a resource reservation protocol (RRP), in this case similar to RSVP, used to assign switch settings to the "Cut-Through Switch." The second transceiver interface is an OFDM signal, in this case with ten subchannels to support cut-through switching. The "Cut-Through Switch" can be set to perform the following functions:

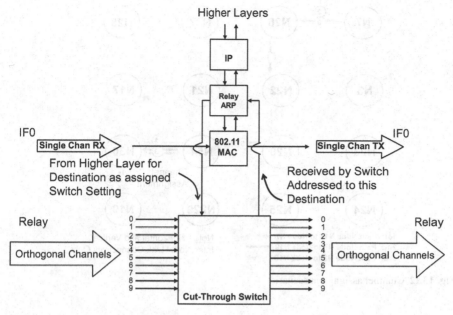

Fig. 15.11 Notional wireless cut-through switch functional diagram

- Cut-through from one incoming channel to one of the other outgoing channel (acting as a cut-through relay)
- Forward to the higher-layers a flit or packet destined for this destination
- Forward a packet from the higher layers with destination assigned to cut-through to an outgoing channel

The assumption given is that the OFDM subchannels can transmit on some subchannels and receive on the others simultaneously.

In section *Unidirectional Noninterfering Channels*, noninterfering unidirectional channels were presented. The channel assignment scenario depicted in Fig. 15.12 shows the channel assignments made to support three traffic flows. Channels are provided through frequency multiplexing with spatial reuse. The transmit power settings for each node provide only horizontal and vertical neighbor link connectivity. Each node has the cut-through functionality in Fig. 15.11.

Given that the internodal link topology is provided by the routing protocol, an interference graph is calculated for each internodal link channel assignment. These graphs are calculated to ensure that a channel assigned to support an internodal link causes limited interference to other internodal link channel assignments. Because we have to provide unidirectional links, the interference graph is a directional graph and it is based upon the following two-hop algorithm:

- "N" transmitting node
- "A" neighbor to "N" all links (N,A)
- "B" neighbor to "A" all links ((N,A),B)

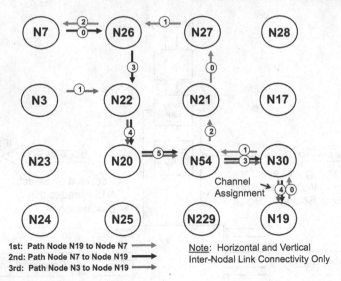

Fig. 15.12 Channel assignment scenario

- Conflicts

  - Any (_,N) pair: a node can not receive on a channel on which it transmits
  - Any (A,_) pair: a neighbor node can not transmit on that channel (possible collision)
  - Any (B,_) pair: hidden-node, possible loop could develop

Once the interference graph is calculated, the remaining noninterfering channel assignments per internodal link must be tracked. This information is based upon the internodal link channel assignments that have already been made to support existing paths in the network. Every time a channel assignment is made, the channels available list per link must be recalculated based upon the updated link channel assignments and the interference graph. All nodes in the network must be able to refer to a current version of this list to make channel assignments.

For example in Fig. 15.12, Path Node 19 to Node 7 has been established with the channel assignments made for each internodal link and the RRP begins to establish the Path Node 7 to Node 19. Channel 0 is assigned to the internodal link Node 7 → Node 26 and Channel 3 is assigned to internodal link Node 26 → Node 22. Why can't Channel 0 be assigned to link Node 22 → Node 20? Because it could interfere with link Node 7 → Node 26. Also it would cause a loop because Node 26 also receives on Channel 0 and cuts-through any packet destined for Node 19. This would result in a loop that would continuously generate duplicate packets destined for Node 19. It is also the reason why Channel 1 cannot be assigned for this link. Why can't Channel 2 be assigned for link Node 22 → Node 20? Because it would conflict with link Node 54 → Node 21. Node 21 also receives on Channel 2, which is in interference range of Node 22's transmission. Channel 3 can not be chosen because Node 22 receives on this channel. These checks would be made by referencing the link channel assignment list. More complex loops would also require loop detection algorithms that require knowledge of all neighbor node link channel assignments. Regardless, Channels 4–9 are available for assignment for this link. Which channel should be chosen? Should it just be the next available one (Channel 4) or one of the others? Could this decision have an impact on available link channel assignments in the future? Yes, it could. If there is knowledge of the ordering of paths to be established and torn down (along with QoS requirements), optimal channel assignment schedules possibly could be calculated to support the paths.

Now, suppose that this channel assignment scenario had more paths already established, to the point that link Node 22 → Node 20 had no channels available (a real possibility). The RRP could attempt to route through another neighbor node with an available channel assignment for that link. If routing through another neighbor node was not possible, it could send an error message to the previous node and try another route. Avoiding routing loops is another issue that would have to be dealt with. These are a few features that must be part of the RRP design. Another prospect should be considered. Could there have been a more optimal channel assignment algorithm or schedule that could have avoided this state of "no channel assignments available for a link" from occurring?

The above example dealt with establishing a path. What about tearing a path down when it is no longer needed? Consider another example using Fig. 15.12. Suppose all three paths in the Channel Assignment Scenario have been established and then the application at Node 3 notifies the RRP that the Path Node 3 to Node 19 is no longer needed. Node 3 would begin by releasing its own resources to support the path, along with releasing the channel assignment of Channel 1 to support the link Node 3 → Node 22. Upon the release of Channel 1 for this link, the channel assignments available list per link is recalculated based upon the updated current link channel assignments. Node 22 then releases the resources that supported the path, but it does not release the channel assignment for link Node 22 → Node 20, because Path Node 7 to Node 19 is still being supported. To realize this, Node 22 must be aware of all paths each link channel assignment is supporting. This means that path dependencies for each link channel assignment according to cut-through destination must be tracked.

Scheduling. The previous two paragraphs touched on this subject. The goal is to try to schedule resources over time optimally to maximize utilization (throughput) in the network. What are the resources that need to be scheduled? They are channels, bandwidth, and buffer space. Channels are constrained not only by the hidden node problem but also by each node's interference range. Bandwidth and buffer space are constraints per node for each channel. When you run out of any of these three resources, no more additional traffic flows can be supported until nonconflicting resources are freed up. If the traffic requirements are known up front or able to be predicted, an optimal schedule of resources could possibly be calculated.

15.5 Directions for Future Research

To understand the research deficiencies in improving the performance of WMNs, we must summarize the work as a whole and determine if it makes a contribution to each of the methods that serve as the WMN architecture building blocks shown in Fig. 15.13. Figure 15.14 presents each area of related work presented in this section and graphically shows each of the building blocks it contributes to (the blocks darkened in black).

Below is a narrative explanation of Fig. 15.14:

- Channel Assignment: Work on centralized and decentralized channel assignment algorithms are presented in [19–23]. The assignment of orthogonal noninterfering channels is essential because this allows for the actual allocation of bandwidth for WMNs. Ensuring that the channels are noninterfering based upon allowable SINR is a challenge and continues to be an active research area. The idea is to get the maximum channel use and reuse to optimize spectrum efficiency.

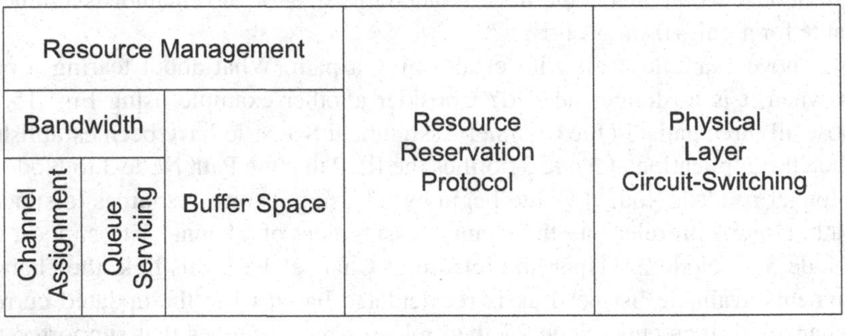

Resource Management				
Bandwidth			Resource Reservation Protocol	Physical Layer Circuit-Switching
Channel Assignment	Queue Servicing	Buffer Space		

Fig. 15.13 Building blocks for an improved performance wireless mesh network architecture

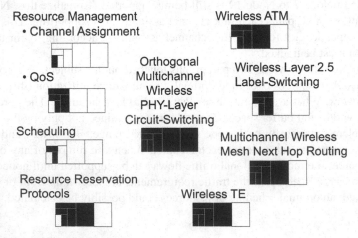

Fig. 15.14 Summary of current work

- QoS: Much mature QoS research techniques are available as a result of wired MPLS and ATM implementations. [11, 12, 24–26] Initial wireless ATM QoS solutions are presented in [27]. QoS models for MANETs were implemented in a multiple access omnidirectional wireless broadcast environment. Buffer space is a controllable nodal resource [28–32]. But bandwidth in this environment is much more difficult to allocate because of neighbor and two-hop neighbor nodes also competing for the same medium along with interference nodes transmitting within interference range.

- Scheduling: Scheduling is a method used to divide up bandwidth by determining when various competing queues of traffic can be serviced/transmitted and for how long. Using scheduling as method to deal with the interference problem especially in these low-latency WMNs using multiple channels is an active research area.

- Resource Reservation Protocols: Considerable research has been done on wireless resource reservation protocols that reserve resources using existing QoS mechanisms to allocate buffer space and bandwidth [17,24,28,30,36,37,40–44]. They do not allocate bandwidth through channel assignment. They also were not designed to support circuit-switching at the physical layer. These features are essential to low-latency WMNs.

- Wireless ATM: Though the RDRN [27,33,39,45,46] did use a RRP similar to INSIGNIA, using high-speed radios and directional antennas for their backbone links and low-speed radios with scheduling for their local links, ATM technology provides switching at layer 2 and not at the physical layer. But layer-2 solutions still have inherent internodal delay that have increased internodal delay compared to PHY layer switching solutions.

- Wireless Layer 2.5 Label-Switching: These protocols modified the contention-based MAC layer to implement label-switching at layer 2.5 (not the physical layer) [47–50]. Because of this they to have inherent internodal delay.

- Multichannel Wireless Mesh Next Hop Routing: With regard to channel assignment, Hyacinth assigned different channels to provide bidirectional (not unidirectional) internodal links and still used the 802.11 MAC layer [20,21]. The use of channel assignment was to reduce contention while still using the existing MAC. Also, interface switching was not implemented in the physical layer.

- Wireless Traffic Engineering: A multichannel 802.11-based WMN was used to extend wired service to train riders [54,55]. The authors used a combination of node placement, channel assignment and routing to reduce interference in this contention-based wireless environment. Packets were relayed using a store-and-forward method at each intermediate node. Published work on practical applications of Wireless Traffic Engineering is lacking and will continue to be an active research area as wireless traffic engineering mechanisms are developed to properly control and optimize the allocation of RF media.

- Orthogonal Multichannel Wireless PHY-Layer Circuit-Switching: Comments on Ramanathan's and Tchakountio's work [2,6] are below.

Notice that only one work contributes to each of the building blocks, Ramanthan and Tchakountio [6]. In [2], Ramanathan proposes that the conventional IP stack will not be able to adequately support high-bandwidth requirements of MANETs. His approach is to use circuit-switching (virtual cut-through, wormhole routing) to support such requirements. He provides a more detailed explanation of his approach in Ramanathan et al. [6]. He proposes using orthogonal channels at each node to provide circuit-switched paths from end-to-end. To support this he developed a path set up mechanism, path access control, PAC, and a switching mechanism, relay oriented physical layer, ROPL. PAC is an RSVP-like mechanism for reserving resources (floor) multiple hops at a time. ROPL is a cut-through relay that pipelines bits through the receive and transmit chains at the physical layer (a cut-through relay mechanism).

All other work contributes one or more portions of the architecture, but none as a whole. The lack of tying all of these methods into a single architecture is what has been missing and is being explored now.

15.6 Conclusion

In this chapter, we have considered the question – why do 802.11-based WMNs provide poor performance? To better understand the source of this poor performance, we identified the shortcomings associated with the architecture and protocols of these contention based networks. Such networks make use of a shared radio frequency (RF) broadcast medium and commonly employ an omni-directional antenna. This creates many challenges particularly because of the unpredictable nature of RF channels with regard to reception and interference range. Together these issues make it more difficult to share/allocate bandwidth compared to wired-optical network architectures.

We then considered the question – how might we combine the advantages of packet switching with the long proven methods of circuit-switching to implement traffic engineering to reduce variance in end-to-end delay? To address the development of such low-latency WMNs, we described an architecture that exploits multiple orthogonal channels, virtual cut-through and wormhole switching, physical layer circuit switch design, and reservation protocols. Such an architecture has the potential to reduce nodal processing delay and interference, which can improve utilization and performance (i.e., delay and jitter) for WMNs.

15.7 Terminologies

1. *Cut-through switching*. A low-latency switching method where a receiving node immediately begins transmitting (forwarding) the packet header and payload after the packet header is read and the next hop is determined. This implies that the initially received part of the packet is forwarded before the remaining portion of the packet is received.

2. *Label switching.* A low-latency method of forwarding packets through an intermediate router based upon a short path label header and router quick look-up table to determine the packet's outgoing interface.
3. *Latency.* The difference between the time a packet, frame or flit departs a sending node and arrives at a destination node.
4. *Quality of service.* The traffic requirements for a traffic flow to support a session. It can be defined in terms of bandwidth, end-to-end delay, jitter (delay variance), packet loss, and priority or class with regard to traffic from other sessions.
5. *Resource reservation protocol (RRP).* A protocol required to implement traffic engineering by reserving the nodal and link resources of the network to meet the QoS requirements of the traffic.
6. *Traffic engineering (TE).* The use of QoS mechanisms to achieve the goal of improving network performance (increased throughput, decreased delay) and make better use of network resources by better matching them with traffic demands.
7. *Unidirectional noninterfering channels.* A communications channel that can support a one-way inter-nodal connection between two adjacent wireless nodes without disruption caused by other transmitting nodes.
8. *Wireless local area network (LAN).* A wireless single-hop, last quarter-mile to one-mile network, consisting of an access point, usually to act as a gateway to wired services and wireless workstations.
9. *Wireless mesh network (WMN).* A peer-peer, self-organizing, wireless multihop network. Its topology can be relatively static or dynamic.
10. *Wormhole switching.* A low latency switching method similar to cut-through switching except that the packet data payload is divided into smaller payload elements called flits. Each flit has a significantly smaller header (sometimes referred to as a label) than the packet header. Each flit is forwarded immediately after the smaller header is processed and the next hop is determined.

15.8 Questions

1. Based on the description of the work by Li, Blake, De Couto et al. [1], multihop performance is limited both by intrinsic properties of RF propagation and also by the 802.11 MAC protocol. What attributes of an 802.11-based wireless mesh network lead to poor performance? What attributes of RF propagation lead to poor performance?
2. Given the data rate for each of the nodes in the six node chain topology shown in Fig. 15.15 is D, what is the effective data rate across the six node chain for a continuous stream of packets departing from node A with a destination of node B in terms of D? Each of the nodes have the interference range = reception range (one hop distance) as shown in the diagram. Also, each node broadcasts omnidirectionally on the same signal channel, using a contention-based MAC protocol such as 802.11.

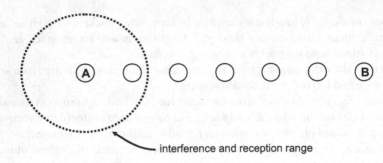

interference and reception range

Fig. 15.15 Six node topology (Question 2)

3. What are some of the attributes of a single-channel wireless omnidirectional networking architecture versus a wired architecture that make it more of a challenge to implement a label-switching protocol?
4. Calculation of reception range and maximum interference distance: Suppose transmitting node x has a transmit power of 0.12 mW (milliwatts) and is using qam-16 with a data rate of 36 Mbps with no ECC with a $SINR_{min}$ of 18.5 dB. Assume $P_{thermal\ noise}$ is −95 dBm. Assume transmitter and receiver gains at both 1.0. The transmit frequency is 2.4 GHz.

   a. What is the maximum reception range to an intended receiver, node y, assuming the free space path loss model?
   b. Assume that the distance between node x, the transmitter, and node y, the intended receiver, is the maximum reception range as calculated in part a minus 1 m. If an interfering node, node z, is also transmitting with a transmit power of 0.12 mW (milliwatts), what is the maximum interference range of node z so that it can interfere with node y's reception from node x (given the initial transmit power and modulation scheme of node x)?
   c. How does interference range affect channel reuse in any network topology?

5. Given the following multichannel wireless node functional design (Fig. 15.16), and WMN topology and state (Fig. 15.17); what channel assignments can be made for the following ordered full-duplex paths with only five orthogonal channels (Chan 0–4):

   First N14 → N13
   Second N17 → N22
   Third N10 → N25

Multichannel node functional design description (see Fig. 15.16):

- IF0 is a single-channel omnidirectional broadcast interface used only by the routing and resource reservation protocols. Consideration of the attributes of this interface does not factor into this question.
- Relay consists of 5 orthogonal channels (0.4) used to support circuit-switched (cut-through) data paths

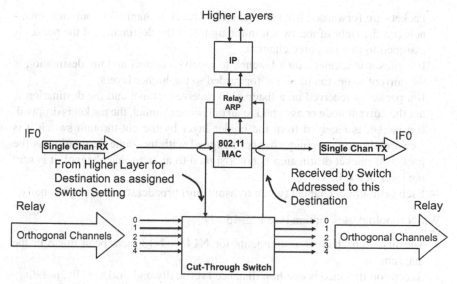

Fig. 15.16 Multichannel wireless node functional design (Question 5)

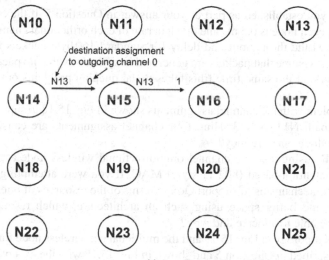

Note: Reception distance is one hop allowing for only horizontal and vertical inter-nodal link connectivity only. Also, interference distance equals reception distance.

Fig. 15.17 Wireless mesh network topology (Question 5)

– The cut-through switch is set up to explicitly listen on certain receive channels (to the left of the switch in the figure) to support paths by the reservation protocol.

- Packets are forwarded from the listened to receive channels to outgoing channels (on the right of the switch in the figure) if the destination of the packet is assigned to that outgoing channel.
- If a packet is received on a listened to receive channel and the destination is the current node, the packet is forwarded to the higher layers.
- If a packet is received on a listened to receive channel and the destination is not the current node or assigned to an outgoing channel, the packet is dropped.
- If a packet is received from the higher layer by the cut-through switch it is forwarded to an outgoing channel assigned with the same destination as the packet. If packet destination is not assigned to an outgoing channel, it is sent out interface IF0.
- Each orthogonal channel (when transmitting) broadcasts omni-directionally.

WMN topology description (see Fig. 15.17):

- The first two channel assignments for N14 → N13 have been made in the diagram.
- Reception distance is one hop distance (vertically and horizontally, not diagonally). Interference is also one hop distance (vertically and horizontally, not diagonally).

6. Given the paths you established as part of your answer in Question 5, if the cut-through delay at each node is 0.5 ms and the data rate of each orthogonal channel is 1 Mbps, what is the throughput and delay for standard 160 byte packets for each of the paths. Assume that packets are generated in such a way that no packet arrives at any node at the same time (this takes nodal queuing problems out of the question).

7. What is the problem with the channel assignments shown in Fig. 15.18 to support the full-duplex path N14 → N13. Hint: four channel assignments are causing loops to occur; which ones are they?

8. Consider a WMN using a single channel omnidirectional wireless node architecture using contention-based (802.11 style) MAC. If you were attempting to implement nodal mechanisms to support QoS in terms of the resources of internodal bandwidth and buffer space using such an architecture, which resource would you not be able to allocate and why?

9. Given the WMN topology in Fig. 15.19 and the multichannel wireless node functional design described in Question 5 and shown in Fig. 15.16 with the exception that there are only four orthogonal channels available (channels 0–3), establish the following full-duplex paths in the order below so that a QoS max delay of six hops is not exceeded:

    First N11 → N24
    Second N18 → N21
    Third N10 → N25

- Given a packet payload size of 300 bytes, header size of 32 bytes and a data rate of 54 Mbps, what is the delay experienced from sender to destination

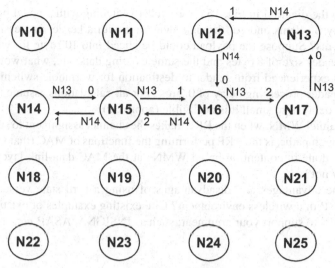

Fig. 15.18 Channel assignments to support full-duplex path N14 → N13 (Question 7)

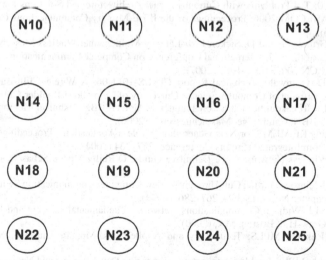

Fig. 15.19 Wireless mesh network topology (Question 9)

according to the diagram in Fig. 15.6 for packet switching with a nodal processing delay of 10 ms and cut-through switching with a header processing delay of 0.5 ms? Suppose the payload could be sliced into 100 byte flit with a reduced header size of 8 bytes and the same existing data rate, what would be the delay experienced from sender to destination for wormhole switching with a flit-header processing delay of 0.2 ms? For all calculations assume the propagation delay is so small it is negligible (approx 0).

- In a multichannel WMN, when the RRP makes the channel assignment to support cut-through paths, is the RRP performing the functions of MAC (that was traditionally done in contention-based WMNs at the MAC data-link layer)? Why or why not?
- What are the advantages and disadvantages of using a hard state versus a soft state RRP in a wireless environment? Cite existing examples of existing wireless RRPs to support your argument; such as INSIGNA, ASAP, etc.

References

1. J. Li, C. Blake, D. De Couto, H. Lee, and R. Morris, Capacity of Ad Hoc Wireless Networks. Mobile Computing and Networking. 61–69 (2001).
2. R. Ramanathan, Challenges: A Radically New Architecture for Next Generation Mobile Ad Hoc Networks. MobiCom 05: Proceedings of the 11th Annual International Conference on Mobile Computing and Networking. 132–139 (2005).
3. R. Jain, Internet 3.0: Ten Problems with Current Internet Architecture and Solutions for the Next Generation. MILCOM 2006: Proceedings of the IEEE Military Communications Conference. 1–9 (2006).
4. R. McTasney, D. Grunwald, and D. Sicker, Low-Latency Multichannel Wireless Mesh Networks. Proceedings of the 16th International Conference on Computer Communications and Networks 2007 ICCCN 2007. 1082–1087 (2007).
5. M. Buettner and D. Grunwald, Technical Report CU-CS-1011-06: A Wireless Flit-Based OpNET Model. Department of Computer Science University of Colorado at Boulder (2006).
6. R. Ramanthan and F. Tchakountio, Ultra Low Latency MANETS BBN Technical Memorandum No. TM-2023. BBN, Cambridge, Massachuessettes (2006).
7. C. Nguyen, Vibrating RF MEMS for Next Generation Wireless Applications. Proceedings of the IEEE 2004 Custom Integrated Circuits Conference. 257–264 (2004).
8. M. Gast, 802.11 Wireless Networks: The Definitive Guide. O'Reilly Media, Sebastabpool, California (2005).
9. P. Kermani and L. Kleinrock, Virtual Cut-Through: A New Computer Communication Switching Technique. Computer Networks 3(4), 267–286, (1979).
10. A. Leon-Garcia and I. Widjaja, Communications Networks: Fundamental Concept and Key Architectures. McGraw-Hill, Boston, MA, (2000).
11. B. Davie and Y. Rekhter, MPLS: Technology and Applications. Morgan Kaufman, New York (2000).
12. I. Minei and J. Lucek, MPLS-Enabled Applications: Emerging Developments and New Technologies. Wiley, West Sussex (2006).
13. C. Perkins, Ad Hoc Networks, Addison-Wesley, New York (2001).
14. V. Park, The Temporally-Ordered Routing Algorithm (TORA). http://www3.ietf.org/proceedings/97dec/slides/manet-tora/index.htm (1999). Accessed 5 March 2007.
15. J. Garcia-Luna-Aceves, M. Spohn, and D. Beyer, IETF MANET Working Group Draft – Source Tree Adaptive Routing (STAR) Protocol. http://tools.ietf.org/html/draft-ietf-manet-star-00 (1998). Accessed 3 March 2007.

16. P. Jacquet, P. Mulethaler, T. Clausen, A. Laouiti, A. Qayyum, and L. Viennot, Optimized link state routing protocol for ad hoc networks. Proceedings of the 5th IEEE Multi Topic Conference. 1–7 (2001).

17. S. Thomas, IP Switching and Routing Essentials: Understanding RIP, OSPF, BGP, MPLS, CR-LDP, and RSVP-TE. Wiley, New York (2002).

18. A. Greenberg, IP Network Traffic Engineering. http://www.nanog.org/mtg-0002/ppt/green/index.htm (2000). Accessed 24 January 2007.

19. S. Ramanathan and E. Lloyd, Scheduling Algorithms for Multihop Radio Networks. IEEE/ACM Transactions on Networking. 166–177 (1993).

20. A. Raniwala, K. Gopalan, and T. Chiueh, Centralized Channel Assignment and Routing Algorithms for Multi-channel Wireless Mesh Networks. SIGMOBILE Mobile Computing and Communications Review. 50–65 (2004).

21. A. Raniwala and T. Chiueh, Evaluation of a Wireless Enterprise Backbone Network Architecture. Proceedings of the 12th Annual IEEE Symposium on High Performance Interconnects 2004. 98–104 (2004).

22. X. Ma and E. Lloyd, Evaluation of a Distributed Broadcast Scheduling Protocol for Multihop Radio Networks. IEEE Military Communications Conference 2001 Communications for Network-Centric Operations. 998–1002 (2001).

23. G. Kulkarni, V. Raghunathan, M. Srivastava, and M. Gerla, Channel Allocation in OFDMA-based Wireless Ad-Hoc Networks. Advanced Signal Processing Algorithms, Architectures, and Implementations XII. (2002) doi:10.1117/12.453813.

24. P. Mohopatra, J. Li, and C. Gui, QoS in Mobile Ad Hoc Networks. IEEE Wireless Communications. 44–52 (2003).

25. E. Valeroso and M. Alam, Adaptive Resource Scheduling Strategies and Performance Analysis of Broadband Networks. LCN'96: Proceedings of the 21st Annual IEEEE Conference on Local Computer Networks. 305 (1996).

26. F. Faucher, T. Nadeau, A. Chieu, W. Townsend, D. Skalecki, and M. Tatham, IETF Internet Draft: Requirements for Support of Diff-Serv-aware MPLS Traffic Engineering (2000).

27. J. Evans, K. Shanmugan, G. Minden, V. Frost, and G. Prescott, Rapidly Deployable Radio Network (RDRN) – Phase II Final Report ITTC-FY2003-1380-15 (2002).

28. H. Xiao, W. Seah, A. Lo, and K. Chua, A Flexible Quality of Service Model for Mobile Ad-Hoc Networks. 2000 IEEE 51st Vechicular Technology Conference Proceedings 445–449 (2000). doi:10.1109/VETECS.2000.851496

29. H. Yan and H. Abdel-Wahab, HQMM: A Hybrid QoS Model for Mobile Ad Hoc Networks. 11th IEEE Symposium on Computers and Communications 2006 194–200 (2006). doi:10.1109/ISCC.2006.85

30. S. Lee and A. Campbell, INSIGNIA: In-Band Signaling Support for QoS in Mobile Ad Hoc Networks. Proceedings of the 5th International Workshop on Mobile Multimedia Communication (1998).

31. A. Taha, H. Hassanein, and H. Mouftah, Integrated Solutions for Wireless MPLS and Mobile IP: Current Status and Future Directions. Canadian Conference on Electrical and Computer Engineering 1463–1466 (2004).

32. S. Vijayarangam and S. Ganesan, QoS Implementation for MPLS Based Wirelss Networks. ASEE Conference 02 1–9 (2002).

33. R. Sanchez, J. Evans, G. Minden, V. Frost, and K. Shanmugan, RDRN: A Rapidly Deployable Network – Implementation and Experience. International Conference on Universal Personal Communications 1998 93–97 (1998).

34. F. Liu, Z. Zeng, J. Tao, Q. Li, and Z. Lin, Achieving QoS for IEEE 802.16 in Mesh Mode. http://zlin.ba.ttu.edu/pdf/CSI-79.pdf (2005) Accessed 5 March (2007).

35. M. Cao, W. Ma, Q. Zhang, X. Wang, and W. Zhu, Modeling and Performance Analysis of the Distributed Scheduler in IEEE 802.16 Mesh Mode. MobiHoc '05: Proceedings of the 6th ACM International Symposium on Mobile Ad Hoc Networking and Computing (2005). doi:10.1145/1062689.1062701

36. R. Braden, L. Zhang, S. Berson, S. Herzog, and S. Jamin, RFC 2205 – Resource ReSerVation Protocol (RSVP) – Version 1 Functional Specification (1997).

37. S. Blake, D. Black, M. Carlson, E. Davies, Z. Wang, and W. Weisss, RFC 2475 – An Architecture for Differentiated Services (1998).

38. I. Chlamtac, M. Conti, and J. Liu, Mobile Ad Hoc Networks: Imperatives and Challenges. Ad Hoc Networks 13–64 (2003).

39. S. Radhakrishnan, V. Frost, and J. Evans, Quality of Service for Rapidly Deployable Radio Networks. Proceedings of the 33rd Annual Hawaii International Conference on System Sciences 2000 11 (2000).

40. J. Xue, P. Stuedi, and G. Alonso, ASAP: An Adaptive QoS Protocol for Mobile and Ad Hoc Networks. IEEE Proceedings on Personal, Indoor and Mobile Radio Communications 2616–2620 (2003).

41. E. Carlson, C. Prehofer, C. Bettstetter, H. Karl, and A. Wolisz, A Distributed End-to-End Reservation Protocol for IEEE 802.11-based Wireless Mesh Networks. IEEE Journal on Selected Areas in Communications 1–10 (2006).

42. H. Liu and D. Raychaudhuri, Label Switched Multi-path Forwarding in Wireless Ad-Hoc Networks. PerCom Workshops 248–252 (2005). doi:10.1109/PERCOMW.2005.42

43. R. Nagarajan and E. Ekici, Flexible MPLS Signaling (FMS) for Mobile Networks. CC 2004 – IEEE International Conference on Communications 4321–4325 (2004).

44. Y. Chi-Hsiang, H. Mouftah, and H. Hassanein, Signaling and QoS Guarantees in Mobile Ad Hoc Networks. IEEE International Conference on Communications 3284–3290 (2002).

45. S. Bush, S. Jagannath, R. Sanchez, J. Evans, V. Frost, and S. Shanmugan, Rapidly Deployable Radio Networks Network Architecture (1997).

46. S. Bush, S. Jagannath, R. Sanchez, J. Evans, G. Minden, S. Shanmugan, and V. Frost, Wireless Networks (1997). http://dx.doi.org/10.1023/A:1019117603571.

47. A. Acharya, A. Misra, and S. Bansal, A Label-Switching Packet Forwarding Architecture for Multi-hop Wireless LANs. Proceedings of the 5th ACM International Workshop on Wireless Mobile Multimedia 33–40 (2002).

48. A. Acharya, A. Misra, and S. Bansal, High-Performance Architectures for IP-Based Multihop 802.11 Networks. IEEE Wireless Communications 22–28 (2003).

49. V. Untz, M. Heusse, F. Rousseau, and A. Duda, Lilith: An Interconnection Architecture Based on Label Switching for Spontaneous Edge Networks. International Conference on Mobile and Ubiquitous Systems 146–151 (2004).

50. V. Untz, M. Heusse, F. Rousseau, A. Duda, On Demand Label Switching for Spontaneous Edge Networks. FDNA '04: Proceedings of the ACM SIGCOMM Workshop on Future Directions in Network Architecture 35–42 (2004).

51. W. Dally, Performance Analysis of k-Ary n-Cube Interconnection Networks. IEEE Transactions on Computers 775–785 (1990).

52. W. Dally, Virtual-Channel Flow Control. IEEE Transactions on Parallel Distributed Systems 194–205 (1992).

53. L. Peh and W. Dally, Flit-Reservation Flow Control, International Symposium on High-Performance Computer Architecture 73–84 (2000).

54. F. De Greve, F. De Turck, I. Moerman, and P. Demeester, Design of Wireless Mesh Networks for Aggregating Traffic of Fast Moving Users. MobiWac '06: Proceedings of the International Workshop on Mobility Management and Wireless Access 35–44 (2006).

55. F. Greve, B. Lannoo, L. Peters, T. Leeuwen, F. Quickenborne, D. Colle, F. Turck, I. Moerman, M. Pickavet, B. Dhoedt, and P. Demeester, FAMOUS: A Network Architecture for Delivering Multimedia Services to FAst MOving USers. Wireless Personal Communications 281–304 Kluwer Academic, Hingham, MA (2005).

56. D. Grunwald, D. Sicker, T. Brown, and P. Mathys, NSF NeTS-FIND Proposal: Radio Wormholes for Wireless Label Switched Mesh Networks. Department of Computer Science and Department of Electrical and Computer Engineering at the University of Colorado at Boulder (2006).

Chapter 16
WiMAX Metro Area Mesh Networks: Technologies and Challenges

Ahmed Iyanda Sulyman and Hossam Hassanein

Abstract MIMO-OFDM technology has emerged as a compelling high-speed solution for the next-generation wireless networks. The IEEE 802.16 standard-based WiMAX system will deploy MIMO-OFDM technology in the broadband wireless access (BWA) and backhaul markets that otherwise depended mostly on proprietary solutions. Standard-based solutions result in inexpensive devices and encourage large-scale deployments, bringing down the cost of the technology to end users, yet making it profitable for service providers and equipment manufacturers. Of the two deployment modes specified in the WiMAX system, mesh mode is currently optional while point-to-multipoint mode is mandatory. In this chapter, we present an overview of the PHY and medium access control (MAC) layer technologies deployed in the WiMAX system and examine the prospects and challenges of mesh operations using them. One of the main impediments for mesh operation in the WiMAX system is that network operators operating the system in licensed spectrum are not keen to provide separate radio channels for access and mesh relay services, as this reduces the numbers of users serviced per spectrum allocation. We discuss in this chapter, an interesting alternative approach that uses the concept of MIMO-multiplexing relaying at each mesh node to provide different links for the access and mesh relaying services on the same radio channel. This approach is cost-effective, and encourages more widespread WiMAX mesh network deployments.

16.1 Introduction

Worldwide Interoperability for Microwave Access (WiMAX) is a new broadband wireless standard offering interesting wireless alternatives to a number of traditional wired and wireless technologies suited to its long-range wireless broadband

H. Hassanein (✉)
School of Computing, Queens University, Kingston, ON, Canada K7L 3N6
e-mail: hossam@cs.queensu.ca

S. Misra et al. (eds.), *Guide to Wireless Mesh Networks*, Computer Communications and Networks, DOI 10.1007/978-1-84800-909-7_16,
© Springer-Verlag London Limited 2009

connectivity. WiMAX is the commercialization of the IEEE 802.16 standard, an evolving standard initiated by an IEEE working group, named 802.16, formed in 1998 and tasked with the goal of producing a new metropolitan area networking (MAN) protocol. In June 2004 following a series of activities, the working group won approval for the latest wireless MAN standard known as IEEE 802.16–2004 standard [10], specifying the physical layer (PHY) and medium access control (MAC) protocols for fixed broadband wireless access (BWA). Unlike its wireless local area networks (WLAN) complement, the new wireless MAN targets long-range broadband applications and thus chooses features and technologies to address this need [1,5]. WiMAX forum is an alliance of network operators, equipment manufacturers, handset vendors, chip makers and other players in the Telecom industry. The forum was formed in April 2001 and tasked with the responsibility of promoting conformance and interoperability of products based on the IEEE 802.16 standards. To this end, it establishes testing labs, defines and conducts interoperability tests, and awards vendor systems a "WiMAX Certified" label upon successful testing of their products. Thus the WiMAX forum takes after the approach pioneered by the Wi-Fi alliance for the IEEE 802.11 WLAN systems, to ensure Worldwide interoperability of WiMAX Certified equipments and it is hoped that this will stimulate similar mass deployments recorded by the WLAN industry.

Fixed wireless solution is the genesis of the WiMAX technology, targeting the so called "last mile" wireless access, as well as the "middle mile" microwave backhaul applications. Two operating modes have thus been specified in the WiMAX system: Point-to-multipoint (PMP) mode, where a backbone of base stations connected to the public network services many fixed subscriber stations situated in the coverage zone of each base station, and mesh mode, where WiMAX mesh networking protocol is used to route data from source to destination via multiple relaying over WiMAX mesh nodes communicating in point-to-point modes. Mesh networking mode is currently an optional feature, with many benefits for the WiMAX network operators when supported. Perhaps the most compelling of these benefits is the potential to use mesh relaying to provide a cost-effective range extension beyond the coverage of a base station. There are two main mesh architectures: Infrastructure mesh and ad hoc or client mesh. In infrastructure mesh, customer premises equipments (CPE) do not relay packets, only the WiMAX base stations have packet relaying capabilities. Backhaul mesh networks are examples of such infrastructure mesh. Client mesh-enabled WiMAX network on the other hand allows both the CPE and the base stations to relay packets and thus brings the full benefits of mesh architecture to the WiMAX system. Infrastructure mesh has the advantages of better security, network performance predictability, and easier management over the client mesh. As the WiMAX technology gets widespread support from major merchant chip vendors such as Intel, Nokia and Motorola, and with the participation of most of the major radio access network (RAN) vendors, strong economic motivation for the provisioning of mobile and personal broadband services that will extend broadband access to users in transit, spurred a new flavor of the WiMAX system. This ignited a series of activities in the IEEE 802.16 working group culminating in the ratification of the IEEE 802.16e-2005 standard in December 2005, and its eventual publication in February 2006 [11]. The IEEE 802.16e-2005 standard addresses the PHY and MAC layer changes necessary to support mobility, and is referred to as the mobile WiMAX extension of the WiMAX technology. Devices manufactured to this standard will support both fixed and mobile services. Mobile WiMAX will allow users to take broadband access with them every where they go, enabling "internet in the pocket" and other mobile data applications, in addition to mobile voice, video, and multimedia services. Personal and ubiquitous broadband services enabled by mobile WiMAX will drive high demand for the WiMAX industry, reminiscent of the effects of cellular system on the telephone industry. As WiMAX-enabled hand-held devices become more pervasive and their technologies mature, the case for engaging them in mesh relay operations will become stronger. WiMAX mesh network-enabled handhelds and CPEs deployed in homes and offices, will soon be widespread enough to cover significant portions of cities and metropolis, and could be employed to provide cost-effective mesh relay links for users outside the range of existing base stations. Network operators and service providers would then take advantage of the long-range broadband connectivity, and the reliability of such WiMAX mesh links, to deploy metro-scale mesh networks where WiMAX mesh nodes provide the core links, connecting WiMAX, Wi-Fi and other broadband users to the core network.

In this chapter, we examine the prospects and challenges of metro-scale WiMAX mesh networks—both fixed and mobile network deployments. A number of articles have discussed the protocols and architectures for wireless mesh networks [2,4]. No article however has yet examined the WiMAX technologies to discuss mesh operations using them. Therefore unlike existing articles on mesh networks, here we focus on the technologies deployed in the WiMAX system and examine the plight of mesh operations using them. Avenues for realizing WiMAX mesh operations that is cost-effective and comparable to the PMP mode using these technologies are then explored. One of the main impediments so far for WiMAX mesh network deployments is that WiMAX network operators are not keen to dedicate half or more of the costly radio spectrum to providing capabilities for the optional mesh mode, which is a requirement if the current dual-radio and multi-radio wireless mesh networks approach are employed in the WiMAX system. For example, the dual-radio broadband solutions developed by Motorola and other vendors use 4.9 GHz radio for mesh relay, and 2.4 GHz radio for access service [15]. Here we discuss an interesting alternative to this approach that uses the concept of MIMO-multiplexing [7] at each mesh node to provide the multiple links for access and mesh relay services on the same radio channel. This is a cost-effective solution for network operators operating the WiMAX system over licensed spectrum, and its deployment can help pave way for a possible convergence of electronic circuit designs for the PMP and mesh modes in the WiMAX system, encouraging wider-scale WiMAX mesh network deployments.

16.2 Background

16.2.1 WiMAX Deployment Modes

There are two deployment modes supportable by WiMAX devices, PMP and mesh modes. PMP mode provides access services, while mesh mode provides backhaul and client mesh services. Application of access services, using PMP modes at WiMAX base stations, is fairly diverse. These include delivery of broadband to enterprises and homes, which can be used for high-speed internet access, cable TV distributions, fiber optic roll-offs, etc. Backhaul service on the other hand is focused around the provisioning of high-capacity long-haul wireless links for transporting voice and data from cellular towers and Wi-Fi hotspots. These services are currently provided using a combination of ADSL (asymmetric digital subscriber line), cable, leased lines, and a number of proprietary solutions developed by the vendors in the respective market. WiMAX provides an opportunity for an efficient and cost-effective solution for these services. Applications of client mesh services include information delivery systems for public safety services (police, fire departments, etc.), information delivery systems for public transport services, and public internet access among others. All WiMAX certified products will support the PMP modes as mandatory functionality, while mesh supports are currently optional. One of the primary reasons for this distinction is that the PMP and mesh modes currently use non-compatible MAC layer protocols and different PHY frame structures. Figure 16.1 a and b illustrate respectively the frame structures for the PMP and mesh modes. In the PMP mode, the downlink (DL) and uplink (UL) bursts transmissions are detached, while in the mesh mode DL/UL separation is achieved using time division duplexing (TDD). Thus vendors hoping to provide both access and mesh capabilities on their devices (base station, BS, and subscriber stations, SS) are faced with the difficult tasks of cost-effectively implementing these two systems together on their units.

16.2.2 WiMAX Mesh Networks

Architectures: There are two main mesh architectures that can be employed in the WiMAX system: Infrastructure and client mesh [2]. Figure 16.2 illustrates the infrastructure and client mesh architectures for a metro-scale WiMAX mesh network, where WiMAX mesh nodes (BS, CPEs, etc.) provide the wide-area mesh links connecting WiMAX, Wi-Fi and other broadband users to the core network. There are two broad realizations for mesh links in the WiMAX system, depending mainly on the antenna technology employed: *logical* and *physical* mesh. Logical mesh use omnidirectional antennas at each mesh node to form logical links to the neighboring devices. These links are considered logical because the hardware configuration does not change for different links to neighboring devices. Physical or directed mesh on the other hand is a form of mesh where substantially directional antennas are used to

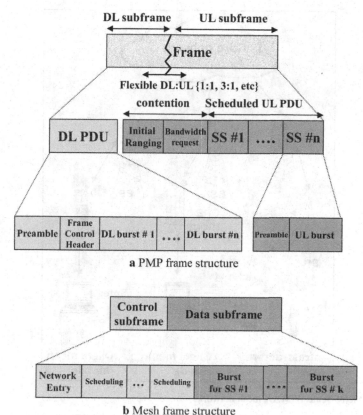

Fig. 16.1 PHY frame structures for PMP and mesh operations

create physical links. Mesh nodes may steer an antenna toward other nodes, or it may use beamforming to create a narrow beam directed toward intended nodes to create the physical mesh links. These links are considered physical because the hardware configuration changes (via beamforming or switching for example) to form different links with different neighbors. The narrow beams employed in the physical mesh links significantly improve the carrier-to-interference (C/I) and improve frequency reuse, which gives high spectral efficiency, but at the expense of increased complexity. While the logical mesh is well specified in the 802.16 standard, physical mesh is comparatively still being developed. Data exchanges among mesh nodes using the logical mesh is based on the concept of neighborhood, where a reference mesh node identifies other nodes in its immediate neighborhood (one-hop neighbors), extended neighborhood (two-hop neighbors), and multi-hop neighbors, as illustrated in Fig. 16.3. This neighborhood definition is then used to realize suitable multi-hop data exchange protocols [2, 6]. Our focus in this article is not on the details of those data routing/scheduling protocols, but rather on the technologies behind the mesh relaying operations in the WiMAX system.

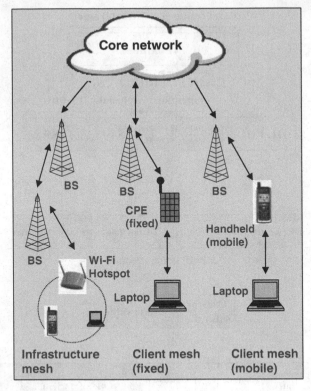

Fig. 16.2 Metro-scale WiMAX mesh networks

Data forwarding options: There are two broad data relaying (or forwarding) options applicable at mesh nodes in WiMAX mesh networks: Amplify-and-forward and Decode-and-forward options. In the amplify-and-forward option, a mesh node simply amplifies the received RF signal and then forwards it, without having any knowledge about the content of the signal. This relaying option puts less processing burden on the mesh nodes and therefore is often preferable when complexity is an issue. In the decode-and-forward option, mesh nodes use appropriate digital signal processing techniques to process the received waveform digitally. The resulting digital signal is then decoded to retrieve the original message bits, and re-encoded again for retransmission (or forwarding). This relaying option adds extra processing complexities and delays at mesh nodes, and allows the relaying nodes access to the content of the transmitted signal, raising some security concerns. Nodes using the amplify-and-forward options are also sometimes referred to as RF repeaters, while those using the decode-and-forward options are known as PHY and MAC repeaters as the increased signal processings required for the decoding process before forwarding involve significant PHY and MAC-layer operations. In the following, we examine the main features and technologies deployed in the WiMAX system closely, and discuss the prospects and challenges of these mesh

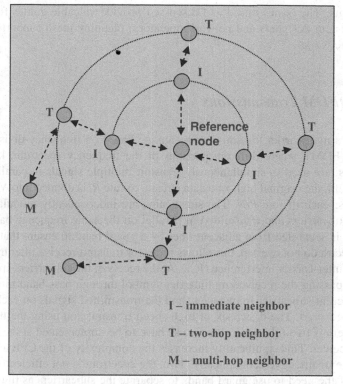

I – immediate neighbor

T – two-hop neighbor

M – multi-hop neighbor

Fig. 16.3 Logical mesh architecture

relaying operations using those WiMAX features and technologies. The discussions are applicable to both the logical and physical mesh, except where otherwise stated.

16.3 Thoughts for Practitioners: Technologies and Challenges

In this section, we present an overview of the PHY and MAC layer technologies deployed in the WiMAX system and examine the prospects and challenges of mesh operations using these technologies. One of the main impediments for mesh operation in the WiMAX system is that network operators operating the system in licensed spectrum are not keen to provide separate radio channels for access and mesh relay services, as this reduces the numbers of users serviced per spectrum allocation. We discuss in this section, an interesting alternative approach that uses the concept of MIMO-multiplexing relaying at each mesh node to provide different links for the access and mesh relaying services on the same radio channel. This approach is cost-effective, and encourages more widespread WiMAX mesh network deployments. The issues raised and discussed provide valuable guides for practitioners (system designers and network operators) planning mesh supports in their WiMAX networks.

16.3.1 OFDM Transmissions

OFDM is a multicarrier transmission technique based on frequency-division multiplexing (FDM), where different portions of the frequency spectrum, known as subcarriers, are used to simultaneously transmit multiple signals in parallel. In an FDM system, the original high-rate data stream, of rate R, is divided into N low-rate substreams, each of rate R/N. The substreams are independently modulated onto different subcarriers and transmitted in parallel on the same frequency band. Each subcarrier is separated from adjacent ones by a guard band to ensure that the subcarrier spectra do not overlap, this allows room for transmit/receive filter transitions and avoid inter-carrier interference (ICI). At the receiver, the subcarriers are demodulated by passing the received multicarrier symbol through pass-band filters tuned to the different subcarrier frequencies, and the transmitted signals on each subcarrier are recovered. The drawback of high-speed transmission using the traditional FDM system is two fold: One, filter banks have to be implemented at the transmitter and receiver. This significantly increases the complexity of the CPE and makes the scheme unattractive economically. Two, the spectrum is not efficiently utilized because of the need to use guard bands to separate the subcarriers as illustrated in Fig. 16.4. OFDM system addresses these two problems while keeping the benefits of multicarrier transmissions, by choosing the N subcarriers orthogonal (perpendicular in mathematical sense) to one another. The orthogonality of the subcarriers allows them to be spaced much closer to one another than in the FDM system. Thus

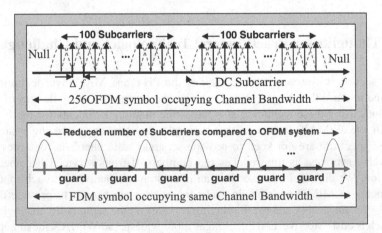

Fig. 16.4 Illustration of 256OFDM and FDM systems

the required bandwidth compared to an FDM system is greatly reduced, resulting in high spectral efficiency for the OFDM system. At the transmitter, N orthogonal waveforms can be generated using an N-point inverse discrete Fourier transform (IDFT). The subcarriers are then modulated with the data to obtain the time-domain OFDM symbol to be transmitted. Before the OFDM symbol is transmitted, a cyclic prefix (CP) is inserted in the time-domain OFDM waveform. The CP is the last part of the OFDM symbol, copied and appended to the beginning of the waveform. This helps to combat inter-symbol interference (ISI). ISI is eliminated if the length of the CP is at least equal to the maximum spread of the multipath channel. At the receiver, the CP is first removed and the data transmitted on each subcarrier is extracted by sampling the spectrum of the received OFDM symbol at Δf spacings. This operation can be accomplished using the discrete-time Fourier transform (DFT). In practice, inverse fast Fourier transform (IFFT) and fast Fourier transform (FFT), i.e., IFFT/FFT pair, are used for implementing OFDM systems as they provide efficient and easy way to generate large number of orthogonal subcarrier waveforms, and are mathematically equivalent to the IDFT/DFT pair. FFT chips are commercially available at low price-performance points, making technologies relying on them cheap, efficient, and therefore attractive. In Fig. 16.4, we illustrate an example of 256OFDM in 802.16-2004. The OFDM symbol consists of 192 data subcarriers, 8 pilot subcarriers and 56 nulls (28 nulls on one side, 27 on the other, and one null used as DC subcarrier). In its most basic form, each data subcarrier could be just on or off to indicate a one or zero bit of information. However, either phase shift keying (PSK) or quadrature amplitude modulation (QAM) is typically used to modulate the data subcarriers in order to increase the data rate. Pilot subcarriers are used for channel estimations, while null subcarriers are used as guard bands and DC subcarriers.

Example 16.1. OFDM symbols of length $4\,\mu s$ each, containing $0.8\,\mu s$ of CP, are transmitted over a wireless channel with root-mean-square (RMS) delay spread $\sigma_\tau = 500\,ns$, and maximum excess delay $\tau_{max} = 1\,\mu s$. (a) Estimate how many samples in the received OFDM symbol sequence contains ISI (assuming a clock rate of $20\,MHz$), (b) suggest adequate length of CP required to eliminate ISI in the received OFDM symbol sequence.

Solution:

(a)

- Time-domain sampling rate for received OFDM symbol $= 20\,MHz$
- Number of samples per OFDM symbol $= 20 \times 10^6$ samples $s^{-1} \times 4 \times 10^{-6}\,s = 80$ samples
- Number of samples per OFDM symbol corrupted with ISI $= 20 \times 10^6$ samples $s^{-1} \times \tau_{max} = 20$ samples

(b)

- RMS delay spread, σ_τ gives estimate of the average delay spread of the channel. However some multipath components (reflections) will arrive delayed

Fig. 16.5 Illustration of channel spread and the resulting ISI when CP is less than the channel spread

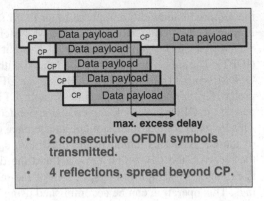

longer than σ_τ. Maximum excess delay τ_{max}, gives time of arrival of the last significant multipath component. Thus, no multipath component will arrive after τ_{max} for a given OFDM symbol transmitted over the channel, as illustrated in Fig. 16.5 [14].

• Therefore to avoid ISI, CP $\geq \tau_{max} \geq 1\,\mu$s.

Challenges for mesh relaying using OFDM transmissions. WiMAX is a high-power (long-range) transmission system. Therefore high power amplifier (HPA) used in the WiMAX system will be operated at high-efficiency operating point close to the saturation point in order to optimize the power efficiency of the system. OFDM time domain waveform however consists of varying signal levels within a symbol duration T. This is the composite signal representing the addition of large number of subcarriers, each with different amplitude and phase. When such signals are passed through HPA operating at high-efficiency operating point, severe nonlinear distortions are introduced in the signals. To reduce the level of nonlinear distortions introduced in the transmitted signal, amplifier operating points are usually reduced (or backed off) from the peak-efficiency operating point (saturation point) [16, 18]. High amplifier backoff results in less distortion but inefficient use of the power amplifiers. For mobile nodes, this translates to a significant reduction in battery life because RF amplifier stage accounts for 60–70% of the total DC power consumptions of modern communication devices. Low backoff on the other hand optimize the power consumption, but signals are more severely distorted. In OFDM transmissions, a combination of low backoff and appropriate signal processing techniques to reduce the peak-to-average power ratio (PAPR) of the time domain OFDM waveform before amplification is used in practice [6, 16]. Depending on the PAPR reduction technique employed however, transmit filters at the output of the HPA, required to reduce out-of-band spurii to legal level, restores the peaks in the amplified OFDM symbols. Thus PAPR reduction may be needed at each mesh node before the symbol is re-amplified and forwarded, as illustrated in Fig. 16.6. For OFDM transmissions in WiMAX mesh networks therefore, relaying nodes would be required to do some

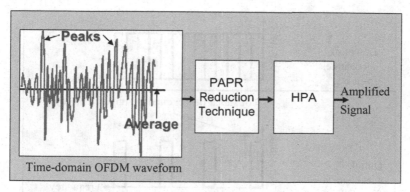

Fig. 16.6 Amplification of OFDM symbols at mesh nodes

indepth processings, beyond the level of the simple amplify-and-forward, before they can carry out the forwarding operation. This tends to limit relaying options in WiMAX mesh network to the more complex decode-and-forward relaying method at each mesh node. Thus security issues associated with this relaying option must be given proper attention in WiMAX mesh networks.

16.3.2 Subchannelization

The access method for 802.16–2004 OFDM PHY is TDMA, while 802.16e-2005 uses the concept of suchannelization to employ OFDM transmission as a multiple access scheme. Subchannelization is a tool designed to improve the uplink (UL) performance and/or balance the UL and downlink (DL) budgets in mobile network deployments where government regulation and need for cost-effective CPEs cause link budget to be asymmetrical –with the UL range/resources very limited. However the benefits of subchannelization in fixed network deployments were also realized and it was subsequently included in the 802.16–2004. Sixteen subchannels were chosen for the 256-OFDM PHY for fixed access. With 192 data subcarriers in the system, each subchannel consists of 12 data subcarriers. Users are assigned one or more subchannels at any time instant based on need and resources availability. Increased UL budget is achieved because a subscriber station (SS) can concentrate its power into a few subcarriers in the assigned subchannel, achieving a power gain (hence link budget gain) per subcarrier of 3 dB every time the number of excited data subcarrier is halved. For the 802.16e-2005 for mobile access, OFDMA (orthogonal FDM/multiple access) option, 32 subchannels were chosen. With these configurations, link budget gains of $10 \log_{10}(16) = 12\,\text{dB}$ and $10 \log_{10}(32) = 15\,\text{dB}$ are achieved in the UL for the fixed and mobile applications respectively (though implementation losses will reduce the gains realized in practice), if one subchannel is assigned per user. In fixed access networks, the maximum number of subchannels that can be assigned to a SS in the UL can be fixed in order to provide some

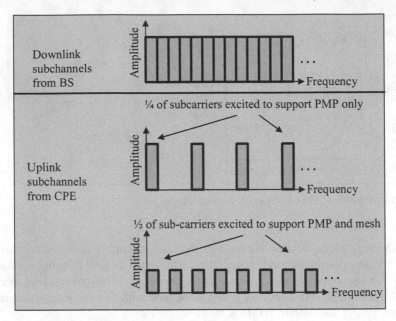

Fig. 16.7 Illustration of subchannelization for PMP and mesh supports

guarantee on subchanneling gain [6]. For mobile access networks however, link asymmetry warrants more flexible DL/UL resources usage. Subchannels assigned to user thus has to be variable, from the smallest to the highest available, in order to allow users to effectively tradeoff mobility for performance gains. In high mobility conditions, using small subchannels provide gains that can help overcome the harsh channel conditions due to user mobility, whereas in low mobility conditions more subchannels can be assigned to increase throughput. This concept can also be extended to mesh relay supports. When relaying functionality is to be supported, extra subchannels to provide the transmission links required for this operation can be assigned to the mesh node as illustrated in Fig. 16.7. Using this approach however, CPE will tradeoff mesh capabilities for access services throughput, but the advantage is that mesh relaying can then be implemented as a supportable functionality on top of the PMP mode. Also, it should be mentioned that this option for mesh support is applicable only to the decode-and-forward mesh relaying option, and may require that the transmissions of bursts for SSs, in Fig. 16.1 be aligned for the access and mesh relay services, which can be realized given ongoing efforts for compatible PMP and mesh modes.

Example 16.2. Estimate the subchanneling gain for a 256-OFDM system, with 16 subchannels in the system, if 4 subchannels are assigned to each subscriber station.

Solution: subscriber station excites $4/16 = 1/4$ of subchannels. Therefore, subchanneling gain $= 10\log_{10}(4) = 6\,\text{dB}$.

16.3.3 OFDMA and SOFDMA

In an OFDM-based system, all subcarriers in an OFDM symbol is used by a single user. In OFDMA however, many users share the FFT carrier space in an OFDM symbol, each user transmitting on the sub-channel(s) assigned to the user. Thus OFDMA uses OFDM transmission for multiplexing of data streams for multiple users in the DL and for multiple-access in the UL, using frequency separation of users. Scalable OFDMA (SOFDMA) brings additional benefits over OFDMA as it scales the size of the FFT to the channel bandwidth in order to keep the subcarrier spacing (and consequently OFDM symbol period) constant across different channel bandwidth. In the IEEE 802.16e, FFT sizes supported are 2,048, 1,024, 512, and 128. System vendors can choose small or large FFT size depending on the intended channel size. The benefits of SOFDMA in mesh networks are similar to those in the PMP networks. By fixing the subcarrier spacing and symbol period, the basic units of physical resources (time and frequency) is fixed. Therefore, the impact to higher layers is minimized when implementing systems with different channel bandwidth. This reduces system complexity and helps to facilitate interoperability for systems implemented in different regions. The scheme also helps to retain robustness over multipath ISI for different channel bandwidth, and results in higher spectrum efficiency for wide channels or cost reduction for narrow channels. Subchanneling is available in both 256OFDM and SOFDMA and can be exploited for "seamless" mesh support as explained above. However the added benefit of SOFMDA lies in the scalable (fixed) FFT size it uses which facilitates uniform implementation of such mesh support across different systems.

16.3.4 Diversity Techniques, MIMO and Adaptive Antenna Systems

Diversity: Diversity technique is a method by which the receiver is provided with multiple copies of the transmitted signal, each of them received over independently fading wireless channel. Diversity technique rely on the fact that with M independently fading replicas of the transmitted signals available at the receiver, the probability of an error detection is improved to p^M, where p is the probability that each signal will fade below a usable level. The link error probability is thus improved without increasing the transmitted power. Recently, the use of diversity technique at the transmitter side also gained wide attentions, and has resulted in the consideration of the more general case of multiple transmit-multiple receiving antennas or MIMO systems. Space-time codes, a coding system combined with MIMO antenna, then gained wide attention.

MIMO systems: The two options for MIMO transmissions in the WiMAX standard are space-time codes and multiplexing. For space-time codes, both space-time trellis codes and Alamouti space-time block codes are specified. However, it is the Alamouti space-time block codes that has yet been implemented by vendors due to its reduced complexity (even though space-time trellis code has better link performance improvements). In the Alamouti scheme designed for two transmitting antennas, a pair of symbol is transmitted at a time instant, and a transformed version of the symbols are transmitted in the next time instant. At the receiver, the decoder detects the four symbols transmitted over two time slots and processes them to obtain 2-branch diversity gain. Thus the Alamouti scheme achieves full diversity, with a rate-1/2 code. For the multiplexing option, the multiple antennas are used for capacity increase. In this option, original high-rate stream is partitioned into N low-rate substreams and each substream is transmitted in parallel over the channel, using different antennas. If there are enough scatters between the transmitter and the receiver, the channel delay for the multiple copies of the transmitted signals received for each of the substreams at the receiver, will be adequate to allow MIMO detection algorithms like zero-forcing, MMSE (minimum mean-square error), or V-BLAST (vertical Bell labs Layered Architecture for space-time codes), to separate the substreams. Thus the link capacity (theoretic upper-bound on the throughput) is increased linearly with min (N, L). The link capacity is given by

$$C \rightarrow m \log_2\left(1 + \text{SNR}\right), \quad m = \min\left(N, L\right) \tag{16.1}$$

where N is the number of transmit and L is the number of receiving antennas. When MIMO transmission is combined with OFDM signaling (MIMO-OFDM), an exciting high-speed solution is achieved.

Example 16.3. Compare the link capacity of the following MIMO-multiplexing systems deployed in WiMAX networks: System A with $N = 8$ transmitting and $L = 2$ receiving antennas, and System B with $N = 4$ transmitting and $L = 4$ receiving antennas.

Solution: For system A: $C \rightarrow 2 \log_2(1 + \text{SNR})$, whereas for system B: $C \rightarrow 4 \log_2(1 + \text{SNR})$. Thus, system B has higher link capacity despite having fewer numbers of combined transmitting and receiving antennas.

MIMO-multiplexing relaying for WiMAX mesh networks: One of the main challenges for metro-scale mesh network deployments is related to capacity scaling. In a full mesh mode, a mesh node act as a mesh router as well as a client access (data) node. Therefore the link capacity is split between mesh relaying and client access services. In the first generation of mesh products, one radio is used for both the relaying and client access services. The performance of such system thus degrades severely as more nodes enter the system, due to more congestions and contentions at each relaying node. It turns out that the relaying traffics take most of the available bandwidth because of the broadcast nature of the relaying protocols, starving the access service. The later generations (second and third generations) of mesh products address this problem by providing separate radios for the relaying and access services. For example in the Wi-Fi WLAN system, typical dual-radio configuration use one radio channels, 2.4 GHz radio, for local access and another radio channel, 5 GHz radio, for mesh relaying. Multi-radio system use one radio for local access and two or more radios for relaying, with the multiple relaying channels providing QoS differentiation. Since the mesh interconnection is performed by separate radios operating on different channels, local wireless access is not affected by mesh forwarding and can run at full speed. This significantly improved bandwidth and latency performance of both mesh relaying and access services. However for the WiMAX system where operations in the licensed spectrum have been actively pursued [11], this implies cutting the over all system capacity in half for the dual-radio system since a service provider must divide its allocated spectrum into two, one for access links, and the other for mesh relay links. Also for the multi-radio system, it translates to cutting the capacity by $1/K$, where K is the number of multiple links (relay and access) created. This is one of the major challenges facing mesh supports for systems operating in licensed spectrum. Here we illustrate the use of MIMO-multiplexing option at each mesh node to address this challenges. MIMO-multiplexing option creates multiple links on the same channel. For example, a 2×2 multiplexing system can provide dual-multiplexing links on the same channel at each mesh node, one for transmitting data streams for client access and the other for transmitting data streams for relaying service as illustrated in Fig. 16.8. At any time instant during a receive mode, the mesh node is capable of simultaneously receiving mesh relay data stream from source node S_1 as well as client access data stream from source node S_2. Also during a transmit mode, the mesh node is capable of simultaneously transmitting mesh relay data stream to destination node D_1 and client access data stream to destination node D_2 as illustrated in the figure. Similarly, a 4×4 multiplexing system can provide quadruple-multiplexing links on the same channel at each mesh node, one for client access and the other three links for providing differentiated QoS relaying services. Figure 16.8 illustrates the capacity achievable at each mesh node using such multiplexing relaying. 2×2 ($N = 2, L = 2$) MIMO-multiplexing system doubles the link capacity, which can be split between access and mesh relaying services, while 4×4 system almost quadruple the link capacity. Notice that the result displayed in this figure assumes independent data streams on each antenna, which can belong to different sources such as access and mesh relaying services. Therefore, MIMO-multiplexing wireless mesh network provides an interesting alternative to the conventional dual- and multi-radio wireless mesh network, where multiple (access and mesh relaying) links are created using costly radio channels. Using the MIMO-multiplexing mesh relaying approach, the overall capacity of the WiMAX system with access and mesh support is similar to that of the access-only WiMAX system using single antenna at each node, with the optional MIMO-multiplexing at each mesh node providing the extra capacity required to provide the optional mesh forwarding functionality when needed. Such system architecture is cost-effective, it facilitates convergence of PMP (access) and mesh mode component designs, and thus encourages more widespread WiMAX mesh network deployments.

Adaptive antenna system (AAS): For mesh relaying involving mobile nodes, adaptive antenna technologies can be used to track the location of the intended user as they roam about the mesh network. AAS is a beamforming technique that adaptively directs the directions of the beams in the electromagnetic wavefront to the intended user device, while directing the nulls to other unintended user devices. This operation results in directed mesh links, with strong link budget gains to support

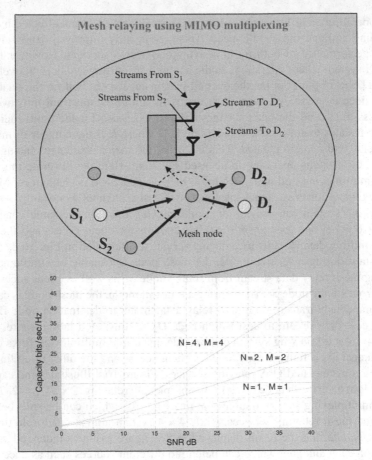

Fig. 16.8 MIMO-multiplexing relaying at WiMAX mesh nodes

mesh services. The potential drawback for AAS however is that beams tend to get blurred (or scatter) in Non LOS propagation conditions, which is typically envisaged by most WiMAX mesh network deployments. AAS however does gives significant link budget gains in LOS conditions.

16.3.5 *Forward Error Correcting Codes and Automatic Repeat Request*

Forward error correcting code (FEC) allows the WiMAX MAC layer to detect errors introduced during the transmissions of frames over the air link. FEC is usually combined with interleaving to spread the errors over many transmission blocks, maximizing the chance that the FEC is able to correct the error in each block. There are three methods of FEC specified in the WiMAX system; Reed-Solomon concatenated with convolutional code (RS-CC), block turbo code (BTC), and convolutional turbo code (CTC). RS-CC is mandatory, while BTC and CTC are made optional due to their complexity, even though they provide 2–3 dB better coding gain than RS-CC.

When the errors in a MAC SDU block is beyond the correcting capability of the FEC, an automatic repeat request (ARQ) algorithm is invoked to effect the retransmission of the lost or corrupted block. 802.16 MAC uses a simple sliding window-based approach, where the transmitter can transmit upto a negotiated number of ARQ blocks (MAC SDU blocks) without receiving acknowledgement (ACK). The receiver sends ACK or NACK (negative ACK) to indicate which MAC SDU blocks was received or lost, and the transmitter slides the window forward for every ACK received. When ACK is not received (or NACK received), the transmitter keeps transmitting until the end of the current window, after which it re-transmits the unacknowledged or lost MAC SDU blocks. This is the selective repeat (SR) protocol. For the 802.16e, a hybrid ARQ (H-ARQ) has been included as an optional feature. There are three types of H-ARQ, classified based on the manner in which they handle the re-transmissions. Type I H-ARQ re-transmits lost or unacknowledged blocks using chase combining in which the old erroneous block is stored at the receiver and compared with the re-transmitted copy. This helps to increase the probability of successful decoding at the FEC block during the re-transmission attempts. Type II/III H-ARQ use incremental coding rate to ensure successful decoding at the FEC block during the re-transmission attempts. These ARQ and H-ARQ schemes are optimized for single-hop transmissions (or PMP mode). For multi-hop wireless relaying applications in WiMAX mesh networks, hop-wise use of the cyclic redundancy check (CRC) in these ARQ schemes can be explored [12]. In ARQ schemes, a transmitted packet is accompanied with CRC bits attached to it from the source node, to detect errors in the received packet. Thus each mesh node in a WiMAX mesh network first check the CRC bits of the packet it receives, to confirm error-free reception, before forwarding the packet. If the CRC bits indicate that the received packet is erroneous, the packet can be discarded. This is termed the zero-retransmission policy [12]. However, a number of other re-transmission policies can be adopted, depending on latency and re-transmission overhead considerations. These include the infinite re-transmission policy where the transmitting mesh node re-transmits the packet repeatedly until the transmission is successful, the finite re-transmission policy (tagged truncated ARQ) where the packet is discarded only after jth transmission failure, and a number of probability-based re-transmission policies [12]. Eventually, only error-free copies of the packet get relayed to the destination. If the packet is correctly detected by the destination node, with ACK feedback at the source node, the source node slides the window forward and continues to transmit other packets as in the SR protocol used in the PMP mode. If all relaying links are bad at any particular transmission time, the destination will not receive the transmitted packet, and after a timeout period expires it will send a NACK message back to the source node via its relay links. In such transmission failure scenarios, there are two re-transmission options that could be employed. A relaying node which has correct copy of the packet can re-forward it, or the source node can be requested to re-transmit the packet. If amplify-and-forward relaying option is used, H-ARQ and FEC parameters used at the source nodes will be maintained at mesh relaying nodes. However if decode-and-forward relaying option is used, these parameters can be modified during the re-transmissions at the source as well as relaying nodes, to enhance the chance of successful re-transmissions.

16.3.6 Adaptive Modulations, Power and Coding Rate Control

Adaptive modulation allows the WiMAX MAC layer to adjust the signal modulation rate depending on the channel or radio link quality [3, 17]. When the channel quality is good, the MAC layer chooses the highest modulation rate, e.g., 64QAM, giving the system the highest throughput. When the channel quality degrades, the MAC layer reduces the modulation rate, e.g., 16QAM, reducing the throughput. In practice, adaptive modulation and coding rate control are used in conjunction with power control. When a link degradation arises, the transmitted power is first increased to provide link budget gain, until it reaches the maximum permitted. If the received signal quality does not improve, then the coding rate is reduced. Extra redundancy is added to provide more coding gain for better error correction performance. If the received signal quality still does not improve, then the modulation rate is reduced as a last resort (as this significantly affects the throughput than others). Similar (reverse) process is also followed when link quality appreciates. For WiMAX mesh networks using the amplify-and-forward relaying option, mesh relaying cannot exploit adaptive modulation technology because relaying nodes are not able to decode the contents of the received OFDM symbols to retrieve the modulated data and re-modulate them at higher or lower rate, in order to increase or reduce the transmission rate (or throughput) of the mesh streams in response to link quality condition. However for mesh networks using the decode-and-forward relay option, adaptive modulation and coding rate control can benefit the mesh relaying operation as mesh nodes can decode the mesh data streams and adjust the coding and modulation rate, depending on the forwarding link quality. For example in Fig. 16.9 a relay node R decodes a data stream originally transmitted from the source node S using 16QAM modulation, and re-modulates the data stream using 64QAM as it has good channel quality to the destination node D that can support this modulation rate. This results in fast and efficient use of the mesh links. In general for M-QAM modulation, the number of bits transmitted on each subcarrier n of an OFDM symbol can be approximated as:

$$r_n = \log_2(1 + \gamma_n P_n), \tag{16.2}$$

where γ_n = channel-gain-to-noise ratio for the subcarrier, and P_n = transmitted power.

Power control is applicable in WiMAX mesh operations in two ways: One, when mesh nodes are relaying data, they are regulated to transmit only the minimum

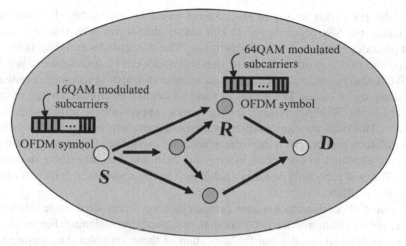

Fig. 16.9 Adaptive modulation at mesh nodes

required power to achieve successful reception at the receiver. Two, when mobile mesh nodes do not have mesh relay (or access service) data to transmit or receive, they go on sleep modes to save battery life.

Example 16.4. Estimate the achievable data rate increase (bits transmitted per OFDM subcarrier) for the second hop transmission in a two-hop relay system, given that the channel-gain-to-noise ratio per subcarrier for the second hop transmission roughly doubles the channel-gain-to-noise ratio per subcarrier for the first hop transmission.

Solution: For the first hop transmission: $r_n^1 = \log_2(1 + \gamma_n P_n)$, whereas for the second hop transmission we have: $r_n^2 = \log_2(1 + 2\gamma_n P_n)$. Thus,

$$r_n^2 - r_n^1 = \log_2(1 + 2\gamma_n P_n) - \log_2(1 + \gamma_n P_n)$$
$$\approx \log_2\left(\frac{2\gamma_n P_n}{\gamma_n P_n}\right) = \log_2(2) = 1 \, \text{bits s}^{-1} \text{Hz}^{-1}. \tag{16.3}$$

Thus, doubling the channel-gain-to-noise ratio roughly adds extra bits s^{-1} Hz^{-1}, assuming constant transmitted power.

16.3.7 Other Issues and Challenges

Scheduling and resource allocations: WiMAX standard specifies two forms of scheduling for the mesh mode: centralized and distributed scheduling [13]. In centralized scheduling, a base station, BS, controls all channel accesses. A subscriber station, SS, with data to transmit requests bandwidth from the base station upon which the BS grants access to the channel based on availability. In distributed scheduling, the SSs negotiate bandwidth among themselves with similar request-grant procedure as in centralized scheduling. The distributed scheduling is further divided into two, coordinated scheme in which nodes elect a coordinator to provide BS-like functionality, and uncoordinated scheme in which scheduling decision is shared among all nodes. The details of these scheduling disciplines are left unstandardized in the WiMAX system, giving vendors opportunity to differentiate their products. This presents some challenges for mesh operations among WiMAX nodes from different vendors, with different scheduling disciplines implemented and the nodes are expected to interrelate to carry out multihop data scheduling in a mesh mode. This will especially be fairly challenging in client mesh where BS infrastructure is not available.

Resource allocation mechanisms pursue efficiency in resource usage interms of power consumption, interference rejection, or capacity provisioning. For mesh networks, the need to balance out the allocation of these variables (i.e., ensure fairness) among mesh nodes in the network while at the same time provide nodes the needed resources for mesh relaying operations when and as needed, posses another challenges.

Handoff considerations: A cell in a mesh network is a collection of mesh nodes that are associated with the same coordinated scheduler (distributed and/or centralized), and a mesh network is the collection of cells, where nodes may arbitrarily associate with cell of choice. Handoff procedures transfer the radio resources of mobile mesh nodes from the coverage of one cell to another, as they roam about the mesh network. Both hard handoffs, where a mobile node's current connection with the serving cell is broken before new connection is established, and soft handoffs, where the mobile node is connected to the new cell before breaking connection with the current serving cell, are supportable [19]. Handoff consideration in mesh networks faces the challenges of lack of comparable signal strength to ensure successful handoff for all the different varieties of nodes that may be involved (e.g., low-powered Wi-Fi and high-powered WiMAX nodes). Also, since handoff consideration in mesh operation is not based only on PHY channel quality but also on the upper-layer metrics such as congestions at the network layer, handoff latencies are very high compared to the PMP mode as the computations of these metrics involve several updates at mesh nodes.

16.4 Directions for Future Research

OFDM technologies and mesh networks: As discussed in Sects. 16.3.1, 16.3.2, and 16.3.6, decode-and-forward relaying option for the OFDM-based transmission in WiMAX mesh network allows more efficient use of transmit amplifier, facilitates the use of subchanneling for mesh supports, and optimize mesh relay link utilization through the use of adaptive modulations. Thus future research works is expected to give significant attention to decode-and-forward relaying option in OFDM-based transmission systems, with the aim of addressing issues such as security, privacy, etc., in order to ensure that performance quality adequate for the need of WiMAX systems is maintained when using this relaying option.

MIMO technologies and mesh networks: Antenna technology is central to the success of wireless networks, especially the WiMAX system where superior link quality performance is targeted. In Sect. 16.3.4, we have explored various MIMO antenna technologies specified in the WiMAX system namely: space-time codes and MIMO-multiplexing. We then introduced the concept of MIMO-multiplexing relaying for mesh support in the WiMAX system. This scheme provides cost-effective means of providing mesh support on top of access service in WiMAX networks operating over licensed bands. Given that this solution for mesh support is attractive from a cost perspective, it is expected that future research will explore the idea in a wider sense, such as its applications in IEEE 802.16j where the principle can be employed to provide different simultaneous relay links to different destinations on the same radio channel, in a dedicated relay system. Practical implementational issues associated with this solution is also expected to be subject of future works.

Apart from MIMO antenna system, the use of AASs (or beamforming) in WiMAX network has also gained significant attention recently. While future research will continue to address traditional problems such as angle-spread, scattered beams, etc., that reduce the gains of beamforming solution in schemes like directed mesh, there will also be significant interest on the use of beamforming-based space-division multiple access (SDMA) for capacity increase both in access and centralized mesh configurations in WiMAX networks [9]. Thus, future research works on beamforming technologies in WiMAX mesh networks will also feature significant exploration of the beamforming solutions.

Scheduling and resource allocations: As explained in Sect. 16.3.7, resource allocation mechanisms pursue efficiency in resource usage in terms of power consumption, interference rejection, or capacity provisioning. Future research on scheduling and resource allocations in WiMAX mesh is also expected to address the issue of optimization of the use of these variables at WiMAX nodes supporting both access and mesh services, such that an existing performance bench mark for access service is not significantly violated when providing resources for mesh supports. This would be a network-wide optimization problem.

16.5 Conclusions

This chapter discusses metro-scale mesh networks using the WiMAX standard. The technologies deployed in the WiMAX system are closely examined to discuss the prospects and challenges of mesh operations using these technologies. We highlight the challenges facing mesh operations with OFDM signaling; and the prospects of using subchannelization and MIMO-multiplexing for mesh networks. It is shown that MIMO-multiplexing relaying at WiMAX mesh node can be used to provide different links for access and mesh relaying services on the same radio channel. For

WiMAX systems operating in licensed spectrum where the use of the conventional dual- and multi-radio mesh networks will result in reduced numbers of users per spectrum allocation, MIMO-multiplexing mesh relaying becomes an attractive and cost-effective solution that can encourage more widespread WiMAX mesh network deployments. Other technologies such as ARQ, adaptive modulation and coding, and power control are also discussed and the challenges presented by multi-hop mesh relaying operations using these WiMAX technologies are enumerated.

Acknowledgments The authors would like to thank Dr. Najjah Abu-Ali of UAE University, and Dr. Mohamed Ibnkahla of Queen's University, for their comments on earlier drafts of this chapter.

16.6 Questions

1. Using Fig. 16.1, explain the network entry procedures for WiMAX subscriber stations supporting both access and mesh services.
2. (a) Given an FDM transmission system with guard band 2 kHz, and subcarrier channel bandwidth of 4 kHz each, how many subcarriers can be accommodated in a system with a total of 9 MHz spectrum available. (b) If OFDM transmission with frequency-bin spacing of $\Delta f = 2$ kHz is employed for this system, estimate how many subcarriers can be accommodated in the system.
3. Given two WiMAX systems, system A with 10 MHz channel, and system B with 5 MHz channel, use illustrations to explain the benefits in employing SOFDMA rather than OFDMA, for these systems.
4. For a WiMAX system employing the 256-OFDM with 16 subchannels per symbol, evaluate the link budget gain for an uplink transmission with: (a) all subchannels excited, (b) half of the subchannels excited, (c) quarter of subchannels excited.
5. A WiMAX operator uses subchanneling to provide access and mesh relaying services multiplexed on a 256-OFDM transmission, where half of the excited subchannels in every OFDM symbol transmitted are dedicated to mesh relay services. Assuming that $1/4$ of the subcarriers in each OFDM symbol transmitted in this system are normally excited for an access-only service, but the operator increases this to half for combined access and mesh supports, in order to keep the access throughput constant. Estimate the loss in access link budget for this approach.
6. A WiMAX equipment manufacturer implemented 2-branch receiver diversity for the uplink transmission in their WiMAX system, alongside subchanneling solution where one subchannel is assigned per user for the OFDMA option of 2,048 subcarriers per OFDM symbol (with 32 subchannels in the system). The company claims that practical tests reveal that these two solutions yield a total link budget gain of 12 dB. Analyze this claim theoretically, and comment on the reasons for any discrepancies.

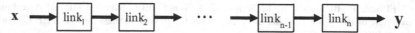

Fig. 16.10 n Cascaded channels

7. Explain the main distinguishing feature of OFDM and OFDMA options in the WiMAX system.

8. Wireless channel capacity is fundamentally limited by SNR. For the single input-single output (SISO) system, the capacity is given by $C = \log_2(1 + \text{SNR})$ bits s^{-1} Hz^{-1}, while for the MIMO system the capacity is given by $C \approx N \log_2(1 + \text{SNR})$ bits s^{-1} Hz^{-1}, where N is the smaller of the number of transmitting and receiving antennas. (1) For the SISO system, estimate how much the transmitted power must be increased to go from 1 to 11 bits s^{-1} Hz^{-1}. (2) Design an $N \times N$ MIMO system to achieve same increase in capacity without increasing the transmitted power.

9. Given that MIMO-multiplexing relaying option is used for mesh support in a WiMAX network deployment. Each wireless hop transmission in the network employs MIMO-multiplexing system with configuration $M = 4$, $L = 4$, where M is the number of transmitting and L is the number of receiving antennas. Suppose mesh relaying services, when supported, are assigned N (where $N \leq M, L$) antennas throughout the network. Let $N = 2$ (i.e., 2×2 MIMO-multiplexing links dedicated to mesh relaying services), (a) estimate the achievable access link capacity per hop, when mesh supports are provided, (b) calculate the achievable access link capacity when mesh supports are not provided, (c) estimate the data rate loss to access service per hop, with mesh supports.

10. For the binary symmetric channel with multi-hop relaying over n cascaded channels [8], shown in Fig. 16.10, the error-rate performance of the system is given by $P_n = \frac{1}{2}[1 - \prod_{j=1}^{n}(1 - 2p_i)]$, where p_i is the error probability of link i, $i = 1, \ldots, n$. What effect does multi-hop relaying has on the error-rate performance of MIMO-multiplexing system (assuming that relaying nodes employ decode-and-forward relaying option)?

References

1. Z. Abichar, Y. Peng, J. M. Chang, WiMAX: The emergence of wireless broadband, *IEEE IT Prof.*, 8(4):44–48, 2006.
2. I. F. Akyildiz, X. Wang, A survey on wireless mesh networks, *IEEE Radio Commun.*, 43(9):23–29, 2005.
3. L. Badia, A. Baiocchi, A. Todini, S. Merlin, S. Pupolin, A. Zanella, M. Zorzi, On the impact of physical layer awareness on scheduling and resource allocation in broadband multicellular IEEE 802.16 systems, *IEEE Wireless Commun.*, 14(1):36–43, 2007.
4. R. Bruno, M. Conti, E. Gregor, Mesh networks: Commodity Multihop ad hoc networks, *IEEE Commun. Mag.*, 43(3):123–131, 2005.

5. C. Eklund, R. B. Marks, K. L. Stanwood, S. Wang, IEEE Standard 802.16: A technical overview of the wireless MAN air interface for broadband wireless access, *IEEE Commun. Mag.*, 40(6):98–107, 2002.
6. C. Eklund et al., *WirelessMANInside the IEEE 802.16 Standard for Wireless Metropolitan Networks*, IEEE Press, New York, 2006.
7. G. J. Foschini, G. D. Golden, R. A. Valenzuela, P. W. Wolniansky, Simplified processing for high spectral efficiency wireless communication employing multi-element arrays, *IEEE J. Sel. Areas Commun.*, 17(11):1841–1852, 1999.
8. H. D. Goldman, R. C. Sommer, An analysis of cascaded binary communication links, *IRE Trans. Commun. Syst.*, 291–299, 1962.
9. C.-H. Hsu, Uplink MIMO–SDMA optimisation of smart antennas by phase–amplitude perturbations based on memetic algorithms for wireless and mobile communication systems, *IET Commun.*, 1(3):520–525, 2007.
10. IEEE Standard for Local and Metropolitan Area Networks – Parts 16: Air Interface for Fixed Broadband Wireless Access System, 2004.
11. IEEE Standard for Local and Metropolitan Area Networks – Parts 16: Air Interface for Fixed and Mobile Broadband Wireless Access System Amendment 2: Physical and Medium Access Control Layers for Combined Fixed and Mobile Operation in Licensed Bands, 2006.
12. T. Issariyakul, E. Hossain, Analysis of end-to-end performance in a multi-hop wireless network for different hop-level ARQ policies, *Proc. Globecom'04*, 3022–3026, 2004.
13. D. Kim, A. Ganz, Fair and Efficient Multihop Scheduling Algorithm for IEEE 802.16 BWA Systems, *Proc. IEEE Int'l Conf. Broadband Netw.*, 2:895–901, 2005.
14. L. Litwin, M. Pugel, The principles of OFDM, *RF Signal Processing*, 30–48, 2001 (available online at: rfdesign.com).
15. MOTOROLA North America, Motorola Inc., 2004, www.motorola.com/mesh.
16. R. Nee, A. Wild, Reducing the peak-to-average power ratio of OFDM, *IEEE VTC'98*, 3:2072–2076, 1998.
17. M. Settembre, M. Puleri, S. Garritano, P. Testa, R. Albanese, M. Mancini, V. L. Curto, Performance analysis of an efficient packet-based IEEE 802.16 MAC supporting adaptive modulation and coding, *Proc. IEEE International Symposium Computer Networks*, 11–16, 2006.
18. A. I. Sulyman, M. Ibnkahla, Performance analysis of nonlinearly amplified M-QAM signals in MIMO channels, *Proc., IEEE ICASSP'04*, Montreal, Canada, 2004.
19. H.-Y. Wei, S. Kim, S. Ganguly, R. Izmailov, Seamless handoff support in wireless mesh networks, *Proc. IEEE Workshop Operator-Assisted (Wireless Mesh) Community Networks*, 1–8, 2006.

Chapter 17
Scheduling and Call Admission Control
A WiMax Mesh Networks View

Daniel Câmara and Fethi Filali

Abstract This chapter discusses the problem of providing call admission control (CAC), scheduling and band reservation for wireless networks. It presents the importance of such procedures focusing mainly on WiMax mesh mode networks. The chapter also classifies some of the most known proposals presented in the literature to solve the scheduling and CAC problems for this kind of network. Differently of some other standards, in the IEEE 802.16 standard the scheduling and CAC procedures are mandatory. No node in the network can communicate, even in the mesh mode, without having the transmission previously scheduled. In this way scheduling becomes one of the most important processes to achieve spectral efficiency and, in consequence, to increase the network capacity.

17.1 Introduction

In the last years wireless mesh networks (WMN) have been attracting a huge amount of attention from both, academia and industry. Indeed, WMN is now emerging as a promising technology for broadband wireless access [2,6]. One of the main reasons for this sudden popularity of WMN is their inclusion in many of the IEEE wireless standards and in special the IEEE 802.16 [22]. The addition of the mesh mode to the IEEE 802.16 standard brought a series of advantages for these networks. Among them we can cite nonline-of-sight (NLOS) capacity, higher network reliability, scaling, throughput and availability [35].

However, to become really useful and valuable for the applications running on top of them, the WMN must to provide some level of quality of service (QoS). To fulfill this requirement, mainly for WMN environments, radio resource management (RRM) techniques play a major role [1]. RRM is the term used to identify a series

D. Câmara (✉)
EURECOM 2229 Route des Crêtes, BP 193-F-06560, Sophia-Antipolis, France
e-mail: daniel.camara@eurecom.fr

S. Misra et al. (eds.), *Guide to Wireless Mesh Networks*, Computer Communications and Networks, DOI 10.1007/978-1-84800-909-7_17,

of strategies and algorithms employed to optimize the use of the radio spectrum and wireless networks limited resources. RRM techniques include frequency and/or time channel allocation, transmission power, access to base stations, handover criteria, modulation schemes, error coding schemes [47]. On behalf of [1] RRM policies, along with the network planning and air interface design, in deep, determine the QoS network performance at both individual user and network level.

This chapter focuses on the problem of providing call admission control (CAC), scheduling and band reservation for the mesh mode of IEEE 802.16 networks [22, 29] also known as WiMax networks. Although these mechanisms are mandatory for IEEE 802.16 networks, the standard just specifies the signaling protocols and messages structure. The transmission scheduling control algorithm is left undefined. This makes the standard open to accommodate extensions and improvements. However, this also may lead, in the future, to incompatibilities among vendors' proprietary solutions.

For future readings, among many other works related to this one, we may highlight the survey presented by Kuran and Tugcu [29] in general emerging broadband wireless technologies. For a survey in general mesh networks the Alkydiz et al. work [2] presents a good overview on many aspects of the mesh networks, discussing how these aspects affect the entire network stack. The problem of CAC mechanisms in general is discussed in [1]. A broad view of the problem of distributed medium access control for mesh networks can be found in [12]. Zhao presents consistent view of the problem of distributed coordination in mesh networks in [39]. For a deep discussion, more specifically for 802.16 mesh networks centralized scheduling algorithms, see [15]. In [36] Redana and Lott present an analysis of the overhead caused by the control messages on the IEEE 802.16 mesh mode and show that, for multihop networks, the centralized approach have a better performance than the distributed one. For an analysis of the times involving the phases of the distributed scheduler mode see [7] and [9].

The remaining of this chapter is organized as follows: Section 17.2 explains better what is CAC and scheduling. After that, Section 17.3 presents an overview of the CAC and scheduling process for WiMax mesh mode networks. Section 17.5 presents a possible classification for CAC and scheduling proposals and classifies some of the most well known proposals of the literature in accordance to the proposed taxonomy. Sections Ideas to Consider and Open Issues, presents, respectively, some of the most interesting techniques of the previously classified approaches and some possible directions for future researches on the field.

17.2 Background

This section presents a deeper discussion of what is CAC and scheduling, showing the importance of such mechanisms for the performance of networks in general and, in special, for the WiMax mesh mode.

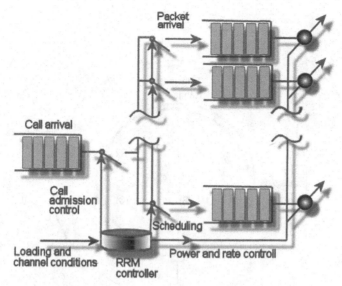

Fig. 17.1 The radio resource management model [1]

As shown in the Fig. 17.1, the CAC procedure is responsible for granting/denying access to the network. The decision of which connections are accepted and which one are not, is based on predefined criteria, taking into account the network status and the requirements of new calls. The admitted calls are then controlled by other mechanisms of the RRM, such as the schedule. The schedule is the RRM process that decides which one is the best moment to grant bandwidth for the admitted calls.

Just considering only throughput, ignoring any other QoS parameter, the scheduling problem is proven to be an NP-hard problem for multihop wireless networks [24, 27]. This means that if the number of nodes, or links, in the WMN increases it becomes computationally impossible to find the optimal scheduling solution. So, in this context, suboptimal scheduling solutions, with lower complexity, are acceptable and even desired for mesh environments.

CAC and scheduling play a central role in the WiMax networks and it is not only because they are mandatory, their importance is far beyond that. They indeed provide a number of important features to the network. Among such features we can highlight: network signal quality, call blocking, dropping probabilities, control of packet delay and transmission rate guarantee. CAC and scheduling mechanisms have been extensively studied for both wired and wireless networks. However, because the intrinsic characteristics of the medium, the application of these techniques for wireless environments is much more challenging than for wired ones.

We need to remember that, by principle, the wireless medium is a broadcast one where, at any time, a number of different stations are addressing the channel concurrently. The main problem with this is that, if concurrent transmissions occur in the same carrier frequency at the same time, this may result in mutual destruction of the transmitted signals. Unfortunately the interference range is greater than the

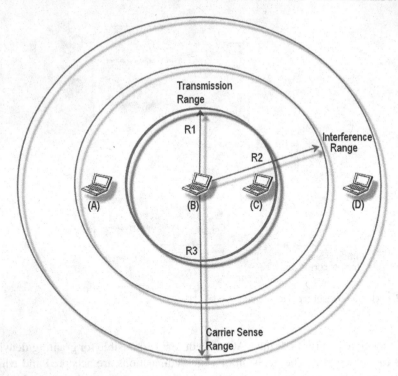

Fig. 17.2 Different ranges in the nodes communication

transmission one. The receiver can only decode or sense the message if the signal-to-interference-and-noise-ratio (SINR) is above some level. For example, in Fig. 17.2, node D can have its signal jammed by the signal sent from B to C and may not be able to actually decode the signal. The interference range means that any transmission made from A, which is in the interference range, can damage the signal between B and C. These different ranges can lead to a number of different scenarios, among them the hidden and exposed node problems, common in IEEE 802.11 networks. For WiMax networks, scheduling and CAC are the techniques used to avoid the interference problems. However, regardless the claims that WiMax networks are free from such problems, Zhu and Lu [48] show they can also occur in WiMax environments.

17.2.1 Reasons to Use CAC and Scheduling

Among the main reasons to use CAC and scheduling schemes we have guarantee of the signal quality, guarantee of transmission rates, decreasing in call dropping probability, possibility to observe packet level parameters, maximization of revenues, prioritization of services and fairness in the medium access.

- *Signal Quality*: CAC schemes guarantee the signal quality once they ensure that a new connection will only be accepted if the network can afford it. The scheduling intends to organize the nodes and decrease the network interference.
- *Transmission Rate*: CAC schemes ensure that the network can offer, at least, the minimum rate required by a given communication and the scheduling observes that the promised transmission rate is really achieved.
- *Call Dropping Probability*: Dropping an ongoing call is normally much more troublesome, from the user point of view, than blocking or delaying a new call. In this way CAC mechanisms are normally used as a control switch to limit new calls in favor of ongoing calls or handoffs.
- *Packet-Level Parameters*: CAC schemes can be used to evaluate if a new call will damage the network performance observing packet-level QoS parameters, e.g., packet delay, delay jitter and throughput. The scheduling may also use such information to improve the quality of the connection.
- *Revenue-Based CAC*: Each new call in the network may bring some kind of revenue to the network. CAC schemes may be used to evaluate such benefits and costs for new connections and decide which calls are more interesting to accept and keep.
- *Prioritize Some Services/Classes*: Some classes of services may have priority over others. CAC schemes can, for example, be used to give priority for traffics that represent better revenues for the network operators. The schedule can also beneficiate such traffics in the resources allocation in detriment of others.
- *Fair Resource Sharing*: Even seeming contradictory, regarding the two previous items, the fairness exists if it is based in some predefined parameters and observed among traffics in the same classes and among different classes.

17.2.2 Aspects to Observe

There are some aspects that good CAC and scheduling schemes should observe. Among the most important ones we have: channel utilization, fairness, end-to-end delay, throughput and QoS support.

- *Channel Utilization*: The greater the channel utilization the better, once it represents the fraction of time used to transmit user data packets in a given period.
- *Fairness*: Traffic flows, with the same QoS level, should gain equal chances to use the wireless medium. However, mainly in highly loaded situations, internal scheduling polices may lead to unfairness. This, as a network behavior, is normally undesirable and should be avoided as much as possible.
- *End-to-End Delay*: This aspect refers to the elapsed time between the generation of a packet at the source station and the correct reception of the packet at the final destination station. The delay performance relays on protocol capabilities of avoiding collision and exploiting spatial reuse. It relays also on the protocol efficiency of channel access and achieved fairness.

- *Throughput*: Throughput is the volume of user data transferred between two stations in a given period. Throughput is one of the most widely used performance metrics. The schedule and CAC algorithms are considered better than others if they help to increase the throughput.
- *QoS Support*: CAC and scheduling schemes are considered as part of the MAC layer protocols. However, they should be able to understand and consider the QoS preferences of the upper layers flows, guaranteeing their specific requirements, such as throughput, packet loss ratio (PLR), packet delay and jitter requirements.

17.2.3 Scheduling Types of Service

The IEEE 802.16 standard defines five different scheduling types of services: unsolicited grant service (UGS), real-time polling service (rtPS), extended real-time polling service (ertPS), nonreal-time polling service (nrtPS), and best effort (BE). Table 17.1 summarizes the main characteristics of these five types of services.

Unsolicited Grant Service – UGS: Designed to support real time data streams where packets are generated in a fixed data rate. For example, VoIP connections without silence suppression. The mandatory QoS parameters for this service are Maximum Sustained Traffic Rate, Maximum Latency, Tolerated Jitter, Uplink Grant Scheduling Type and Request/Transmission Policy. Once the rate is constant, if present, the Minimum Reserved Traffic Rate parameter should has the same value as the Maximum Sustained Traffic Rate parameter, once the data rate is constant. The grants for this service are issued periodically and without any explicit request. The main advantage of this is that it eliminates the overhead and the latency of the subscriber station (SS) issuing for new grants for this specific traffic.

- *Real-time Polling Service – rtPS*: The Real-time Polling Service is designed to support the same kind of traffic that UGS does, but with variable data rate, e.g., MPEG video. The mandatory QoS parameters are Minimum Reserved Traf-

Table 17.1 Services and their main parameters and characteristics

Characteristic Scheduling type	Max sustained traffic rate	Min reserved traffic rate	Max latency	Tolered jitter	Traffic priority	Request/ transmission policy	Piggy back request	Bandwidth stealing
UGS	M	O	M	M	X	M	NA	NA
rtPS	M	M	M	O	M	M	A	A
ertPS	M	M	M	M	M	M	A	NA
nrtPS	M	M	X	X	M	M	A	A
BE	M	X	X	X	M	M	A	A

M – Mandatory, O – Optional, X – Not Available, A – Allowed, NA – Not Allowed

fic Rate, Maximum Sustained Traffic Rate, Maximum Latency, Uplink Grant Scheduling Type and Request/Transmission Policy. Differently of the UGS flow, this service offers periodic unicast request opportunities for the SS to adjust the size of its grants.

- *Extended Real-Time Polling Service – ertPS*: The extended rtPS service, introduced latter into the standard [23], is a service based on both UGS and rtPS. For ertPS the flow has some amount of resource reserved in an unsolicited grant way, but the allocation may change if the SS requests for that. In other words, the allocation is dynamic and depends on the needs of the SS, but when set it works as the UGS type. The key service information elements are the Maximum Sustained Traffic Rate, Minimum Reserved Traffic Rate, Maximum Latency and Request/Transmission Policy. The extended rtPS is designed to support real-time service flows that generate variable size data packets on a periodic basis, such as Voice over IP services with silence suppression.

- *Non-real-time Polling Service – nrtPS*: The nrtPS is designed to support delay-tolerant data streams consisting of variable-sized data packets that require variable data grant on regular basis. File transfer protocol (FTP) is an example of application that could use this kind of service. The mandatory QoS parameters for this scheduling service are Minimum Reserved Traffic Rate, Maximum Sustained Traffic Rate, Traffic Priority, Uplink Grant Scheduling Type and Request/Transmission Policy. The advantage of this kind of service is that it can support data streams even in very saturated network conditions. The mesh base station (BS) provides SS the opportunity to request bandwidth using unicast and contention period. In addition, piggyback request opportunities are also available.

- *Best Effort Service – BE*: Best Effort service intend to be used for any other kind of traffic that does not have any significant QoS requirements and that can be handled on a space-available basis, e.g., http and e-mail traffic. The mandatory QoS service flow parameters for this scheduling service are Maximum Sustained Traffic Rate, Traffic Priority and Request/Transmission Policy.

17.3 WiMax Mesh Mode Overview

The WiMax mesh mode, introduced in the standard by the IEEE 802.16a amendment [21], supports two different physical layers: WirelessMAN-OFDM^{TM}, operating in a licensed band, and WirelessHUMAN^{TM}, operating in an unlicensed band. Both of them use 256 point FFT OFDM TDMA/TDM for channel access and operate in a frequency band below 11 GHz.

Even though the standard permits both time division duplex (TDD) and frequency division duplexing (FDD) as access scheme, for the mesh mode only the TDD is allowed [22]. This means that the uplink and downlink transmissions share the same frequencies and, doing so, they must to occur at different times. However, for IEEE 802.16j, the relay networks upcoming part of the standard some people proposed the use of FDD [42].

The Mesh frame is divided into control and data subframes. There are two types of control subframes: schedule control and network control subframe. The network control subframe provides basic functionality for network entry and topology management. The schedule control subframe controls the transmissions. The scheduling is done negotiating minislots ranges for the traffic demands of each link. All the communications are done in terms of the links established among nodes. All data transmissions between two nodes are done through one link and the QoS is provisioned over links on a message by message basis. Upper layer protocols are in charge of the traffic classification and flow regulation.

17.3.1 Scheduling Policies

In Mesh mode all transmissions must to be scheduled, not even the Mesh BS can transmit without having its transmission coordinated with other nodes [22]. To organize the medium access, the standard defines three different schedule mechanisms: coordinated centralized scheduling, coordinated distributed scheduling and uncoordinated distributed scheduling. These three schedule policies can be either used alone or together in the same network.

According some authors the centralized schedule should be used for external traffic and the distributed schedule for intra network traffic [8, 13]. This came from the fact that the centralized schedule trusts in a mesh BS, which is in last instance, a backhaul responsible for act as gateway between the internal and external network traffic. Table 17.2 presents the messages used by the CAC and schedule mechanisms in the WiMax mesh mode.

Table 17.2 Mesh MAC management messages

Message type	Name	Description	Connection mode
39	MSH-NCFG	Mesh network configuration	Broadcast
40	MSH-NENT	Mesh network entry	Basic
41	MSH-DSCH	Mesh network distributed schedule	Broadcast
42	MSH-CSCH	Mesh network centralized schedule	Broadcast
43	MSH-CSCF	Mesh network centralized schedule configuration	Broadcast

17.3.2 Centralized Scheduling

For the centralized scheduling, the mesh BS schedules all network transmissions, even the mesh BS ones. The resource request and the mesh BS assignments are both transmitted during the control portion of the frame. The centralized scheduling coordinates the transmissions and ensures that they are all collision-free. Once the BS has the knowledge of the entire network, it is typically more optimal using the spectrum than the distributed forms. Algorithm 1 [5] defines the downstream transmission ordering for MSH-CSCF, or MSH-CSCH, messages, being the upstream transmission ordering the same, but in the reverse order.

> //Downstream MSH-CSCF or MSH-CSCH messages use the following algorithm
> **Begin** {
>   The mesh BS initiates the frame;
>   Collect the eligible children of the mesh BS, with hop count equal 1;
>   Order them in by their appearance in the most recent MSH-CSCF packet;
>   Transmit in accordance to the established order;
>   **If** (The message does not fit entirely in a subframe)
>   Fragment the message;   **While** (Exists eligible nodes) {
>   Increase the hop count by one;
>   Ordered nodes by their appearance in the MSH-CSCF packet;
>   Transmit in accordance to the established order;
>   **If** (The message does not fit entirely in a subframe)
>   Fragment the message;
>   }//while
>   **If** (A node's order requires it to transmit immediately after receiving)
>   Insert a MinCSForwardingDelay delay;
> }//Begin

Algorithm 1 *Centralized scheduling control transmit order algorithm [5].*

The MSH-CSCH message has two variants, MSH-CSCH Request and MSH-CSCH grant. With the MSH-CSCH Request each node estimates and reports the level of its own upstream and downstream traffic demand to its parent. This demand comprises also the demands reported by the node's children. With the MSH-CSCH Grant the mesh BS propagates down, through the routing tree, the levels of flows and grants to each node in the network. Figure 17.3 shows an example of message flow for the centralized schedule.

All MSH-CSCH Grant messages contain information about all network grants, because all nodes need the complete information for the schedule computation. Upon receiving any message in the current scheduling sequence and assuming that nodes have up-to-date scheduling configuration information, any node is able to compute locally the schedule for all transmissions, including its own. Besides the mesh BS, a node should not transmit any downstream centralized scheduling packet without receiving a MSH-CSCH message from a parent. Also, a node should not send any centralized scheduling packets, if its MSH-CSCF information is outdated.

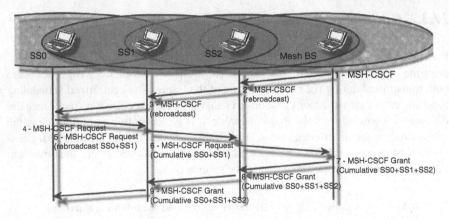

Fig. 17.3 A message flow example for the centralized scheme

In terms of eligibility to send and receive MSH-CSCH messages, all nodes are eligible to retransmit the grant schedule, except those with no children. For transmitting MSH-CSCH grant messages, all nodes with children are eligible. For transmitting MSH-CSCH request messages, all nodes, except the mesh BS are eligible.

17.3.3 Distributed Scheduling

In both distributed scheduling mechanisms, coordinated and uncoordinated, all the stations in the two hop neighborhood must to have their transmissions coordinated to avoid collision. The coordinated distributed scheduling uses the control part of the frame to transmit its own traffic schedule. Both schedule schemas, centralized and distributed, may coexist at the same time at the same network.

The uncoordinated distributed scheduling is a simpler version of the distributed scheduler and may be used for fast ad hoc setup of schedules in a hop-by-hop basis. The uncoordinated schedule is basically an agreement between two nodes and should not cause collision with the data and control traffic scheduled by the coordinated schedules. Both coordinated and uncoordinated distributed scheduling employ a three-way handshake to setup the connection.

The first message in the three-way handshake is a MSH-DSCH request. The transmission is scheduled using a random-access algorithm among the "idle" slots of the current schedule. If the attempt was unsuccessful a random backoff is used to avoid new collisions. Figure 17.4 shows schematically the messages in the distributed schedule three way handshake.

The MSH-DSCH Grant can be issued by any neighbor that listens the MSH-DSCH Request. The grant message contains the list with the subset of the resources awarded. The first granter node may start its grant transmission in the immediately

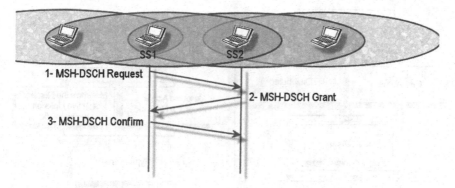

Fig. 17.4 Distributed scheduling three way Hand Shake

following base-channel idle minislot. More than one granter may also respond the request.

The requester node sends the same received MSH-DSCH Grant message in confirmation. Doing this the requester's neighbors became aware of the grant awarded. The grant confirmation is then sent in the first available minislots following the minislots reserved for the grant opportunity of the last potential granter.

17.4 Network Configuration

Two more messages, responsible for create and maintain the network configuration, may be transmitted in the network control sub frame: mesh network configuration (MSH-NCFG) and mesh network entry (MSH-NENT).

A new node that wishes to join the mesh network needs to wait until listen a MSH-NCFG message. When the new node receives this message it is able to establish the synchronization with the mesh network. In truth it should decide which node will be the best sponsor for its communication, so the new node may wait for more than one MSH-NCFG message to arrive. When the sponsor node is chosen, the new node sends though the sponsor a MSH-NENT message to the mesh BS with its registration information. The sponsor node then establishes a quick schedule, through the uncoordinated scheduler process, and communicates it to the new node. The new node confirms the schedule and sends the required security information. Finally, in the last step, the sponsor node grants the new node the access to the network.

17.5 Taxonomy

This section presents a possible classification for the proposed algorithms for CAC and scheduling for IEEE 802.16 networks and also frames some of the most important works of the literature on this classification. Figure 17.5 shows a diagram with

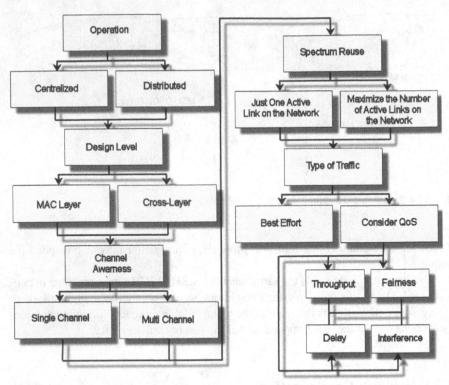

Fig. 17.5 Proposed classification for WiMax mesh mode CAC algorithms

the topics used in the classification and Table 17.3 presents how some important works fits on the classification. The aspects observed are: operation mode, design level, channel awareness, spectrum reuse, type of traffic and QoS observed.

As Fig. 17.5 shows, one proposal can, and indeed should, present more than one of the observed characteristics. It is perfectly possible to have, for example, a proposal that has a centralized approach, with cross layer design, that try to maximize the number of active links and that observe QoS parameters. Actually this is exactly the case of the proposal presented in [43]. However it is important to highlight that the topics presented here are, by no means, an extensive list, in special in what concerns the QoS support aspects. The values present in the classification are just some of the more common used to distinguish the algorithms. Other classifications can be found in [1, 12].

- *Operation Mode*: The operation mode reflects if the proposal focuses in the centralized or distributed mode of the standard. In the centralized approach all the scheduling and CAC decisions are made in the mesh BS. Without a central coordination, distributed approaches are more challenging than centralized ones. All the communications in the IEEE 802.16 networks must to be synchronized. It is important to notice that the synchronization problem is considerably harder in

Table 17.3 Classification of literature proposals

Proposal	Operation mode	Design level	Channel aware	Spectrum reuse	Type of traffic considered	QoS aspects observed
[36]	Distributed	MAC	No	No	No	No
[30]	Centralized	MAC	No	Yes	Yes	Five types of service
[34]	Distributed	MAC	No	No	Yes	Priority channels
[17]	Distributed	MAC	No	Yes	Yes	Yes
[18]	Centralized	MAC	No	Yes	Yes	Yes
[13]	Dist/Central	MAC	No	Yes/No	No	No
[26]	Centralized	MAC	No	Yes	No	No
[41]	Centralized	Crosslayer	No	No	No	No
[20]	Centralized	MAC	No	Yes	No	No
[31]	Centralized	Crosslayer	Yes	Yes	Different rewards for dif. connections	QoS and Non QoS connections
[40]	Centralized	Crosslayer	No	No	Yes UDP and TCP	Yes
[46]	Centralized	Crosslayer	No	No	No	Yes
[11]	Centralized	Crosslayer	No	Yes	No	Yes
[25]	Centralized	MAC	No	Yes	Yes	Yes, all the classes
[8]	Centralized	Crosslayer	No	Yes	Yes TCP and UDP	No
[43]	Centralized	Crosslayer	No	Yes	No	No
[14]	Distributed	MAC	Possible	Yes	No	No

a distributed environment. Both schedules can be running simultaneously inside the network, using different messages and configuration slots. Although this is a standard and expected organization for slots, even explicit in the standard [22], the work of Cheng et al. [12] shows that the avoidance of such division may lead to better performance results.

- *Design Level*: The conventional protocol stack requires different protocol layers to be transparent to each other. This normally leads to simpler and more scalable implementation and operation for protocols. Unfortunately, this design approach does not necessarily lead to an optimum solution for wireless networks [2]. The CAC and the scheduling mechanisms are normally agreed to be part of the protocols from the MAC layer. However, some proposals have interfaces to receive information from other network layers and such information may influence the protocol behavior, in the MAC layer. Because the unreliability and relative vulnerability of the wireless links the crosslayer approach may lead to better results.

- *Channel Awareness*: The channel awareness aspect is related to how the proposal treats and perceives the communication channel. Some approaches treat every communication as occurring in one single communication channel, others allow the communication to be divided into different frequencies. The use of multichannel communication allows more than one communication to occur at the same time, into different frequencies, even among neighbor nodes. The OFDM technology, used in the WiMax mesh mode, permits nodes to transmit different messages into different subcarriers. This makes the scheduling problem more interesting and effective in avoiding collisions and increasing the network capacity. However, the allocation of frequencies makes the scheduling problem even harder. Other point to observe is that to use multifrequency the scheduled channels must to be orthogonal to avoid interference. Considering that, one must be aware that part of the available frequency spectrum is lost.
- *Spectrum Reuse*: Some protocols permit, even incentive, the spectrum reuse as a mean to increase the spectral efficiency. On the other hand, consider possible just one transmission in the whole network at a time, even though the standard permits spectrum reuse.
- *Type of Traffic*: Some protocols make distinction among the kinds of the traffic they are handling, whereas others do not. The differentiation normally targets the possible QoS traffics presented in Sect. 17.2.3.
- *QoS Aspects Observed*: Some scheduling and CAC mechanisms observe QoS aspects to enhance and improve the network behavior. The observed QoS aspects may be in terms of the quality of the flows, e.g., throughput and delay, or may be in terms of fairness of access medium for the calls. We also consider the use of other techniques, such as interference minimization, also as a QoS aspect. Again, a proposal may present more than just one of these aspects.

Table 17.3 presents some of the most known proposals for CAC and scheduling for IEEE 802.16 networks classified in accordance to the proposed taxonomy.

17.5.1 Comparison of Some of the Main Existing Proposals

Each one of the existing proposals has its own objectives and mechanisms, therefore any comparison among the strategies is, in principle, unfair. Some of the works just want to test one aspect of the IEEE 802.16 Mesh mode CAC and scheduling problem, whereas others try to go further and really implement the mechanisms in the terms the standard proposes. Without implementing all proposals and comparing them within the same parameters and conditions it is unlikely that any one can affirm, without any shadow of doubt, which one is the best. Indeed, some works, like [16] that implemented some different proposals, have consistent results comparing the performance of the implemented ones. However, here our purpose is to present a summary of the most relevant ideas to guide future works on this field. The comparison, summarized in Table 17.3, is done in architectural terms and based on the taxonomy purposed on the beginning of the section. This comparison does not intend to show which proposal is the best or even the most complete one. In our opinion it is more interesting here to observe the proposals and try to evaluate how to use all the different techniques they present to improve the network performance.

17.6 Thoughts for Practitioners

As stated previously, no communication is allowed in WiMax networks, if not previously scheduled. This means that, more than correct and well designed, the CAC and scheduling mechanisms must also be fast and computational efficient enough to process all the network traffic. In addition to this, the scheduling problem in multihop networks is proven to be NP-hard [24, 27]. Because this, some optimal techniques normally present also an alternative heuristic, not optimal, to solve the problem [17,31,46]. In the real world, suboptimal solutions may be the only way to apply scheduling and CAC techniques to mesh networks.

The fairness is another interesting issue and, probably, one of the most diverse aspects among the proposed methods. The fairness is in truth an umbrella that accommodates many different definitions. However, it is commonly agreed that some kind of fairness is valuable for the network [19]. A peculiar, although interesting fairness approach, dynamic fairness, is introduced in [8]. The concept of dynamic fairness seems to be more interesting for the link unstable mesh network context, even though in the general case neither hard nor dynamic fairness is welcomed. Other simple and efficient ideas related to fairness, like the establishment of threshold for different class of services presented in [25, 34, 44], can also be interesting and even applicable in conjunction to other different techniques.

Many of the proposed approaches also proved that the interference is a real problem that must be treated carefully. The proposals to handle the interference vary in many senses and can use, for example, a conflict graph [17] or a conflict matrix [26]. For TDMA like approaches the techniques can be the constructing better routes [11,20,36,43] or dividing the spectrum [31].

Mainly for the centralized scheduling, it is agreed by many of the proposals that the creation of a scheduling tree is the best approach [8,11,17,20,30,31,40,41,43]. If we consider the OSI seven layers model [3], the creation of this tree rooted at the mesh BS is routing and, in truth, part of the job of the network layer. In this sense such proposals present a crosslayer design. Such kind of proposal normally presents really good perspectives and seems to be a good direction for new proposals to follow.

The standard itself [22, 23] defines a series of different types of services, presented here in Sect. 17.2.3, to be used by the applications. These services are considered by some approaches [12,30] in conjunction to their particular characteristics. Some of the approaches, more than just consider differentiation among the different services, also consider during the scheduling and CAC a reward for connections [31] or nodes [26] served. One of the main objectives of the CAC and scheduler in these approaches is to maximize the reward of the network. It is important to notice here that this really may provide better quality to the nodes in the privileged classes, but can be very unfair to other classes. We need to keep in mind that the available amount of resources is always the same. Sometimes to present gains some techniques may penalize some users. This must be done really carefully to avoid rash unfairness.

The standard states that the grants, even for centralized approaches, should be done hop by hop. Normally the approaches distribute the grants exactly in this way, but some proposals go little beyond that. In [30], for example, it is proposed that each node should be represented by *n* different virtual nodes, being *n* the number of different services. This intends to make easier the manipulation of the scheduling and the grants distribution among the services and nodes.

17.7 Directions for Future Research

The WiMax mesh mode is a good and valuable approach, but it is still a young part of the IEEE 802.16 standard and presents a lot of room for improvements. In this section we will discuss some topics that, in the best of our knowledge, were not explored deeply enough yet for this kind of networks.

A number of parameters must to be set to reach good protocol performance, e.g., holdoff exponent, periodicity of MSH-NCFG messages. Some consistent work has been done analyzing the network performance, but more works exploring these parameters are needed and surely enough would represent a valuable contribution to the field. The holdoff exponent value, for example, strongly affects IEEE 802.16 performance [35] and not many works have explored this.

The characterization of the traffic distribution on the mesh network is also important, not only for network simulation purposes, but also to be used in the design of newer and better algorithms. Some authors, when analyzing and validating their protocols just use poison or normal distribution to generate traffic. Also in [45] is argued that wide-area network traffic is much better modeled using selfsimilar processes [32]. However, for WMN, the traffic distribution and patterns for the different QoS services is still to be studied, at least in deeper way.

Some works present good results working with orthogonal channel allocation for IEEE 802.11 mesh networks [28]. This kind of technique could be even easier applied in WiMax networks, but, again, little has been done exploring this field. The frequency reuse is another topic that may be important for Mesh networks, and that has been studied for PMP (Point to Multi Point) WiMax networks [4], but not for the mesh mode.

A new working group is studying the problem of relay networks, the IEEE 802.16j, that is a problem very near to the mesh networks one. In the best of our knowledge, up to now no schedule or CAC mechanisms were proposed to such networks. Apart from that, could also be interesting to study the mix of both networks, IEEE 802.16 mesh mode and IEEE 802.16j, for example, adding some relay points in the mesh network [45]. This can open new opportunities for scheduling and routing, where new algorithms can take advantage of the relay characteristics to help the network performance.

So far, in the best of our knowledge, no work on CAC or scheduling for WiMax mesh mode makes use of adaptive power allocation (APA) to decrease the interference in the network. Much more in opposite, some techniques even consider always node at full power transmission [40]. Some work on this field, using APA and CAC mechanisms have been studied for PMP networks [37, 38], but no work addressed it for WiMax mesh networks.

Mixing networks, many different standards address mesh as a valid architectural topology, e.g., IEEE 802.16, IEEE 802.20, but so far no work addressed the interconnection of such standards. Some people explore hierarchical approaches for CAC for CDMA networks [10]. The general idea could be also applied to IEEE 802.16 mesh mode, as well the cluster based reservation, explored in [33].

We observe also that no technique so far considers mobility, even though mobility being a key aspect for WMN. Indeed, there are no guarantees of how the actual methods will behave in face of mobility. New and efficient procedures must be designed to handle handoffs and the constant position changing in the network topology.

Some techniques approach the scheduling and CAC problems using simple heuristics. However, could be interesting to see how to apply more sophisticated artificial intelligence techniques to solve the scheduling problem, once it is a NP-hard one.

Some techniques propose reward for connections schemas, which can be used as indicative of revenue, but up to now no one discussed about the billing in such networks. No one likes to talk about it, but who and how one will pay for the access for WiMax networks and how this will influence the CAC mechanism is not fully comprehended yet. Cheng et al. present in [12] a list of open research issues on CAC mechanisms for wireless networks in general, and truth also valid for mesh networks. A good discussion about important emerging trends and future research issues for CAC mechanisms can also be found in [1].

17.8 Conclusion

This chapter presented an overview of the CAC and scheduling schemes for IEEE 802.16 mesh mode standard. The literature presents many CAC and scheduling algorithms for many different kinds of networks, including WiMax ones. Some of these algorithms are even suited to very specific networks and situations. However, for the general case, the broader and fairer the algorithm the better it is considered, once normally one hopes to use the same algorithm in a broad range of situations.

IEEE 802.16 is still a young standard and CAC and scheduling mechanisms for it are not fixed yet. As this part of the standard is open for different implementations this represents an opportunity for research. Comparisons of different schemes and the proposition of innovative algorithms are always welcomed by those who work on this field.

17.9 Terminologies

1. *CAC – Call admission control.* Comprehends the mechanisms to decide if a new call is accepted or not.
2. *Scheduling.* The decision of when the previously accepted calls will have their share of resources.
3. *RRM – Radio resource management.* identify a series of strategies and algorithms employed to optimize the use of the radio spectrum and wireless networks limited resources. CAC and scheduling are examples of such strategies.
4. *Wireless mesh networks.* Kind of wireless network without a fixed structure and where the nodes that provide access are also wireless nodes.
5. *Mesh BS – Mesh base station.* Network node that has the responsibility of concentrate and organizes the network. Normally is a network backhaul.
6. *Backhaul.* A node that has connections with both wireless and outside network.
7. *IEEE 802.16.* IEEE standard that define the communication for broadband wireless networks.
8. *WiMax – Worldwide interoperability for microwave access.* An initiative to ensure the IEEE 802.16 compatibility and interoperability among different implementations and even promote compatibility with other broadband standards, mainly the HIPERMAN, the European standard.
9. *QoS – Quality of service.* Term used to identify the need for a differentiated kind of traffic from a call. CAC and Scheduling are examples of mechanisms used to ensure that the calls will maintain the desired/requested QoS level.
10. *MAC – Medium access control.* part of the network stack, in the IEEE standards, that defines topology dependent access control protocols.

17.10 Questions

1. Explain what is scheduling and its importance to Mesh networks.
2. Discuss about three reasons to use CAC and scheduling algorithms.
3. Comment three of the points to observe about CAC and scheduling algorithms.
4. Explain, with your own words, each type of service defined in the IEEE 802.16 standard.
5. Consider a medical application that permits a doctor to perform an online non-presential surgery. Which type of service would be more suited for this kind of application and why?
6. It is possible for applications that use BE, as a connection type, to have more bandwidth than those that use other mechanisms?

7. Chose one of the algorithms from Table 17.3, explain it, summarize its main advantages and disadvantages and check its classification in accordance to the taxonomy.
8. Propose a new scheduling algorithm that respects the IEEE 802.16 specification and classify it in the taxonomy.

9. Analyze your algorithm and evaluate its strong and weak points, verify for which kind of traffic and network it fits better.
10. Why if you force all communications in the network to occur in a scheduled way do you tend to avoid problems such as hidden and exposed terminals?

References

1. M.H. Ahmed, Call admission control in wireless networks: a comprehensive survey, Communications Surveys & Tutorials, IEEE, First Quarter 7(1):49–68, 2005.
2. I.F. Akyildiz, X. Wang, W. Wang, Wireless mesh networks: a survey, Computer Networks 47(4):445–487, 2005.
3. Andrew, S. Tanenbaum, Computer Networks, Prentice Hall, 4 edn., 2002.
4. C. Ball, E. Humburg, K. Ivanov, F. Treml, Comparison of IEEE 802.16 WiMax Scenarios with Fixed and Mobile Subscribers in Tight Reuse, IST Summit Dresden, Germany, 2005.
5. D. Beyer, N. van Waes, C. Eklund, Tutorial: 802.16 AC Layer Mesh Extensions Overviews, IEEE 802.16 (document S802.16a-02/30), St Louis, MO, USA, 2002.
6. R. Bruno, M. Conti, E. Gregori, Mesh networks: commodity multihop ad hoc networks. IEEE Communications Magazine 43(3):123–131, 2005.
7. M. Cao, W. Ma, Q. Zhang, X. Wang, Zhu, W., Modelling and Performance Analysis of the Distributed Scheduler, IEEE 802.16 mesh mode, MobiHoc'05 ACM Press, Urbana-Champaign, IL, USA, 2005.
8. M. Cao, V. Raghunathan, P.R. Kumar, A Tractable Algorithm for Fair and Efficient Uplink Scheduling of Multi-hop WiMax Mesh Networks, Proceedings Second IEEE Workshop on Wireless Mesh Networks, Reston, VA, Sep., 2006.
9. M. Cao, W. Ma, Q. Zhang, X. Wang, Analysis of IEEE 802.16 mesh mode scheduler performance, IEEE Transactions on Wireless Communications 6(4), 2007.
10. H.-H. Chen, W.-T. Tea, Hierarchy schedule sensing protocol for CDMA wireless networks – performance study under multipath, multiuser interference, and collision capture effect, IEEE Transactions on Mobile Computing 4(2), 2005.
11. J. Chen, C. Chi, Q. Guo, A Bandwidth Allocation Model with High Concurrence Rate in IEEE 802.16 Mesh Mode, 2005 Asia-Pacific Conference on Communications, Perth, Western Australia, Oct., 2005.
12. H.T. Cheng, H. Jiang, W. Zhuang, Distributed medium access control for wireless mesh networks, Wireless Communications and Mobile Computing 6(6), 2006.
13. S. Cheng, P. Lin, D. Huang, S. Yang, A Sstudy on Ddistributed/Ccentralized Sscheduling for Wwireless Mmesh Network, In Proceeding of the 2006 International Conference on Communications and Mobile Computing, ACM Press, Vancouver, British Columbia, Canada, 2006.
14. C. Cicconnetti, I.F. Akyildiz, L. Lenzini, Bandwidth Balancing in Multi-Channel IEEE 802.16 Wireless Mesh Networks, IEEE Infocom Conference, Anchorage, AK, USA, 2007.
15. P. Djukic, S. Valaee, 802.16 mesh networking, in Handbook of Wimax (S. Ahson, M. Ilyas, eds.), CRC Press, New York, 2007.
16. P. Djukic, S. Valaee, Performance Ccomparison of 802.16 Ccentralized Sscheduling Aalgorithms, Submitted to ACM/Springer Mobile Networks and Applications, Jun., 2007.

17. P. Djukic, S. Valaee, Distributed Llink Sscheduling for TDMA Mmesh Nnetworks, in Proceedings of ICC 2007, Glasgow, Scotland, Jun., 2007.

18. P. Djukic, and S. Valaee, Link Scheduling for Minimum Delay in Spatial Re-use TDMA, 26th Annual IEEE Conference on Computer Communications, Anchorage, AK, USA, May, 2007.

19. V. Gambiroza, B. Sadeghi, E. Knightly, End-to-Eend Performance and Fairness in Multihop Wireless Backhaul Networks, MobiCom'04, ACM, New York, NY, 2004.

20. B. Han, F. Po, Tso, L. idong Ling, W. Jia, Performance Evaluation of Scheduling in IEEE 802.16 Based Wireless Mesh Networks, IEEE International Conference Mobile Adhoc and Sensor Systems (MASS), Vancouver, Canada, Oct., 2006.

21. IEEE Standard for Local and metropolitan area networks – Part 16: Air Interface for Fixed Broadband Wireless Access Systems – Amendment 2: Medium Access Control Modifications and Additional Physical Layer Specifications for 2–11 GHz, IEEE Std. 802.16a, Apr. 2003.

22. IEEE Standard 802.16–2004, IEEE Standard for Local and Metropolitan Area Networks – Part 16: Air Interface for Fixed Broadband Wireless Access Systems, 2004.

23. IEEE Standard 802.16–2004, Part 16: Air Interface for Fixed and Mobile Broadband Wireless Access Systems Amendment 2: Physical and Medium Access Control Layers for Combined Fixed and Mobile Operation in Licensed Bands and Corrigendum 1, IEEE Std 802.16e, Feb. 2006.

24. K. Jain, J. Padhye, V.N. Padmanabhan, L. Qiu, Impact of Interference on Multi-hop Wireless Network Performance, ACM MobiCom, San Diego, CA, USA, 2003.

25. Chi.-H.ong Jiang, T.zu-C.hieh Tsai, Token Bucket Based CAC and Packet Scheduling for IEEE 802.16 Broadband Wireless Access Networks, Consumer Communications and Networking Conference, Las Vegas, NV, USA, Jan., 2006.

26. D. Kim, A. Ganz, Fair and Efficient Multihop Scheduling Algorithm for IEEE 802.16 BWA Systems, 2nd International Conference on Broadband Networks, vol. 2, Oct., 2005.

27. M. Kodialam, and T. Nandagopal, Characterizing the Achievable Rates in Multihop Wireless Networks, ACM MobiCom, San Diego, CA, USA, 2003.

28. M. Kodialam, T. Nandagopal, Characterizing achievable rates in multi-hop wireless mesh networks with orthogonal channels, IEEE/ACM Transactions on Networking 13(4): 2005.

29. S.M. Kuran, T. Tugcu, A survey on emerging broadband wireless access technologies, The International Journal of Computer and Telecommunications Networking, 51(11),: 2007.

30. M.S. Kuran, B. Yilmaz, F. Alagoz, T. Tugcu, Quality of Service in Mesh Mode IEEE 802.16 Networks, SOFTCOM 2006, Split, Croatia, Sep., 2006.

31. S. Lee, G. Narlikar, M. Pal, G. Wilfong, L. Zhang, Admission Control for Multihop Wireless Backhaul Networks with QoS Support, IEEE Wireless Communications and Networking Conference, Las Vegas, NV, USA, 2006.

32. W.E. Leland, On the self-similar nature of ethernet traffic (extended version), IEEE Communications Magazine, 40:1–15, 2002.

33. C.R. Lin, M. Gerla. Adaptive clustering for mobile wireless networks, IEEE Journal on Selected Areas in Communications, 15(7):1265–1275, 1997.

34. F. Liu, Z. Zeng, J. Tao, Q. Li, Z. Lin, Achieving QoS for IEEE 802.16 in Mesh Mode, 8th International Conference on Computer Science and Informatics, Salt Lake City, UT, USA, Jul., 2005.

35. C. Min, R. Vivek, P.R. Kumar, A Tractable Algorithm for Fair and Efficient Uplink Scheduling of Multi-hop WiMax Mesh Networks, Proceedings Second IEEE Workshop on Wireless Mesh Networks, Reston, VA, USA, Sep., 2006.

36. S. Redana, M. Lott, Performance Analysis of IEEE 802.16a in Mesh Operation Mode, IST SUMMIT 2004, Lyon, France, June 2004.

37. B. Rong, Y. Qian, H.-H. Chen, Adaptive power allocation and call admission control in multiservice WiMAX access networks, IEEE Wireless Communications, 14(1), 2007.

38. B. Rong, Y. Qian, K. Lu, Revenue and Fairness Guaranteed Downlink Adaptive Power Allocation in WiMAX Access Networks, 16th IST Mobile and Wireless Communications Summit, Budapest, Hungary, 2007.

39. Z. Rui, Mesh Distributed Coordination Function for Efficient Wireless Mesh Networks Supporting QoS, Communication Networks, RWTH Aachen University, Master Thesis, 2007.

40. H. Shetiya, V. Sharma, Algorithms for routing and centralized scheduling to provide QoS in IEEE 802.16 mesh networks, Proceedings of the 1st ACM workshop on Wireless multimedia networking and performance modeling, Montreal, Quebec, Canada, 2005.
41. H. Shetiya, V. Sharma, Algorithms for Routing and centralized Scheduling in IEEE 802.16 Mesh Networks, IEEE Wireless Communications and Networking Conference, Las Vegas, NV, USA, Apr., 2006.
42. Y. Sun, Dharma Basgeet, Khurram Rizvi, Zhong Fan, Paul Strauch, Dynamic Frame Structure for IEEE802.16j Relaying Transmission to Support Efficient Scheduling, IEEE 802.16, IEEE C80216j-06_224, November, 07, 2006.
43. J. Tao, F. Liu, Z. Zeng, Z. Lin, Throughput enhancement in WiMax mesh networks using concurrent transmission, Wireless Communications, Networking and Mobile Computing 2005, 2, 2005.
44. T.-C. Tsai, C.-H. Jiang, C.-Y. Wang, CAC and packet scheduling using token bucket for IEEE 802.16 networks, Journal of Communications, V1, N2, Academy Publisher, New York, 2006.
45. H. Wang, B. He, D.P. Agrawal, Admission Control and Bandwidth Allocation above Packet Level for IEEE 802.16 Wireless MAN, 12th International Conference on Parallel and Distributed Systems – Vol. 1, Minneapolis, MN, USA, 2006.
46. H-Y. Wei, S. Ganguly, R. Izmailov, Z.J. Haas, Interference-Aware IEEE 802.16 WiMax Mesh Networks, In 61st IEEE Vehicular Technology Conference (VTC 2005 Spring), Stockholm, Sweden, May-Jun., 2005.
47. J. Zander, Radio resource management in future wireless networks: requirements and limitations, IEEE Communication. Magazine., Aug., 35(8), 1997.
48. H. Zhu, K. Lu, Performance of IEEE 802.16 Mesh Coordinated Distributed Scheduling Under Realistic Non-Quasi-Interference Channel, in Proc. of the International Conference on Wireless Networks, Las Vegas, NV, USA, Jun., 2006.

Chapter 18
The Symbiosis of Cognitive Radio and Wireless Mesh Networks

Brent Ishibashi and Raouf Boutaba

Abstract Although wireless mesh networks (WMNs) have quickly been success-fully deployed, the dual usage of wireless communication makes them very resource dependent. Proposed cognitive radio (CR) concepts appear to be a good solution to provide WMNs with additional bandwidth and improved efficiency. In addition, we believe that applying CR to WMN can be very beneficial to CR, speeding the devel-opment and acceptance of the technology.

18.1 Introduction

Wireless mesh networks (WMNs) would, of course, benefit from additional wire-less bandwidth. In fact, a WMN's dual use of wireless communication for both user access and data transit places additional strain on a already scarce resource, com-pared to other wireless networks (such as WLANs). To date, most mesh systems have been designed to use unlicensed spectrum, particularly the 2.4 GHz band used by IEEE 802.11 b/g. As this spectrum is unlicensed, it is also heavily used – not only by other 802.11 devices, but also by a wide range of other devices, including cordless phones, remote controls, and even microwave ovens.

However, obtaining additional spectrum is very difficult. Under the current sys-tem of spectrum allocation, spectrum is strictly allocated, with only a few small pockets that are unlicensed. This leaves two options: either use (along with a large number of technologies and users) the unlicensed spectrum, or obtain (at great expense) spectrum to dedicate specifically to a WMN. For a technology such as mesh, both options are potentially very limiting to the applications where it can be deployed.

R. Boutaba (✉)
David R. Cheriton School of Computer Science, University of Waterloo, 200 University Ave West, Waterloo, ON, Canada N2L 3G1
e-mail: rboutaba@uwaterloo.ca

S. Misra et al. (eds.), *Guide to Wireless Mesh Networks*, Computer Communications and Networks, DOI 10.1007/978-1-84800-909-7_18,
© Springer-Verlag London Limited 2009

There are however indications that the system is changing. Many of the players responsible for spectrum allocation have acknowledged the need for a more advanced, more dynamic system – a system that makes better, more efficient use of available bandwidth. A concept has emerged of a radio system the gathers all available information about its environment, then uses this information to determine the most effective way – when, where, and how – to communicate. This concept is cognitive radio (CR).

Cognitive radio is currently only a concept, an eventual goal for intelligent wireless communication. However, as a general concept, it incorporates many ideas, from many existing research fields. In fact, one of the driving forces behind the rise of the CR concept has been the increased resource demand of new types of wireless networks, including mobile ad hoc networks, and wireless meshes.

However, not only will the development of CR technology benefit WMNs, but WMNs can potentially contribute greatly to both the development and implementation of cognitive radio. Although CR is envisioned as universal wireless technology, not one bound to any particular network structure, specific characteristics of WMNs suggest that WMNs could be a great facilitator of the technology.

As a result, the paths of WMNs and cognitive radio appear to be closely intertwined. In this chapter, we will present an overview of CR work, focusing on how it relates to WMNs.

18.2 Background

18.2.1 Radio Communication

In this chapter, although other types of wireless communication exist (microwave, visible light, acoustics, etc.), our discussion will be confined to radio communication. Radio communication uses a transmitter to encode information and generate radio waves. These electromagnetic waves occurring at low frequencies, in the range of 3 Hz to 30 GHz, propagate through the air. A receiver is used to detect and decode the signal.

The characteristics of radio waves are very important in the wireless world. The transmitter creates a wave with a certain power. However, as the wave travels, it attenuates, reflects, and refracts. The characteristics of transmission are dependent on the frequency used. Although lower frequencies (below $\sim 10\,\text{GHz}$) will pass through some obstacles, higher frequencies require a clear line of sight. All of this complexity makes modeling the wireless environment extremely difficult.

At the receiver, the signal quality must be great enough to allow the signal to be decoded. To properly receive the transmission, a sufficient signal-to-noise ratio (SNR), or signal-to-interference-plus-noise (SINR) must be achieved. Unfortunately, there is a great deal of radiation within the radio spectrum, both naturally occurring and generated by transmissions. The required SNR is dependent on the characteristics of the antenna, receiver, and the encoding scheme used.

In today's world, many radio devices are often operating in close proximity to one another. This makes interference between devices extremely troublesome, as two transmissions may mutually interfere with each other, preventing one or both from being properly received. The devices may be using the same network or technology, or could have entirely different purposes. A system for controlling who uses spectrum – where, when, and how – is required to ensure communication can occur effectively.

Fortunately, the radio spectrum can be shared along three dimensions – frequency, time, and space. The total spectrum space is large relative to the needs (and power capabilities) of individual devices/communications. Therefore, communications only use a small band (frequency range). Different bands can be used simultaneously. Wireless resources are also completely renewable, so individual transmissions can be made one after another. Finally, because of the attenuation properties of signal propagation, spectrum can be reused geographically, if the distance between devices is great enough that signal interference is low enough relative to signal strength (that is, the required SINR is maintained).

18.2.2 Spectrum Allocation

To ensure that wireless spectrum is used and shared effectively, the resource is tightly regulated. Regulatory bodies set out rules on what, where, and how spectrum can be used, and who can use it. The system that has been developed relies principally on frequency-division. Spectrum is sub-divided into frequency bands, and allocated to particular uses or users. Geographical divisions also occur, because of both political borders and regional requirements.

In the USA, the Federal Communications Commission (FCC) [2] is a government agency responsible for regulating wireless spectrum. The canadian radio-television telecommunications commission (CRTC) [3] has similar responsibilities in Canada, and similar agencies exist in other countries. In addition, the international telecommunications union (ITU) [4] and its Radiocommunication subcommittee (ITU-R) is a UN agency responsible for coordinating spectrum allocation worldwide. The ITU-R works to coordinate spectrum allocation internationally, to allow certain technologies to use the same spectral bands throughout most of the world, as well as to avoid major interference problems across international borders.

The current allocation of frequency bands has been arrived at as a result of a number of different methods for allocating spectrum. Many frequencies have been allocated to, or reserved for public service uses (e.g., governmental, military, or emergency services). Some frequencies are allocated because of their historical placement – as technologies are developed, they use a particular frequency. As the technology is adopted, it becomes increasingly difficult to change the allocation, even if technology advances no longer require that band to be used. Certain frequencies have been allocated for open use – these unlicensed bands (such as the 2.4 GHz band used by 802.11 b/g) can be used by any user or technology, as long as certain power rules are met. The Canadian spectrum map can be found in [67].

Most regulatory bodies now favor the spectrum auction as the method of choice for allocating new frequency bands [5]. The FCC has conducted spectrum auctions since 1994, with spectrum licenses granted to the highest bidder. The auction system replaced the previous "best public use" method, where applicants were required to demonstrate that their proposal would deliver the most benefit for the public. After obtaining a spectrum license, the licensee is given exclusive use of that spectrum, subject to the conditions of the license (e.g., location, power constraints).

18.2.3 Spectrum Usage

Although the current system of fixed spectrum allocation and spectrum auctions is straightforward, it suffers from a few problems. Most notably, with the ever-increasing numbers and types of wireless devices, new spectrum is becoming increasingly scarce. Bandwidth is becoming increasingly expensive, and difficult to obtain. However, studies of existing spectrum usage have yielded an interesting result.

Spectrum is vastly under-used. Although certain frequencies, in certain locations, are heavily congested, studies have shown that the overall spectrum is remarkably quiet [6]. For example, measurements were taken at six locations. Overall spectrum usage was only 5.2% (averaged over the six locations), and although certain bands were heavily used in some areas, even the location with the highest occupancy had a total use of only 13.1%. This means that, despite the incredible value of wireless resources, they are to a large extent wasted. This is to be expected to some extent – usage is dependent on need. However, in some cases, overall spectrum use was quite low, despite the fact that certain bands were very heavily used. In these cases, wireless demand was clearly present, confined to a small band while other spectrum is idle.

Because of the historical nature of the allocation system – the long life of spectrum licenses, the current allocation may not be ideal. Many older technologies make inefficient use of their resources. However, there is a large investment in existing technologies, making replacement undesirable.

18.2.4 Change

In 2003, the FCC charged a task force with looking at the way spectrum allocation is performed. The Spectrum Policy Task Force (SPTF) investigated ways to evolve the "command and control" approach to spectrum regulation [7]. In their report, the task force acknowledged the inefficiencies of the current license system. They found that current spectrum policy could not keep up with technology, and identified the need for a new system that allow better use of the existing spectrum resources. In particular, they identified the need for the new system to be more dynamic, responding better to changes in usage and to new technologies.

18.3 Cognitive Radio

As the regulatory agencies were acknowledging the need for a more dynamic system or resource allocation, a concept had emerged within academic literature. This concept incorporated many different ideas from several research fields. Although unrealizable in the short term, it caught on as a unifying vision of how a future radio device might behave. This concept is cognitive radio.

18.3.1 What is Cognitive Radio?

The term "cognitive radio" is generally credited to J. Mitola. It first appeared in 1999, in an article coauthored by G.Q. Maguire [8]. This was followed by Mitola's PhD dissertation in 2000 [9]. The dissertation described a language for describing and communicating the characteristics of a device's radio interface. In the work, he used the term "cognitive radio" to describe a device that used its awareness of its environment to intelligently choose the best parameters to use for its own communications.

The concept of applying intelligence to communication is not a new one. Patterning wireless on the characteristics of human conversation has long been a topic of research. However, the identification of the CR concept is indicative that the underlying technology has reached a point where such a system is becoming realistic. Several key factors point in this direction.

First, there has been an incredible boom in wireless networks and devices. With the near ubiquity of WLAN access, it is easy to forget that the IEEE 802.11 standard is only about a decade old [10]. Even the popularity of cellular phones is relatively recent, even though the first commercial networks were deployed almost thirty years ago (1979). However, today, many locations are serviced by variety of different wireless technologies and service providers.

Second, this boom has increased interest in wireless research. New types of wireless networks, particularly multihop networks such as mobile ad hoc networks (MANETs) and sensor networks have made the scarcity of wireless resources abundantly clear. A huge number of new protocols were proposed to improve the efficiency of communications, especially routing [11] and MAC protocols [12]. In addition, this work revealed a need to consider cross-layer information in protocol design.

Third, technological advances have made software-defined radio (SDR) possible and increasingly capable [13]. SDR allows the behavior of the radio to be controlled by software, rather than in fixed hardware. This goes beyond basic parameter configuration, to allow control over all aspects of the radio interface, including frequency, modulation, power, and medium access control. SDR allows a device to switch between different network technologies, using a single physical radio. SDR focuses on specifying architectures and the wireless interface, an important component for building CR devices [14].

With these factors coming together, it was becoming possible to realistically envision the cognitive radio concept. CR is a future technology, a target towards which research will progress. However, before this goal can be achieved, a large number of issues must be addressed. Many of these issues have been studied in the context of various types of networks, including WMNs, however, all of this work must be brought together within the CR view. Bringing CR to fruition will require developmental work in engineering, architectural design, protocol development, network management, and applications, not to mention overcoming regulatory obstacles.

18.3.2 Key Characteristics of a Cognitive Radio

The development of cognitive radio will need to take advantage of many different technologies to succeed. By its nature, CR must allow new technologies to coexist with current devices. However, as a long-term goal, many of these new technologies are still in their infancy; others may not even have been conceived yet. With this in mind, this section attempts to give a picture of CR as it is currently envisioned.

18.3.2.1 Advanced Interoperability

The CR will take advantage of advanced technologies to have significantly greater capabilities than current radio interfaces. Current work on antenna technologies: antenna arrays, MIMO, and adaptive beam-forming give some idea of what might be expected, with advances in digital signal processing allowing radio devices to gather as much information as possible [15–17]. Ultrawideband radio (UWB) is also a possibility for using the medium without adding significantly to overall interference levels [18]. However, as technologies progress, different devices will have different capabilities, and one of the goals for cognitive radio is a system for all of these devices to operate effectively within the environment. As previously mentioned, at least the first generation of CR devices will have to coexist with existing noncognitive wireless technologies.

18.3.2.2 Frequency Agility

CR devices are envisioned to be highly flexible in the way that they send and receive. In addition to MIMO (Multiple in, multiple out capabilities), they will be frequency agile, being able to dynamically adjust the frequencies and bandwidth of their transmissions. This functionality is envisioned to go well beyond the basic capabilities of SDR however, with the adaptive ability to fill fragmented spectrum holes as required by the current radio environment. In addition, a CR will require a much better ability to detect different types of transmissions, including those spread over a range of frequencies [19].

18.3.2.3 Awareness

This ability to detect transmissions will give the CR a greatly improved ability to gather information about its radio environment. The ability to sense and measure channel conditions, throughout the spectrum, is only the beginning. The CR will rely not only on current information, but will also retain a memory of its environment [20]. It will therefore also need improved systems for maintaining this information.

Not only more aware of its surrounding, a CR will also have an increased level of selfawareness. This includes awareness of its hardware, applications, user characteristics, and particularly its goals. For example, knowledge of the application may provide traffic characteristics and requirements [21].

18.3.2.4 Cognition

The CR uses its awareness of the environment to make decisions on how to best meet its operational goals. The gathered information is analyzed to determine the optimal set of parameters for each communication. The CR must decide what transmission must occur – to whom, on what network – and when and how the transmission will occur. Because of the very large number of variables, both in terms of awareness and decisions, the CR decision-making process will have to be fairly advanced, with the ability to adapt and learn [20].

The collected awareness will be stored to maintain a memory, and modeled to predict future conditions. Prediction may take the form of sophisticated pattern recognition of cycles or trends, or the simple recognition of poor conditions, with the expectation that future conditions will improve. A CR that senses, stores, and uses its awareness effectively should have an advantage over less capable devices. It will develop a better strategy to be used in competition with other nodes [22].

18.3.2.5 Collaboration

By its nature, radio communication is dependent on collaboration. At the minimum, sender and receiver must collaborate, however in reality the open nature of the wireless medium demands that far more nodes must be involved. The CR must consider the interactions between not only different nodes in its network, but in all networks – in fact, the CR's abilities allow it to choose to interact with different networks.

18.3.3 How Cognitive Radio Changes Spectrum Management

The CR's envisioned capabilities differ considerably from any previous radio interface. Its awareness and cognitive abilities allow it to be very flexible and dynamic. Although it would be possible for a CR to operate within the current spectrum rules (i.e., fixed band allocations), the true benefits may come from the combination of CR development and changes to spectrum management.

Open Spectrum is one of the first major works considered to address the cognitive radio concept [23]. An Open Spectrum Policy (OSP) has been proposed so that available spectrum can be more fully used. Recognizing the need for incumbent technologies to continue to function correctly, researchers have proposed different methods for CR devices to use the same licensed frequencies – while still avoiding interference with existing devices either spatially or temporally.

The IEEE 802.22 Working Group is addressing this approach, looking at ways to share the frequencies occupied by broadcast television [24]. Several approaches have been proposed for re-using this spectrum. First, over-the-air TV bands currently have guard bands between the reserved channels. These guard band frequencies are not used – they are designed so that adjacent TV channels do not interfere with each other. As a result, a CR could use these gaps, as long as it could control its signal so that it does not cause any problems for TV receivers. A second option relies on technological improvements giving CRs far greater sensitivity and signal processing ability than existing devices. Cognitive radios can then communicate at transmission powers and ranges that are low enough to avoid interfering with TVs. Third, if a CR can determine when and where there are no users of the primary (incumbent) technology, it may be able to make full use of the spectrum [25]. IEEE has also established a Standards Coordinating Committee on Dynamic Spectrum Access Networks. SCC41, continues the work of the P1900.X standards development committee [26], and is currently developing guidelines for the use of dynamic access throughout the radio spectrum.

Open spectrum illustrates the important ability of cognitive radio to share spectrum with existing technologies. In allowing a CR to use licensed spectrum, it capitalizes on previously wasted bandwidth. The CR must always ensure that it does not interfere with spectrum usage by the primary user. However, even this contravenes current spectrum allocation rules and licenses.

Therefore, changes to spectrum management are required to make cognitive radio a reality [27]. At the minimum, certain spectrum licenses must be made available for spectrum sharing according to known methods, as in the example of 802.22. However, with regulatory agencies considering major changes to spectrum allocation, a dynamic system could better match the flexibility of CR.

Different proposals exist as to what form a more dynamic spectrum allocation system might take. These include shorter-term licenses (and more frequent auctions), licenses allowing for secondary cognitive use while maintaining primary user rights and priority, and a fully dynamic spectrum market. The latter option presents the most flexibility, with the ability to buy, sell, trade, or lease spectrum rights. For example, if a spectrum licensee decides that it will not fully use its bandwidth, it may arrange with another party to temporarily lease the extra resources.

18.4 Applying Cognitive Radio to a WMN

The nature of WMNs makes them prime candidates for applying cognitive radio. In this section, the characteristics of WMNs will be discussed, and the potential benefits of CR considered.

18.4.1 WMN Characteristics

WMNs are designed to provide wireless network access to user devices. However, rather than requiring a wired connection to each access point, mesh access points (MAPs) are interconnected wirelessly. This greatly reduces the cost of deploying the network, and allows additional flexibility in the placement of nodes. User devices communicate with a MAP via an access link. The traffic is then forwarded through the mesh, from MAP to MAP, via transit links. This multihop forwarding delivers traffic to a gateway, nodes within the mesh that possess an additional interface to the Internet. This network structure is depicted in Fig. 18.1. Some traffic may also flow between two WMN users – this peer-to-peer traffic does not need to pass through the gateway, staying within the mesh. However, most traffic is likely to occur between a WMN user and a second endpoint elsewhere on the Internet [28].

Both the access link and the transit links operate via wireless communication. Although the gateway link could also be wireless, most works to date have assumed

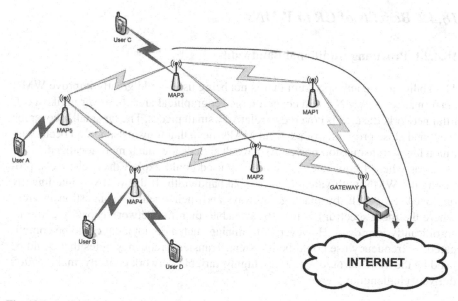

Fig. 18.1 A wireless mesh network

it to be wired. Many works have also considered the access and transit links to operate over separate wireless interfaces and on different channels. Typically, the transit network is the limiting factor in a WMN. Several things contribute to this: first, the wireless medium is openly shared, requiring traffic from multiple links to share the same bandwidth; second, multihop forwarding requires a single traffic packet to be transmitted multiple times to reach its destination; third, the presence of the gateway tends to accumulate traffic in its region, as most traffic flows either to or from the gateway. Therefore, the nodes and links surrounding the gateway not only carry the most traffic, but also interfere with each other so that this large volume of traffic must share the available bandwidth [29].

Added to this is the fact that existing MAC protocols do not make efficient use of the wireless channel. The IEEE 802.11 DCF is most frequently used in WMN works. However, the CSMA mechanism of DCF is designed primarily for use in WLANs. In multihop wireless networks, the floor acquisition model of the RTS-CTS mechanism results in each link requiring a large number of neighboring links to remain silent. This limits the network's ability to re-use the medium and have transmissions occur simultaneously [30].

Some existing WMN products use directional antennas within the transit network to alleviate some of the interference and medium re-use issues. For example [31], uses directional antennas to communicate between MAPs. This, combined with the use of multiple channels, allows multiple links to operate simultaneously. However, even in this case, the throughput capacity of the network is limited by the ability of the gateway to send and receive through its transit interface.

18.4.2 Benefits of CR to WMNs

18.4.2.1 Providing Additional Bandwidth

The ability to use any spectrum that is not being used could greatly improve WMN performance. A WMN could cover a large geographical area, however unlike a cellular network, the area is covered in relatively small pieces. The far smaller coverage areas and close proximity of adjacent MAPs mean that transmissions can occur with much lower transmission power. As a result signals are much more localized.

Areas where frequencies are not being used could exploit these channels, providing the WMN with valuable additional bandwidth. If the WMN is intelligently deployed, especially by placing gateways and their resulting congestion in areas where the most spectrum is usually available, then the network capacity could be significantly improved. However, determining such a deployment could be complicated, as frequency use could be transient. Transient frequency holes that could be used by the WMN would result in a highly variable network capacity, making QoS delivery challenging.

18.4.2.2 Rebalancing the Access and Transit Network Bandwidth

Without requiring full CR capabilities, a WMN could use CR techniques to make better use of the wireless channels available to it. Using a technology such as 802.11, several equal-sized channels can be used (three non-overlapping channels in 802.11b/g). However, if both the access and transit links use the same technology, the transit network has far lower throughput capacity than an individual access link. The access channel is under-used, as the transit network is incapable of handling the total traffic if every MAP fully uses its access link.

Numerous works have considered the use of multiple channels and/or multiple interfaces within a multihop wireless network to increase the network throughput capacity. Multichannel MACs must co-ordinate which nodes should use which channels, and when. One option is to use a fixed control channel [32]. Nodes request resources on the control channel, then switch to an alternate channel for the transmission of data. Other approaches assign home channels to nodes [33]. To contact a particular node, a sender must switch to that node's channel.

Nodes with multiple interfaces can use different channels simultaneously. In [34], each MAP has two transit interfaces. One is used for the uplink (towards the gateway), while the other handles the downlink (away from the gateway). Other schemes may have additional network interfaces (k-NICs) to use additional channels. However, as shown in [35], adding interfaces selectively within the network can yield similar improvements by alleviating the bottleneck.

A similar result can be gained using CR. The ability to dynamically allocate and use frequencies allows for more bandwidth to be allocated in the bottleneck regions. Even without additional bandwidth, a redivisioning of the channels assigned to access and transit would result in an increased capacity, as well as a more complete use of all channels (Fig. 18.2).

18.4.2.3 Changing the Nature of Gateways

If every MAP is a CR node, they already have the capability to use the equivalent of an additional interface. With wireless technologies such as IEEE 802.16 (WiMAX) emerging [36], it could be possible to also use wireless for the gateway link as shown in Fig. 18.3. All WMN nodes would then only require a power connection. It would also allow a much larger number of gateways to be placed within the network,

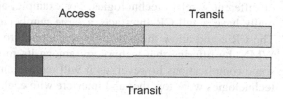

Fig. 18.2 Balancing the access (*dark*) and transit (*light*) networks – wasted access bandwidth is shaded

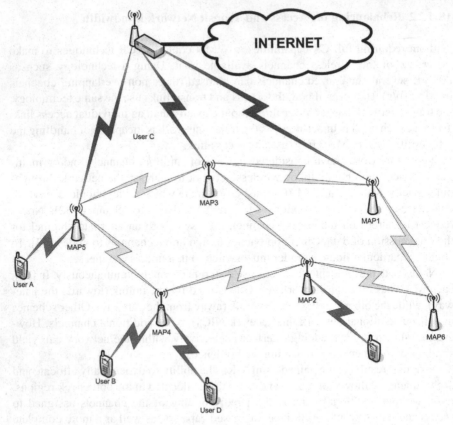

Fig. 18.3 Wireless gateway links feeding mesh gateways

which would reduce both the distance traffic would need to flow, and the number of nodes served by each gateway. This architecture would also allow for greater path redundancy and possibly the opportunity for a node to exploit multiple gateways simultaneously to maximize throughput.

18.4.2.4 Multiple User Technologies

With a CR interface, MAPs could possibly support user devices utilizing a variety of different wireless technologies. For example, although user devices may eventually have a full CR interface, a large number of legacy devices already exist. The CR could allow these devices to continue to be used to connect through the WMN, by offering that technology among its modes of access. This does create some complications however, as it will require an understanding of how different technologies work together and interfere with each other [37].

18.5 CR Research

Clearly there is a lot of work to be done to take cognitive radio from its current vision to a usable technology. As a long-term goal and a combination of a number of components, one of the keys will be to find ways to allow intermediate technological advances to enter mainstream use before the completion of a fully capable cognitive radio. This is particularly true for CR, because of the nature of current spectrum regulations. In this section, we address some of the key research areas for CR, considering the existing work and what advances must be made in order for CR to be realized.

18.5.1 Transmission

Perhaps one of the most difficult areas to predict in terms of future progress are the characteristics of physical antennas and transceivers. It is therefore a major challenge to create a technology such as CR that must adapt and incorporate these technologies into the system. Although current research and developments give an indication of some of the capabilities of the next generation of PHYs, the CR architecture will have to be extremely flexible.

However, researchers are beginning to explore the limits of the capabilities for radio techniques. Works on information theory are starting to develop a picture of how much usable information can be encoded on a channel using different techniques. For example, ultrawideband communication (UWB) spreads a communication over a very wide frequency band [38]. To avoid interfering with other nodes operating with those frequencies, it uses a very low transmission power. Therefore, a UWB receiver can detect the short-range transmissions, while it contributes very little to the overall noise floor experienced by other nodes. Information theory works are finding the capacity regions for these works – the theoretically achievable rates for multiple transmissions within the same frequencies [39, 40].

UWB, directional antennas, multiple-in multiple-out (MIMO), and frequency agility are all examples of current research areas that are likely to become (and in some cases already are) part of next-generation wireless communication systems. As digital signal processing techniques (and processing speed) continue to improve, these technologies will mature. However, at this point it seems likely that different devices will continue to have different capabilities, depending on hardware factors such as processors, memory storage, physical size, and power supply.

18.5.2 Awareness

Building awareness is a key function to cognitive radio. However, there is a huge volume of environmental data that can be collected, and not only must it be collected, it must also be shared with others as appropriate.

In current wireless technologies, data collection is typically very simplistic. Consider two common systems. In a cellular system, medium access control is typically centrally controlled. It is aware only of its own clients, and only requires knowledge of the frequencies it controls. An 802.11-based WLAN is slightly more complicated, with the clients required to sense the medium under the carrier-sense multiple access (CSMA) scheme. Only a limited memory is maintained, as captured by the backoff scheme.

Previous works have shown that extending this awareness even a little can be quite difficult. Even spectrum sensing is much more difficult, when wide spectrum is considered. The dynamic nature of the wireless medium also causes problems. For example, several works on MANETs and WMNs have attempted to evaluate the quality of links between nodes [40, 41]. Small changes in position, the physical environment, or environmental noise, can often cause major changes in the achievable link quality. Therefore, knowledge must exist, not only about the received signal, but also about the source, so that further results can be extrapolated.

Mitola's original work on CR addressed this point. The radio knowledge representation language (RKRL) was designed to allow devices to represent information about their radio characteristics [9]. Sharing this knowledge allows other radios to adapt their own communications, whether to communicate with the device, or to avoid interfering with it. Another concept, Interference Temperature, was proposed as a metric for estimating the cumulative interference energy at a receiver [42]. However, the concept behind this work was determined to be "not a workable concept" and investigation by the FCC was terminated in May 2007. An alternative approach appears in [43], where the spectrum resource is divided into virtual cubes, with the dimensions of the cubes representing time, frequency, and power.

For CR, one special type of awareness involves knowledge about primary spectrum users. The location and activity of these users is important, as CRs must avoid interfering with them. Primary users can be either active (transmitters or transceivers) or passive (receiver only). CR devices can detect active primary users by sensing the medium, although an idle user could be missed. However, passive users are problematic. As receivers (e.g., a television), passive users do not transmit any signal. Therefore, a CR must rely on other information to reveal the presence of a passive user.

This is an excellent example of how a CR must combine information from a variety of sources to understand its complete radio environment. Different approaches could be used for identifying a passive primary user. One option is to maintain a database of user locations and characteristics that CRs must check before using certain frequencies [44]. Frequencies that may have passive users (or even idle active users) would be considered required knowledge for all CR devices. Another option would be to protect a passive user with a simple CR device, responsible for notifying other CR nodes of the user's presence [45].

A CR should be aware of network load and application conditions. This knowledge may be required for several networks – both networks the node is involved in, as well as others that affect the radio environment. Gathering this information could be expensive. Active techniques such as probing may yield better information, however passive techniques have less impact on the network. For a CR, its sensing capabilities should allow it to rely heavily on passive sensing – gathering information by overhearing it.

18.5.3 Sharing Information

Nodes can also gather information by sharing with other cognitive radios. Because of differences in location, configuration, and history, different nodes will have a different set of collected knowledge. Communication between CR devices can extend the knowledge a node has about its environment beyond what it can obtain on its own. Certain information (such as location) may only be attainable based on collaborative sensing, where several nodes share and combine their information.

If we consider location determination, one approach is to incorporate a GPS receiver into each device. However, this has limitations in terms of cost, complexity, size, and usability. Although GPS functionality may eventually be included within the capabilities of a CR – a GPS receiver must have far higher sensitivity than current radios – alternative solutions are being investigated. A number of variants of the problem exist, from locating mobiles in a conventional WLAN [46], to establishing a complete map of positions in an ad hoc network. The amount of information used varies, from received signal strength [47] to simple connectivity [48].

If all nodes can determine signal strength, why would they choose to use only connectivity data? This illustrates one of the problems faced in sharing spectrum data between nodes. Two receivers, depending on their own characteristics, can obtain very different measurements. Therefore, although an individual node may have more detailed information, to combine it, the information must be reduced to a form that is mutually compatible.

In addition, systems must be able to control or limit the information that is exchanged, or else the communication process could overwhelm resources. There is a huge volume of data that could be collected. In fact, even relatively simple exchanges can overwhelm a dynamic network, as has been seen in the propagation of routing information in MANETs [11].

The field of sensor networks has yielded considerable work that is closely related to this problem. Sensor networks are designed to gather information, but inherently filter it as it is communicated to the necessary location within the network [49]. Sensor network protocols are also designed to be lightweight, minimizing the resources required for them to operate. However, many sensor network protocols focus on relaying information to a sink node, whereas communication within a CR-based network could have to be more distributed, depending on the network topology.

18.5.4 Decision Making

The cognitive radio communication process is very different from a conventional wireless interface. The conventional process is very linear, modeled on a protocol stack. For example, consider a basic WLAN device. An application generates packets. Routing determines the destination for the packets, and sends them to the appropriate interface. The network interface uses the medium according to a predetermined MAC protocol (e.g., the 802.11 DCF), then transmits the packet according to the interface's specifications and configuration. By comparison, the CR interface is decidedly nonlinear.

In the CR, all parts of the communication must be decided. This includes what to send, as well as where, when, and how to send it. All aspects of the communication are part of the decision, and are governed by the user preferences and goals – including whether or not to communicate at all. All of these decisions are closely interrelated, and cannot be made independently of one another.

The CR decision-making process can be viewed as the ultimate destination for work on cross-layer protocols. Cross-layer protocols have been popular in recent years for MANETs, where it was realized that considering network conditions such as routing and congestion in medium access decisions can yield improved performance [50, 51]. CR takes this to the extreme, utilizing all available information in all communication-related decisions.

18.5.4.1 What to Send

The decision of what to send is typically beyond the scope of the wireless interface. Data is simply passed to the interface with the expectation that it will be transmitted according to the communication protocol. A CR can decide to communicate now or wait until later, or adjust the traffic according to the characteristics of the network and the wireless resources.

Adaptive applications are an example of network conditions impacting on what to send [52]. A QoS-sensitive voice application might adjust its sound quality (e.g., sampling rate, stereo/mono) to match the available resources of the network. An adaptive web client might reduce image resolutions (or not load images) if it is using a low bandwidth connection.

18.5.4.2 Where to Send it

The flexibility of a CR allows it to connect to different networks. This ability is similar to some existing devices that have multiple interface cards, although the CR will have the ability to connect to any available network. This may be used to allow vertical handoff [53], depending on network coverage, or simply to choose the preferred network. This may be dependent on cost, availability, congestion, or QoS [54]. The CR could even make use of different networks for different traffic types.

In addition to network choice, the CR must also consider the routing of traffic. For single hop wireless networks, this is reasonably simple, although the CR should be able to make full use of mobility management techniques as it moves between access points, to ensure optimal handoffs [55]. For multihop networks, the routing process may be far more complicated. The CR may even consider whether or not a route is established in determining whether or not a particular network should be used.

18.5.4.3 How to Send

The question of how to send depends heavily on the decision of where to send. Depending on the network chosen, many or all of the communication parameters will be predetermined by the technology being used. However, in other cases – for example in a MANET made up of CR nodes, or even just an ad hoc connection between two devices – the communicating nodes have to choose the appropriate parameters based on the current environment.

This is where the full flexibility of cognitive radio appears. The cognitive radio can choose each parameter for the communication to best fit the environment it is in.

18.5.4.4 When to Send

There are two aspects to the question of when to send. There is the question of medium access control. The MAC determines when (combined with how) to access the medium for each communication. We will cover this in more detail in the spectrum management section. On a larger time scale, having knowledge of past conditions and a model of future conditions may create a scenario where best effort traffic is not necessarily sent immediately, as quickly as possible. A CR may predict that conditions will improve, because of more available resources, a lower price, a particular network, etc. In this case, it may choose to delay the communication, waiting for the preferred conditions.

18.6 Thoughts for Practitioners

With the current spectrum allocation system, the medium access method is dependent on the technology deployed in each particular band. Even technologies operating in unlicensed spectrum use specific channels, and then use a MAC protocol to negotiate traffic within that channel. With the importance of the MAC protocol, a large quantity of wireless research has been devoted to improving the MAC protocols, especially the 802.11 DCF. Numerous works have addressed adding priority mechanisms, handling multiple channels, using directional antennas, utilizing cross-layer information, and much more.

However, as the move towards cognitive radio allows, and likely requires, the reconsideration of the allocation system, we now consider the types of system that could emerge, and some of the key issues that must be dealt with.

The simplest approach would likely be to continue with the current system with only minor changes to allow additional dynamic access within spectrum holes. Under the condition that primary systems must be fully protected, cognitive radio usage would be strictly limited to scenarios such as 802.22's use of the TV broadcast bands. Each proposed system could be thoroughly tested within each band under consideration.

However, it appears that regulatory agencies are prepared to implement greater changes than this [56]. Therefore, more dynamic systems are being created. The proposals vary in three major ways. First, on what time-scale should spectrum be allocated? Second, should a centralized or a distributed approach be used? Third, should access be scheduled or contention-based?

For the first question, long allocation periods are good for licensees, particularly if they need to deploy infrastructure. Long periods allow them to invest in their network, with the security that they will have the resource for a certain length of time. However, this creates a system that is less dynamic, and less responsive to change, resulting in inefficiencies. Shorter periods are more flexible and responsive, at the expense of stability.

There are several possible solutions to this question. One option to address this instability would be to use short licenses while giving a priority to renewing an existing license. This would allow an operator to obtain a license and deploy infrastructure, with a reasonable expectation that they can maintain the license as long as the spectrum is adequately used. Another approach would be to have a system of variable-length licenses, with extended licenses being granted as required. In the extreme, with fully cognitive nodes, even a system with fully open spectrum could be envisioned.

The question of centralized versus distributed is dependent on the length of licenses. For very short leases, the fully centralized approach of a single regulatory body would be overwhelmed. Similarly, a highly distributed system would be unnecessary for very long leases. In reality, some type of hybrid or hierarchical system is likely. Consider the spectrum server solution presented in [57]. In this system, clients request resources from a centralized server, which allocates spectrum. Similarly [58], presents a framework for real-time spectrum auctions. To scale, these systems would likely require many servers, with extensive coordination, however they can be effective at avoiding conflicts in spectrum allocations.

The fully distributed case is in fact the question typically considered in medium access control, where individual nodes or links must obtain a spectrum opportunity to communicate. At this scale, spectrum allocation can be very fine, with resources allocated for single flows or even single packets. With allocation handled at this level of precision, no additional MAC protocol would be needed. In [59], a distributed allocation scheme is presented where groups of nodes bargain with each other for spectrum access. With the geographical limitations of wireless communication, distributed decisions become quite natural. However, the difficulty lies in knowing how the decision affects more distant nodes.

Scheduled or contention-based? This is an interesting question for dynamic spectrum. Although the opportunistic nature of contention seems to lend itself naturally to the task, the inefficiency of contention-based approaches seems contrary to cognitive radios goals. However, this inefficiency arises predominantly from the expense of the contention process relative to the length of the communication. Although WLANs typically contend for a single packet transmission opportunity, a more flexible spectrum license could be more feasible.

Scheduled spectrum allocation could be very efficient, with very little spectrum wasted. The challenge lies in creating this schedule, whether centrally, or in a distributed manner [60]. In part, the cost of scheduling resources is dependent on the dynamics of the nodes. As seen in MANETs, high mobility, or highly variable traffic requirements require frequent changes to the schedule, increasing the cost of computation and communication.

One proposal for spectrum management has been to facilitate the concept of a secondary spectrum market [61]. The concept allows the current allocation system and spectrum rights to be maintained, while making better use of wireless resources. Many current licensees object to dynamic spectrum access and the Open Spectrum concept, as it infringes on their spectrum, a resource that they may have acquired at considerable expense. A secondary market would maintain the rights of the primary spectrum license, and in fact allow the licensee to make additional profit from their residual resources [62].

A secondary market involves the primary licensee re-leasing any residual bandwidth that they cannot use. Consider a cellular provider with a set of frequency channels. The provider must have sufficient resources to avoid blocking incoming calls and to keep call dropping due to handoff to a minimum, even during busy periods. During low periods, there is a large amount of residual capacity that goes unused. A secondary market allows the provider to lease this residual capacity to another provider, if required. This can ability can be used to effectively pool resources, leading to a dramatically improved level of QoS [63].

One further area of spectrum management requires consideration. Enforcing spectrum rights is already difficult under the current system, and dynamic spectrum usage further complicates it. Currently, spectrum is protected primarily through regulation and control over radio emitters. In the US, the FCC approves radio-emitting devices only after extensive compliance testing, to ensure they operate as required, without generating harmful interference. Detecting transmissions from unauthorized users is important, as they may impact on QoS and, in the case of intrusions, may present a security risk [64].

Dynamic spectrum necessitates an integrated enforcement solution. Mechanisms for secure devices have been proposed, so that devices were ensured to observe proper channel etiquette to transmit [65]. It has also been suggested that this etiquette could be captured within a channel license, which could be limited in duration to ensure their eventual expiration [66]. In addition, the sensing capabilities of a CR suggest that a distributed approach to detect rogue transmissions could be created [67].

18.7 Directions for Future Research

The development of CR could be a long process, however WMNs present several characteristics that could assist in bringing CR technology, or at least some parts of it, to use much sooner.

18.7.1 Static Core Topology

The relatively static nature of the WMN could greatly simplify CR systems. This effect has already been seen in the development of WMNs from MANETs – by removing the mobility, routing overhead is reduced and the technology becomes feasible. For CR, the fixed network changes the problem of collecting awareness of the network's surroundings. The WMN nodes provide a static frame of reference against which environmental data can be collected. Although mobile users and external interference sources may change throughout the life of the network, the WMN can establish normal values and possibly even identify periodic or predictable behavior.

18.7.2 Spectrum Information Collection

The WMN also presents a distributed infrastructure to collect spectrum data at a large number of locations. Interference levels are most important at the destination rather than the source, so to ensure harmful interference is not created, a CR system will need to be able to check levels at several sites. The presence of many user devices may also assist in the process, as CR-capable devices may assist in this detection process – in essence operating as sensor nodes (Fig. 18.4). It will be important to develop systems for collecting this data and maintaining it in a relevant form.

18.7.3 Traffic Awareness

As the primary traffic pattern in a WMN is focused to and from the gateway, knowledge about network traffic is fairly easy to obtain. The gateway is in a position to learn about traffic either by observation or reservation. Collecting information about other MAP-to-MAP flows in the network will require additional communication if this information is needed for network-wide resource management decisions.

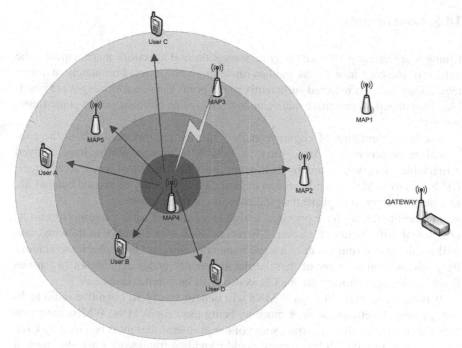

Fig. 18.4 CR devices act as sensors to gauge interference levels

18.7.4 Data Distribution and Decision Making

After the data is collected, it must be shared to ensure that the appropriate nodes have access to it. The structure of the WMN suggests that these decisions can be initially limited to within the mesh infrastructure. Although the WMN may receive sensing data or send control instructions to mobile devices, the processing and decision-making could occur solely within the mesh itself. This could greatly reduce the scope of data distribution.

18.7.5 Spectrum Monitoring and Policing

In order for cognitive radio to be initially accepted by policy-makers, it is extremely important that primary spectrum rights be protected. Using the sensing capabilities of the WMN, the WMN may be able to collaborate to detect users and determine the location of illegal transmissions. For users that violate spectrum policy, the WMN could also play a role in actively policing the action by denying or reducing that user's future service, through fines or pricing premiums, or by reporting the violation to the authorities.

18.8 Conclusions

Changes are coming to wireless communication and spectrum management. The explosive increase in wireless use has made them necessary. Fortunately, wireless technology has also matured sufficiently to the point where such changes are possible. The concept of cognitive radio can now serve as an ultimate goal to guide future research.

The implementation of cognitive radio will play an important role in the continued development and deployment of WMNs. Although WMNs have enjoyed remarkable success, they are, by their nature, very demanding of wireless resources. CR brings to WMNs the potential to exploit a large quantity of unused bandwidth, and the flexibility to improve the efficiency of communication.

In this chapter, we have provided a high-level view of what a cognitive radio is, and how it will operate. However, CR represents a major change in thinking, and will require the development of a large number of different technologies to achieve its goals. Although many of these technologies are under exploration in various fields, their combination into one CR system will be a tremendous task.

It is our view that although WMNs will benefit from CR, cognitive radio technology may benefit as much or more by being used for WMNs. WMNs have certain characteristics that constrain some of the problems that must be faced by CRs. In particular, the fixed infrastructure could provide a framework for gathering and maintaining information about its environment. The WMN structure fits naturally with both centralized and distributed approaches, which can be used in creating and enforcing spectrum allocation policies.

Despite the promise of cognitive radio, it faces many obstacles to obtaining authorization from regulatory bodies. However, we believe that by developing CR in conjunction with WMNs, a deployment, at least within a limited spectrum region, could be achieved much more readily. With both technologies benefiting from this relationship, a successful demonstration could suppress concerns about a new spectrum paradigm, and pave the road for the full realization of cognitive radio.

18.9 Terminologies

1. *Cognition.* The functions and processes of intelligence. These include learning, inference, decision-making, and planning. The development of knowledge.
2. *Cognitive radio.* The accumulation and use of knowledge about all aspects of the wireless environment for the purpose of intelligently deciding how to most effectively exploit wireless resources.
3. *Software defined radio.* A radio that is implemented in software rather than hardware, allowing for a far greater degree of control and re-configurability.
4. *Frequency agility.* The ability of a radio device to use a variety of different radio frequencies.

5. *Unlicensed spectrum*. Radio spectrum frequencies that are allocated for open use. These frequencies can be used by any device as long as they follow certain power and interference constraints.
6. *Open spectrum policy*. Policy advocating an increase in the openness of spectrum allocation. This includes not only the increase of available unlicensed spectrum, but also the opening up of existing licensed spectrum, possibly through cognitive radio techniques.
7. *Regulators*. Decision-making bodies responsible for the management of spectrum resources, e.g., the FCC in the USA or the CRTC in Canada.
8. *Spectrum auction*. One of the common methods for allocation of spectrum resources. A spectrum auction involves the release of a block of frequencies by the regulator. Parties submit bids to obtain a license for those frequencies.
9. *Spectrum license*. The right to dictate the usage of a particular set of frequencies. The license is granted to the licensee by the regulator.
10. *Fixed spectrum allocation*. The current system of spectrum allocation whereby each frequency channel is assigned to a particular purpose. Any changes occur via the regulatory process, usually over very long periods of time.

18.10 Questions

1. Explain how wireless resources can be both scarce and under-used.
2. Wireless spectrum has become a very valuable resource. Compare spectrum to two other natural resources in terms of renewal, scarcity, usage, overuse, etc.
3. List and describe three pros and three cons for an existing spectrum owner to open their resources to CR.
4. Compare current wireless communications to human speech. How are they similar? How are they different?
5. How does cognitive radio change the comparison in Q4?
6. How would using a wireless technology such as WiMax benefit a WMN?
7. Use a simple example to show why a WMN with equal channels dedicated to the access and transit links cannot fully use its available bandwidth.
8. Can one node detect a spectrum violation under the current system?
9. Can one node detect a spectrum violation in a CR system?
10. What systems do you think are necessary for implementing a secondary spectrum market?

References

1. Federal Communications Commission (US). http://www.fcc.gov/
2. Canadian Radio-television and Telecommunications Commission (CRTC). http://www.crtc.gc.ca/
3. International Telecommunications Union (ITU). http://www.itu.int/

4. T.W. Hazlett, The spectrum allocation debate. IEEE Internet Computing 10(5): 68–74 (2006).

5. M.A. McHenry (2005) NSF Spectrum Occupancy Measurements Project Summary. Shared Spectrum Company Report. http://www.sharedspectrum.com/

6. Federal Communications Commission (FCC) (2002) Spectrum Policy Task Force ET Docket no. 02–135.

7. J. Mitola III and G.Q. Maguire Jr, Cognitive radio: Making software radios more personal. IEEE Personal Communications 6(4): 13–18 (1999).

8. J. Mitola, Cognitive radio: An integrated agent architecture for software defined radio. Doctor of Technology, Royal Institute of Technology (KTH), Sweden (2000).

9. EEE 802.11 Working Group (2007) IEEE 802.11–1997: Wireless LAN Medium Access Control (MAC) and Physical Layer (PHY) Specifications.

10. L. Hanzo and R. Tafazolli, A survey of QoS routing solutions for mobile ad hoc networks. IEEE Communications Surveys and Tutorials, 9(2): 50–70 (2007).

11. S. Kumar, V.S. Raghavan and J. Deng, Medium access control protocols for ad hoc wireless networks: A survey, Ad Hoc Networks (Elsevier) 4:326–358 (2006).

12. T. Shono et al. IEEE 802.11 wireless LAN implemented on software defined radio with programmable architecture. IEEE Transactions on Wireless Communications 4(5):2299–2308 (2005).

13. A. Bourdoux et al., Receiver Architectures for Software-defined Radios in Mobile Terminals: the Path to Cognitive Radios. In Proc. of 2007 IEEE Radio and Wireless Symposium (2007).

14. R. Wang et al., Capacity and performance analysis for adaptive multi-beam directional networking. In Proc. of MILCOM 2006 (2006).

15. R. Bhatia and L. Li, Throughput optimization of wireless mesh networks with MIMO links. In Proc. of IEEE Infocom (2007).

16. R. Ramanathan, On the performance of ad hoc networks with beam-forming antennas. In Proc. of 2001 ACM Intl. Symp. On Mobile Ad hoc Networking and Computing (2001).

17. D. Zhang and Z. Tian, Spatial capacity of cognitive radio networks: Narrowband versus ultra-wideband systems. In Proc. of IEEE WCNC (2007).

18. Z. Tian and G.B. Giannakis, A wavelet approach to wideband spectrum sensing for cognitive radios. In Proc. of 1st Cognitive Radio Oriented Wireless Networks and Communications Conference (2006).

19. S. Haykin, Cognitive Radio: Brain-empowered wireless communications. IEEE Journal on Selected Areas in Communications, 23(2):201–220 (2005).

20. P. Sutton, L.E. Doyle and K.E. Nolan, A reconfigurable platform for cognitive networks. In Proc. of 1st Cognitive Radio Oriented Wireless Networks and Communications Conference (2006).

21. J. Zhu and K.J.R. Liu, Dynamic spectrum sharing, a game theoretical overview. IEEE Communications Magazine 45(5):88–94 (2007).

22. E.M. Noam, Taking the next step beyond spectrum auctions: open spectrum access. IEEE Communications Magazine 33(12):66–73 (1995).

23. IEEE 802.22 WG on Wireless Regional Area Networks. http://www.ieee802.org/22/

24. C. Cordeiro et al. IEEE 802.22: the first worldwide standard based on cognitive radios. In Proc. of IEEE DySpAN 2005 (2005).

25. C. Siller and R. Boutaba, Standards – A new challenge for ComSoc. IEEE Communications Magazine 43(8) (2005).

26. D. Maldonado et al., Cognitive radio applications to dynamic spectrum allocation: a discussion and an illustrative example. In Proc. of IEEE DySpAN 2005 (2005).

27. I.F. Akylidiz and X. Wang, A survey on wireless mesh networks. IEEE Communications Magazine 43(9):S23–S30 (2005).

28. J. Jun and M.L. Sichitiu, The nominal capacity of wireless mesh networks. IEEE Wireless Communications 10(5):8–14 (2003).

29. D. Niculescu et al., Performance of VoIP in a 802.11 Wireless Mesh Network. In Proc. of INFOCOM (2006).

30. S.M. Das et al., DMesh: Incorporating directional antennas in multichannel wireless mesh networks. IEEE Journal on Selected Areas in Communications 24(11):2028–2039 (2006).

31. H. Koubaa, Fairness-enhanced multiple control channels MAC for ad hoc networks. In Proc. of IEEE 61st Vehicular Technology Conference (2005).
32. J. Mo, H.W. So and J. Walrand, Comparison of multi-channel mac protocols. In Proc. of the 8th ACM/IEEE International Symposium on Modeling, Analysis and Simulation of Wireless and Mobile Systems (2005).
33. P. Kyanasur and N.H. Vaidya, Routing and interface assignment in multi-channel multi-interface wireless networks. In Proc. of 2005 IEEE Wireless Communications and Networking Conference (2005).
34. B. Aoun and R. Boutaba, Analysis of capacity improvements in multi-radio wireless mesh networks. In Proc. of 63rd IEEE Vehicular Technology Conference (2006).
35. D.T. Chen, On the Analysis of Using 802.16e WiMAX for point-to-point wireless backhaul. In Proc. of 2007 IEEE Radio and Wireless Symposium (2007).
36. X. Jing and D. Raychaudhuri, Spectrum co-existence of IEEE 802.11b and 802.16a networks using reactive and proactive etiquette policies. Journal of Mobile Networking Applications, Springer Science (2006).
37. D. Porrat, Information theory of wideband communications. IEEE Communication Surveys 9(2):2–16 (2007).
38. S.A. Jafar and S. Srinivasa, Capacity limits of cognitive radio with distributed and dynamic spectral activity. IEEE Journal on Selected Areas in Communications 25(3):529–537 (2007).
39. I. Wormsbecke and C. Williamson, On channel selection strategies for multi-channel MAC Protocols in wireless ad hoc networks. In Proc. of WiMob 2006 (2006).
40. D. Aguayo et al., Link-level Measurements from an 802.11b Mesh Network. In Proc. of SIG-COMM (2004).
41. P.J. Kolodzy, Interference temperature: a metric for dynamic spectrum utilization. International Journal of Network Management, Wiley Interscience (2006).
42. I. Akyildiz et al., AdaptNet: An adaptive protocol suite for the next generation wireless internet. IEEE Communications Magazine 42(3):128–136 (2004).
43. M.J. Marcus, Unlicensed cognitive sharing of TV spectrum: the controversy at the Federal Communications Commission. IEEE Communications Magazine 43(5):24–25 (2005).
44. V.K. Varma et al., A beacon detection method for sharing spectrum between wireless access systems and fixed microwave systems. In Proc. of 1993 Vehicular Technology Conference (1993).
45. K. Kaemarungsi, Distribution of WLAN received signal strength indication for indoor location determination. In Proc. of 1st Intl. Symp. On Wireless Pervasive Computing (2006).
46. C. Savarese, J. Rabay, and K. Langendoen robust positioning algorithms for distributed ad-hoc wireless sensor networks. In Proc. of USENIX Technical Annual Conference (2002).
47. N. Sundaram and P. Ramanathan, Connectivity based location estimation scheme for wireless ad hoc networks. In Proc. of Globecom 2002 (2002).
48. I. Akyildiz et al., A survey on sensor networks. IEEE Communications Magazine 40(8):101–114 (2002).
49. L. Jiang and G. Feng, A MAC aware cross-layer routing approach for wireless mesh network. In Proc. of WiCOM (2006).
50. C.J. Merlin and W.B. Heinzelman, A first look at a cross-layer facilitating architecture for wireless sensor networks. In Proc. of 2nd WiMesh Workshop (2006).
51. P.M. Ruiz, J.A. Botia, and A. Gomez-Skarmeta, Providing QoS through machine-learning-driven adaptive multimedia applications. IEEE Transactions on Systems, Man, and Cybernetics 34(3):1398–1411 (2004).
52. E. Stevens-Navarro and V.W.S. Wong, Comparison between vertical handoff decision algorithms for heterogeneous wireless networks. In Proc. of the 63rd IEEE Vehicular Technology Conference (2006).
53. G. Lee et al., A user-guided cognitive agent for network service selection in pervasive computing environments. In Proc. of IEEE PerCom'04 (2004).
54. X. Fu et al., Extended mobility management challenges over cellular networks combined with cognitive radio by using Multi-hop network. In Proc. of SNPD (2007).

55. J.B. Bernthal et al., Trends and precedents favoring a regulatory embrace of smart radio technologies. In Proc. of DySPAN (2007).
56. C. Raman, R.D. Yates, and N.B. Mandayam, Scheduling variable rate links via a spectrum server. In Proc. of IEEE DySpAN (2005).
57. S. Gandhi et al., A general framework for wireless spectrum auctions. In Proc. of IEEE DySPAN (2007).
58. L. Cao and H. Zheng, Distributed spectrum allocation via local bargaining. In Proc. of IEEE SECON (2005).
59. M. Thoppian et al., MAC-layer scheduling in cognitive radio based multi-hop wireless networks. In Proc. of IEEE WoWMoM (2006).
60. J.M. Peha and S. Panichpapiboon, Real-time secondary markets for spectrum. Telecommunications Policy (Elsevier) 28 (2004).
61. F. Capar and F. Jondral, Spectrum pricing for excess bandwidth in radio networks. In Proc. of PIMRC (2004).
62. M. Oner and F. Jondral On the extraction of the channel allocation information in spectrum pooling systems. IEEE Journal on Selected Areas in Communications 25(3):558–565 (2007).
63. M.K. Chirumamilla and B. Ramamurthy, Agent based intrusion detection and response system for wireless LANs. In Proc. of IEEE ICC'03 (2003).
64. W. Xu, P. Kamat and W. Trappe, TRIESTE, A trusted radio infrastructure for enforcing Spectrum Etiquettes. In Proc. of 1st IEEE Workshop on Networking Technologies for Software Defined Radio Networks (2006).
65. J.M. Chapin and W.H. Lehr, Time-limited leases for innovative radios. In Proc. of DySPAN (2007).
66. A.A. Tomko, C.J. Rieser and L.H. Buell, Physical-layer intrusion detection in wireless networks. In Proc. of Military Communications Conference (2006).
67. Industry Canada. Radio Spectrum Allocations in Canada (Chart). http://www.ic.gc.ca/epic/site/smt-gst.nsf/en/h_sf01678e.html.

Chapter 19
Construction and Evaluation of a Wireless Mesh Network Testbed

Alexander Zimmermann, Martin Wenig, and Ulrich Meis

Abstract Wireless mesh networks (WMN) are supposed to provide flexible and high-performance wireless network access for large indoor and outdoor areas, e.g., community networking and metropolitan area networks. However, these claims are mostly substantiated by simulation studies only as real testbeds are inflexible and associated with high maintenance effort. In this work we present a hybrid, i.e., partly real and partly virtualized, WMN testbed. This provides a high degree of realism while still allowing the flexibility known from simulations. In addition to the architectural discussion we present measurement results from our testbed highlighting the optimization potential of small protocol parameter changes.

19.1 Introduction

Over the past years, one approach has received a great deal of attention: *mobile ad hoc networks* (MANET). In spite of massive efforts in research of MANETs, this type of network has not yet seen mass-market deployment. The low commercial penetration of products based on MANET technology is due to the very restricted application scenarios. However, the masses demand for versatile networks providing them *high bandwidth* and *access to the Internet* [3]. To make MANETs useful for the mass-market some changes to the common definition of MANETs are mandatory. By relaxing one of the main constraints of MANETs, "the network is made of user devices only and no infrastructure exists," a new class of networks emerges: *wireless mesh networks* (WMN) [1, 3].

A. Zimmermann (✉)
Department of Computer Science, Informatik 4, RWTH Aachen University Ahornstr. 55, 52074 Aachen, Germany
e-mail: zimmermann@cs.rwth-aachen.de

S. Misra et al. (eds.), *Guide to Wireless Mesh Networks*, Computer Communications and Networks, DOI 10.1007/978-1-84800-909-7_19,

Regarding the research activities in mesh networking, the majority of prior work has relied on simulation based evaluation only because simulation offers a convenient combination of flexibility and controllability. However, simulation suffers from the fact that the gained insights are difficult to transfer to reality since it is too complicated to model the complex nature of wireless multi-hop networks in all details. On the other hand the WMN testbeds that have been deployed, e.g., MIT Roofnet [2] or Microsoft Research's testbed [10] provide a high degree of realism, but are inflexible in terms of scenario creation, repeatability, and controllability.

In this chapter we describe and evaluate *UMIC-Mesh.net*, our alternative approach to study WMNs. UMIC-Mesh.net is characterized by a hybrid architecture, consisting of a real testbed and a virtualized environment. The virtualization allows the *development*, *validation*, and *testing* of software as if it was executed on real mesh nodes, but in a more repeatable and controllable way. The newly developed software can afterwards be executed and evaluated in the real testbed without any modifications. The results and conclusions gained by the evaluation can be easily transferred into the real world, since the testbed provides a high degree of realism. To the best of our knowledge our testbed is the first that combines a real testbed and a virtualized environment to study WMNs.

The performance evaluation of our testbed shows that the choice of parameters of well-known protocols has a big impact on the overall mesh performance. We argue that there has been little research in the community on how to set these parameters, or on analyzing the impact of these parameters on the performance of a mesh network. Especially the impact of parameters of routing protocols and the usage of routing metrics are not sufficiently researched. For example in the case of the optimized link-state routing protocol (OLSR), only few conference papers can be found about parameter tuning in the digital libraries of IEEE [8, 12, 14] and ACM [28]. It is our belief that we need a much richer understanding of the relevance of different protocol parameters to network performance.

The remainder of this chapter is organized as follows. Section 19.2 gives a common definition of WMNs. Based on this, we motivate our solution to investigate WMNs. Section 19.4 presents our testbed in detail. The performance evaluation is presented in Sect. 19.5. Previous work is surveyed in Sect. 19.6. The thoughts for practitioners are given in Sect. 19.7. Section 19.8 outlines the directions for future research. Finally, in Sect. 19.9 we draw some conclusions.

19.2 Wireless Mesh Networks

Several different definitions of WMNs and their relation to MANETs have been proposed in literature which, in some parts, even contradict each other [1, 24]. Therefore, a definition of what shall be understood by the term WMN in this work is presented in the following together with main characteristics of WMNs.

19.2.1 System and Network Architecture

Figure 19.1 depicts a hierarchical and layered architecture that integrates various approaches and, thus, helps to identify the main parts of a WMN. This view is more general than that usually presented insofar as that other approaches often leave out some layers, e.g., they consider only clients without routing functionality [3].

The top level consists of *backbone mesh gateways* connected to the Internet by wire and providing wireless Internet access to the second level entities, the so-called *backbone mesh routers*. Both, gateways and routers are installed at certain fixed positions and form the wireless, meshed backbone of the WMN. The wired Internet connections are not considered to be part of the WMN, i.e., the WMN itself is completely wireless. The lowest level contains mobile user devices, the *mesh clients*. They are subdivided into two groups: *routing mesh clients* and *non-routing mesh clients*. Routing mesh clients communicate among each other in a multi-hop fashion, thus forming a MANET with gateways connected to the WMN backbone. Non-routing mesh clients associate with mesh routers in the same way as conventional IEEE 802.11 clients do with wireless access points (AP).

The hierarchy achieved by the distinction between clients and routers promotes the utilization of multiple radios for mesh routers and gateways, separating the traffic in the backbone from the one of the clients. Routing and configuration tasks are

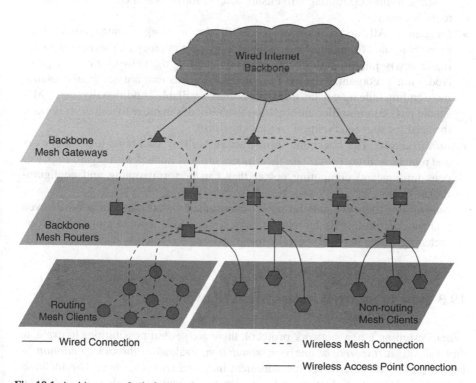

Fig. 19.1 Architecture of wireless mesh networks

assigned to mesh routers in order to unburden mesh clients that are probably power constrained because of their inherent mobility. Furthermore, the WMN has got a spontaneous and dynamic character as mesh clients can leave at any time and new clients may arrive at any time.

19.2.2 Network Characteristics

The presented architecture of a WMN leads to several characteristics that are quite general and many of them also hold for other views of WMNs.

- *Wireless*: WMNs must cope with the limited transmission range and the potentially high loss rates due to packet collision and fading of the wireless channel during the transmission.
- *Multi-hop*: WMNs use multi-hop routing, i.e., a WMN node forwards traffic generated by nodes which are not within its direct transmission range.
- *Redundancy*: The wireless backbone of a WMN provides redundant links between backbone mesh routers, backbone mesh gateways, and mesh clients, which makes the WMN more resistant against node or link failures.
- *Mobility*: A mobile client can change the mesh router it connects to and the dynamic multi-hop routing will ensure that the traffic is still correctly forwarded to its destination.
- *Dynamics*: All nodes have to establish the network spontaneously (*self-organizing*) and to maintain their connectivity continuously (*self-healing*). Leaving or newly joining nodes cause topology changes the network must adapt to. Nodes must reorganize their routes, invalidate paths that are not available anymore and include new paths that have become available. Additionally, the WMN should pass configuration information to new nodes in order to reduce or remove the need for user intervention (*self-configuring*).
- *Infrastructure*: The backbone infrastructure built by backbone mesh gateways and routers is almost static. As these nodes are less limited with regard to power consumption and computing power they can perform routing and configuration tasks.
- *Integration*: Light-weight and power-constrained clients can join a WMN even without contributing to the routing service. Therefore, a broad range of devices, including whole networks, can be integrated into the wireless backbone.

19.3 How to Study Wireless Mesh Networks

When designing a new network protocol, there are several possibilities to evaluate and validate it: *theoretical analysis, simulation*, evaluation through *emulation* or *virtualization*, and the direct measurement in a *real world testbed*. The methods differ in their level of abstraction in relation to the real application. A theoretical analysis has the highest level of abstraction and, in descending order, is followed by simulation, emulation, virtualization, and finally reproduction in a real world testbed. The use of simplifying quantitative models leads thereby to a deviating behavior of the experimental setup. The more parameters remain unconsidered in such a model, the more inaccurate the evaluation will be [18].

19.3.1 Theoretical Analysis

Performing a theoretical analysis means to use mathematical models to evaluate the network performance. Queuing theory is one of the most common mathematical tools in network performance studies. Unfortunately, theoretical analysis of WMNs is very difficult, since the mathematical constructs get very complex for realistic considerations; useful mathematical tools do not exist.

19.3.2 Simulations

A simulation environment offers a high degree of control and repeatable results to the researcher. This is especially useful when studying highly distributed networks like MANETs or WMNs. During the study of such a network, typically few parameters are varied while most remain fixed. This allows to study the effects of certain parameters on the network performance. Simulation studies are very flexible and the related costs are normally low.

However, a simulation study has also its disadvantages. The simulation environment is typically an abstraction of the reality and therefore contains many simplifications. In the case of mobile and wireless networks, which have a very complicated and dynamic environment, the simulation environments are far from being *realistic*. This leads to results that do not fit with real-world measurements.

19.3.3 Emulation

Emulation is a hybrid study environment that consists of two parts: existing hardware and real network layers or parts, and a simulated environment. Which elements are real and which are simulated depends on the study goals and may differ considerably. However, with emulation it is possible to increase the quality of the study environment by making it more realistic.

An important advantage of emulation environments over simulation environments is the possibility of validation against real traffic. The advantage of emulation environments over real world experiments is the possibility of scaling to larger topologies by multiplexing simulated elements on physical resources, e.g., network interfaces [5].

19.3.4 Virtualization

In general, virtualization environments can be classified into three classes. The first class is the *system virtualization* with the virtual machine monitor (VMM) *inside* the host system. The virtual machine (VM) simulates the complete hardware, allowing an unmodified operating system (OS) for a completely different CPU to be run. The second class of virtualization is also a type of system virtualization, but in contrast to the previous one, the VMM is *underneath* the host OS. Thus, the VMM runs directly on the hardware. In general this allows multiple OSes to run, unmodified, at the same time. The third class of virtualization is the *operating system-level virtualization*. It virtualizes a physical server at the OS level, enabling multiple isolated and secure virtualized servers on a single physical server. The guest OS is the same OS as the host system, since the same OS kernel is used to implement the guest environments.

Beside the technical aspects the virtualization offers an adequate tool to evaluate communication protocols. With the aid of virtualization, it is possible to create several VMs on a single host system. Each VM can run a separate OS and hence represents an entire computer system. By coupling several VMs over the network, it is possible to create a whole virtual network of VMs. The most important advantage of virtualization is that the software development can be done on real machines with a real OS, and tested on the virtual network of VMs.

19.3.5 Real Testbeds

The seemingly best environment to study network protocols and to conduct experiments in is a real testbed. Typically, this is done by prototype implementations. The results and conclusions can be easily transferred to reality, since not only the prototype, but also the testbed represents a high degree of realism.

However, in the case of distributed and mobile networks, it is very difficult to conduct experiments. The researcher has only limited control over the environment, since there are many influences from the study environment, e.g., interference with production networks. Experiments are typically difficult to repeat, and the experiment setups are restricted in size as well as in complexity. It is also very expensive to conduct experiments in the real world from the hardware point of view as well as from labor intensity. Last but not least, these kind of experiments are limited to existing technologies.

19.3.6 Summary

The upper part of Table 19.1 summarizes the evaluation of the presented methods for studying WMNs. The following parameters are taken into account:

Table 19.1 Overview of the characteristics of environments for wireless networks

Characteristics	Environments				
	Theoretical analysis	Simulation	Emulation	Virtualization	Real testbed
Applicability	Poor	Low	Middle	High	High
Repeatability	–	High	Low	Low	Poor
Controllability	High	High	Middle	Middle	Poor
Maintainability	–	High	Middle	Middle	Poor
Scenario creation	–	Simple	Middle	Middle	Complex
Scalability	–	High	Middle	Middle	Low
Duration	–	Variable	Real-time	Real-time	Real-time
Cost	–	Low	Middle	Middle	High
Application	–	Low	High	High	High
Transport	–	Low	Middle	High	High
Network	–	Low	Middle	High	High
Data link	–	High	Middle	Middle	High
Physical	–	High	Middle	Low	High

- *Applicability*: Evaluates the degree of transferability of the results, and conclusions into the real world.
- *Repeatability*: Rates how straightforward the repetition of a given experiment in that study environment is.
- *Controllability*: Assesses the degree of control the researcher has over the study environment.
- *Maintainability*: Describes the ability to maintain the environment, i.e., how much effort is necessary to keep the system runnable.
- *Scenario creation*: Describes the freedom in creating different experiment scenarios in terms of network topology, the number of nodes, etc.
- *Scalability*: Assesses the feasibility of large scale experiments with respect to the number of nodes, etc.
- *Duration*: Describes the experiment time. Variable means that experiments can be conducted over long periods of time. In contrast, realtime means that experiments are conducted in real-world time.
- *Cost*: Evaluates the cost of experiments. The cost is related to hardware and software costs.

In the case of theoretical analysis we have evaluated only two categories. The other categories do not restrict the environment, since it depends heavily on the modeling capabilities of the researcher, e.g., scalability is not an issue here. In summary, we argue that a *theoretical analysis* of a complete WMN is not possible, but can only be done for particular components of the network. This environment provides a high degree of control and abstraction and at the same time a poor applicability of the results and conclusions. The *simulation* combines low cost with high flexibility for different types of network studies. The most important disadvantage is the limited applicability of results to the real world. The *virtualization* provides a healthy tradeoff between maintainability, scalability, and applicability. From our point of view, virtualization has some inherent advantages by allowing the development and testing of software as if it was executed on backbone mesh routers, but in a more repeatable and controllable way. It is easy to port the software to the real nodes of the testbed afterwards.

The highest degree of applicability and therefore transferability of results, conclusion, and system environment is given in the case of *real testbeds*. The main disadvantage of this environment is its low scalability and the complexity in experiment scenario generation.

When designing experiments to study performance parameters of WMNs it is important to have an idea which degree of realism can be expected from the study environments. In the lower part of Table 19.1 we have summarized them with respect to networking layers. This helps to determine which of the environments provides the researcher with the required degree of realism. The *theoretical analysis* approach does not provide any realistic instances of the network layers. In contrast a *real testbed* provides realism on all layers. *Simulation* typically provides a high degree on the data link-and physical layer. The upper layers are typically simplified. The degree of realism in *emulation* depends heavily on the parts which are represented by real hard-and software. In the case of *virtualization* the upper layers are real, since the virtualized machine and the OS provide all necessary functionalities. However, if VMs are coupled via a network the physical layer may have low realism, if both VMs are run on the same physical computer.

19.4 UMIC-Mesh.net: A Hybrid Testbed for WMNs

In this section we present our project *UMIC-Mesh.net*. The aim of this project is twofold. From the scientific point of view, the goal is to build a large and scalable WMN to pursue various networking studies. Considering real applications, the goal is to provide the members of the Computer Science Department with an easy way to get network access.

19.4.1 Motivation for a Hybrid Testbed

As we have seen in Sect. 19.3, there are different possibilities to study wireless and mobile networks. Before we present our solution to study WMNs, we will shortly review the software development process to point out the advantages of our approach. Basically, as known from software engineering, the *software development process* is iterative. It comprises *developing, distributing*, and *testing*. In the developing step the realization of new network protocols and tools starts. Implementation and debugging are made. Subsequently, in the distributing step, the installation of the implementation and its validation are done to ensure a correct distribution among all testbed nodes. In the final step, the functionality and performance of the implementation have to be tested and evaluated respectively. In addition, if any failure occurs in one of these steps, debugging information has to be collected and analyzed. Preferably, the environment for developing and testing should be as close to reality as possible generally achieved best by utilizing real hardware and standard software.

All in all, the iterative software development process is a complex and labor-intensive undertaking. In particular, the distribution of new software versions is a challenging task, since new versions have to be distributed among all nodes of the testbed frequently. A *hybrid testbed* that combines a *real testbed* and a *virtualized environment* offers a solution to the problems mentioned above. The hybrid testbed allows us to benefit from the advantages of a real testbed but, at the same time, to avoid its disadvantages.

The separate tasks of the software development process are distributed over the virtualized environment and the real testbed. On one hand, the virtualized environment takes over the *developing* task (implementation and debugging), the *validation*, and the *functionality testing*. On the other hand, *performance evaluation* takes place in the real testbed, since a high degree of realism is of great concern to this step. Thus, there is no need for an accurate emulation of a wireless medium, as for example in [16], since the virtualized environment is used only for software development and functionality testing. Network performance evaluation is solely done in the real testbed. However, for scenario creation or testing special cases some emulation might be of help (see Sect. 19.4.3).

To illustrate the benefit of the virtualized environment for the software development process consider for example that booting a real mesh router may take up to 2 min whereas booting the kernel of a virtualized machine only takes about 20 s. The same holds for compiling. As the real mesh routers are equipped with hardware which is appropriate for routing and forwarding only, compiling on the mesh routers is hardly bearable. In contrast, the virtualized environment may come up with high-end hardware to enable fast cycling through the software development process.

Besides a more efficient software development process a hybrid testbed offers a more flexible *scenario creation* as well. As virtualization allows fast and flexible setup of an almost arbitrary number of nodes, scenarios with many or few nodes, potentially leaving and joining the network, are easy to achieve. Moreover, even complicated scenarios in which we realize arbitrary WMN architectures (see Sect. 19.2.1) are possible by defining which mesh routers operate as gateways or which mesh clients have their routing functionality enabled. Thus, any mesh protocols, e.g., routing protocols or channel assignment algorithms, can be easily evaluated in different mesh architectures.

19.4.2 System and Network Architecture

Figure 19.2 depicts the general system and network architecture of the UMIC-Mesh.net testbed. In accordance with the previous considerations that a hybrid

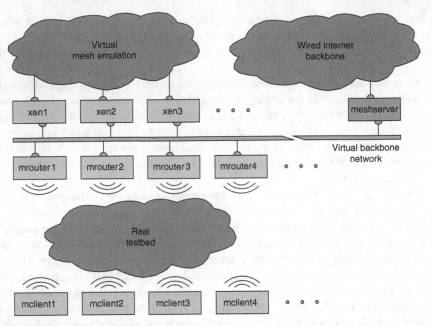

Fig. 19.2 Architecture of the UMIC-Mesh.net

testbed presents an appropriate environment to study WMNs, the testbed is realized by using two different components: a virtualized environment and a real testbed.

A main disadvantage of both real testbeds and virtualized environments is the high maintenance effort. Especially, if there is a failure during an ongoing performance evaluation, the distribution of a new, corrected implementation is still labor-intensive. To minimize this effort a central server, the so-called *meshserver*, is integrated into the testbed. It offers *source* and *drain* functionalities. The most important service of the source functionality is the provision of an OS to all nodes via the network. Therefore, the basic setup is the same in each node of the hybrid testbed. The nodes may even share the same kernel including modules and drivers. Another important service provided by the meshserver is the Internet access, which is required for the backbone mesh gateways. The drain functionality incorporates kernel, system message and Simple network management protocol (SNMP) message logging, gathering measurement results of the real and emulated mesh networks, etc. The central logging enables a quick detection of any problem in the testbed.

To interconnect the virtualized environment and the real testbed all mesh routers and virtual machines are connected by a common wired network, the *virtual backbone network*. The term "virtual" emphases the fact that this network is *solely* used for booting and configuring the attached nodes as well as for the audit trail processing. That means, the clients in the real testbed cannot use it for their data traffic. Thus, their data is forwarded in a multi-hop fashion via the wireless network interfaces.

19.4.3 Testbed Realization

The testbed is set up at the Department of Computer Science at the RWTH Aachen University. The routers are deployed in two four-story buildings which are interconnected through a ground floor. All mesh routers are placed in different offices at different floors. The placement of the routers was unplanned, i.e., no interference measurements were done. The only constraint was to have a completely connected mesh. Unlike wireless-friendly cubicle environments, our buildings have rooms with floor-to-ceiling walls and solid wood doors. Thus, in order to realize a completely connected mesh, three routers were distributed across every floor. The nodes are in fixed locations and did not move during the experiments reported here.

19.4.3.1 Hardware

The real testbed part consists of 51 identical mesh routers. Each backbone mesh router consists of a single board computer (SBC), two identical IEEE 802.11a/b/g wireless network interfaces based on the *Atheros AR5213 XR* and two omni-directional antennas. The SBC is an *ALIX.3C2* board [20]. While the first network interface card (NIC) is reserved for router-to-router communication, the second one handles the router-to-client communication. In order to clearly separate these communication types, the first NIC operates in IEEE 802.11 g channel 1 and the second one in channel 11. Both cards transmit at 100 mW and operate in a non-standard independent basic service set (IBSS) mode *ahdemo* [19]. This solves the tendency of the IBSS mode to form partitions which have different basic service set identifiers (BSSID) despite having the same network identifier. Such partitions made it impossible to reliably operate the WMN with the IBSS mode. All mesh routers share the same extended service set identifier (ESSID) pair, that is one ESSID for channel 1 and one ESSID for channel 11.

The virtualized environment consists of 7 Core 2 Duo PC with 2 GB RAM. With this amount of RAM it is possible to run about 10 VMs per host.

19.4.3.2 Software

One goal of the UMIC-Mesh.net implementation is to achieve a central configuration. For this reason a single OS image, a standard Ubuntu Linux distribution, is provided to all nodes via network by using the *network file system* (NFS) protocol. To enable network booting we deploy a combination of *EtherBoot* and *PXELinux*. The vital parts of this process are to get an IP configuration and a kernel to boot. The central audit trail processing is realized by a combination of logging (syslog-ng) and monitoring (NET-SNMP).

To implement the WMN architecture in the real testbed we employ the WLAN driver *madwifi-ng* [19]. The nodes in the virtualized environment are driven by the VMM *XEN* [29]. To emulate the multi-hop behavior in the virtual testbed a combination of the packet filtering and advanced routing features [15] of the Linux kernel are used. At the core, we deploy a virtual private network (VPN) on top of our wired virtual backbone. For this, the *generic routing encapsulation* (GRE) [11] tunneling protocol is utilized. It emulates a broadcast medium on top of an existing network by using a multicast address for its broadcast traffic. To control the communication between all participants of the network, standard packet filtering as provided by *iptables* is employed. There is no need to emulate a wireless medium in the virtualized environment, since it is used only for software development and functionality testing and not for performance evaluation. However, if the functionality testing requires a wireless medium behavior, e.g., wireless packet loss or additional delay, it can be realized with *NetEm* [13].

Currently, the *DYMO* [25] and *OLSR* [27] routing protocols are employed. We made this choice since these protocols are typical representatives of the two routing philosophies in MANETs: reactive and proactive routing.

19.5 Performance Evaluation

Obviously, the employed hard- and software significantly affects the performance of a WMN. Similarly, the configuration parameters of the deployed software and protocols have a big impact on the overall mesh performance. However, there has been little research in the community on how to set these parameters, or on analyzing the impact of these parameters on the performance of a mesh network. Especially the impact of routing parameters and the choice of routing metrics are not sufficiently examined.

To demonstrate the importance of a study regarding these parameters we discuss two scenarios concerning the OLSR routing protocol [6]. We focus our research on an excerpt of parameters instead of presenting an extensive evaluation to show that small changes may have a big impact on the overall mesh performance. In particular we evaluate the influence of OLSR HELLO and topology control (TC) message emission intervals. As performance metrics we use throughput, average hop count, and average packet loss.

19.5.1 Optimized Link-State Routing Protocol

Before we discuss the measurement results in detail we will first give a rough introduction on how OLSR works. OLSR is a link-state routing protocol that periodically advertises the links in the network. It optimizes the link advertisement process by reducing the amount of advertised links and the number of nodes advertising them. HELLO messages are periodically sent by all participating nodes to become aware of one-hop and two-hops neighbors. To make the sensing of links more robust to small changes in wireless connectivity a node computes an additional link quality value for each discovered neighbor. The link hysteresis decides on the validity of the link state of a detected neighbor. Multipoint relay (MPR) nodes are selected by each node in the network (called MPR Selector) as the minimum set of one-hop neighbors that allow reaching every two-hop neighbor throughout the nodes in the MPR set. MPRs are the only nodes generating and forwarding TC messages which advertise the links between MPRs and MPR Selectors. Nodes calculate their routing tables based on their local neighborhood tables and the topology information received by the TC messages.

The OLSR.org implementation uses the loss rate of OLSR messages to calculate the link quality instead of using link layer information (e.g., signal-to-noise ratio or the number of link layer retransmits).

19.5.2 Methodology

Table 19.2 shows the plot of our performance evaluation. There are four parameter sets. In this Scenario we varied the OLSR HELLO and TC message emission intervals. This study aims at establishing a better understanding of the impact of these parameters on the end-to-end performance.

To reflect the different wireless characteristics we conducted measurements over a number of different paths (see Table 19.3). We picked these particular paths as they exemplify different challenges for wireless transmissions. The two buildings E1 and E2 each consist of two shifted aisles. The aisles are connected via a staircase. Paths which connect two aisles therefore have to pass this area consisting mainly of

Table 19.2 Scenario Parameter Sets

Parameter set	802.11g Broadcast rate	Routing metric	OLSR emission interval	
			HELLO (s)	TC (s)
1	$1\,\text{Mb s}^{-1}$	Hop Count	2	5
2	$1\,\text{Mb s}^{-1}$	Hop Count	4	5
3	$1\,\text{Mb s}^{-1}$	Hop Count	4	10
4	$1\,\text{Mb s}^{-1}$	Hop Count	6	10

Table 19.3 Selected paths for measurements

Path	Building	Floor	Aisle	Side
$3 \rightarrow 1$	E1	4–4	Same	Same
$26 \rightarrow 25$	E2	1–1	Same	Opposite
$30 \rightarrow 31$	E2	2–2	Different	Same
$34 \rightarrow 27$	E2	3–1	Same	Same
$26 \rightarrow 2$	E2–E1	1–4	Different	Same
$34 \rightarrow 16$	E2–E1	3–1	Same	Same

concrete walls. Connections on the same aisle and same floor merely are obstructed by drywalls. The two buildings are connected via two routers on the ground floor only. Paths between these buildings therefore are relatively long and always have to pass this bottleneck.

Each measurement consists of a *ping* test and a *flowgrind* (see Sect. 19.5.3) test. The ping test yields the packet loss ratio and the average hop count, the flowgrind test measures TCP throughput. The ping test sends 50 maximum transmission unit (MTU) sized packets (1464 bytes) at a rate of five packets per second. The flowgrind test runs a simple one-way bulk-data TCP flow lasting 15 s with the TCP congestion control NewReno TCP.

To cope with environmental influences (e.g., other WLAN sources, Bluetooth devices, etc.) the following approach was taken: one measurement was run sequentially over all mentioned paths. This pattern was repeated for 30 times to take a decent sample. After such a sampling run, the parameter set was changed, OLSR was restarted and allowed to even out the new routing for 10 min. A certainly better approach for coping with environmental influences would be to allow more frequent changes in the parameter set, e.g., to cycle through the parameter sets for each path. But as OLSR needs time at the magnitude of 10 min to follow a change in its parameters, this approach is unfortunately intractable.

The setup of the mesh routers was as follows: vanilla Linux kernel 2.6.16.19, madwifi-ng 0.9.3 [19], and OLSR 0.5.0 from the OLSR.org project [27]. All software was configured as default except for the settings shown in Table 19.2. A further exception was that we set $tcp\_no\_metrics\_save = 1$ to prevent the kernel from gathering long term performance statistics which could cause test runs to influence each other.

19.5.3 Flowgrind

To evaluate the TCP end-to-end performance we deploy our measurement tool *flowgrind* [26]. Flowgrind features some unique characteristics which are of use when exploring the idiosyncrasies of WMNs. Most important, it allows to split the data and control connection, i.e., for negotiating test setup and transmission of statistics flowgrind may use a different connection (over a potentially different route) than for the actual test data. This is beneficial to ensure test robustness, especially when testing over noisy links, as the amount of data traversing the control connection is non-negligible. In the UMIC-Mesh.net testbed the virtual backbone is exploited to handle this control connection.

Moreover, flowgrind constantly reports on not only throughput but also on round trip time, inter-arrival time and a number of Linux kernel variables revealing the status of TCP connections (e.g., congestion window, slow start threshold, number of bytes unacknowledged in the network, etc.). For all other features which are of less use for this work we refer to [26].

19.5.4 Measurement Discussion

In this measurement we analyzed the influence of the OLSR HELLO and TC message emission intervals. Figure 19.3 shows the results of the throughput measurements. It is quite obvious that the standard parameter set leads to poorer performance compared to all other parameter sets. The highest performance gain is achieved by doubling the HELLO interval from 2 to 4 s. This performance gain increases with the hop count. For example the throughput between the nearby nodes 26 and 25 increased by 13% whereas the throughput of the long path 30 → 31 increased even by 120%. Moreover, as seen in Fig. 19.4 the average packet loss shows a significant decrease for all paths. Further increase of the HELLO interval does not show the same decisive effect. One has to conclude that the OLSR messages cause a non-negligible amount of interference and thereby lower the overall performance.

Figure 19.5 shows that even the average slightly decreases, especially for 3 → 1. It is likely that the lesser protocol overhead causes the link quality of some links to raise above the hysteresis threshold. Figure 19.5 further shows a bias towards less deviating hop counts for longer emission intervals. This seems natural as fewer OLSR messages presumably result in fewer route changes. It is important to note that all 51 nodes, i.e., *not* just the nodes on the considered paths, contribute to the routing overhead.

Hence, the OLSR message emission intervals in WMNs have quite some optimization potential. However, the maximum possible gain by prolonging these intervals is not clear and remains subject to further studies.

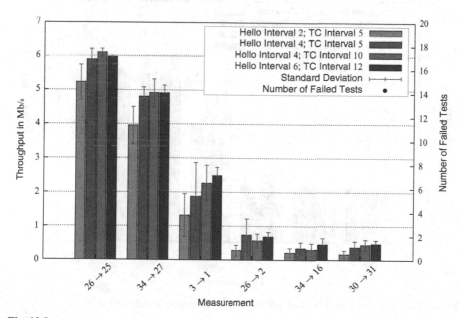

Fig. 19.3 Average throughput per path for different OLSR message emission intervals

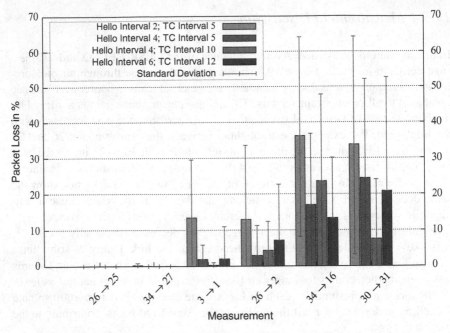

Fig. 19.4 Average packet loss per path for different OLSR message emission intervals
Performance Evaluation!Packet Loss

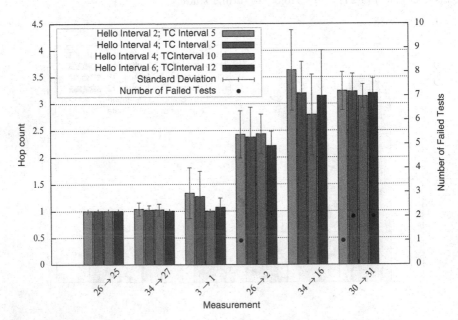

Fig. 19.5 Average hop count per path for different OLSR message emission intervals

19.6 Related Work

In this section we present some WMN testbed projects. We will discuss their goals and the software and hardware on which the mesh nodes are based on. The discussed list of projects is not exhaustive. The objective of this section is to show the various interpretations of WMN and the different implementations rather than to compare the performance evaluations conducted in the different testbeds. As Table 19.4 shows the testbeds may vary fairly and thereby comparison is hardly possible anyway.

19.6.1 Existing Testbeds

The *MIT Roofnet project* [2] consists of 37 nodes based on PCs running Linux and the Click modular router [17]. Each node has an IEEE 802.11b NIC and an omnidirectional antenna. All nodes run on the same channel. The goal of the project is to provide Internet access to students. There are a total of four Internet gateways. The Roofnet mesh routers distribute IP addresses via DHCP. The employed routing protocol is *SrcRR*, which is similar to dynamic source routing (DSR). An own header is used to carry IP packets, thus introducing the roofnet layer (RL).

Microsoft Research works on a community mesh network. A community mesh network allows the residents of a neighborhood to share existing Internet gateways. The focus is on capacity and range enhancement, multi-path multi-hop routing [9], and recently on feasibility studies of mesh networks for all-wireless offices [10].

The *University of California at Santa Barbara* runs a mesh network project called *MeshNet* [22]. In MeshNet each router has two Linksys WRT54G wireless devices. One device is used for routing within the WMN and the other for managing the router. The routers run OpenWRT and a modified version of the ad hoc on-demand distance vector (AODV) routing protocol that uses a reliability-based routing metric instead of minimal hop count.

Table 19.4 Overview of wireless mesh network testbed projects

Project	Nodes	802.11	Software	Routing Layer	Routing Protocol	Roaming	Config	MANET
MIT Roofnet	37	b/g	Linux, CMR	RL	SrcRR	–	×	–
Microsoft	21	a/b/g	Windows CE	MAC	MCL	–	×	×
USCB MeshNet	25	a/b/g	OpenWRT	IP	MCL	–	×	–
Purdue	32	a/b/g	–	IP	AODV	–	×	–
Georgia Tech	15	b/g	–	–	AODV,OLSR	–	–	–
Carleton Univ.	??	a/g	μClinux	IP	–	–	×	–
Hyacinth	10	a	Windows XP	–	OLSR	×	×	–
UMIC-Mesh.net	51	a/b/g	Linux	IP	DYMO,OLSR	×	×	×

The *Purdue University* has a WMN project called *Mesh@Purdue* with 32 nodes [21]. The routers are small form-factor desktops equipped with two wireless and one wired NIC. The latter one is used for management purposes. The goal of the project is first and foremost to provide Internet access. Besides, the research group works on modifications of AODV and OLSR.

The *Georgia Institute of Technology* runs a WMN project with 15 nodes [1]. The goal of the project is to study various performance metrics of WMNs, e.g., the effects of inter-router distance, backbone placement and clustering. Furthermore, existing protocols are re-investigated to review their performance in the testbed.

The *Carleton University* runs a WMN project in which each mesh router is equipped with two wireless NICs [4]. One is used for the communication within the WMN among the wireless mesh routers and the other is used for the communication with the clients. The WMN provides Internet access to the clients. The mesh routers are running μClinux and a quality of service (QoS) enhanced OLSR [7] as the routing protocol. To provide clients with addresses DHCPv6 is used. Thus, the WMN deploys only IPv6. If clients need to access an IPv4 network, e.g., the Internet, the packets are tunneled.

Hyacinth [23] is the WMN project at the *State University of New York*. Each mesh node is a small form-factor PC running Windows XP and is equipped with three IEEE 802.11a NICs. The nodes obtain an unique IP address from a global DHCP server, which is placed in the wired network. Each mesh node acts also as a local DHCP server and can assign IP addresses to mobile stations. To this end each mesh node receives an IP address range from the global DHCP server. Furthermore, roaming of mobile stations is supported, since all mesh nodes act also as a home/foreign agent like in Mobile IP (MIP).

19.6.2 Summary

The testbeds are summarized in Table 19.4. All information originates from the respective project websites. The *routing* column denotes at which network layer the routing is done and which protocol is used. A checkmark in the *roaming* column indicates that mesh clients can move around through the mesh without loosing connection. *Config* shows the capability to configure clients (and possibly routers) automatically. Finally, the column *MANET* is checked if the architecture is able to integrate independent MANETs into its routing.

Currently, the largest testbeds in terms of numbers of nodes are MIT Roofnet with 37 nodes and UMIC-Mesh.net with 51 nodes. All testbeds deploy IEEE 802.11 compliant technology. This holds for the communication among the mesh routers as well as for the communication between these routers and the mesh clients. Four projects perform routing at IP level, three projects at MAC level, the remaining projects do not reveal how they route. AODV, DYMO, DSR, and OLSR or variants are used as routing protocols. Most projects do not allow roaming of mesh clients; only Hyacinth and UMIC-Mesh.net do. Similar, independent integration of MANETs is only supported by two projects, Microsoft's Mesh Connectivity Layer (MCL) and UMIC-Mesh.net.

19.7 Thoughts for Practitioners

When contemplating the idea of running a WMN testbed there are a few things to consider. First of all, you should realize that besides covering the purchasing costs you will have to deal with the labor intensive tasks of installation and maintenance. Should you aim for a hybrid testbed like the UMIC-Mesh.net these tasks will be reduced but not eliminated.

Depending on your requirements on the testbed's topology, the planning phase can also take a non-negligible amount of time and resources. However, the most important decision you are going to be faced with is most likely the choice of node hardware, i.e., which platform, what kind of non-wireless connections (LAN, USB), how much processing power, and last but not least what kind of wireless NIC and how many of them should the nodes have. How programmable the wireless NICs are will determine what kind of experiments you can run in your testbed, i.e., down to which layer of the protocol stack you can influence node configuration.

Finally, an important decision will be whether you are going to install the nodes at fixed locations and rely on a permanent wired network connection. This decision largely depends on your scientific interests but will greatly affect your maintenance efforts. Requiring a permanent wired connection reduces your flexibility in node placement and raises installation costs but will greatly ease monitoring, software distribution and measurement as we have outlined in Sect. 19.4 about the UMICmesh.net. If done right, it will also ensure that node configurations will always be synchronized, thereby eliminating a major source of error in measurements.

19.8 Directions of Future Research

The performance evaluation shows that multi-hop wireless networks have big potential. However, as the evaluation shows further there are subtle difficulties when configuring WMNs. In our measurements we could show major performance gains coming from very small parameter set tunings. In detail, we showed that simply changing the routing control message emission intervals has non-negligible effects on the overall performance of the WMNs. Further effort is needed to investigate the trade-off between slow network state propagation and message overhead. Ideally a routing protocol would adapt its emission intervals to the network conditions.

19.9 Conclusions

Nowadays, the study of wireless multi-hop networks is mainly based on simulation. Although simulations provide researchers with many advantages like low cost, flexibility, and controllability they also have disadvantages. The prime disadvantage is that simulation results may have limited applicability in reality as simulations oversimplify. Maximum applicability is attainable by real word testbeds. However, they are expensive, do not scale, and are complicated to maintain. A possible solution to this dilemma is to build a hybrid testbed like UMIC-Mesh.net. It consists of a wireless mesh networking testbed and a virtualized environment. The former ensures a high degree of realism while the latter provides us with a flexible environment for fast protocol development and variable scenario creation.

19.10 Terminologies

1. *Backbone mesh gateway*. A mesh router with wired Internet connection.
2. *Backbone mesh router*. A mesh router that is installed at a fixed location and is therefore part of the backbone.
3. *Emulation*. Analysis of a system by duplicating its operation with a different and usually simpler and/or smaller system.
4. *Flowgrind*. A network testing tool that generates and analyzes TCP streams.
5. *Hybrid testbed*. A hybrid testbed combines a classical network testbed with virtualized nodes.
6. *Mesh client*. A node that's driven by a mesh user, possibly taking part in mesh routing (routing mesh client) or not (non-routing mesh client).
7. NetEm. NetEm is a Linux kernel extension that allows the adjustment of the characteristics of network links; e.g., by introducing artificial loss.
8. *Simulation*. Analysis of a system by simulating its components and their interactions, usually with the help of a simulator.
9. *Software development process*. The software development process comprises all the steps from the intent to develop a piece of software up to its realization.
10. *Testbed*. A network testbed is a collection of nodes that can communicate with each other and is used to analyze the behavior of the network or parts thereof.
11. *Theoretical analysis*. Analysis of a system with mathematical tools.
12. *Virtualization*. Abstraction of the physical resources of a single system to allow several virtual systems to run on one actual system and use these resources concurrently.
13. *Virtualized environment*. A system that acts in a virtualized environment is presented with an abstract notion of the actual resources of the underlying system; thereby these resources can be reused by several virtual systems.
14. *XEN*. XEN is a virtualization system and hypervisor originally developed by the University of Cambridge.

19.11 Acronyms

ACM	Association for computing machinery
AODV	Ad hoc on-demand distance vector routing protocol
AP	Access point
BSSID	Basic service set identifier
CPU	Central processing unit
DHCP	Dynamic host configuration protocol
DHCPv6	Dynamic host configuration protocol for IPv6
DSR	Dynamic source routing
DYMO	Dynamic MANET on-demand routing protocol
ESSID	Extended service set identifier
GRE	Generic routing encapsulation
IBSS	Independent basic service set
IEEE	Institute of electrical and electronics engineers
IP	Internet protocol
IPv4	Internet protocol version 4
IPv6	Internet protocol version 6
MAC	Medium access control
MANET	Mobile ad hoc network
MCL	Mesh connectivity layer
MIP	Mobile IP
MPR	Multipoint relay
MTU	Maximum transmission unit
NFS	Network file system
NIC	Network interface card
NID	Network identifier
OLSR	Optimized link-state routing protocol
OS	Operating system
PC	Personal computer
QoS	Quality of service
RAM	Random access memory
RL	Roofnet layer
SBC	Single board computer
SNMP	Simple network management protocol
TC	Topology control
TCP	Transmission Control Protocol
VM	Virtual machine
VMM	Virtual machine monitor
VPN	Virtual private network
WLAN	Wireless local area network
WMN	Wireless mesh network

19.12 Questions

1. Which parts belong to the infrastructure of a wireless mesh network?
2. What is the difference between Routing and non-routing clients?
3. Name at least three characteristics of WMNs.
4. Name at least two possibilities to study WMNs and one advantage and disadvantage.
5. Name two routing protocols applicable to WMNs.
6. What are some of the benefits of a hybrid testbed?
7. What are the basic types of virtualization?
8. What can be the benefits of a central server in a wireless testbed where all nodes are connected by wire?
9. What do our OLSR measurements with varying message emission intervals show?
10. What are the benefits of the flowgrind measurement tool?

References

1. I. F. Akyildiz, X. Wang, and W. Wang. Wireless mesh networks: a survey. Computer Networks, 47(4):445–487, 2005.
2. J. Bicket, D. Aguayo, S. Biswas, and R. Morris. Architecture and Evaluation of an Unplanned 802.11b Mesh Network. In Proceedings of the 11th Annual ACM/IEEE International Conference on Mobile Computing and Networking (MobiCom'05), pp. 31–42. ACM Press, 2005.
3. R. Bruno, M. Conti, and E. Gregori. Mesh networks: commodity multihop ad hoc networks. IEEE Communications Magazine, 43(3):123–131, 2005.
4. Carleton University. Wireless Mesh Networking. URL http://kunz-pc.sce.carleton.ca/MESH/index.htm.
5. M. Carson and D. Santay. NIST Net: a linux-based network emulation tool. ACM SIGCOMM Computer Communication Review, 33(3):111–126, 2003.
6. T. Clausen and P. Jacquet. The Optimized Link State Routing Protocol (OLSR). RFC 3626, October 2003.
7. Communications Research Centre. CRC OLSR. URL http://www.crc.ca/en/html/manetsensor/home/software/software.
8. S. Demers and L. Kant. MANETs: Perfrormance Analysis and Management. In Proceedings of the 25th IEEE Military Communications Conference (MILCOM'06). IEEE Communications Society Press, 2006.
9. R. Draves, J. Padhye, and B. Zill. Routing in Multi-Radio, Multi-Hop Wireless Mesh Networks. In Proceedings of the 10th Annual ACM/IEEE International Conference on Mobile Computing and Networking (MobiCom'04), pp. 114–128. ACM Press, 2004.
10. J. Eriksson, S. Agarwal, P. Bahl, and J. Padhye. Feasibility Study of Mesh Networks for All-Wireless Offices. In Proceedings of the 4th International Conference on Mobile Systems, Applications and Services (MobiSys'06), pp. 69–82. ACM Press, 2006.
11. D. Farinacci, T. Li, S. Hanks, D. Meyer, and P. Traina. Generic Routing Encapsulation (GRE). RFC 2784, March 2000.
12. C. Gomez, D. Garcia, and J. Paradells. Improving Performance of a Real Ad-hoc Network by Tuning OLSR Parameters. In Proceedings of the 10th IEEE International Symposium on Computers and Communications (ISCC'05), pp. 16–21. IEEE Computer Society Press, 2005.

13. S. Hemminger. Network Emulation with NetEm. In Proceedings of the 6th Australia's National Linux Conference (LCA'05). 2005.
14. Y. Huang, S. N. Bhatti, and D. Parker. TUNING OLSR. In Proceedings of the 17th Annual IEEE International Symposium on Personal, Indoor and Mobile Radio Communications (PIMRC'06). IEEE Communications Society Press, 2006.
15. B. Hubert, T. Graf, G. Maxwell, R. van Mook, M. van Oosterhout, P. B. Schroeder, J. Spaans, and P. Larroy. LARTC – Linux Advanced Routing & Traffic Control. URL http://lartc.org/.
16. G. Judd and P. Steenkiste. Using Emulation to Understand and Improve Wireless Networks and Applications. In Proceedings of the 2nd USENIX Symposium on Networked Systems Design and Implementation (NSDI'05), pp. 203–216. 2005.
17. E. Kohler, R. Morris, B. Chen, J. Jannotti, and M. F. Kaashoek. The Click modular router. ACM Transactions on Computer Systems, 18(3):263–297, 2000.
18. D. Kotz, C. Newport, R. S. Gray, J. Liu, Y. Yuan, and C. Elliott. Experimental Evaluation of Wireless Simulation Assumptions. In Proceedings of the 7th ACM International Symposium on Modeling, Analysis and Simulation of Wireless and Mobile Systems (MSWiM'04), pp. 78–82. ACM Press, 2004.
19. Madwifi Project. Madwifi – Multiband Atheros Driver for Wireless Fidelity. URL http://madwifi.org/.
20. PC Engines. WRAP router platform (Version WRAP.2E). URL http://www.pcengines.ch.
21. Purdue University. Purdue University Wireless Mesh Network Testbed. URL https://engineering.purdue.edu/MESH.
22. K. N. Ramachandran, K. C. Almeroth, and E. M. Belding-Royer. A Framework for the Management of Large-scale Wireless Network Testbeds. In Proceedings of the 1st Workshop on Wireless Network Measurements (WiNMee'05). 2005.
23. A. Raniwala and T. Chiueh. Architecture and Algorithms for an IEEE 802.11-Based Multi-Channel Wireless Mesh Network. In Proceedings of the 24th Annual Joint Conference of the IEEE Computer and Communications Societies (INFOCOM'05), vol. 3, pp. 2223–2234. IEEE Communications Society Press, 2005.
24. S. Ruffino, P. Stupar, T. Clausen, and S. Singh. Connectivity Scenarios for MANET. Internet Draft, July 2005.
25. P. M. Ruiz and F. Ros. DYMOUM – A DYMO implementation for real world and simulation. URL http://masimum.dif.um.es/?Software:DYMOUM.
26. RWTH Aachen University. UMIC-Mesh.net – A hybrid Wireless Mesh Network Testbed. URL http://www.umic-mesh.net.
27. The Olsr.org Project. OLSRd – The olsr.org OLSR daemon. URL http://www.olsr.org/.
28. P. E. Villanueva-Pena, T. Kunz, and P. Dhakal. Extending Network Knowledge: Making OLSR a Quality of Service Conducive Protocol. In Proceedings of the 2006 International Conference on Wireless Communications and Mobile Computing (IWCMC'06), pp. 103–108. ACM Press, 2006.
29. XenSource, Inc. XEN – The Xen virtual machine monitor. URL http://www.xensource.com/.

Index